ITALIAN PHYSICAL SOCIETY

PROCEEDINGS

OF THE

INTERNATIONAL SCHOOL OF PHYSICS « ENRICO FERMI »

Course LIV

edited by R. Gatto
Director of the Course

VARENNA ON LAKE COMO
VILLA MONASTERO
2nd - 14th AUGUST 1971

Developments in High-Energy Physics

1972

ACADEMIC PRESS • NEW YORK AND LONDON

SOCIETA' ITALIANA DI FISICA

RENDICONTI

DELLA

SCUOLA INTERNAZIONALE DI FISICA «ENRICO FERMI»

LIV CORSO

a cura di R. GATTO
Direttore del Corso

VARENNA SUL LAGO DI COMO
VILLA MONASTERO
2-14 AGOSTO 1971

Sviluppi della fisica delle alte energie

1972

ACADEMIC PRESS • NEW YORK AND LONDON

ACADEMIC PRESS INC.
111 FIFTH AVENUE
NEW YORK 3, N. Y.

United Kingdom Edition
Published by
ACADEMIC PRESS INC. (LONDON) LTD.
24/28 OVAL ROAD, LONDON N. W. 1

Library of Congress Catalog Card Number: 72-12210

PRINTED IN ITALY

INDICE

R. Sexl – Introduction to general relativity.

S. A. Bludman – Equation of state of ultradense relativistic matter.

A. Salam – Nonpolynomial Lagrangians, infinities and gravity theory.

Preface.

R. Gatto

Istituto di Fisica dell'Università - Roma

In the summer of 1953 the International School of Physics of the Italian Physical Society was inaugurated in Varenna with a course, directed by G. Puppi, dealing with Cosmic Rays and Elementary Particles. Professor Powell in his beautiful lectures gave a list of the existing particles at the time. The hyperons listed were Λ^0 and $\Omega^{\pm}$, the latter decaying into $\mathcal{N}+\pi^{\pm}$ with a Q-value of 130 MeV, and therefore it should not be confused with what is nowdays called the Ω-hyperon. Heavy mesons had been reported and called $K^{\pm}$, $\tau^{\pm}$, χ and θ^0. The leptons were $\mu^{\pm}$, $e^{\pm}$, and ν. Except for the distinction between ν_e and ν_{μ} the number of leptons has not increased since that time—against the advice of theoreticians.

The subsequent Varenna School, in the summer of 1954, had E. Fermi among the teachers, who gave lectures on pion- and nucleon-physics, dealing with experiments carried out at accelerators. The CERN project of a proton-synchrotron was illustrated by Amaldi, Adams, Citron and Hine.

Those Varenna schools were perhaps remarkable historical events. The number of summer schools has now everywhere increased, perhaps as rapidly as the number of resonant baryon- and meson-states. Also the entire field of physics has rapidly expanded and a wide spectrum of specialized topics is dealt with each year at the Varenna School.

The present book contains the lectures given at the LIV course during the first half of August 1971. The course was entitled « Developments in High Energy Physics » and designed to deal with the topics of more recent interest in strong-, weak- and electromagnetic-interactions, rather than with a single specialized field—as with the older courses in Varenna.

Professor G. Toraldo di Francia, President of the Italian Physical Society, gave all his support to the success of the School. We owe to him and to the Council of the Italian Physical Society our best thanks. Professor Menotti acted with great enthusiasm as Scientific Secretary. Professor Germanà took care of all organizational details with great competence. The secretaries worked hardly in typing and reproducing the preliminary lecture notes during the school. Unfortunately a tragic accident saddens our memory of this particular Varenna course: one of our secretaries, Miss NUNZIA LOPERFIDO died by car accident while driving back from Varenna.

The authors of the contributions contained in this volume do not need presentations. They are all well-known physicists. We only want to thank them for their unvaluable collaboration. The students were very active and contributed to lively round-table discussions and with seminars. We thank them for their participation. To P. PAPALI and to the staff of the Italian Physical Society, who assisted in the preparation of this volume, goes our acknowledgement for their help.

1. R. Ferrari
2. G. Immirzi
3. F. Nicolò
4. G. C. Rossi
5. D. Amati
6. A. Di Giacomo
7. F. Pempinelli Boiti
8. R. R. Gatto
9. V. Alessandrini
10. R. Dashen
11. M. L. Goldberger
12. P. Menotti
13. B. Drevillon
14. A. Ferrer
15. P. Goddard
16. G. Puglierin
17. M. Evrard
18. L. Maiani
19. A. R. White
20. R. Vergara Caffarelli
21. U. E. Schroeder
22. P. D. Mannheim
23. J. Olmos
24. E. Reya
25. V. P. Seth
26. K. Schilcher
27. I. C. Zuravk
28. P. Marcolungo
29. A. Din
30. R. Nobili
31. L. Lugiato
32. B. Desplanques
33. R. A. Leo
34. S. Pallua
35. R. Collina
36. R. Aldrovandi
37. S. Rock
38. W. C. Ng
39. P. Furlan
40. S. Zerbini
41. A. Ballestrero
42. M. Boiti
43. B. Loiseau
44. J. Provost
45. P. Dondi
46. H. Gottlieb
47. J. H. Danskin
48. A. Rothery
49. G. Altarelli
50. G. Soliani
51. J. Gomis Torné
52. E. Galli
53. K. Schlüpmann
54. G. Marchesini
55. K. Konishi
56. M. Magg
57. M. Ciafaloni
58. H. Strubbe
59. H. Saller
60. G. Cicuta
61. L. Hughston
62. J. Becker
63. G. Bella
64. L. Ryder
65. R. Barbieri

SOCIETÀ ITALIANA DI FISICA

SCUOLA INTERNAZIONALE DI FISICA « E. FERMI »

LIV CORSO - VARENNA SUL LAGO DI COMO - VILLA MONASTERO - 2 - 14 Agosto 1971

Multiperipheral Models and High-Energy Processes (*).

M. L. GOLDBERGER

Joseph Henry Laboratories, Princeton University - Princeton, N. J.

During the past few years there has been a great deal of activity among theorists and experimentalists trying to understand collision phenomena at high energies. In particular there has been a growing awareness of the richness of multiparticle production processes initiated either by virtual photons or hadrons. In some cases we have reached the limits of available accelerators (such as in certain of the deep-inelastic electron experiments at SLAC and must wait upon critical tests of theoretical speculations for results from the intersecting storage rings at CERN and from the 500 GeV accelerator at the National Accelerator Laboratory. Many theorists, myself included, have their necks fairly well extended on rather simple minded models for a class of high-energy processes which are based on the old idea of multiperipheralism and closely related concepts of Regge theory. Preliminary results from CERN seem to indicate (happily perhaps for physics) that real life is much more complicated. Nevertheless, I shall in these lectures tell about work on multiperipheral models that has been carried out at Princeton during the past year by ABARBANEL, CHEW, SAUNDERS and me. Some of this material has been or will be published elsewhere. Where there is duplication, I have tried to make the presentation more transparent than is customary in *The Physical Review* where the style tends to be one of punitive pedagogy. There is, in addition, new material which my colleagues may or may not choose to associate themselves with.

Two years ago at Erice I gave a series of lectures on multiperiphalism which have now been published and I can recommend them to you without reservation for background reading. There have been a number of improvements in technology and in physical ideas since that time and it is these that I shall address here.

(*) Research sponsored by the U.S. Atomic Energy Commission under Contract AT(30-1)-4159.

1. – Introduction to the ABFST model.

1'1. *Notation, kinematics and integral equation.* – The basic idea of the ABFST model is that the absorptive part of the elastic-scattering amplitude can be computed from a multiparticle-production amplitude which factors in a way to be seen below. This makes it possible to write down a recursion relation for the absorptive part connecting amplitudes corresponding to, say, n-particle production to those for $n-1$ particles. This enables one to write a kind of integral equation for the absorptive part which has been exhaustively studied.

We shall deal (almost) exclusively with the very simplest case of, spin-zero neutral particles of mass m, which when they collide may produce other particles of various masses that need not concern us at the moment. Our notational conventions are as follows: For an elastic-scattering process in which $m_a + m_b \to m_a + m_b$ (all particles spinless) we relate the barycentric-scattering amplitude $f(s, \cos\theta)$ to the (dimensionless) invariant amplitude $M(s, t)$ by

$$f(s, \cos\theta) = \frac{M(s, t)}{8\pi\sqrt{s}},$$

where

$$s = (p_a + p_b)^2,$$

$$t = -2k^2(1 - \cos\theta),$$

$$4sk^2 = \Delta(s, m_a^2, m_b^2),$$

with θ the scattering angle, p_a, p_b the initial four-momenta ($p_a^2 = m_a^2$, $p_b^2 = m_b^2$) and the triangle function $\Delta(a, b, c) = a^2 + b^2 + c^2 - 2ab - 2ac - 2bc$. The elastic differential cross-section is

$$\frac{d\sigma_e}{dt} = \frac{1}{16\pi\Delta(s, m_a^2, m_b^2)} |M(s, t)|^2,$$

and the total cross-section σ is

$$\sigma = \frac{4\pi}{k} \operatorname{Im} f(s, 1),$$

$$\sigma = \frac{1}{\Delta^{\frac{1}{2}}(s, m_a^2, m_b^2)} \operatorname{Im} M(s, 0),$$

$$\sigma \equiv \frac{A(s, 0)}{\Delta^{\frac{1}{2}}(s, m_a^2, m_b^2)},$$

with $A(s, t)$ being the absorptive part of the elastic amplitude $M(s, t)$. The elastic unitarity relation takes the form

$$\operatorname{Im} M(s, t) = \frac{1}{16\pi^2 \Delta^{\frac{1}{2}}} \int \frac{\mathrm{d}t_1 \mathrm{d}t_2 M^*(s, t_2) M(s, t_1)}{[-\Delta(t, t_1, t_2) + (t t_1 t_2)/k^2]^{\frac{1}{2}}} .$$

The region of integration is that for which the bracket in the denominator is positve.

Our model may be defined by a recursion relation based on the assumption that the multiparticle-production amplitude factors into a product of unspecified blobs connected by single-particle propagators. The absorptive part of the elastic amplitude for n blobs is then related to that for $n-1$ blobs by

$$A_n(P, K, Q) = \frac{2}{(2\pi)^4} \int \mathrm{d}^4 P' \, V(P, P', Q) S(P', Q) A_{n-1}(P', K, Q) ,$$

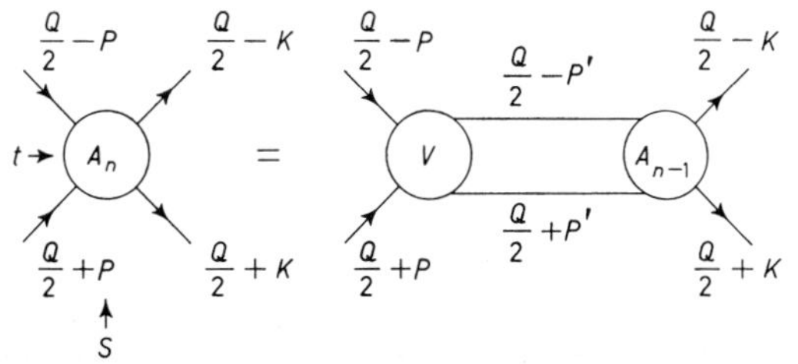

Fig. 1. – Recursion relation for ABFST model.

where the momentum vectors are shown in Fig. 1. The quantity $S(P', Q)$ is the product of the upper and lower single-particle propagators:

$$S(P', Q) = \frac{1}{m^2 - (Q/2 + P')^2} \, \frac{1}{m^2 - (Q/2 - P')^2} .$$

The s-channel total-energy squared is $s = (P - K)^2$ and the overall momentum-transfer squared is $t = Q^2$. The quantity V is the specific input to the model. In the simplest case of single-particle exchange we have

$$V(P, P', Q) = \pi g^2 \delta[(P - P')^2 - m_0^2] .$$

In the most widely studied model we take V to be some appropriate off-mass-shell extension of the absorptive part of elastic π-π scattering; in this case the particles of mass m are also pions. Without stopping for the moment to talk about off-shell effects, we indicate what V on shell is. Obviously one can diagonalize the recursion relation in t-channel internal quantum numbers

(I-spin, say) so that the t-channel «potential» $V^{I}t$ must be obtained by crossing from the s-channel π-π scattering. Thus we have (on-shell!)

$$V^{I_t}(s, t) = \sum_{I_s} \beta_{I_t, I_s} \frac{1}{16\pi^2 \Delta^{\frac{1}{2}}(s, m^2, m^2)} \iint \frac{\mathrm{d}t_1 \mathrm{d}t_2 \, M^*_{I_s}(s, t_2)\, M_{I_s}(s, t_1)}{\{-\Delta(t, t_1, t_2) + (tt_1t_2)/(s/4 - m^2)\}^{\frac{1}{2}}},$$

where β_{I_t, I_s} is the crossing matrix. When we take up our physical model in detail we'll discuss the off-shell question.

Returning now to formalism, we convert the recursion relation for A_n into something that looks like an integral equation by defining the total absorptive part $A(P, K, Q)$ by

$$A(P, K, Q) = \sum_{n=1}^{\infty} A_n(P, K, Q)\,,$$

with $A_1 \equiv V$. Then summing the recursion relation from $n = 2$ to ∞ we find

$$A(P, K, Q) = V(P, K, Q) + \frac{2}{(2\pi)^4} \int \mathrm{d}^4P'\, V(P, P', Q) S(P', Q) A(P', K, Q)\,.$$

Needless to say, this equation is smart enough to recognize that if V has threshold (such as m_0^2 in the ladder model or $4m^2$ in the π-π scattering potential) there will be no more terms in the iterative solution of the above equation than are allowed by energy conservation. The above version of the original recursion relation is what is conventionally called ABFST integral equation. Note that it can also be written as

$$A(P, K, Q) = V(P, K, Q) + \frac{2}{(2\pi)^4} \int \mathrm{d}^4P'\, A(P, P', Q) S(P', Q)\, V(P', K, Q)\,,$$

since these two forms lead to precisely the same (finite number of terms) iterative solution. It is important to notice that one cannot restrict attention to on-shell quantities in these equations because of the integrations over P'; the on-shell values of the variables P, K are given by the relations (for a process $m_a + m_b \to m_a + m_b$)

$$P^2 + \frac{Q^2}{4} = m_a^2\,, \qquad P \cdot Q = 0\,,$$

$$K^2 + \frac{Q^2}{4} = m_b^2\,, \qquad K \cdot Q = 0\,.$$

1'2. *Variational principle for* $A(P, K, Q)$. – We shall make considerable use of variational methods in our discussion, so we note that one may write a stationary expression for

$$\mathscr{A}(P, K, Q) \equiv A(P, K, Q) - V(P, K, Q)\,,$$

namely

$$\mathscr{A}(P, K, Q) =$$

$$= \frac{(2/(2\pi)^4)\int \mathrm{d}^4P'\, A(P, P')S(P')V(P', K)\int \mathrm{d}^4P'\, V(P, P')S(P')A(P', k)}{\int \mathrm{d}^4P'\, A(P, P')S(P^{\iota})A(P', K) - 2/(2\pi)^4\int \mathrm{d}^4P'\int \mathrm{d}^4P''\, A(P, P')S(P')V(P', P'')S(P'')A(P'', K)}\,,$$

where on the right we have suppressed the dependence on Q which should appear in every factor: $A(P, P', Q) \to A(P, P')$, $S(P', Q) \to S(P')$, etc.

1·3. *Diagonalization of the* ABFST *equation.* – In order to simplify the analysis as much as possible it is very helpful to take advantage of all possible symmetry to reduce the four-dimensional ABFST equation to equations of lower dimension. This procedure has been described in great detail by ABARBANEL and SAUNDERS and we shall take over their results without going into the straightforward but rather involved calculations.

We begin with the simplest case of forward scattering $t = Q^2 = 0$, and imagine that we express all quantities as functions of scalar variables rather than four-vectors. We choose them as

$$\begin{aligned} P^2 &= u\,, & s &= (P - K)^2\,,\\ K^2 &= v\,, & s' &= (P' - K)^2\,,\\ P'^2 &= u'\,, & s_0 &= (P - P')^2\,, \end{aligned}$$

and make frequent use of hyperbolic angles defined by

$$\begin{aligned} \cosh \Theta(u, v) &= -\frac{P\cdot K}{\sqrt{uv}} = \frac{s - u - v}{2\sqrt{uv}}\,,\\ \cosh \Theta'(u', v) &= -\frac{P'\cdot K}{\sqrt{uv}} = \frac{s' - u' - v}{2\sqrt{u'v}}\,,\\ \cosh \Theta_0(u, u') &= -\frac{P\cdot P'}{\sqrt{uu'}} = \frac{s_0 - u - u'}{2\sqrt{uu'}}\,. \end{aligned}$$

Evidently s is the total $(\text{energy})^2$ in the s-channel and s_0 is the corresponding quantity for each blob described by V. The variables u, v will be assumed negative throughout the analysis in spite of the fact that one eventually wants them positive on the mass shell. This greatly simplifies the kinematics since it constrains u' also to be negative. The symmetry of the equation for $t = 0$

is well known to be $O_{3,1}$, so that it is reasonable to use the harmonics of this group to affect the diagonalization. We define a « partial-wave amplitude » $A_\lambda(u, v)$ by

$$A_\lambda(u, v) = \int_{L^2}^{\infty} \mathrm{d}s \exp[-(\lambda+1)\Theta(u, v)]A(s, u, v) .$$

Here L^2 is the minimum mass associated with the potential V. By introducing this integration over all s we hereby convert the ABFST recursion equation into a honest integral equation. The amplitude $A_\lambda(u, v)$ will be analytic in the complex λ-plane to the right of $\mathrm{Re}\,\alpha_m$ where, for fixed u, v, $A(s, u, v)$ for large s grows no faster than s^{α_m}. Using the fact that $\mathrm{d}s = 2\sqrt{uv}\sinh\Theta\,\mathrm{d}\Theta$ we may invert the above projection, which is nothing more than a Laplace transform of $2\sqrt{uv}\exp[-\Theta]\sinh\Theta\ A(s, u, v)$ to obtain

$$A(s, u, v) = \int_{c-i\infty}^{c+i\infty} \frac{\mathrm{d}\lambda}{2\pi i} \frac{\exp[+(\lambda+1)\Theta]A_\lambda(u, v)}{2\sqrt{uv}\sinh\Theta} \approx \left(\frac{1}{\sqrt{uv}}\right)^{\lambda+1} \int_{c-i\infty}^{c+i\infty} \frac{\mathrm{d}\lambda}{2\pi i} s^\lambda A_\lambda(u, v) ,$$

where $c > \mathrm{Re}\,\alpha_m$ and the second form is that appropriate for large s.

We now apply this projection to the forward equation where we introduce invariants as variables, namely

$$A(s, u, v) = V(s, u, v) + \frac{2}{(2\pi)^4} \int_{L^2}^{(\sqrt{s}-L)^2} \mathrm{d}s' \int_{L^2}^{(\sqrt{s'}-L)^2} \mathrm{d}s_0 \int_{u_-}^{u_+} \mathrm{d}u' \frac{V(s_0, u, u')A(s', u', v)}{J(m^2-u')^2}$$

with

$$\frac{1}{J} = \int \mathrm{d}^4P'\,\delta[(P-P')^2-s_0]\,\delta[P'^2-u']\,\delta[(P'-K)^2-s'] ,$$

and $u_\pm$ are functions of s, s_0, u, u', v which are determined by the δ-functions. The transformed equation is

$$A_\lambda(u, v) = V_\lambda(u, v) + \frac{1}{16\pi^3(\lambda+1)} \int_{-\infty}^{0} \mathrm{d}u' \frac{V_\lambda(u, u')A_\lambda(u', v)}{(m^2-u')^2} ,$$

where

$$V_\lambda(u, u') = \int_{L^2}^{\infty} \mathrm{d}s_0 \exp[-(\lambda+1)\Theta_0(u, u')]\,V(s_0, u, u') .$$

It is important to note that in this equation, the phase-space integral J^{-1} has been treated exactly. Furthermore, in contrast to older treatments we have not first taken a large-s limit before transforming; this means that we do not lose control over the scale of the inhomogeneous term and consequently can discuss the magnitude of A and not just the location of singularities associated with solutions of the homogeneous equation.

The corresponding diagonalization of the ABFST equation for $t \neq 0$ is somewhat more complicated. Since the momentum transfer Q is held fixed and $t = Q^2 < 0$, we use the symmetry associated with the little group of the spacelike vector Q, namely $O_{1,2}$. We introduce scalar variables and also decompose all 4-vectors into parts parallel to Q and orthogonal to Q. Thus we write

$$P^2 = u\,, \qquad z = P\cdot Q/\sqrt{ut}\,, \qquad y = \frac{-\tilde{P}\cdot\tilde{K}}{\sqrt{\tilde{P}^2\tilde{K}^2}}\,,$$

$$K^2 = v\,, \qquad z = K\cdot Q/\sqrt{vt}\,, \qquad y_0 = \frac{-\tilde{P}\cdot\tilde{P}'}{\sqrt{\tilde{P}^2\tilde{P}'^2}}\,,$$

$$P'^2 = u'\,, \qquad z' = P'\cdot Q/\sqrt{u't}\,, \qquad y' = \frac{-\tilde{P}'\cdot\tilde{K}}{\sqrt{\tilde{P}'^2\tilde{K}^2}}\,,$$

where vectors designated by $\sim$ are defined by

$$\tilde{N} = N - \frac{N\cdot Q}{Q^2}Q\,.$$

The y's are in turn related to energies s, s_0 and s' by

$$y = \frac{\cosh\Theta(u, v) - z\zeta}{\sqrt{(1-z^2)(1-z^2)}}\,,$$

$$y_0 = \frac{\cosh\Theta_0(u, u') - zz'}{\sqrt{(1-z^2)(1-z'^2)}}\,,$$

$$y' = \frac{\cosh\Theta'(u', v) - z'\zeta}{\sqrt{(1-z'^2)(1-\zeta^2)}}\,,$$

and the quantities $\cosh\Theta$, etc. are given as before in terms of s, u, v. Note that in contrast with the forward case, u, v, u' are no longer simply (mass)^2. For example, in a general process where $m_a + m_b \to m_a' + m_b'$

$$P^2 + \frac{Q^2}{4} = u + \frac{t}{4} = \frac{m_a^2 + m_a'^2}{2}\,,$$

$$K^2 + \frac{Q^2}{4} = v + \frac{t}{4} = \frac{m_b^2 + m_b'^2}{2}\,.$$

The first step then is to introduce these scalar variables and a Jacobian that transforms the integration d^4P' into one over u', z', y_0 and y'. We anticipate that the group theory will enable us to integrate over the variables y_0, y' that specify the orientation of the vector P' in the three-dimensional space orthogonal to Q. The appropriate transform is

$$A_l(u, z; v, \zeta; t) \equiv \int_{L^2}^{\infty} ds\, Q_l(y) A(s; u, z; v, \zeta; t) ,$$

where, of course

$$ds = 2\sqrt{uv(1-z^2)(1-\zeta^2)}\, dy .$$

If we similarly define

$$V_l(u, z; v\zeta; t) = \int_{L^2}^{\infty} ds\, Q_l(y)\, V(s; u, z; v, \zeta; t) ,$$

$$V_l(u, z; u'z'; t) = \int_{L^2}^{\infty} ds_0 Q_l(y_0)\, V(s_0; u, z; v, \zeta; t) ,$$

we find

$$A_l(u, z; v, \zeta; t) = V_l(u, z; v, \zeta; t) + \frac{1}{16\pi^4}\int_{-\infty}^{0} du' \int_{-1}^{+1} \frac{dz'}{\sqrt{1-z'^2}} V_l(u, z, u', z'; t) \cdot$$

$$\cdot A_l(u', z'; v, \zeta; t)/[(m^2 - u' - t/4)^2 - u't'z'^2] .$$

The inversion formula to recover $A(s, \dots)$ reads

$$A(s, \dots) = \int_{c-i\infty}^{c+i\infty} \frac{dl(2l+1)}{2\pi i} A_l(\dots) P_l(y) \left[2\sqrt{uv(1-z^2)(1-\zeta^2)}\right]^{-1} .$$

A remark is in order about the connection between $A_l(u, z; v, \zeta; t)$ and $A_\lambda(u, v)$ in the limit of $t \to 0$. There is a considerable delicacy in this question. Although the l-plane singularities must arrange themselves in such a way as to be consistent with the λ-plane over $t \to 0$, it is not true that A_l in any sense approaches A_λ. If one imagines computing $A(s, \dots, t)$ from A_l and then passing to the limit $t \to 0$, one must find the same $A(s, \dots)$ as obtained by inverting directly at $t = 0$ using $A_\lambda(u, v)$. There is, after all, no singularity in the physical amplitude at $t = 0$. Any skeptics about the subtlety of this whole issue are invited to try the following exercise: Compare the first Fredholm approximations to the forward equation for $A_\lambda(u, v)$ to the $t \to 0$ limit in the same approximation $A_l(u, z;\ v\zeta;\ t)$ for a model in which $V(P, P', Q) = \pi g^2 \delta[(P-P')^2 - m_0^2]$. Thus one is concerned with

$$D(\lambda) \approx 1 - \frac{g^2}{16\pi^2}\frac{1}{\lambda+1}\int_{-\infty}^{0} du\, \frac{\exp[-(\lambda+1)(u, u)]}{(m^2-u)^2},$$

$$D(l) \approx 1 - \frac{g^2}{16\pi^3}\int_{-\infty}^{0} du \int_{-1}^{+1} \frac{dz}{\sqrt{1-z^2}}\, Q_l\left[1 - \frac{m_0^2}{2u(1-z^2)}\right]\frac{1}{(m^2-u)^2},$$

where

$$\cosh\Theta(u, u') = \frac{m_0^2 - u - u'}{2\sqrt{uu'}}.$$

These two things are *not* equal. For example, in the limit $m = 0$

$$D(\lambda) = 1 - \left(\frac{g}{4\pi m_0}\right)^2 \frac{2}{\lambda(\lambda+1)(\lambda+2)},$$

$$D(l) = 1 - \left(\frac{g}{4\pi m_0}\right)^2 \frac{1}{l(l+1)}.$$

1·4. *A particular model.* – One can imagine that it might be convenient to split the potential into a low-energy and a high-energy part. This is particularly appropriate in a model which is based on units consisting of π-π scattering that we have referred to earlier. Then there is a low-energy resonance region and a high-energy part which can hopefully be described in terms of Regge behavior. It has been found in earlier work that the low-energy potential is much larger than the high-energy tail although the latter plays a crucial role in determining the structure in the angular-momentum plane near $l = 1$.

To see how the division is made, consider first the potential for $t = 0$. We write

$$V_\lambda(u, v) = V_\lambda^R(u, v) + V_\lambda^P(u, v),$$

where

$$V_\lambda^R(u, v) = \int_{4m^2}^{s^*} ds\, \exp[-(\lambda+1)\Theta(u, v)]\, V(s, u, v),$$

$$V^R(u, v) = \int_{s^*}^{\infty} ds\, \exp[-(\lambda+1)\Theta(u, v)]\, V(s, u, v).$$

Note that we have specialized to the π-π model in writing the lower limit of integration as $4m^2$. The energy s^* would be chosen to be above my prominent resonances where $V(s, u, v)$ (at least on-shell) would be expected to be smooth. We remark that in this model, on-shell, $V(s, t = 0)$ takes the form

$$V^{I_t}(s, t = 0) \sum_{I_s} \beta_{I_t, I_s} \Delta^{\frac{1}{2}}(s, m^2, m^2) \sigma_{I_s}^{\text{elastic}}(s) ,$$

where we have again explicitly shown the internal quantum numbers I_t, I_s.

It is obvious that an exactly analogous separation of the potential can be made for $t \neq 0$.

Let us now study in a little more detail how the high-energy part of the potential is constructed. For simplicity let us assume that for $s > s^*$, the elastic amplitude $M(s, t)$ can be represented by a simple Regge pole, temporarily suppressing the dependence on the variables u, v, $z\zeta$. We write

$$M_{I_s} = \mathscr{M}_{I_s}(t) s^{\alpha_0(t)} .$$

Strictly speaking we should use something like $P_{\alpha_0}(y)$ but this is an affectation that will in no way affect our results provided s^* is large (say more than 3 $(\text{GeV})^2$). We recall that

$$\operatorname{Im} M_{I_s} = A_{I_s}(s, 0) = \frac{1}{16\pi \Delta^{\frac{1}{2}}(s, u, v)} \int_{-4k^2}^{0} dt |M_{I_s}(s, t)|^2 \approx \frac{1}{16\pi s} \int_{-\infty}^{0} dt |M_{I_s}(s, t)|^2 ,$$

where we imagine that $s \gg u, v$; the error made by replacing the lower limit by $-\infty$ is exponentially small according to most reasonable models for $M_{I_s}(s, t)$. Now for the purpose of our high-energy potential we have

$$V_\lambda^P = \int_{s^*}^{\infty} ds \exp[-(\lambda + 1)\Theta] V(s) ,$$

$$V_\lambda^P \approx (\sqrt{uv})^{\lambda+1} \int_{s^*}^{\infty} ds\, s^{-(\lambda+1)} V(s) ,$$

and supplying the appropriate crossing matrix we find

$$V_\lambda^{P, I_t} = (\sqrt{uv})^{\lambda+1} \sum_{I_s} \frac{\beta_{I_t, I_s}}{16\pi} \int_{-\infty}^{0} dt \frac{|\mathscr{M}_{I_s}(t)|^2}{\lambda - (2\alpha_0(t) - 1)} (s^*)^{2\alpha_0(t) - 1 - \lambda} .$$

The corresponding formula for $t \neq 0$ is obtained by using the elastic unitarity relation for $t \neq 0$ in the high-energy limit, namely

$$A_{I_s}(s, t) \simeq \frac{1}{16\pi^2 s} \int \mathrm{d}t_1 \int \mathrm{d}t_2 \frac{\mathscr{M}^*_{I_s}(s, t_1)\mathscr{M}_{I_s}(s, t_2)}{[-\Delta(t, t_1, t_2)]^{\frac{1}{2}}} ;$$

next we transform with $Q_l(y)$ which we approximate as

$$Q_l(y) \sim \frac{\sqrt{\pi}\,\Gamma(l+1)}{\Gamma(l+\frac{3}{2})} \left[\sqrt{uv(1-z^2)(1-\zeta^2)}\right]^{l+1} s^{-(l+1)} ,$$

having written

$$y \approx \frac{s}{2\sqrt{uv(1-z^2)(1-\zeta^2)}} ,$$

which is consistent with the assumption $s \gg u, v$. We find then for $V_l^{P,t}$ the result

$$V_l^{P,I_t}(t) = \sum_{I_s} \frac{\beta_{I_t,I_s}}{16\pi^2} [\sqrt{uv(1-z^2)(1-\zeta^2)}]^{l+1} \frac{\sqrt{\pi}\,\Gamma(l+1)}{\Gamma(l+\frac{3}{2})} \cdot$$

$$\cdot \iint \frac{\mathrm{d}t_1 \mathrm{d}t_2}{[-\Delta(t, t_1, t_2)]^{\frac{1}{2}}} \frac{\mathscr{M}^*_{I_s}(s, t_1)\mathscr{M}_{I_s}(s, t_2)}{l-\alpha_0(t_1)-\alpha_0(t_2)+1} (s^*)^{-l-1-\alpha_0(t_1)-\alpha_0(t_2)} .$$

We note in passing that since V_λ^R and $V_l^R(t)$ are defined by finite integrals whose integrands are analytic in λ or $l > -1$ the only singularities in these planes for the whole potential arise from the high-energy tails. These in turn are (logarithmic) branch points associate with the vanishing of the denominators $\lambda - 2\alpha_0(t) + 1$ or $l - \alpha_0(t_1) - \alpha_0(t_2) + 1$. This is the well-known AFS branch point which can be quite close to 1 if $\alpha_0(0) \sim 1$.

1·5. *Off-shell behavior.* – We turn now briefly to the question of off-shell behavior of the potentials V_λ and V_l. The very simplest procedure would be to adopt the on-shell value of $V(s, t)$. The rationale for this is that behind the whole model which assumes such a steep decrease of amplitudes on momentum transfer that the most important effects are those of the particle propagators which involve the (small) pion mass. Thus if the propagators alone give sufficient convergence to keep the kernel of our integral equation, which is $V \cdot S(u, z)$, Fredholm we are quite happy with this naive rule. Such is the case for V_λ^R and V_l^R because, as can easily be seen, for any finite value of s^*, for large u, $V^R \sim (\sqrt{-u})^{-l-1}$ or $(\sqrt{-u})^{-\lambda-1}$. Thus, for example, if we were to approximate the low-energy region by a single p-wave resonance at $s = m_0^2$ (the ρ-meson, say)

$$V_l^R \sim \left[1 + \frac{t/2}{(m_0^2/4) - m^2}\right] Q_l \left[\frac{\cosh\theta - z\zeta}{(1-z^2)(1-\zeta^2)}\right],$$

where $\cosh\theta = (m_0^2 - u - v)/2\sqrt{uv}$ and

$$V_\lambda^R \sim \exp[-(\lambda+1)\theta(u, v)].$$

The t-dependent factor in V_l^R is just $\cos\theta_s = 1 + (t/2k^2)$ with $k^2 = s/4 - m^2$ evaluated at $s = m_0^2$.

Unfortunately this simplest off-shell extrapolation fails for the high-energy potential because of the explicit appearance of the factor $(\sqrt{uv})^{\lambda+1}$ in V_λ^P and V_R^P. This *cannot* be cured simply by doing the high-energy trasform more accurately. One must rely on the off-shell damping provided by the factors $\mathscr{M}_{I_s}(t)$ when these in turn are computed from the *solution* of our equations. This is the first time self-consistency has entered the picture. To see how this comes about let us consider the contribution to M_{I_s} from an isolated I_t = singlet Regge pole with even signature and trajectory $l = \alpha_0(t)$. We know from general principles that near such a singularity the solution of the ABFST equation will have the form

$$A_l^{I_t=1}(u, z; u', z'; t) \approx \left[\sqrt{uu'(1-z^2)(1-z'^2)}\right]^{\alpha_0(t)+1} \frac{b(u, z, t)b(u', z', t)}{l - \alpha_0(t)},$$

where we have isolated conventional relative momentum factors. Using the asymptotic form (for large s) of the inversion formula

$$A(s, \ldots) = \int \frac{dl(2l+1)A_l}{2\pi i} \frac{P_l[(\cosh\Theta(u, u') - zz')/\sqrt{(1-z^2)(1-z'^2)}]}{2\sqrt{uv(1-z^2)(1-z'^2)}},$$

in which we replace

$$P_l[\;] \to \frac{\Gamma(l+\frac{1}{2})}{\sqrt{\pi}\,\Gamma(l+1)} \left[\frac{s}{\sqrt{uv(1-z^2)(1-z'^2)}}\right]^l,$$

we find

$$A^{I_t=1}(s, u, z; u', z'; t) \to \frac{\Gamma(\alpha_0(t)+\frac{3}{2})}{\sqrt{\pi}\,\Gamma(\alpha_0(t)+1)} b(u, z, t)\, b(u', z', t)\, s^{\alpha_0(t)}.$$

From this, introducing a signature factor

$$-\frac{\exp[-i\pi\alpha_0(t)] + 1}{2\sin\pi\alpha_0(t)} = \frac{1}{2}\left(i - \operatorname{ctg}\frac{\pi\alpha_0(t)}{2}\right),$$

and a factor N^{-1} with N the multiplicity of the regular representation of SU_n to account for crossing from the singlet t-channel pole to any s-channel state, we have for M_{I_s} the result

$$M_{I_s}(s, u, z;\ u', z';\ t) \sim \beta(u, z;\ t)\beta(u', z';\ t)s^{\alpha_0(t)},$$

where

$$\beta(u, z;\ t) = \left\{\frac{1}{N}\left(i - \operatorname{ctg}\frac{\pi\alpha_0(t)}{2}\right)\frac{\Gamma(\alpha_0(t)+\frac{3}{2})}{2\sqrt{\pi}\,\Gamma(\alpha_0(t)+1)}\right\}^{\frac{1}{2}} b(u, z, t)\,.$$

This $M_{I_s}(s, \ldots)$ is the input we need to compute the high-energy part of our potentials. Thus in addition to the troublesome factor of $(\sqrt{uv})^{l+1}$ we must expect further u, v-dependence arising from the residue functions β. There is one further slightly tricky point involved in the actual calculation. Consider the evaluation of $V_\lambda^P(u, u')$. In the implied integration over the momentum transfer t, we have to do with an s-channel scattering amplitude for a process $P+(-P')\to q_1+q_2$, where $P^2=u$, $P'^2=u'$ whereas $q_1^2=q_2^2=m^2$ and it is necessary to express the variables u, z, u', z' in our expression given above for M_{I_s} in terms of the quantities u, u' that are to emerge in $V_\lambda^P(u, u')$ as well as in terms of the t we are integrating over. The notation used in writing M_{I_s} was that for the original nonforward process shown in Fig. 1, so we must bring this in accord with the process under consideration. If we write $P=(P-Q/2)+Q/2$, $q_1=-(P-Q/2)+Q/2$, then the «u» occuring in our M_{I_s}, call it $\bar{u}$, should be $\bar{u}=(P-Q/2)^2$ and z, call it $\bar{z}=(P-Q/2)\cdot Q/\sqrt{ut}$. Now $q_1^2=m^2=(P-Q)^2$, so that we find

$$\bar{u}=\frac{u+m^2}{2}-\frac{t}{4}, \qquad \bar{z}=\frac{u-m^2}{2\sqrt{\bar{u}t}}\,.$$

Similarly we find that u', z' should in M_{I_s} be replaced by $\bar{u}'$, $\bar{z}'$ with

$$\bar{u}'=\frac{u'+m^2}{2}-\frac{t}{4}, \qquad \bar{z}'=\frac{u'-m^2}{2\sqrt{\bar{u}'t}}\,.$$

Now we must ask whether it is reasonable to expect significant damping from the residue functions. The answer is that if we imagine an iterative process whereby one starts out with only a low-energy kernel which is taken from on-mass-shell quantities with no dependence on u, v, one finds an output residue function $\beta(\bar{u}, \bar{z};\ t)$ that for large $\bar{u}$ goes like

$$\beta(\bar{u}, \bar{z}, t) \sim (-\bar{u})^{-\alpha(t)-1}\,.$$

This is a quite general feature of the ABFST equation and follows simply from the fact that for finite s, the variable y grows for large u like $\sqrt{-u}$ and thus $Q_l(y) \sim (\sqrt{-u})^{-l-1}$ and in turn $A_l \sim (\sqrt{-u})^{-l-1}$; since $\beta \sim A/(\sqrt{-u})^{l+1}$ by definition we see the advertised result. The upshot of this is that in the interesting neighborhood of the AFS branch point, $V_\lambda^P(u, v) \sim (-u)^{-\alpha_{\max}-1}$ and our kernel will be square integrable in a large region of the λ-plane. Self-consistency is thus seen to play an important role in making our separation into high- and low-energy potentials.

1'6. *Solution of the* ABFST *equation.* – Even in the simplest model, that of simple ladders, the ABFST equation is extremely resistant to analytic solution. The one known exact solution to the ladder model is the physically uninteresting case of zero-mass exchange at zero momentum transfer. To see why this is the case, we note that

$$V(s, u, v) = \pi g^2 \delta(s - m_0^2),$$

defines the general ladder problem and

$$V_\lambda(u, v) = \exp[-(\lambda + 1)\theta(u, v),$$

with $\cosh\theta = (m_0^2 - u - v)/2\sqrt{uv}$. We note that

$$\exp[-\theta] = \frac{m_0^2 - u - v - \Delta^{\frac{1}{2}}(m_0^2, u, v)}{2\sqrt{uv}} \to \frac{-u-v-|u-v|}{2\sqrt{uv}} \quad \text{as } m_0^2 \to 0,$$

$$= \sqrt{\frac{v}{u}} \qquad \text{for } u < v,$$

$$= \sqrt{\frac{u}{v}} \qquad \text{for } u > v.$$

This quasi-separability of the kernel enables one to reduce the integral equation to a hypergeometric equation that can be easily solved. The result is

$$A_\lambda(u, v) = \pi g^2 \frac{\Gamma(\beta + \lambda + 1)\Gamma(\lambda - \beta + 2)}{\Gamma(\lambda + 1)\Gamma(\lambda + 2)} f\left(-\frac{u_<}{m^2}\right) f\left(-\frac{m^2}{u_>}\right),$$

where $u_<$ ($u_>$) mean the smaller (larger) of u and v and

$$f(z) = (z)\frac{\lambda + 1}{2}(1 + z)^\beta {}_2F_1(\beta, \beta + \lambda + 1; \lambda + 2; -z),$$

$$\beta = \frac{1}{2} - \sqrt{\frac{1}{4} + \left(\frac{g}{4\pi m}\right)^2}.$$

There are poles at

$$\lambda = -n - \frac{3}{2} \pm \sqrt{\frac{1}{4} + \left(\frac{g}{4\pi m}\right)^2}, \qquad n = 0, 1, 2, \dots,$$

with residue R given by

$$R = \pi g^2 \frac{n!\,\Gamma(2\lambda+3)}{\Gamma(2\lambda+n+3)\Gamma(\lambda+2)\Gamma(\lambda+1)} \left[\frac{\sqrt{uv}}{(m^2-u)(m^2-v)}\right]^{\lambda+1} \cdot C_n^{\lambda+\frac{3}{2}}\left(\frac{u+m^2}{u-m^2}\right) C_n^{\lambda+\frac{3}{2}}\left(\frac{v+m^2}{v-m^2}\right),$$

where the C_n^v are Gegenbauer polynomials. This result implies that the eigenfunctions of the kernel defined by

$$\frac{g^2}{16\pi^2}\frac{1}{\lambda+1}\int_{-\infty}^{0} \mathrm{d}u' \exp\left[-(\lambda+1)\theta(u,u')\right]\frac{b_\mu(u')}{(m^2-u')^2} = \mu b_\mu(u)$$

are

$$b_\mu(u) = \left[\frac{m\sqrt{-u}}{m^2-u}\right]^{\lambda+1} C_n^{\lambda+\frac{3}{2}}\left(\frac{u+m^2}{u-m^2}\right),$$

and the corresponding eigenvalues μ are

$$\mu = \frac{(g/4\pi m)^2}{(\lambda+n+1)(\lambda+n+2)}.$$

Setting $\mu = 1$ we find the previously noted Regge poles. These results were kindly provided by SAXTON from his Princeton thesis.

It is possible to find approximate but quite accurate solutions to both the forward and nonforward ladder equation into the limit where the exchange mass m_0^2 is much larger than m^2. This is a much more realistic case if we think of m_0 as being the ρ-meson and m being the pion. We consider first the forward equation and observe that if $m_0^2 \gg (-u)$, $(-u')$:

$$\exp\left[-\theta(u,u')\right] = \frac{m_0^2-u-u'-\Delta^{\frac{1}{2}}(m_0^2,u,u')}{2\sqrt{uu'}},$$

$$= \frac{2\sqrt{uu'}}{m_0^2-u-u'+\Delta^{\frac{1}{2}}(m_0^2,u,u')},$$

$$\approx \frac{\sqrt{uu'}}{m_0^2-u-u'},$$

$$a) \qquad \approx \frac{m_0^2\sqrt{uu'}}{(m_0^2-u)(m_0^2-u')}\,,$$

$$b) \qquad \approx \frac{\sqrt{uu'}}{\sqrt{m_0^2-2u}\sqrt{m_0^2-2u'}}\,.$$

Both of the approximations a) and b) lead to separable kernels and hence trivially soluble integral equations. The kernel a) is always smaller than kernel b) for all negative u, u'. Further, the exact kernel is larger than a) and smaller than b). Since

$$\frac{\mathrm{d}}{\mathrm{d}l}\,\frac{\exp[-(l+1)\theta]}{l+1}<0\,,$$

we can conclude that the eigenvalue of any of these kernels, call them $\mu(l)$, $\mu_a(l)$, $\mu_b(l)$ are decreasing functions of l. Since the Regge poles correspond to the solution of the homogeneous equation, setting the μ's $=1$ determines the poles. This then implies that for a given value of the coupling g^2, the leading Regge poles for a), b) and exact are in the relation

$$\alpha_a(0)<\alpha(0)<\alpha_b(0)\,.$$

The eigenfunction provided by kernel a) might be expected to be more accurate than that for b) since approximation a) has the correct behavior for both small and large values of $(-u)$ whereas b) is correct only for small $(-u)$.

In general, for a separable kernel for our forward-scattering equation in the ladder model we write

$$\exp[-(\lambda+1)\theta(u,u')]=f_\lambda(u)f_\lambda(u')\,.$$

The forward equation becomes

$$A_\lambda(u,v)=\pi g^2 f_\lambda(u)f_\lambda(v)+\frac{g^2}{16\pi^2}\frac{f_\lambda(u)}{(\lambda+1)}\int\limits_{-\infty}^{0}\mathrm{d}u'\,\frac{f_\lambda(u')}{(m^2-u')^2}A_\lambda(u',v)\,,$$

which is instantaneously soluble:

$$A_\lambda(u,v)=\frac{\pi g^2 f_\lambda(u)f_\lambda(v)}{1-(g^2/16\pi^2)(1/(\lambda+1))\int_{-\infty}^{0}\mathrm{d}u\,(f_\lambda^2(u)/(m^2-u)^2)}\,.$$

There are three approximations that we have studied: Two are the ones labeled a), b) above, and the third, c), is one in which the denominator function which appears in our solution coincides with the first Fredholm approximation. The latter is of course given by

$$D(\lambda) = 1 - \frac{g^2}{16\pi^2}\frac{1}{\lambda+1}\int_{-\infty}^{0} du \frac{\exp[-(\lambda+1)(u,u)]}{(m^2-u)^2}.$$

We have then, obviously, the following f_λ's:

a) $f_\lambda = \left[\dfrac{m_0\sqrt{-u}}{m_0^2-u}\right]^{\lambda+1}$,

b) $f_\lambda = \left[\dfrac{\sqrt{u}}{\sqrt{m_0^2-2u}}\right]^{\lambda+1}$,

c) $f_\lambda = \exp\left[-\dfrac{(\lambda+1)}{2}\theta(u,u)\right] = \left[\dfrac{2(-u)}{m_9^2-2u+m_0\sqrt{m_0^2-4u}}\right]^{((\lambda+1)/2)}.$

These f_λ's are, of course, the eigenfunctions $\psi_\mu(u)$ of our separable kernels, and the eigenvalues (μ) defined by

$$\frac{g^2}{16\pi^2}\frac{1}{(\lambda+1)}\int_{-\infty}^{0} du' \frac{f_\lambda(u)f_\lambda(u')}{(m^2-u')^2}\psi_\mu(u') = \mu\psi_\mu(u'),$$

are given by

$$\mu(\lambda) = \frac{g^2}{16\pi^2}\frac{1}{\lambda+1}\int_{-\infty}^{0} du \frac{f_\lambda^2(u)}{(m^2-u)^2}.$$

The Regge poles, corresponding to the vanishing denominator function, are determined by $\mu(\lambda) = 1$.

The quantities $\mu(\lambda)$ for all three cases can be evaluated explicitly and we record them more for completeness rather than beauty ($\eta = m^2/m_0^2$):

$$\mu^{(a)}(\lambda) = \left(\frac{g}{4\pi m_0}\right)^2 \frac{\Gamma(\lambda+1)\Gamma(\lambda+2)}{\Gamma(2\lambda+4)}\left[\frac{2}{1+\eta}\right]^2 {}_2F_1\left(1, \frac{3}{2}; \lambda+\frac{5}{2}; \left(\frac{1-\eta}{1+\eta}\right)^2\right),$$

$$\mu^{(b)}(\lambda) = \left(\frac{g}{4\pi m_0}\right)^2 \frac{2}{(\lambda+1)(\lambda+2)} {}_2F_1(2, 1; l+3; 1-2\eta).$$

Although it can be written down in terms of ${}_3F_2$-functions, the case c) corresponding to the trace approximation is very ugly. We give only the expansion in powers of η:

$$\mu^{(c)}(\lambda) = \left(\frac{g}{4\pi m_0}\right) \cdot$$

$$\cdot\left[\frac{2}{\lambda(\lambda+1)(\lambda+2)} - \frac{\pi}{\sin\pi\lambda}\eta^\lambda + \frac{24}{\lambda(\lambda+1)(\lambda+2)(\lambda+3)(1-\lambda)}\eta + \dots\right].$$

In the limit $m^2 \to 0$, which can be taken only if $\lambda > 0$, we find with $(g/4\pi m_0)^2 = \gamma$

$$\mu^{(a)}(\lambda) = \gamma \frac{\Gamma(\lambda)\Gamma(\lambda+1)}{\Gamma(2\lambda+2)} ,$$

$$\mu^{(b)}(\lambda) = \gamma \frac{1}{2^\lambda \lambda(\lambda+1)} ,$$

$$\mu^{(c)}(\lambda) = \gamma \frac{2}{\lambda(\lambda+1)(\lambda+2)} .$$

Exact numerical integration in this case shows that to get $\mu(\lambda) = 1$ at $\lambda = 1$ requires a value of $\gamma = 5.3$; for the above three approximations we have *a*) $\gamma = 6$, *b*) $\gamma = 4$, *c*) $\gamma = 3$, so that approximation *a*) is not bad for the eigenvalue calculation.

Before turning to the nonforward case we note some mathematical results which are amusing and even useful. The first has to do with the Mellin transform of the ladder kernel, the transform being taken in the exchange mass. This is the fastest way to generate the expansion of $\mu^{(c)}(\lambda)$ given above and if necessary to find the series for large values of m^2/m_0^2. Consider

$$I(s) = \int_0^\infty \mathrm{d}x\, x^{-s-1} \left[\frac{x - u - u' - \Delta^{\frac{1}{2}}(x, u, u')}{2\sqrt{uu'}}\right]^{\lambda+1} ,$$

where we have written $x = m_0^2$ and the quantity in brackets is just $\exp[-\theta(u, u')]$. It is relatively straightforward to show that

$$I(s) = [\sqrt{uu'}]^{-s} x_0^{1+s+\lambda} \frac{\Gamma(-s)\Gamma(\lambda+1+s)}{\Gamma(\lambda+1)} {}_2F_1(s, \lambda+s+1; \lambda+2; x_0^2) ,$$

where $x_0 = (u_>/u_<)^{\frac{1}{2}}$ and $u_>$ $(u_<)$ is the larger (smaller) of u, u'; note that since u, u' are negative, $x_0 < 1$. The next observation is that the ladder kernel can be written as an integral in which the dependence on the three variables m_0^2, u, u' is separated:

$$\frac{1}{\lambda+1} \exp[-(\lambda+1)\theta(u, u')] = 2m_0 \int_0^\infty \mathrm{d}y\, K_1(m_0 y)\, J_{\lambda+1}(y\sqrt{-u})\, J_{\lambda+1}(y\sqrt{-u'}) .$$

This interesting representation was derived by SHAFFER. It can be utilized effectively in the calculation of the second Fredholm approximation to the forward ABFST equation. We recall that the Fredholm form for the solution of the equation

$$\psi(x) = \varphi(x) + \int \mathrm{d}y\, K(x, y)\psi(y) ,$$

with a symmetric kernel may be written as

$$\psi(x) = -\int dy\,[\delta D/\delta K(x, y)]\varphi(y)/D\,,$$

where the Fredholm denominator D is given by

$$D = \det|1 - K| = \exp[\mathrm{Tr}\ln(1 - K)] \approx 1 - \mathrm{Tr}\,K - \tfrac{1}{2}[\mathrm{Tr}\,K^2 - (\mathrm{Tr}\,K)^2] - \dots .$$

In our model

$$K = \frac{g^2}{16\pi^2}\frac{1}{\lambda+1}\exp[-(\lambda+1)\theta(u, u')]\frac{1}{(m^2 - u')^2}$$

and

$$\mathrm{Tr}\,K^2 = \left[\frac{g^2}{16\pi^2}\frac{1}{\lambda+1}\right]^2 \int_{-\infty}^{0} du \int_{-\infty}^{0} du' \frac{\exp[-2(\lambda+1)\theta(u, u')]}{(m^2 - u)^2(m^2 - u')^2}\,.$$

It is unnecessary to explicitly symmetrize K in evaluating D. Using our representation we find

$$\mathrm{Tr}\,K^2 = \frac{4m_0}{\lambda+1}\left(\frac{g^2}{16\pi^2}\right)^2 \int_0 dy\,K_1(m_0 y)\left[\int_{-\infty}^{0} du\,\frac{J_{2\lambda+2}(y\sqrt{-u})}{(m^2 - u)^2}\right]^2.$$

The u-integral can be done but it is too scary to reveal. If however we set $m = 0$, everything is trivial and we find

$$\mathrm{Tr}\,K^2 = \frac{1}{\lambda+1}\left(\frac{g^2}{16\pi^2 m_0^2}\right)^2 \left[\frac{2}{\lambda(\lambda+1)(\lambda+2)}\right]^2.$$

Remembering that for $m = 0$,

$$\mathrm{Tr}\,K = \frac{g^2}{16\pi^2 m_0^2}\,\frac{2}{\lambda(\lambda+1)(\lambda+2)}\,,$$

we find (writing $\gamma = (g/4\pi m_0)^2$),

$$D = 1 - \gamma\frac{2}{\lambda(\lambda+1)(\lambda+2)} + (\gamma^2/2)\left[\frac{2}{\lambda(\lambda+1)(\lambda+2)}\right]^2 \frac{\lambda}{\lambda+1}\,.$$

Note that if $D = 0$ at $\lambda = 1$, $\gamma = 6$.

We turn now to a discussion of the nonforward ABFST equation in the ladder approximation. The equation takes the form

$$A_l(u, z; v, \zeta; t) = \pi g^2 Q_l \left[\frac{\cosh\theta(u, v) - z\zeta}{\sqrt{(1-z^2)(1-\zeta^2)}}\right] +$$

$$+ \frac{g^2}{16\pi^3}\int_{-\infty}^{0} du' \int_{-1}^{+1} \frac{dz'}{\sqrt{1-z'^2}} Q_l \left[\frac{\cosh\theta(u, u') - zz'}{\sqrt{(1-z^2)(1-z'^2)}}\right] S(u', z', t) A_l(u', z'; v, \zeta; t)\,,$$

where again

$$\cosh\theta(u, u') = \frac{m_0^2 - u - u'}{2\sqrt{uu'}}\,,$$

$$S(u, z, t) = \frac{1}{(m^2 - u - t/4)^2 - tuz^2}\,.$$

The separability of the kernel Q_l in an approximate sense will be assumed just as in the forward equation

$$Q_l \left[\frac{\cosh\theta(u, u) - zz'}{(1-z^2)(1-z'^2)}\right] = F_l(u, z) F_l(u', z')\,.$$

The solution to our equation is then

$$A_l(u, z; v, \zeta; t) = \frac{\pi g^2 F_l(u, z) F_l(v, \zeta)}{1 - (g^2/16\pi^2)\int_{-\infty}^{0} du \int_{-1}^{+1} dz/(\sqrt{1-z^2})\, F_l^2(u, z) S(u, z, t)}\,.$$

Note that in this approximation, the entire t-dependence of A comes from the denominator function which we shall refer to as $D(l, t)$. The Regge poles correspond to $D(l, t) = 0$.

Choices for the function $F_l(u, z)$ are suggested by considering large values of the exchange mass m_0^2 which means, of course, not only $m_0^2/m^2 \gg 1$ but also m_0^2 larger than the relevant values of $-u$, $-u'$. Then $\cosh\theta(u, u') \gg 1$ and we may then use the asymptotic form for Q_l. Further, since the z's are less than or equal to one, we drop the zz'-term compared to $\cosh\theta$. Thus

$$Q_l \to \frac{\sqrt{\pi}\,\Gamma(l+1)}{\Gamma(l+\frac{3}{2})} \frac{[\sqrt{uu'(1-z^2)(1-z'^2)}]^{l+1}}{[m_0^2 - u - u']^{l+1}}\,.$$

The z-dependence is already separable while the generation of a separable approximation for the u-dependence is exactly the same as the problem already solved in connection with the forward equation. We write

$$F_l(u, z) = \sqrt{C_l}[\sqrt{1-z^2}]^{l+1} f_l(u),$$

where

$$C_l = \frac{\sqrt{\pi}\Gamma(l+1)}{\Gamma(l+\frac{3}{2})}$$

and

$$f_l^{(a)}(u) = \left[\frac{m_0\sqrt{-u}}{m_0^2 - u}\right]^{l+1},$$

$$f_l^{(b)}(u) = \left[\frac{\sqrt{-u}}{\sqrt{m_0^2 - 2u}}\right]^{l+1}.$$

We can obviously make an approximation which makes $D(l, t)$ coincide with the first Fredholm approximation by choosing

$$F_l(u, z) = \left\{Q_l\left[1 - \frac{m_0^2}{2u(1-z^2)}\right]\right\}^{\frac{1}{2}}.$$

For all three of these approximations the z integration in $D(l, t)$ may be carried out explicitly and we are led to the following expressions for $D(l, t)$:

$$D^{(a,b)}(l, t) = 1 - \frac{g^2}{16\pi^2}\frac{1}{l+1}\int_{-\infty}^{0} \mathrm{d}u \frac{f_l^2(u)}{(m^2-u-t/4)^2}\,{}_2F_1\left(\frac{1}{2}, 1; l+2; \frac{ut}{(m^2-u-t/4)^2}\right),$$

$$D\quad(l, t) = 1 - \frac{g^2}{16\pi^2}\int_0^{\infty}\frac{\mathrm{d}x}{x} Q_l\left(1+\frac{m_0^2}{2x^2}\right)\frac{1}{\sqrt{x^2+m^2(z^2+m^2-t/4)}}.$$

The first expression is to be used with either of the above two separable approximations $f_l^{(a)}$, $f_l^{(b)}$. The second is what corresponds to the first Fredholm (or trace) approximation. Note that $D_l^{(a,b)}(t=0)$ coincides precisely with the corresponding expressions derived for the forward equation. As we noted before, the trace approximation does *not* go over into the forward trace approximation as $t \to 0$.

The above expressions cannot be evaluated very neatly or easily although it is straightforward to generate a power series in t. Since we shall encounter the same problem in another connection in the next chapter we shall not stop to discuss the matter here.

We conclude this introductory Part with a remark about the problem of continuing our solutions onto the mass shell. We recall that we have consistently assumed that the quantities u, v were negative. There is no reason to believe there should be any fundamental problem associated with going to the positive values implied by going to the mass shell provided the external masses are not such as to involve us in anomalous thresholds. In the case of forward scattering where we have $m_a + m_b \to m_a + m_b$, the physical values of u and v are $u = m_a^2$, $v = m_b^2$. For nonforward scattering we have $u = m_a^2 - t/4$, $v = m_b^2 - t/4$. In this case, for large $(-t)$, u, v can become very large and positive. The whole problem of continuation to positive u, v has not been too widely studied. It has been shown by SAXTON in the zero-mass exchange case, that so long as u, v are different from the internal mass m^2 the solutions for negative u, v are, in fact, valid for all u, v. We should note that our various approximate solutions based on making the various kernels separable have some problems associated with them in so far as analyticity in t is concerned. It is easy enough to see that $D(l, t)$ has only the expected branch point at $t = 4m^2$ and thus that (in general) the trajectories determined by $D(l, t) = 0$ will share this behavior. But the residues of the poles are quite crazy. It is well known that Regge residues should have a branch point also only for $t = 4m^2$ whereas something like $[m_0^2 - u]^{-l-1} = [m_0^2 - m_a^2 + t/4]^{-l-1}$ will have strange, quite unphysical branch points. Thus our approximate solutions cannot be correct in the whole t-plane although they might be reasonable predictors of trajectories and even quite accurate for small t.

2. – Variational principle for ABFST equations and applications.

2'1. *The variational principle.* – We have already called attention to a stationary expression for the absorptive part of the nonforward elastic scattering amplitude $A(P, K, Q)$. This is not a particularly useful expression because one is dealing with a four-dimensional space and because one requires trial functions valid for all values of the variables $(P+K)^2$, P^2, K^2, $P\cdot Q$, $K\cdot Q$, Q^2. It is considerably easier to use the partially diagonalized nonforward equation or the one-dimensional forward equation and formulate a variational basis for these. We shall see in this Section how the variational principle may be used to extract much useful information from the multiperipheral equations.

The integral equations have the formal structure

$$A_{\gamma\gamma'} = V_{\gamma\gamma'} + \sum_{\gamma''} V_{\gamma\gamma''} S_{\gamma''} A_{\gamma''\gamma'},$$

or writing this as a matrix equation

$$A = V + VSA = V + ASV.$$

The indices γ refer to u, z in the nonforward case or to u for the forward one; V is the transform of the basic elements of the multiperipheral chain and S is the propagator. The summation includes the measure of the integration variables as well as various constant factors. We shall make these all explicit later on. A stationary expression for A which is a slight modification of the familiar Schwinger variational principle is easily found. Defining $A - V \equiv \mathscr{A}$, we may readily verify that

$$\mathscr{A}_{\gamma\gamma_0} = \frac{(ASV)_{\gamma\gamma_0}(VSA)_{\gamma\gamma_0}}{(ASA)_{\gamma\gamma_0} - (ASVSA)_{\gamma\gamma_0}}$$

is stationary under arbitrary variations of A about the solution of the equations $A = V + VSA$ or $A = V + ASV$. The great power of the variational principle is that it enables one to work in the neighborhood of a solution of the homogeneous equation $A = VSA$ which determines the spectrum of singularities in the l and λ planes.

2·2. *Trajectory slope in the* ABFST *model.* – Our first application will be to the calculation of the slope of a Regge trajectory at $t = 0$ in terms of the solution of the much simpler ABFST equation for $t = 0$. Basically, what we shall do is to take the $t = 0$ solution as a trial function and systematically expand everything else to first order in t. Since we use a variational method, the errors will be of order t^2 and thus we can calculate the slope *exactly*. We treat in detail the simple ladder model and later comment on generalizations to more realistic kernels.

In the ladder model the nonforward equation takes the form

$$A_l(u, z; v, \zeta; t) = \pi g^2 Q_l\left[\frac{\cosh\theta(u, v) - z\zeta}{\sqrt{(1-z^2)(1-z^2)}}\right] + \frac{g^2}{16\pi^3}\int_{-\infty}^{0} du' \int_{-1}^{+1} \frac{dz'}{\sqrt{1-z'^2}} \cdot$$

$$\cdot Q_l\left[\frac{\cosh\theta(u, u') - zz'}{\sqrt{(1-z^2)(1-z'^2)}}\right] S(u', z', t) A(u', z'; v, \zeta; t) ,$$

where we have used our standard notation

$$\cosh\theta(u, u') = \frac{m_0^2 - u - u'}{2\sqrt{uu'}} ,$$

$$S(u, z, t) = \frac{1}{(m^2 - u - t/4)^2 - tuz^2} .$$

The traslation to the abstract notation in our variational principle is

$$\gamma \;= (u, z)\,, \qquad \gamma_0 = (v, \zeta)\,,$$

$$V_{\gamma\gamma'} = \pi g^2 Q_l\left[\frac{\cosh\theta(u, u') - zz'}{\sqrt{(1-z^2)(1-z'^2)}}\right],$$

$$\sum_{\gamma'} \;= \frac{1}{16\pi^4}\int_{-\infty}^{0}\mathrm{d}u'\int_{-1}^{+1}\frac{\mathrm{d}z'}{\sqrt{1-z'^2}}\,.$$

The Regge poles $l = \alpha(t)$ correspond to the vanishing of the denominator of the variational principle or in other words to those values of l for which the homogeneous equation $A = VSA$ is satisfied. In such a neighborhood, the dependence of A on the indices γ, γ_0 factors, so that $A_{\gamma\gamma_0} = a_\gamma a_{\gamma_0}$. We shall take as a general trial function such a factorized form, so that the variational principle now looks like

$$\mathscr{A}_{\gamma\gamma_0} = \frac{\left(\sum\limits_{\gamma'} V_{\gamma\gamma'}S_{\gamma'}a_{\gamma'}\right)\left(\sum\limits_{\gamma'} a_{\gamma'}S_{\gamma'}a_{\gamma_0}\right)}{\left(\sum\limits_{\gamma'} a_{\gamma'}S_{\gamma'}a_{\gamma'}\right) - \left(\sum\limits_{\gamma'\gamma''} a_{\gamma'}S_{\gamma'}V_{\gamma'\gamma''}S_{\gamma''}a_{\gamma''}\right)}\,.$$

Note that the denominator is now a number and not a matrix. Writing this out explicitly for our ladder model, we have

$$\mathscr{A}_l(u, z;\, v, \zeta;\, t) = \frac{(g^4/16\pi^2)\,I_l(u, z, t)\,I_l(v, \zeta, t)}{I_0(l, t) - (g^2/16\pi^3)\,I_2(l, t)}\,,$$

where

$$I_l(u, z, t) = \int_{-\infty}^{0}\mathrm{d}u'\int_{-1}^{+1}\frac{\mathrm{d}z'}{\sqrt{1-z'^2}}\,Q_l\left[\frac{\cosh\theta(u, u') - zz'}{\sqrt{(1-z^2)(1-z'^2)}}\right]S(u', z', t)\,a_l(u', z', t)\,,$$

$$I_0(l, t) \;= \int_{-\infty}^{0}\mathrm{d}u'\int_{-1}^{+1}\frac{\mathrm{d}z'}{\sqrt{1-z'^2}}\,a_l^2(u', z', t)\,S_l(u', z', t)\,,$$

$$I_2(l, t) \;= \int_{-\infty}^{0}\mathrm{d}u\int_{-1}^{+1}\frac{\mathrm{d}z}{\sqrt{1-z^2}}\int_{-\infty}^{0}\mathrm{d}u'\int_{-1}^{+1}\frac{\mathrm{d}z'}{\sqrt{1-z'^2}}\,a_l(u, z, t)\,S(u, z, t)\,\cdot$$

$$\cdot\, Q_l\left[\frac{\cosh\theta(u, u') - zz'}{\sqrt{(1-z^2)(1-z'^2)}}\right]S(u', z', t)\,a_l(u', z', t)\,.$$

To proceed now with the slope calculation we make an explicit ansatz for the trial function, namely

$$ {}_l(u, z, t) = \sqrt{C_l}\,[\sqrt{1-z^2}]^{l+1} f_l(u)\,, $$

where C_l is a convenient (but, of course, irrelevant) constant, the factor $\left[\sqrt{1-z^2}\right]^{l+1}$ is the same threshold behavior we noted in the previous chapter and $f_l(u)$ will be specified shortly. The crucial assumption for the moment is that it is independent of t. We concentrate our attention on the denominator.

Consider first

$$ I_0(l, t) = \int_{-\infty}^{0} \mathrm{d}u\, f_l^2(u)\, C_l \int_{-1}^{+1} \mathrm{d}z \frac{[\sqrt{1-z^2}]^{2l+1}}{(m^2-u-t/4)^2 - tuz^2}\,. $$

With C_l chosen to be

$$ C_l = \frac{\sqrt{\pi}\,\Gamma(l+1)}{\Gamma(l+\frac{3}{2})}\,, $$

the z-integral is one we had already encountered and conquered earlier and we define

$$ S_l(u, t) \equiv C_l \int_{-1}^{+1} \mathrm{d}z \frac{[\sqrt{1-z^2}]^{2l+1}}{(m^2-u-t/4)^2 - tuz^2} = $$

$$ = \frac{\pi}{l+1} \frac{1}{(m^2-u-t/4)^2}\, {}_2F_1\left(\frac{1}{2}, 1; l+2; \frac{ut}{(m^2-u-t/4)^2}\right). $$

With this notation

$$ I_0(l, t) = \int_{-\infty}^{0} \mathrm{d}u\, f_l^2(u)\, S_l(u, t)\,. $$

To deal with the double integral $I_2(l, t)$ we make use of a brilliant formula found by SAUNDERS, SAXTON and TAN:

$$ Q_l\left[\frac{\cosh\theta - zz'}{\sqrt{(1-z^2)(1-z'^2)}}\right] = \sum_{n=0}^{\infty} \frac{[\Gamma(l+1)]^2\, 2^{2l+1} n!}{\Gamma(n+2l+2)} [\sqrt{(1-z^2)(1-z'^2)}]^{l+1}\cdot $$

$$ \cdot\, C_n^{l+1}(z)\, C_n^{l+1}(z') \exp[-(l+1+n)\theta]\,, $$

where the $C_n^{l+1}(z)$ are Gegenbauer polynomials defined by

$$ \frac{1}{(1-2tz+t^2)^{l+1}} = \sum_{n=0}^{\infty} C_n^{l+1}(z)\, t^n\,. $$

When this expansion is put into I_2 along with our trial function, we see that the two z-integrals factor into $S_l^n(u, t)S_l^n(u', t)$, where

$$S_l^n(u, t) = \int_{-1}^{+1} dz \frac{[\sqrt{1-z^2}]^{2l+1} C_n^{l+1}(z)}{(m^2 - u - t/4)^2 - tuz^2} =$$

$$= \frac{\pi a^n}{(m^2 - u - t/4)^2} \frac{\Gamma(2l+2+n)}{2^{2l+1}\Gamma(l+1)\Gamma(l+n+2)} {}_2F_1\left(\frac{n+1}{2}, \frac{n}{2}+1; n+l+2; a^2\right)$$

with

$$a = \frac{\sqrt{ut}}{m^2 - u - t/4}, \qquad n \text{ even}, \qquad S_l^n = 0, \qquad n \text{ odd}.$$

For $n = 0$

$$S_l^0 = C_l^{-1} S_l(u, t).$$

Since we are ultimately to be concerned with terms no higher than first order in t we need not use this complicated formula except for $n = 0$; the leading t-dependence is $t^{n/2}$ and n is 2, 4, ... Thus the product $S_l^n(u, t)S_l^n(u', t)$ is proportional to t^n and the terms corresponding to $n > 0$ may be dropped. That this is true may also be seen without evaluating the integral for S^n explicitly by noticing that

$$\int_{-1}^{+1} dz [\sqrt{1-z^2}]^{2l+1} C_n^{l+1}(z) = 0, \qquad n \neq 0.$$

We are now in position to complete the evaluation of $I_2(l, t)$ to the requisite (first) order in t. We note that

$$\frac{2^{2+1}[\Gamma(l+1)]^2}{(\Gamma 2l+2)} = \frac{\sqrt{\pi}\Gamma(l+1)}{\Gamma(l+\frac{3}{2})} = C_l,$$

and that we had cleverly supplied a factor of $\sqrt{C_l}$ in our trial function. Hence from the $n = 0$ term in the Q_l expansion we find

$$I_2(l, t) = \int_{-\infty}^{0} du \int_{-\infty}^{0} du' f_l(u) S_l(u, t) \exp[-(l+1)\theta(u, u')] S_l(u', t) f_l(u').$$

We may now write the denominator of the variational principle, call it

$D(l, t)$, as

$$D(l, t) = \int_{-\infty}^{0} du\, f_l^2(u) S(u, t) - $$
$$- \frac{g^2}{16\pi^3} \int_{-\infty}^{0} du \int_{-\infty}^{0} du' f_l(u) S_l(u, t) \exp[-(l+1)\theta(u, u')] S_l(u', t) f_l(u') .$$

This is our fundamental quantity and it is important to emphasize that it is accurate up to and including terms linear in t.

It is worth noting that if one makes, ab initio, the approximation

$$Q_l \left[\frac{\cosh\theta - zz'}{\sqrt{(1-z^2)(1-z'^2)}}\right] \approx C_l [\sqrt{(1-z^2)(1-z'^2)}]^{l+1} \exp[-(l+1)\theta]$$

in the integral equation for $A_l(u, z;\ v, \zeta;\ t)$ [corresponding to the term $n = 0$ in the Saunders, Saxton, Tan expansion] one may derive a one-dimensional integral equation for the nonforward problem. This comes about because the z-dependence is separable. The deduction of the equation is completely straightforward and we give only the result

$$A_l(u, z;\ v, \zeta;\ t) = C_l [\sqrt{(1-z^2)(1-z^2)}]^{l+1} B_l(u, v, t) ,$$

$$B_l(u, v, t) = \pi g^2 \exp[-(l+1)\theta(u, v)] + $$
$$+ \frac{g^2}{16\pi^3} \int_{-\infty}^{0} du' \exp[-(l+1)\theta(u, u')] S_l(u', t) B_l(u', v, t) ,$$

where $S_l(u, t)$ is, as before, given by

$$S_l(u, t) = \frac{\pi}{l+1} \frac{1}{(m^2 - u - t/4)^2}\, {}_2F_1\left(\frac{1}{2}, 1;\ l+2, \frac{ut}{(m^2 - u - t/4)^2}\right) .$$

Note that as $t \to 0$, the equation satisfied by $B\ (u, v, 0)$ is precisely the forward scattering ABFST equation in the ladder model:

$$B_l(u, v, 0) = \pi g^2 \exp[-(l+1)\theta(u, v)] + $$
$$+ \frac{g^2}{16\pi^2(l+1)} \int_{-\infty}^{0} du' \frac{\exp[-(l+1)\theta(u, u')] B_l(u', v, 0)}{(m^2 - u')^2} .$$

It would be interesting to study the equation for $B(u, v, t)$ in detail since it is certainly more tractable than the general two-dimensional nonforward equation.

A variational principle for $\mathcal{B}_l(u, v, t)$ defined by

$$\mathcal{B}_l(u, v, t) = B_l(u, v, t) - \pi g^2 \exp[-(l+1)\theta(u, v)],$$

is given by

$$\mathcal{B}_l(u, v, t) = \frac{(g^4/16\pi^2) I_1(u, v, t) I_1(u, v, t)}{I_0(u, v, t) - (g^2/16\pi^3) I_2(u, v, t)},$$

where

$$I_1(u, v, t) = \int_{-\infty}^{0} du' \exp[-(l+1)\theta(u, u')] S_l(u', t) B_l(u', v, t),$$

$$I_1(u, v, t) = \int_{-\infty}^{0} du' B_l(u, u', t) S_l(u', t) \exp[-(l+1)\theta(u', v),$$

$$I_0(u, v, t) = \int_{-\infty}^{0} du' B_l(u, u', t) S_l(u', t) B_l(u', v, t),$$

$$I_2(u, v, t) = \int_{-\infty}^{0} du' \int_{-\infty}^{0} du'' B_l(u, u', t) S_l(u', t) \exp[-(l+1)\theta(u', u'')] S_l(u'', t) B_l(u', v, t).$$

If we now restrict attention to trial functions of the form $B_l(u, v, t) = b_l(u, t) b_l(v, t)$, we find for $\mathcal{B}_l$ the expression

$$\mathcal{B}_l(u, v, t) = \frac{(g^4/16\pi^2) I_l(u, t) I_l(v, t)}{I_0(l, t) - (g^2/16\pi^3) I_2(l, t)}$$

with

$$I_l(u, t) = \int_{-\infty}^{0} du' \exp[-(l+1)\theta(u, u')] S_l(u', t) b_l(u', t),$$

$$I_0(l, t) = \int_{-\infty}^{0} du\, b_l^2(u, t) S_l(u, t),$$

$$I_2(l, t) = \int_{-\infty}^{0} du \int_{-\infty}^{0} du' b_l(u, t) S_l(u, t) \exp[-(l+1)\theta(u, u')] S_l(u', t) b_l(u', t).$$

Finally we note that if $b_l(u, t) = f_l(u)$ we recover the same denominator function $D(l, t)$, we had been led to when we used the exact kernel.

We turn now to a study of this denominator function. Our object is to determine the Regge trajectory $l(t)$ from the requirement $D(l, t) = 0$. We choose the trial function $f_l(u)$ to be the solution of the homogeneous forward equation corresponding to an eigenvalue $l = l_0$. Specifically we set $f_l(u) = f_0(u)$ where

$$f_0(u) = \frac{g^2}{16\pi^3} \int_{-\infty}^{0} du' \exp[-(l_0 + 1)\theta(u, u')] S_0(u') f_0(u') ,$$

where

$$S_0(u) = S_{l_0}(u, 0) = \frac{\pi}{l_0 + 1} \frac{1}{(m^2 - u)^2} .$$

We next expand $S_l(u, t)$ to first order in t and in $l - l_0 = \Delta l$ and also $\exp[-l-1]\theta(u, u')$ about l_0:

$$S_l(u, t) \approx S_0(u) + tS_0'(u) - \frac{\Delta l}{l_0 + 1} S_0(u) ,$$

$$\exp[-(l+1)\theta] = \exp[-(l_0 + 1)\theta][1 - \Delta l\theta] ,$$

where

$$S_0'(u) \equiv S_{l_0}'(u, 0) = \frac{\pi}{2(l+1)(l+2)} \frac{1}{(m^2 - u)^3} \left[l + 1 + \frac{m^2}{m^2 - u} \right] .$$

Note that $S_0'(u) > 0$ for all $u \leqslant 0$. Finally we substitute these expansions into

$$D(l, t) = \int_{-\infty}^{0} du\, f_0^2(u) S_l(u, t) - $$
$$- \frac{g^2}{16\pi^3} \int_{-\infty}^{0} du \int_{-\infty}^{0} du' f_0(u) S_l(u, t) \exp[-(l+1)\theta(u, u')] S_l(u', t) f_0(u') ,$$

and choose Δl so that $D(l_0 + \Delta l, t) = 0$. We make heavy use of the equation satisfied by $f_0(u)$ and find

$$\Delta l = t \frac{\displaystyle\int_{-\infty}^{0} du\, f_0^2(u) S_0'(u)}{\displaystyle\int_{-\infty}^{0} du \frac{f_0^2(u) S_0(u)}{l+1} + \frac{g^2}{16\pi^3} \int_{-\infty}^{0} du \int_{-\infty}^{0} du' f_0(u) S_0(u) \exp[-(l_0+1)\theta(u, u')]\theta(u, u') S_0(u') f_0(u')} .$$

Fortunately, this unpleasant looking denominator can be written in an interesting and simple way. We write the forward-equation relation

$$\int_{-\infty}^{0} du\, f_0^2(u)\, S_0(u) = \frac{g^2}{16\pi^3} \int_{-\infty}^{0} du \int_{-\infty}^{0} du'\, f_0(u)\, S_0(u) \exp\left[-(l_0+1)\theta(u,u')\right] S_0(u')\, f_0(u')\,;$$

this can be regarded as a stationary principle for the coupling constant g^2. We now differentiate the whole expression with respect to l_0, regarding g^2 as a function of l_0. We find

$$\frac{1}{g^2}\frac{\partial g^2}{\partial l_0} = \\ = \frac{\int_{-\infty}^{0} du\, f_0^2 S_0 + \frac{g^2}{16\pi^3}\int_{-\infty}^{0} du \int_{-\infty}^{0} du'\, f_0(u)\, S_0(u) \exp\left[-(l_0+1)\theta(u,u')\right]\theta(u,u')\, S_l(u')\, f_0(u')}{\int_{-\infty}^{0} du\, f_0^2 S_0},$$

so that our result for Δl may be written as

$$\Delta l = t g^2 \frac{\partial l_0}{\partial g^2} \frac{\int_{-\infty}^{0} du\, f_0^2(u)\, S_0'(u)}{\int_{-\infty}^{0} du\, f_0^2(u)\, S_0(u)}\,.$$

This is our final, *exact* expression for the slope of the Regge trajectory in the ladder model. [We shall note later on that essentially the same result obtains for the most general multiperipheral model.] There are two features that are worth noting. First, the slope $\Delta l/t$ is positive, since $\delta l_0/\partial g^2$ is positive (as can be seen from the relation we used to simplify the original formula) and, as previously noted, $S_0'(u) > 0$. Second, the factor $g^2\,\partial l_0/\partial g^2$ is the coefficient of the $\ln s$ term in the mean multiplicity

$$\langle n \rangle = g^2 \frac{\partial l_0}{\partial g^2} \ln s + \text{const}\,.$$

The appearance of this same factor in our slope evaluation is a surprise and does *not* seem very obvious. As we shall see shortly, it survives when we extend our result to more general multiperipheral models. So far, no direct, model-independent interpretation has been found for the ratio of integrals in the Δl formula though it is obviously some sort of mean inverse $(\text{mass})^2$.

It is possible to give a rather simpler derivation of the slope formula. Imagine that the $t = 0$ equation has been solved with the coupling constant g^2

adjusted so that the eigenvalue l takes on some preassigned value $l(0, g_0^2)$, where g_0^2 is the required coupling constant. Next, insist that $D(l, t) = 0$ for the same numerical value of l for all t. Evidently this would require a t-dependent coupling constant so that

$$l(t, g^2) = l(0, g^2) .$$

Again, working to first order in t, we write $g^2 = g_0^2 + \Delta g^2$ and expand $l(t, g^2)$:

$$\begin{aligned} l(t, g^2) &\approx l(t, g_0^2) + \Delta g^2 \frac{\partial l(t, g_0^2)}{\partial g_0^2} \\ &\approx l(t, g_0^2) + \Delta g^2 \frac{\partial l(0, g_0^2)}{\partial g_0^2} , \end{aligned}$$

where the last line follows from the fact that Δg^2 is already linear in t. Thus we have

$$\Delta l = l(t, g_0^2) - l(0, g_0^2) = -\Delta g^2 \frac{\partial l(0, g_0^2)}{\partial g_0^2} .$$

To compute Δg^2, we go into our expression for $D(l, t)$, expand the propagator $S_l(u, t)$ to first order in t, write $g^2 = g_0^2 + \Delta g^2$ and demand that Δg^2 be chosen to make $D(l, t) = 0$. We obtain in this way

$$\Delta g^2 = -g_0^2 t \frac{\int_{-\infty}^{0} \mathrm{d}u \, f_0^2(u) S_0'}{\int_{-\infty}^{0} \mathrm{d}u \, f_0^2(u) S_0} ,$$

substituting this into our expression for Δl leads to the previously given formula for Δl.

There are not many cases in which the eigenfunctions $f_0(u)$ and corresponding eigenvalues l_0 are known. In fact, the only one seems to be the case of zero exchange mass, $m_0^2 = 0$ in which case (writing l_0 simply as l) we have

$$f_0 = \left[\frac{m\sqrt{-u}}{m^2 - u}\right]^{1+l} ,$$

$$l = -\frac{3}{2} + \frac{1}{4} + \left(\frac{g}{4\pi m}\right)^2 ,$$

as we noted in Sect. **1**. We find

$$\Delta l = \frac{t}{2m^2} \frac{(l+1)(l+2)^2}{(2l+3)(2l+5)} .$$

Just to get a slight feel for the more realistic situation corresponding to a large exchange mass (*i.e.* $m_0^2 \gg m^2$), we take for f_0 the solution corresponding to one of our separable models, namely

$$f_0 = \left[\frac{m_0\sqrt{-u}}{m_0^2 - u}\right]^{l+1}.$$

The eigenvalue condition given in Sect. **1** was

$$1 = \left(\frac{g}{4\pi m_0}\right)^2 \frac{\Gamma(l+1)\Gamma(l+2)}{\Gamma(2l+4)} \left(\frac{2}{1+\eta}\right)^2 {}_2F_1\left(1, \frac{3}{2}; l+\frac{5}{2}; \left(\frac{1-\eta}{1+\eta}\right)^2\right),$$

where $\eta = m^2/m_0^2 \ll 1$. The slope formula may be written as

$$\Delta l = tg^2 \frac{\partial l}{\partial g^2} \frac{1}{2m^2} \frac{[-((l+1)/2)\eta(\partial/\partial\eta) + (\eta^2/6)(\partial^2/\partial\eta^2)]I(l, \eta)}{(l+2)I(l, \eta)},$$

where $I(l, \eta)$ is $(l+1)$ times the same integral that determines the eigenvalue l:

$$I(l, \eta) = \eta^l \int_0^\infty \mathrm{d}x \frac{x^{l+1}}{(1+x)^2(1+\eta x)^{2l+2}} =$$

$$= \frac{\Gamma^2(l+2)}{\Gamma(2l+4)} \left(\frac{2}{1+\eta}\right)^2 {}_2F_1\left(1, \frac{3}{2}; l+\frac{5}{2}; \left(\frac{1-\eta}{1+\eta}\right)^2\right).$$

This is a rather nasty function to expand in powers of η if you want to consider integral values of l. The reason is that the standard analytic continuation formulae that take you from argument z to $1-z$ break down for integral l. The best way to do the calculation is to take a Mellin transform of $I(l, \eta)$ in the η variable and invert. One finds thereby the representation

$$I(l, \eta) = \int_{c-i\infty}^{c+i\infty} \frac{\mathrm{d}s}{2\pi i} \frac{\Gamma(s+2)\Gamma(-s)\Gamma(l-s)\Gamma(l+s+2)}{\Gamma(2l+2)} \eta^s,$$

where, we assume $l \geqslant 0$, $-2 < c < 0$. To get the series in ascending powers of η, close the contour to the right. We find

$$I(l, \eta) = \frac{\Gamma(l)\Gamma(l+2)}{\Gamma(2l+2)} + \Gamma(-l)\Gamma(l+2)\eta^l - \frac{2\Gamma(l-1)\Gamma(l+3)}{\Gamma(2l+2)}\eta -$$

$$-2(l+2)\Gamma(-l-1)\Gamma(l+3)\eta^{l+1} + \frac{3\Gamma(l-2)\Gamma(l+4)}{\Gamma(2l+2)}\eta^2 + \dots .$$

The disease of integral l's is manifest and one must pass to the limit with care. For example,

$$I(1,\eta)=\tfrac{1}{3}+\tfrac{14}{3}\eta+2\eta\ln\eta+17\eta^2+12\eta^2\ln\eta+\dots.$$

Exact computer evaluation of $I(1,\eta)$ shows that the above expansion gets bad if $\eta\geqslant0.1$. In order to get the slope it is necessary to calculate

$$\frac{\mathrm{d}}{\mathrm{d}l}\frac{I(l,\eta)}{l+1},$$

and this is quite unpleasant. At $l=1$ we find

$$\left.\frac{\mathrm{d}}{\mathrm{d}l}\frac{I(l,\eta)}{l+1}\right|_{l=1}=-\frac{4}{9}+\frac{\eta}{3}\left(\pi^2-\frac{35}{3}\right)+\frac{1}{2}\eta\ln^2\eta+\eta^2(3\ln^2\eta+2\ln^2\eta)+\dots.$$

Putting it all together we obtain for $\Delta l/t$ at $l=1$ the following values:

η	$m_0(2\,\Delta l/t)$
0.01	0.85
0.02	0.75
0.10	0.52

It is clear that our utilization of the variational method to calculate the slope of the Regge trajectory is not crucially dependent upon the precise form of the kernel in the multiperipheral model. We write the general equation in the form

$$A_l(u,z;v,\zeta;t)=\gamma V(u,z;v,\zeta;t)+\frac{\gamma}{16\pi^4}\int_{-\infty}^{0}\mathrm{d}u'\int_{-1}^{+1}\frac{\mathrm{d}z'}{\sqrt{1-z'^2}}\cdot$$
$$\cdot V_l(u,z'u',z';t)\,S(u',z',t)\,A_l(u',z';v,\zeta;t)\,.$$

We proceed exactly as in the ladder model to formulate the variational principle with γV replacing $\pi g^2 Q_l$ and proceed to make the same formal anzatz for the trial function. In the variational denominator the integral $I_0(l,t)$ appears exactly as before and we have the formal denominator $I_0(l,t)-(\gamma/16\pi^4)I_2(l,t)$, but $I_2(l,t)$ is now given by a new expression which we write as

$$I_2(l,t)=C_l^2\int_{-\infty}^{0}\mathrm{d}u\int_{-\infty}^{0}\mathrm{d}u'\,f_0(u)\left[\int_{-1}^{+1}\mathrm{d}z'\,[\sqrt{1-z^2}\,]^{2l+1}S(u,z,t)\cdot\right.$$
$$\left.\cdot\int_{-1}^{+1}\mathrm{d}z'\,[\sqrt{1-z'^2}]^{2l+1}S(u',z',t)\,S_l(u,z;u',z';t)\right]f_0(u')\,,$$

where

$$\mathscr{V}_l(u, z; u', z'; t) = \frac{V(u, z; u', z'; t)}{C_l[\sqrt{(1-z^2)(1-z'^2)}]^{l+1}}.$$

Now we recall from Sect. **1** that $V(u, z; u' z'; t)$ is the $Q_l(Z)$ transform of the basic kernel $V(P, P', Q)$ and we may, of course, use our Q_l expansion to effect the transform. We first write $V(P, P', Q) = \overline{V}(s_0, u, z; u', z'; t)$ and find

$$\mathscr{V}(u, z; u', z'; t) = \frac{1}{C_l}\sum_{n=0}^{\infty} \frac{\Gamma^2(l+1)2^{l+1}\, n!}{\Gamma(n+2l+2)} C_n^{l+1}(z)\, C_n^{l+1}(z') \cdot$$

$$\cdot \int_{L^2}^{\infty} ds_0\, \overline{V}(s_0, u, z; u', z'; t) \exp[-(l+1+n)\Theta],$$

where

$$\cosh\Theta = \frac{s_0 - u - u'}{2\sqrt{uu'}}.$$

If it should be the case that $\overline{V}(s_0, u, z; u', z'; t)$ is well approximated by $\overline{V}(s_0, u, 0; u', 0; t)$ only the $n = 0$ term in the expansion of $\mathscr{V}$ would be important in the slope calculation for the same reason as in the ladder model, and then the formal structure of $D(l, t)$ is very much the same as before. The only change is that in place of $\exp[-l-1]\theta(u, u')$ as a kernel, one would have

$$\mathscr{V}_l(u, u', t) \equiv \int_{L^2}^{\infty} ds_0\, \overline{V}(s_0, u, 0; u', 0; t) \exp[-(l+1)\Theta].$$

If we define the trial function to be the solution of

$$f_0(u) = \gamma \int_{-\infty}^{0} du'\, \mathscr{V}_l(u, u', 0)\, S_0(u')\, f_0(u'),$$

the slope formula now becomes

$$\frac{\Delta l}{t} = \gamma \frac{\partial l}{\partial \gamma} \frac{\int_{-\infty}^{0} du\, f_0^2 S_0' + \int_{-\infty}^{0} du \int_{-\infty}^{0} du'\, f_0(u)\, S_0(u)\, \mathscr{V}_l'(u, u', 0)\, S_0 f_0(u')}{\int_{-\infty}^{0} du f_0^2 S_0(u)},$$

where

$$\mathscr{V}'_l = \frac{\partial}{\partial t}\mathscr{V}_l(u, u', t)\Big|_{t=0}.$$

It is clearly possible to treat any specific case but hard to write neat but explicit formulae for an abstract $V_l(u, z;\ u', z', t)$.

As a simple illustration we consider the spin-one exchange ladder model with our simplest off-shell prescription which merely multiplies the spin-zero kernel by $(1+2t/m_0^2)$. This leads to

$$\frac{\Delta l}{t} = g^2\frac{\partial l}{\partial g^2}\left[\frac{\int_{-\infty}^{0} \mathrm{d}u\, f_0^2 S'_0}{\int_{-\infty}^{0} \mathrm{d}u\, f_0^2 S_0} + \frac{2}{m_0^2}\right].$$

We are currently studying the trajectory slope when the high-energy tail of the potential is included along with the resonance low-energy part.

2·3. *The multiperipheral model and diffractive dissociation.* – We take up now the use of the variational method in a treatment of the high-energy tail of the potential and its relation to diffractive dissociation. We discussed in Sect. **1** the break up of the potential into low- and high-energy parts and now wish to see how the discrete spectrum of simple poles associated with the low-energy potential in the λ or l planes are perturbed by the addition of the high-energy tail. We shall confine our attention almost entirely to the forward scattering problem for the remainder of this Section

We recall that the general forward equation for the absorptive part of the elastic scattering imagined diagonalized in internal t-channel quantum numbers takes the form (for the « partial waves »)

$$A_\lambda(u, v) = V_\lambda(u, v) + \frac{1}{16\pi^3}\frac{1}{\lambda+1}\int_{-\infty}^{0}\mathrm{d}u'\,\frac{V_\lambda(u, u')A_\lambda(u', v)}{(m^2-u')^2}.$$

Thus in the abstract notation

$$A_{\gamma\gamma_0} = V_{\gamma\gamma_0} + \sum_{\gamma'} V_{\gamma\gamma'}S_{\gamma'}A_{\gamma'\gamma_0},$$

the index γ stands for u, γ_0 for v and $S_{\gamma'}$ for $(m^2-u')^{-2}$; the sum $\sum_{\gamma'}$ includes the factor $[16\pi^3(\lambda+1)]^{-1}$ as well as $\int_{-\infty}^{0}\mathrm{d}u'$. The variational principle has the same formal structure as before, namely

$$'A'_{\gamma\gamma_0} = V_{\gamma\gamma_0} + \frac{(ASV)_{\gamma\gamma_0}(VSA)_{\gamma\gamma_0}}{(ASA)_{\gamma\gamma_0} - (ASVSA)_{\gamma\gamma_0}},$$

where the variationally computed A which we have designated by $'A'$ becomes the true answer if the trial A's on the right are solutions of

$$A = V + VSA = V + ASV .$$

We now make two basic assumptions to develop our perturbation theory: 1) when the full potential is replaced by its low energy, or resonance part, V^R, one has a reasonable first approximation to the true solution and 2) that we work in a region of the λ-plane where the largest eigenvalue $\mu(\lambda)$ of the approximate kernel $V^R S$ is near unity with the smaller eigenvalues being well separated. This latter assumption has the following implication. If we define the eigenvalues and eigenfunctions of the truncated kernel by the equation

$$V^R S|\psi_\mu\rangle = \mu|\psi_\mu\rangle ,$$

which is an abbreviation for

$$\frac{1}{16\pi^3}\frac{1}{\lambda+1}\int_{-\infty}^{0} \mathrm{d}u' \frac{V^R(u, u')\psi_\mu(u')}{(m^2 - u')^2} = \mu\psi_\mu(u) ,$$

the solution of $A^R = V^R + V^R S A^R$ may be written as

$$A^R = \sum_\mu \mu \frac{|\psi_\mu\rangle\langle\psi_\mu|}{1-\mu} ,$$

or less symbolically as

$$A^R(u, v) = \sum_\mu \mu \frac{\psi_\mu(u)\psi_\mu(v)}{1-\mu} ;$$

we have assumed that the ψ_μ are normalized according to

$$\langle\psi_\mu|S|\psi_\nu\rangle = \delta_{\mu\nu} .$$

In the neighborhood of the presumed isolated singularity near $\mu = 1$

$$A^R \approx \frac{|\psi_\mu\rangle\langle\psi_\mu|}{1-\mu} .$$

We shall return to this formula shortly but for now press on to the variational calculation for the complete problem.

By our assumptions it is reasonable to take as a trial function for $'A'$ the dyadic $|\psi_\mu\rangle\langle\psi_\mu|$ where from now on we mean by μ the largest eigenvalue of the resonance kernel $V^R S$. We have then

$$'A' = V + \frac{VS|\psi_\mu\rangle\langle\psi_\mu|SV}{\langle\psi_\mu|S|\psi_\mu\rangle - \langle\psi_\mu|SVS|\psi_\mu\rangle},$$

which with our agreement upon normalization becomes

$$'A' = V^R + V^P + \frac{(\mu + V^P S)|\psi_\mu\rangle\langle\psi_\mu|(\mu + SV^P)}{1 - \mu - \langle\psi_\mu SV^P S|\psi_\mu},$$

where V^P is the high-energy part of the potential defined in Part 1. The superscript P is intended to suggest that the pomeron trajectory will be the dominant contributor to the potential V^P.

This is our principal result. It shows the denominator as involving a function $\mu(\lambda)$ which is analytic for $\operatorname{Re}\lambda > -1$ and the term involving V^P which contains the infinite AFS branch point. Although the numerator function plays an important role in the general theory, we shall here concentrate only on the denominator in the particular case where the internal quantum number $I_t = 1$, because this is the quantity, as we shall see, which is intimately connected to diffractive dissociation.

In order to be prepared to make contact with various physical processes, we translate our eigenfunctions into the more familiar residue functions of high-energy elastic amplitudes. Adding the λ-plane label we have first

$$A_\lambda^R(u, v) \sim \frac{\psi_\mu(u)\,\psi_\mu(v)}{1 - \mu} = \frac{\psi_{\alpha_R}(u)\,\psi_{\alpha_R}(v)}{\lambda - \alpha_R},$$

where α_R is the largest value of λ corresponding to $\mu(\lambda) = 1$, and

$$\psi_{\alpha_R}(u) = k^{\frac{1}{2}}\psi_\mu(u)\big|_{\mu=1},$$

with

$$k = -\frac{\mathrm{d}\mu^{-1}}{\mathrm{d}\lambda_{\mu=1}}.$$

Now suppose that for large s the s-channel absorptive part corresponding to internal quantum number I_s is

$$A_{I_s}^R(s, u, v) = \tilde{\beta}^R(u)\tilde{\beta}^R(v)s^{\alpha_R},$$

then since the crossing matrix from singlet I_t to any I_s state is N^{-1}, the t-channel singlet A^R is

$$A^R(s,\, u,\, v) = N\tilde{\beta}^R(u)\tilde{\beta}^R(v)s^{\alpha_R}\,.$$

On the other hand, from our expression for $A^R_\lambda(u,\, v)$ in terms of ψ_{α_R} we find from the inversion formula

$$A^R(s,\, u,\, v) \sim \int\limits_{c-i\infty}^{c+i\infty} \frac{d\lambda}{2\pi i}\, \frac{s^\lambda}{(\sqrt{uv})^{\lambda+1}}\, A^R_\lambda(u,\, v) = \frac{\psi_{\alpha_R}(u)\,\psi_{\alpha_R}(v)}{(\sqrt{uv})^{\alpha_{R+1}}}\, s^{\alpha_R}\,.$$

Thus

$$\psi_{\alpha_R}(u) = \sqrt{N}(-u)^{\alpha_{R+1/2}}\tilde{\beta}^R(u)\,.$$

It should be noted that $\tilde{\beta}^R(u)$ is the vertex function for a « particle » of $(\text{mass})^2 = u$ and the Regge pole of zero mass with intercept $\alpha(t=0) = \alpha_R$.

The potential $V^{\text{P}}(u,\, v)$ may be computed from the general formula given in Chapter 1. We have

$$V^{\text{P}}_\lambda(u,\, v) = (\sqrt{uv})^{\lambda+1}\, \frac{N}{16\pi} \int\limits_{-\infty}^{0} dt\, \frac{|\beta(u,\, m^2,\, t)|^2\, |\beta(v,\, m^2,\, t)|^2}{\lambda - 2\alpha_{\text{P}}(t) + 1}\, (s^*)^{2\alpha_{\text{P}}(t)-1-\lambda}\,,$$

where the three arguments of the vertex functions now refer to the three masses of the vertex legs: u, v, to the external « particles », m^2 the mass of the pions in our π-π elastic unitarity calculation of V and t is the usual momentum transfer in that collision, carried by the pomeron $\alpha_{\text{P}}(t)$.

We now put all the pieces together and evaluate the denominator of the variational expression, $D(\lambda)$, near $\mu = 1$ or equivalently near $\lambda = \alpha_R$:

$$D(\lambda) = 1 - \mu - \langle\psi_\mu|SVS|\psi_\mu\rangle = \frac{1}{k}[\lambda - \alpha_R - k\,\langle\psi_\mu|SVS|\psi_\mu\rangle]\,,$$

and

$$\varrho_{\text{P}}(\lambda) \equiv k\langle\psi_\mu|SVS|\psi_\mu\rangle = \frac{1}{16\pi} \int\limits_{-\infty}^{0} dt\, \frac{(s^*)^{2\alpha_{\text{P}}(t)-1-\lambda}}{\lambda - 2\alpha_{\text{P}}(t) + 1}\, G^2_{\text{P}}(t)\,,$$

where

$$G_{\text{P}}(t) = \frac{1}{16\pi^3}\, \frac{N}{\alpha_{R+1}} \int\limits_{-\infty}^{0} du(-u)^{\alpha_{R+1}}\, \frac{|\beta(u,\, m^2_\pi,\, t)|^2\, \tilde{\beta}^R(u)}{(m^2 - u)^2}\,.$$

If we assume that $G_P(t)$ is analytic near $t=0$ and that $\alpha_P(t)$ is monotonical rising, it is clear that $\varrho(\lambda)$ has a logarithmic branch point at $\lambda=\alpha_c\equiv\equiv 2\alpha_P(0)-1$. It is not difficult to show that

$$\varrho(\lambda)=\frac{1}{16\pi}\frac{1}{2\alpha'_P[\tau(\lambda)]}G_P^2[\tau(\lambda)]\ln\frac{1}{\lambda-a_c}+\text{a function regular near }\lambda=\alpha_c\,,$$

where $\tau(\lambda)$ is determined by

$$2\alpha_P[\tau(\lambda)]-1=\lambda\,,$$

at the branch point $\lambda=\alpha_c$, $\tau=0$ and we see that the strength of the branch point is determined by the important parameter

$$\eta_P\equiv\frac{1}{16\pi}\frac{1}{2\alpha'_P(0)}G_P^2(0)\,.$$

We shall not stop to discuss the implications of this result for the fine structure of the spectrum of $D(\lambda)$. It can be shown that if $\alpha_P(0)=1-a_P$, we expect η_P to be of the same order as a_P. Unless energies are so high that $\ln s\gg\eta_P^{-1}$ this fine structure will not show up and the combined effect of the singularities near $\lambda=1$ including both poles and the branch point will look very much like a single pole at $\lambda=\alpha_R$.

We turn now to the important influence that the parameter η_P has on certain diffractive dissociation experiments.

3. – Triple-pomeron vertex and diffractive dissociation.

3'1. *Experiment and general structure of the problem.* – The term diffractive dissociation is usually used in connection with a process in which particles A and B collide with A simply recoiling with a momentum transfer $(-t)^{\frac{1}{2}}$ but retaining its integrity while B breaks up into a mass of things with a large total mass $(s')^{\frac{1}{2}}$. It is also customary to picture this as involving an exchange of vacuum quantum numbers between A and B and thus (at sufficiently high total energy $s=(p_A+p_B)^2$) the exchange of a pomeron trajectory. This is shown in Fig. 2.

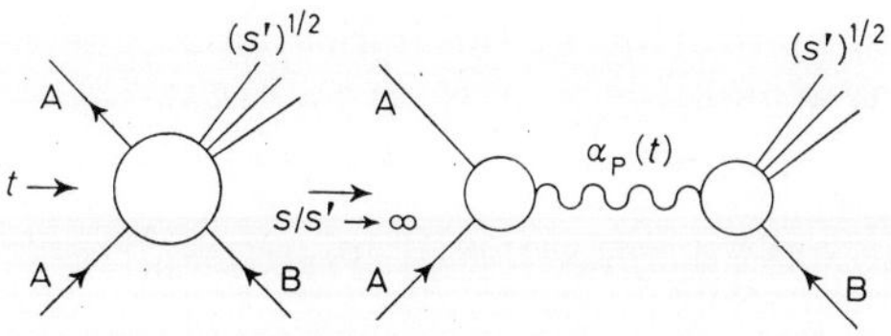

Fig. 2. – Diffractive dissociation of B.

If we imagine summing over all the particles composing the mass $(s')^{\frac{1}{2}}$ at definite s'—what is called a single-particle inclusive experiment—we expect further that the cross-section $d\sigma/d|t|ds'$ for the process $A+B \to A+$ (anything

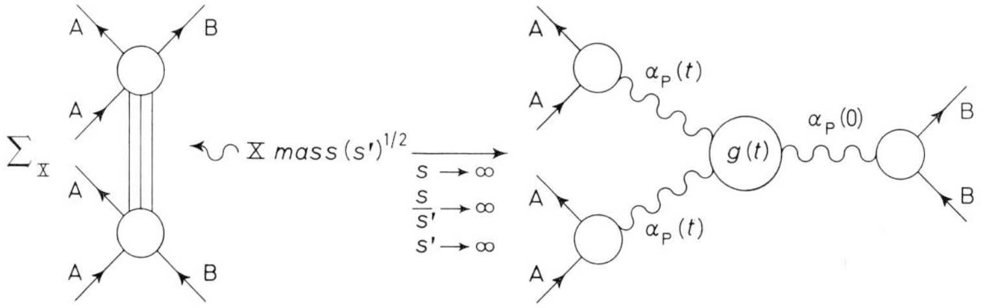

Fig. 3. – Cross-section for diffractive dissociation.

of total mass $(s')^{\frac{1}{2}}$) will be described by Fig. 3. The second form of the cross-section is what is expected in the kinematic regime where s' is itself so large that it too may be described by pomeron exchange in the process $\alpha_P(t) + B \to \to \alpha_P(t) + B$. The quantity $g(t)$ is what we shall call for the time being the triple-pomeron vertex (in Sect. **3**.3. we introduce a more general vertex). Many people have discussed this reaction and have noted that it is expected to have a characteristic energy-dependence $(s/s')^{2\alpha_P(t)}(s')^{\alpha_P(0)}$ and have recognized the significance of the triple-pomeron vertex in this connection. What we have shown and what will be described here is the connection between the quantity $g_P(t)$ and our parameter η_P which enters the Reggeon « propagator » $D(\lambda)$ we have just treated in Sect. **2**.2.

It is evident that we could study a process like $A+B \to C+$ (anything of mass $(s')^{\frac{1}{2}}$) in which case the trajectory above called $\alpha_P(t)$ would generally carry nonvacuum quantum numbers; $\alpha_P(0)$ would still appear however from the sum over particles that are not observed.

We have already commented on the expected energy-dependence of the relevant differential cross-section and we now specify all the normalizations quite precisely. In order to save space we shall no longer always indicate that we are working in the limit $s \to \infty$, $s' \to \infty$ and $s/s' \to \infty$. We write for the spin-averaged differential cross-section

$$s^2 \frac{d\sigma_{AB}}{d|t|ds'} = \frac{1}{16\pi} |\beta_{PAA}(t, m_A^2)|^2 \tilde{\beta}_{PBB}(0, m_B^2) g_P(t) \left(\frac{s}{s'}\right)^{2\alpha_P(t)} (s')^{\alpha_P(0)},$$

where the absolute normalization of the indicated Regge residues are fixed by giving the elastic A-B scattering and total cross-sections, namely

$$s^2 \frac{d\sigma_{AB}^{el}}{d|t|} = \frac{1}{16\pi} |\beta_{PAA}(t, m_A^2)|^2 |\beta_{PBB}(t, m_B^2)|^2 s^{2\alpha_P(t)},$$

$$s\sigma_t^{AB} = \tilde{\beta}_{PAA}(0, m_A^2) \tilde{\beta}_{PBB}(0, m_B^2) s^{\alpha_P(0)}.$$

To see the connection between $g_P(t)$ and the parameter η_P we consider a double diffractive process for which in the pomeron exchange between A and B both A and B dissociate into large masses $(s'')^{\frac{1}{2}}$ and $(s')^{\frac{1}{2}}$, respectively, as shown in Fig. 4.

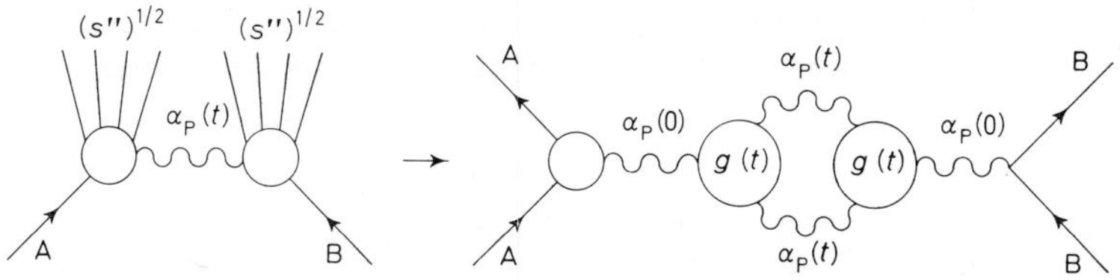

Fig. 4. – Double diffractive dissociation in limit $s\to\infty$, $s'\to\infty$, $s''\to\infty$ and $(s/s's'')\to\infty$.

With our chosen normalization we have for the spin-averaged cross-section for this process

$$s^2\,\frac{\mathrm{d}\sigma_{AB}}{\mathrm{d}s'\,\mathrm{d}s''\,\mathrm{d}|t|}=\frac{1}{16\pi}\,\tilde{\beta}_{PAA}(0,\,m_A^2)\,g_P^2(t)\,\tilde{\beta}_{PBB}(0,\,m_B^2)\left(\frac{s}{s'\,s''}\right)^{2\alpha_P(t)}(s'\,s'')^{\alpha_P(0)}\,.$$

It is very tempting to identify Fig. 4 with a single-loop modification of the pomeron propagator. Within the framework of the multiperipheral model developed in Sect. **2** this would appear to be related to the denominator function $D(\lambda)$ encountered there where the « bare propagator » $(\lambda-\alpha_R)^{-1}$ is corrected to first order by the high-energy tail V^P; recall that we had found $D(\lambda)\sim\lambda-a_R-\eta_P\ln(\lambda-a_c)^{-1}$.

Because of this author's inability to understand adequately the extensive work by GRIBOV and his co-workers on Reggeon calculus where such ideas of propagator modifications are widely discussed, we shall resort to an explicit calculation of the single diffractive dissociation process within the framework of the multiperipheral model we have developed.

3'2. *Evaluation of* $g_P(t)$ *in the multiperipheral model.* – In order to simplify writing let us ignore all internal quantum number questions. The single-particle inclusive cross-section may be inferred from the forward integral equation

$$A(P,\,K)=V(P,\,K)+\frac{2}{(2\pi)^4}\int\mathrm{d}^4P'\,\frac{V(P,\,P')\,A(P',\,K)}{(m^2-p'^2)^2}\,,$$

when we recall that $V(P,\,P'')$ is computed from elastic unitarity according to the expression

$$V(P,\,P')=\frac{(2\pi)^4}{2}\int\frac{\mathrm{d}^4q_1}{(2\pi)^3}\,\delta_+(q_1^2-m^2)\int\frac{\mathrm{d}^4q_2}{(2\pi)^3}\,\delta_+(q_2^2-m^2)\cdot$$
$$\cdot|M(q_1,\,q_2;\,P,\,-P')|^2\,\delta(q_1+q_2-P+P')\,,$$

where $M(q_1, q_2; P, -P')$ is the elastic π-π amplitude for the process $P+(-P')\to \to q_1+q_2$; the minus P' results simply from our original sign conventions. (See Fig. 5).

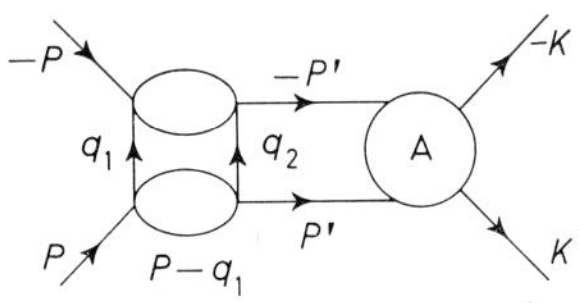

Fig. 5. – Structure of the multiperipheral model.

To obtain the desired differential cross-section for the process $P+(-K)\to \to q_1+$anything,we simply suppress the integration over d^4q_1 and multiply by the usual flux factor $\Delta^{\frac{1}{2}}(s, P^2, K^2)$ to find

$$\Delta^{\frac{1}{2}}(s, P^2, K^2)\, d\sigma/d^4q_1\delta_+(q_1^2-m^2)(2\pi)^{-3} =$$
$$= \frac{1}{(2\pi)^3}\int d^4P'\, \frac{\delta_+[m^2-(P-P'-q_1)^2]}{[m^2-P'^2]^2}\,|M(q_1, P-P'-q_1;\; P, -P')|^2 A(p', K).$$

Now to get into the interesting region we must assume that the subenergy $(P-P')^2$ is so large that we can approximate the matrix element M by its pomeron-pole contribution, namely to write it as

$$M\sim\beta(P^2, (P-q_1)^2)\beta((P-q_1)^2, P'^2)[(P-P')^2]^{\alpha((P-q_1)^2)},$$

where α means α_P from now on. We introduce scalar variables $s=(P-K)^2$, $s'=(P-q_1-K)^2$, $s''=(P'-K)^2$, $t=(P-q_1)^2$, $u'=P'^2$. Note that s' is total energy of all the produced particles beside the distinguished one, q_1. We next assume that the important values of s, s', s'' are large and evaluate all quantities in this limit. It is important, however to retain the ratio s''/s' because it can become of order unity. We set

$$A(P', K)=\tilde{\beta}(u', 0)\tilde{\beta}(0, K^2)(s'')^{\alpha(0)}.$$

The hardest part of the calculation is the evaluation of $(P-P')^2$ in terms of the scalars and an azimuthal angle. For the benefit of fanatics who want to check everything we remark that we formally carried out the transformation to scalar variables in the frame where $P-q_1-K$ is at rest and actually used $q_2= =P-q_1-P'$ as an integration variable rather than P' and took as a polar axis $\boldsymbol{P}-\boldsymbol{q}_1$. We find then

$$(P-P')^2=2\sqrt{ut}\,\frac{s}{s'}\left\{\cosh q-\sqrt{\frac{t}{u'}}\,x-\left[1-2x\sqrt{\frac{t}{u'}}\cosh q+\frac{t}{u'}x^2\right]^{\frac{1}{2}}\cos\varphi\right\},$$

where

$$x = s''/s' ,$$

$$\cosh q = \frac{m^2 - u' - t}{2\sqrt{u't}} ,$$

and φ is the azimuthal angle which enters in the angle between $\boldsymbol{q}_1$ and $\boldsymbol{q}_2$ in the chosen frame. The introduction of scalar variables leads most naturally to an integral of the form

$$\int_{4m^2}^{s'} \mathrm{d}s''/s' \int_{-\infty}^{u_+} \mathrm{d}u' \int_0^{2\pi} \mathrm{d}\varphi ,$$

where $u_+ = ux - m^2 x/(1-x)$; to get the answer given below, we introduce x in place of s'' and interchange the x and u'_+ integrations.

From this point on one needs only adequate patience with factors to deduce that in the high-energy regime, $s \to \infty$, $s' \to \infty$, $s/s' \to \infty$:

$$s^2 \frac{\mathrm{d}\sigma}{\mathrm{d}s'\mathrm{d}|t|} \to \frac{|\beta(P^2, t)|^2}{(16\pi^2)^2} \left(\frac{s}{s'}\right)^{2\alpha(t)} (s')^{\alpha(0)} \tilde{\beta}(0, K^2) \int_{-\infty}^{0} \mathrm{d}u' \frac{|\beta(t, u')|^2 \tilde{\beta}(u', 0)}{(m^2 - u')^2} \varkappa(t, u') ,$$

where

$$\varkappa(t, u') = \left[2\sqrt{u't}\right]^{2\alpha(t)} \int_0^{\exp[-q\sqrt{u'/t}]} \mathrm{d}x x^{\alpha(0)} \frac{1}{\pi} \int_0^{\pi} \mathrm{d}\varphi \cdot$$

$$\cdot \left\{\cosh q - \sqrt{\frac{t}{u'}}\, x - \left[1 - 2x\sqrt{\frac{t}{u'}} \cosh q + \frac{t}{u'} x^2\right]^{\frac{1}{2}} \cos\varphi\right\}^{2\alpha(t)} .$$

This quantity can be drastically simplified by the substitution

$$z = \frac{\cosh q - \sqrt{t/u'}\, x}{\sinh q} .$$

We find

$$\varkappa(u', t) = \left[2\sqrt{u't}\right]^{2\alpha(t)} \left[\sqrt{\frac{u'}{t}}\right]^{\alpha(0)+1} [\sinh q]^{2\alpha(t)+1} \cdot$$

$$\cdot \int_1^{\operatorname{ctgh} q} \mathrm{d}z [\cosh q - z \sinh q]^{\alpha(0)} \frac{1}{\pi} \int_0^{\pi} \mathrm{d}\varphi \, \{z - \sqrt{z^2 - 1} \cos\varphi\}^{2\alpha(t)} =$$

$$= \left[2\sqrt{u't} \sinh q\right]^{2\alpha(t)} \left[\sqrt{\frac{u'}{t}}\right]^{\alpha(0)+1} \sinh q \int^{\operatorname{ctgh} q} \mathrm{d}z \, [\cosh q - z \sinh q]^{\alpha(0)} P_{2\alpha(t)}(z) ,$$

where we have used a standard representation for $P_{2\alpha(t)}(z)$. Finally we note that the z integration may be carried out explicitly (GRADSHTEYN and RYZHIK, 7.137, No. 9, p. 802) to give

$$\varkappa(u', t) = [2\sqrt{u't}\sinh q]^{2\alpha(t)}\left[\sqrt{\frac{u'}{t}}\right]^{\alpha(0)+1}\Gamma(\alpha(0)+1)\,P_{2\alpha(t)}^{-\alpha(0)-1}(\operatorname{ctgh} q)\,.$$

It is clear that all the members of the pseudoscalar multiplet will contribute the same amount to our cross-section, so that we must multiply by a factor of N. If we now substitute this into our differential cross-section we find

$$s^2\frac{d\sigma}{ds'\,d|t|} = \frac{|\beta(P^2, t)|^2}{(16\pi^2)^2}\left(\frac{s}{s'}\right)^{2\alpha(t)}(s')^{\alpha(0)}\tilde{\beta}(0, K^2)\,N\Gamma(\alpha(0)+1)\cdot$$
$$\cdot\int_{-\infty}^{0} du'\,\frac{|\beta(t, u')|^2\tilde{\beta}(u', 0)}{(m^2-u')^2}\,[2\sqrt{u't}\sinh q]^{2\alpha(t)}\left[\sqrt{\frac{u'}{t}}\right]^{\alpha(0)+1} P_{2\alpha(t)}^{-\alpha(0)+1}(\operatorname{ctgh} q)\,.$$

Comparison with the formula for single diffractive dissociation given given in Sect. **3·1.** shows that $g_P(t)$ is given by

$$g_P(t) = \frac{N\Gamma(\alpha(0)+1)}{16\pi^3}\int_{-\infty}^{0} du'\left(\sqrt{\frac{u'}{t}}\right)^{\alpha(0)+1}[2\sqrt{u't}\sinh q]^{2\alpha(t)}\cdot$$
$$\cdot|\beta(t, u', m^2)|^2\tilde{\beta}(0, u', u')\,P_{2\alpha(t)}^{-\alpha(0)-1}(\operatorname{ctgh} q)\,,$$

where we have added the additional-mass label to the vertex functions. A diagrammatic representation of the calculation we have just carried out is shown in Fig. 6.

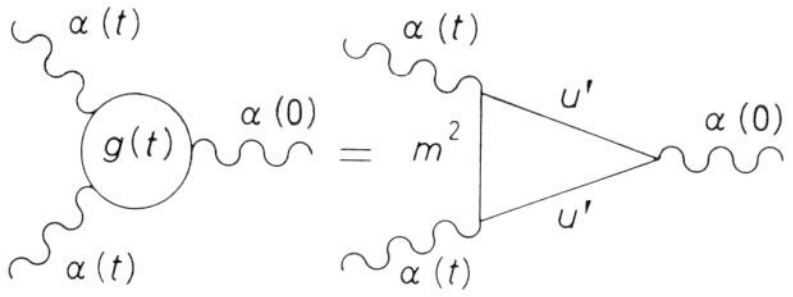

Fig. 6. – Triple-Regge vertex in multiperipheral model.

The t-dependence of $g_P(t)$ is quite complicated, occurring as it does in $\cosh q$, the vertex function $\beta(t, u', m^2)$ and in $\alpha(t)$. One question of critical importance is the value of $g_P(t=0)$. We remember that $\cosh q = (m^2-u'-t)/2\sqrt{ut'} \to$ $\to (m^2-u')/2\sqrt{u't}$ for small t and using

$$P_\nu^\mu(z) \to 2^{\frac{\mu}{2}}(z-1)^{-\mu/2}/\Gamma(1-\mu)\,, \qquad z\to 1\,,$$

we find

$$g_P(0) = \frac{N}{16\pi^3[\alpha(0)+1]} \int_{-\infty}^{0} du' \frac{(-u')^{\alpha(0)+1}}{(m^2-u')^2} (m^2-u')^{\alpha(0)-1} |\beta(0, u', m^2|^2 \beta(0, u', u') .$$

If we assume that $\alpha(0) \approx 1$, we have

$$g_P(0) = \frac{N}{32\pi^3} \int_{-\infty}^{0} du' \frac{(-u')^2}{(m^2-u')^2} |\beta(0, u', m^2)|^2 \tilde{\beta}(0, u', u') .$$

To the extent that α_R is sufficiently close to $\alpha(0)$ and thus nearly one, to the same perturbative approximation, we see that $g_P(0)$ is the same as the quantity $G_P(0)$ introduced in Sect. **2**˙2, in connection with the evaluation of $D(\lambda)$. In other words, the same dimensionless parameter η_P,

$$\eta_P = \frac{g_P^2(0)}{16\pi} \frac{1}{2\alpha'(0)} ,$$

controls single diffractive dissociation and the pomeron fine structure in the λ-plane.

The fact that $g_P(0)$ is different from zero means that this is to avoid a violation of the Froissart bound, $\alpha(0) < 1$. Of course we have done only a model calculation. We have actually evaluated η_P using some of the separable models of Sect. **1** and find $\eta_P \approx 0.02$. There has been at least one calculation based on a dual resonance model which leads to $g_P(0) = 0$, however some of the other predictions of the same model are so bizarre that the calculation should probably not be taken seriously.

3˙3. *Multiplicity as a function of momentum transfer.* – A question of some physical interest is the manner in which the mean multiplicity of high-energy collisions varies as a function of momentum transfer in a collision. An understanding of this question might be useful for building some geometrical intuition about the importance of various regions of impact parameter space—the natural space conjugate to momentum transfer.

The very concept of multiplicity as a function of momentum transfer is not easy to define. Some work is in progress on this matter which will not be reported here, but a somewhat appealing (to the author) definition is provided by the consideration of the single-particle exclusive reaction we have just given. Reference to Fig. 2 shows that we are dealing with a process in which a «particle» (albeit a Reggeon) transmits momentum $(-t)^{\frac{1}{2}}$ from particle A to B resulting in the production of a number of particles of total mass $(s')^{\frac{1}{2}}$. Suppose that there were n particles produced in that collision. It would be natural to write

$$\Delta^{\frac{1}{2}}(s, P^2, K^2)\mathrm{d}\sigma_n/\mathrm{d}^4q_1\,\delta_+(q_1^2-m^2)(2\pi)^{-3} = B_n(P, P-q_1, K)\,,$$

or in terms of scalar variables, in the large s limit

$$s^2\frac{\mathrm{d}\sigma_n}{\mathrm{d}s'\,\mathrm{d}|t|} = \frac{1}{16\pi^2}B_n(s, s', t, P^2, K^2)\,,$$

in terms of which we define $\langle n(t, s)\rangle$ by

$$\langle n(t, s)\rangle = \frac{\int_{4m^2}^{s}\mathrm{d}s'\sum_{n=1}^{\infty} nB_n(s, s', t, P^2, K^2)}{\int_{4m^2}^{s}\mathrm{d}s'\sum_{n=1}^{\infty} B_n(s, s', t, P^2, K^2)}\,.$$

The denominator of this quantity is, of course, just the inclusive cross-section, aside from constant factors, but we need some kind of dynamical model to compute the numerator. We turn to our detailed model.

In our model based on π-π scattering we produce two mesons in each elementary unit of the multiperipheral chain. Thus the natural quantity B_n that we used, namely

$$B_n(s, s', t, P^2, K^2) = \frac{|\beta(P^2, t)|^2}{32\pi^3 s'}\int_{4m^2}^{s'}\int_{-\infty}^{u_+}\mathrm{d}u'\frac{|\beta(t, u')|^2}{(m^2-u')^2}[(q_1+q_2)^2]^{2\alpha(t)}A_n(s'', u', K^2)\,,$$

in reality corresponds to the production of $2n+1$ particles, the index n on A_n referring to the number of units in the chain.

We remark that the quantity $N(s, u, v)$ defined by

$$N(s, u, v) = \sum_{n=1}^{\infty} A_n(s, u, v)\,,$$

can be computed by quadratures from $A(s, u, v)$

$$N(s, u, v) = A(s, u, v) + \frac{2}{(2\pi)^4}\int \mathrm{d}^4P'\frac{A(P, P')A(P', K)}{[m^2-P'^2]^2}\,.$$

This can be deduced from the fact that A_n satisfies the recursion relation

$$A_n(P, K) = \frac{2}{(2\pi)^4}\int \mathrm{d}^4P'\frac{A_{n-m}(P, P')A_m(P', K)}{[m^2-P'^2]^2}\,,$$

where $1 \leqslant m \leqslant n-1$. Summing over m and over n from 1 to ∞ leads to our formula for $N(s, u, v)$. It is natural to define the mean number of basic units $\langle n(s, u, v)\rangle$ by

$$\langle n(s, u, v)\rangle = \frac{N(s, u, v)}{A(s, u, v)} .$$

A simple way to evaluate $N(s, u, v)$ is to transform it in the same way we did for the forward equation. We write

$$N_\lambda(u, v) \equiv \int\limits_{4m^2}^{\infty} ds \exp[-(\lambda+1)\theta(u, v)]N(s, u, v) ,$$

where $\cosh\theta(u, v) = (s-u-v)/2\sqrt{uv}$ and find

$$N_\lambda(u, v) = A_\lambda(u, v) + \frac{1}{16\pi^3(\lambda+1)} \int\limits_{-\infty}^{0} du' \frac{A_\lambda(u, u')A_\lambda(u', v)}{(m^2-u')^2} .$$

Next we introduce the eigenfunctions of the kernel of the equation for A analogous to that introduced in Sect. **2**·2. Thus we write

$$A_\lambda(u, v) = \sum_\mu \frac{\psi_\mu(\lambda, u)\psi_\mu(\lambda, v)\mu(\lambda)}{1-\mu(\lambda)} ,$$

where

$$\frac{1}{16\pi^3(\lambda+1)} \int\limits_{-\infty}^{0} du \frac{\psi_\mu(\lambda, u)\psi_\nu(\lambda, u)}{(m^2-u)^2} = \delta_{\mu\nu} .$$

We find then for $N_\lambda(u, v)$:

$$N_\lambda(u, v) = \sum_\mu \frac{\psi_\mu(\lambda, u)\psi_\mu(\lambda, v)\mu(\lambda)}{[1-u(\lambda)]^2} .$$

For simplicity assume that we can consider to sufficient accuracy only the largest value of μ and thus drop the sum over μ. Further, imagine that only one of the values of λ defined by $\mu(\lambda)=1$ is important for large s; call that $\lambda = \alpha(0)$. Thus

$$A(s, u, v) \approx \int \frac{d\lambda}{2\pi i} \frac{s^\lambda}{(\sqrt{uv})^{\lambda+1}} \frac{\psi(\lambda, u)\psi(\lambda, v)}{1-\mu(\lambda)} \mu(\lambda) = \left[\frac{1}{(-d\mu/d\lambda)} \frac{\psi(\lambda, u)\psi(\lambda, v)}{(\sqrt{uv})^{\lambda+1}}\right]_{\lambda=\alpha(0)} s^{\alpha(0)},$$

$$N(s, u, v) = \left[\frac{1}{(-d\mu/d\lambda)^2} \frac{d}{d\lambda}\left\{\frac{\psi(\lambda, u)\psi(\lambda, v)}{(\sqrt{uv})^{\lambda+1}} \mu(\lambda)s^\lambda\right\}\right]_{\lambda=\alpha(0)} .$$

This shows that

$$\langle n(s, u, v)\rangle = -\left(\frac{\mathrm{d}\mu}{\mathrm{d}\lambda}\right)^{-1}_{\lambda=\alpha(0)} \ln s + \text{const}\,.$$

It is elementary to show that if we introduce a coupling constant g^2 into the interaction [so that the eigenvalue condition $\mu = 1$ is replaced by $g^2\tilde{\mu} = 1$] $(-\,\mathrm{d}\mu/\mathrm{d}\lambda)^{-1} = g^2(\mathrm{d}\lambda/\mathrm{d}g^2)$ which is an expression for the coefficient of the log that we have used before.

Since we are doing an exploratory calculation let us make the whole model very explicit and take one of our separable models for $A_\lambda(u, v)$:

$$A_\lambda(u, v) = \frac{\pi g^2[(m_0^2\sqrt{uv})/(m_0^2 - u)(m_0^2 - v)]^{\lambda+1}}{1-\mu(\lambda)}\,,$$

where the value of $\mu(\lambda)$ was given in Sect. **1** and need not concerns us. Then we have (again keeping only the largest $\lambda = \alpha(0)$)

$$A(s, u, v) = \pi g^2 \frac{m_0^2}{(m_0^2 - u)(m_0 - v)}\left(-\frac{\mathrm{d}\mu}{\mathrm{d}\lambda}\right)^{-1} x \left[\frac{m_0^2 s}{(m_0^2 - u)(m_0^2 - v)}\right]^{\alpha(0)},$$

$$N(s, u, v) = g^2 \frac{m_0^2}{(m_0^2 - u)(m_0^2 - v)}\left(\frac{\mathrm{d}\mu}{\mathrm{d}\lambda}\right)^{-2} \frac{\mathrm{d}}{\mathrm{d}\lambda}\left[\frac{m_0^2 s}{(m_0^2 - u)(m_0^2 - v)}\right] =$$

$$= \pi g^2 \frac{m_0^2}{(m_0^2 - u)(m_0^2 - v)}\left(\frac{\mathrm{d}\mu}{\mathrm{d}\lambda}\right)^{-2}\left[\frac{m_0^2 s}{(m_0^2 - u)(m_0^2 - v)}\right]^{\alpha(0)} \ln \frac{m_0^2 s}{(m_0^2 - u)(m_0^2 - v)}\,,$$

and finally

$$\langle n(s, u, v)\rangle = \left(-\frac{\mathrm{d}\mu}{\mathrm{d}\lambda}\right)^{-1}_{\lambda=\alpha(0)} \ln \frac{m_0^2 s}{(m_0^2 - u)(m_0^2 - v)}\,,$$

and this should be multiplied by 2 to be consistent with our original model of a π-π kernel. The only real virtue in this explicit model is the fact that the precise scale of all quantities is fixed.

It is clear that with either the single eigenfunction or the separable model

$$\langle n(t, s)\rangle = 1 + 2\left(-\frac{\partial\mu}{\partial\alpha(0)}\right)^{-1} \frac{(\partial/\partial\alpha(0)\bar{B}(s, t)}{\bar{B}(s, t)}\,,$$

where

$$\bar{B} = \int_{4m^2}^{s} \mathrm{d}s' \sum_n B_n(s, s', t, P^2, K^2)\,.$$

In order to simplify the writing, let us assume that the vertex functions β are strongly peaked about small values of their arguments, so that we neglect

u', v compared to m_0^2, but not compared to the small mass m^2. Then we have in the first place

$$\langle n(s, u, v)\rangle = 2\left(-\frac{\partial\mu}{\partial\alpha(0)}\right)^{-1}\ln\frac{s}{m_0^2},$$

where we have supplied the factor of two mentioned above. Reference to the formulae of the preceeding Section shows that in the stated approximation

$$\bar{B} = \left(\frac{g}{4\pi m_0}\right)^2\left(-\frac{\partial\mu}{\partial\alpha(0)}\right)^{-1}|\beta(P^2, t)|^2 s\left(\frac{s}{m_0^2}\right)^{\alpha(0)}\frac{1}{(\alpha(0)+1)(\alpha(0)-2\alpha(t)+1)}\cdot$$

$$\cdot\int_{-\infty}^{0}du'\frac{|\beta(t, u')^2(-u')^{\alpha(0)}}{(m^2-u')^2(m^2-u'-t)^{\alpha(0)+1-2\alpha(t)}},$$

and finally

$$\langle n(s, t\rangle = 1 + 2\left(-\frac{\partial\mu}{\partial\alpha(0)}\right)^{-1}\ln\frac{s}{m_0^2} + C(t)/D(t),$$

where

$$C(t) = 2\int_{-\infty}^{0}du\,\frac{|\beta(t, u)^2(-u)^{\alpha(0)+1}}{(m^2-u)^2(m^2-u-t)^{\alpha(0)-2\alpha(t)+1}}\cdot\left(-\frac{\partial u}{\partial\alpha(0)}\right)^{-1}\cdot$$

$$\cdot\left\{\ln\frac{(-u)}{m^2-u-t} + \frac{\partial}{\partial\alpha(0)}\left[\frac{1}{\alpha(0)+1}\,\frac{1}{\alpha(0)-2\alpha(t)+1}\right]\right\},$$

$$D(t) = \int_{-\infty}^{0}du\,\frac{(-u)^{\alpha(0)+1}|\beta(t, u)|^2}{(m^2-u)^2(m^2-u-t)^{\alpha(0)-2\alpha(t)+1}}.$$

The significance of this result, if any, is that the logarithmic term in $n(s, t)$ is the same as that found in the usual definition of multiplicity and the t-dependence appears only in an energy-independent term. This type of limiting behavior is frequently found in the multiperipheral model.

3·4. *The general triple-Regge vertex.* – The final topic we shall address is related to the general triple-Regge vertex. We have previously discussed in Subsects. **3·1.** and **3·2.** of this Section a special case of such a vertex where the masses of two of the legs are $(-t)^{\frac{1}{2}}$ whereas the third was at zero. The more general case can in principle be related to an appropriate kinematic limit of a six-point function as shown below in Fig. 7. If we imagine that the external particles are all spinless, our six-point function depends on $8(3N-10)$ scalar variables which may be taken as the three sub-energies related to the three Regge poles, the three t-values, and two Toller angles relating the orientation in some frame of the planes of, say, C, C' and A, A' with respect to that of B, B'. A slightly less general quantity arises quite naturally in the high-energy limit of the nonforward absorptive amplitude in our ABFST model which corresponds to requiring than the three-momentum of A' equals that of C. This reduces the number of variables by three t's. There is then no room for Toller-angle-dependence in the vertex. This matter will be discussed more extensively elsewhere.

Consider again our ABFST model based on a π-π scattering kernel just as in Sect. **3**˙2. but now for nonforward scattering. The kinematics are shown in Fig. 8.

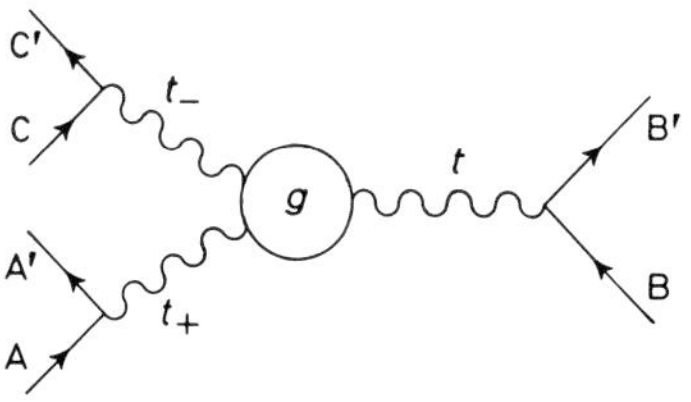

Fig. 7.

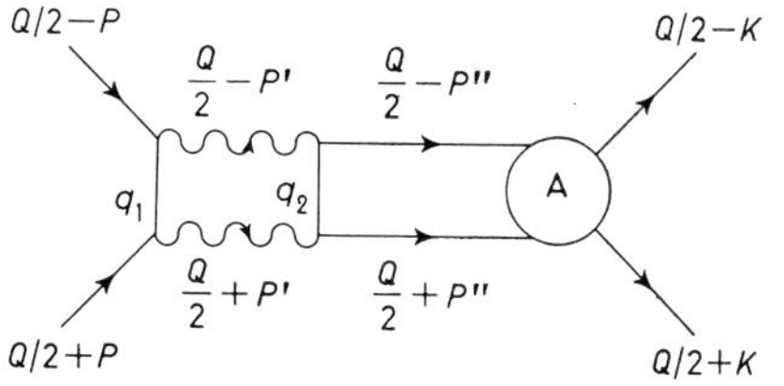

Fig. 8.

Fig. 7. – General triple Regge vertex.

Fig. 8. – Kinematics for model calculation.

The nonforward absorptive part (aside from the two-particle intermediate state) is given by

$$A(P, K, Q) = \int d^4P'\, \delta[(P-P')^2 - m^2]\beta(t_+, m^2, t'_+)\beta^*(t_-, m^2, t'_-)B(P, P', K, Q)\,,$$

where

$$B(P, P', K, Q) = \frac{d^4P''}{(2\pi)^6}[(P'-P'')^2 - m^2]\beta(t'_+, m^2, t''_+)\beta^*(t'_-, m^2, t''_-)\cdot \\ \cdot [(P'-P'')^2]^{\alpha(t'_+)+\alpha(t'_-)} S(P', Q)A(P'', K, Q)\,,$$

where we have introduced the following notation:

$$t_\pm = \left(\frac{Q}{2} \pm P\right)^2, \qquad t'_\pm = \left(\frac{Q}{2} \pm P'\right)^2, \qquad t''_\pm = \left(\frac{Q}{2} \pm P''\right)^2,$$

$$S(P'', Q) = \frac{1}{[m^2 - ((Q/2) + P'')^2][m^2 - ((Q/2) - P'')^2]}\,.$$

The various factors follow from our general formalism of the model in which we write

$$A(P, K, Q) = \frac{2}{(2\pi)^4}\int d^4P''\, V(P, P'', Q)\, S(P'', Q)\, A(P'', K, Q)\,,$$

$$V(P, P'', Q) = \frac{(2\pi)^4}{2}\int \frac{d^4q_1}{(2\pi)^3}\int \frac{d^4q_2}{(2\pi)^3}\, \delta_+(q_1^2 - m^2)\,\delta_+(q_2^2 - m^2)\cdot$$
$$\cdot\,\delta(P - P'' - q_1 - q_2)\, M_+(P, -P'', Q)\, M_-^*(P, -P'', Q)\,,$$

and have assumed

$$M_\pm(P, -P'', Q) = \beta(t_\pm, m^2, t'_\pm)\beta(t'_\pm, m^2, t''_\pm)[(P - P'')^2]^{\alpha(t'_\pm)}\,.$$

We shall also take under the integral defining B

$$A(P'', K, Q) = \tilde{\beta}(t''_+, t''_-, t)\tilde{\beta}(t, t^0_+, t^0_-)[(P'' - K)^2]^{\alpha(t)}\,,$$

where $t = Q^2$ and $t^0_\pm = ((Q/2) \pm K)^2$.

The B-amplitude is the quantity we desire and we see that except for some factors it is essentially the same thing we defined for the forward equation in the previous Section.

To proceed with the evaluation of B it is useful to introduce some scalar variables in place of our four-vectors. We define

$$s = (P - K)^2\,, \qquad u = P^2\,, \qquad u' = P'^2\,, \qquad u'' = P''^2\,, \qquad v = K^2\,,$$

$$s' = (P' - K)^2\,, \qquad z = \frac{P\cdot Q}{\sqrt{ut}}\,, \qquad z' = \frac{P'\cdot Q}{\sqrt{u't}}\,, \qquad z'' = \frac{P''\cdot Q}{\sqrt{u''t}}\,,$$

$$s'' = (P'' - K)^2\,,$$

$$\cosh\theta = \frac{-P\cdot K}{\sqrt{uv}} = \frac{s - u - v}{2\sqrt{uv}}\,, \qquad \cosh\theta' = \frac{-P'\cdot K}{\sqrt{u'v}} = \frac{s' - u - v}{2\sqrt{uv}}\,,$$

$$\cosh\theta'' = \frac{-P''\cdot K}{\sqrt{u''v}} = \frac{s'' - u'' - v}{2\sqrt{u''v}}\,, \qquad \cosh\theta_0 = \frac{-P''\cdot P}{\sqrt{uu''}} = \frac{(P - P'')^2 - u - u''}{2\sqrt{u''u}}\,,$$

$$x' = \frac{\cosh\theta}{\cosh\theta'} \simeq \sqrt{\frac{u'}{u}\,\frac{s}{s'}} \gg 1\,,$$

$$x'' = \frac{\cosh\theta''}{\cosh\theta'} \simeq \sqrt{\frac{u'}{u''}\,\frac{s''}{s'}} \leqslant 1\,.$$

To isolate the triple-Regge vertex we must be in a regime where $s\to\infty$, $s'\to\infty$, $s/s'\to\infty$, $s''/s' \leqslant 1$ but $s'' \gg$ all relevant masses. Although it surely could be avoided we shall use sometimes some intermediate variables that appear in the

Erice lectures and some work of SILVERMAN and TAN, namely in place of z', z'', $\eta' = P'\cdot Q = \sqrt{u't}z'$, $\eta'' = P''\cdot Q = \sqrt{u''t}z''$. Then the relevant Jacobian for the transformation to the scalars s'', u'', z'' becomes

$$J = \int d^4P''\,\delta[(P'-P'')^2 - m^2]\,\delta[P''^2 - u'']\,\delta(P''\cdot Q - \eta'')\,\delta[(P''-K)^2 - s''] = \frac{1}{2s'}\frac{(L)}{L^{\frac{1}{2}}},$$

where

$$L = (u't - \eta'^2)y^2 - 2\left\{(u' + u'' - m^2)\frac{t}{2} - \eta'\eta''\right\}y + (u''t - \eta''^2),$$

and $y \equiv s''/s'$. Thus we have for B the following:

$$B = \frac{1}{2(2\pi)^6}\int\limits_{4m^2}^{s'}\frac{ds''}{s'}\int du''\int d\eta''\,\frac{K(t'_\pm, t''_\pm, t)}{L^{\frac{1}{2}}}\,[(P-P'')^2]^\gamma (s'')^{\alpha(t)},$$

where the limits are determined by $L > 0$ and we have lumped into K all those things we can never integrate over, namely

$$K = \beta(t'_+, m^2, t''_+)\beta^*(t'_-, m^2, t''_-)S(P'', Q)\tilde{\beta}(t''_+, t''_-, t)\tilde{\beta}(t, t^0_+, t^0_-);$$

note that

$$\gamma = \alpha(t'_+) + \alpha(t'_-), \qquad u'' = \frac{t''_+ + t''_-}{2} - \frac{t}{4}, \qquad \eta'' = \frac{t''_+ - t''_-}{2}.$$

The next (and quite frightening) task is the computation of $(P-P'')^2 \equiv \equiv \sum u + u'' + 2\sqrt{uu''}\cosh\theta_0$ in terms of the scalars. This was the hardest part of the corresponding calculation in Sect. **3**·2. and now it is really unpleasant. One must equate to zero the Gram determinant composed of the scalar products of the five four-vectors in the problem: P, P', P'', K, Q. In the limit under discussion the 5×5 determinant boils down to a quadratic equation for $\cosh\theta_0$ which is

$$(\cosh\theta_0)^2 - 2x'\cosh\theta_0\{\cosh q_2 - z'z'' - x''(1-z'^2)\} + \\ + x'^2\{(\cosh q_2 - zz')^2 - (1-z^2)(1-z'^2)\} = 0,$$

where $\cosh q_2 = -P\cdot P''/\sqrt{uu''} = (m^2 - u' - u'')/2\sqrt{u'u''}$ and we recall $x' \sim s/s' \gg 1$, $x'' \sim s''/s' \sim 1$. Since

$$\frac{s'}{s}\Sigma \simeq 2\sqrt{uu'}\,\frac{\cosh\theta_0}{x'},$$

we find in terms of $y \equiv s''/s'$ the following formula for Σ:

$$\frac{s'}{s}\Sigma = 2\sqrt{u'u''}\cosh q_2 - 2\sqrt{u'u''}\,z'z'' + 2u'y(1-z'^2) \pm [4u'^2(1-z'^2)y^2 + \\ + 4\{2\sqrt{u'u''}\cosh q_2 - 2\sqrt{u'u''}\,z'z''\}u'(1-z'^2)y + 4u'u''(1-z'^2)(1-z''^2)]^{\frac{1}{2}}.$$

The appearance of this sign ambiguity is a reflection of the fact that the mapping from the space of P'' to the space of the invariants is 2 to 1 and not 1 to 1. We shall have to take the sum of the contributions from the two values of Σ. This point is discussed in detail by SILVERMAN and TAN.

It is useful to introduce some new variable invented in Erice:

$$\omega'^2 = -u'(1-z'^2) = -u' - \beta'^2, \qquad \beta' = \sqrt{-u'}\,z' = \eta'/\sqrt{-t},$$

$$\omega''^2 = -u''(1-z''^2) = -u'' - \beta''^2, \qquad \beta'' = \sqrt{-u''}\,z'' = \eta''/\sqrt{-t},$$

$$u't - \eta'^2 = -t\omega'^2,$$

$$L^{\frac{1}{2}} = \sqrt{-t\omega''^2}\left[1 - \frac{2y\varrho}{\omega''^2} + \frac{\omega'^2}{\omega''^2}y^2\right],$$

$$\varrho = \frac{m^2 - u' - u''}{2} - \sqrt{u'u''}\,z'z'' = \frac{m^2 + \omega'^2 + \omega''^2 + (\beta' - \beta'')^2}{2},$$

$$\frac{s'}{s}\Sigma = 2\left[\varrho - \omega'^2 y \pm \omega'\omega''\sqrt{1 - \frac{2y\varrho}{\omega''^2} + \frac{\omega'^2}{\omega''^2}y^2}\right].$$

Now we put it all together:

$$B = \frac{1}{1(2\pi)^6}\int_{-\infty}^{\infty}\mathrm{d}\eta''\int_{-\infty}^{\eta''^2/t}\mathrm{d}u''\,K(t'_\pm, t''_\pm, t)\int\frac{\mathrm{d}s''}{s'}\,\frac{(s'')^{\alpha(t)}[(P-P'')^2]^\gamma}{L^{\frac{1}{2}}} =$$

$$= \frac{1}{2(2\pi)^6}\left(\frac{s}{s'}\right)^\gamma (s')^{\alpha(t)}\int_{-\infty}^{\infty}\mathrm{d}\beta''\int_0^\infty \mathrm{d}\omega''\int_0^{y_-}\mathrm{d}y\,\frac{y^{\alpha(t)}[\Sigma_+^\gamma + \Sigma_-^\gamma]}{[(1-(y/y_-))(1-(y/y_+))]^{\frac{1}{2}}},$$

where

$$\Sigma_\pm = 2\left[\frac{y_+ + y_-}{2} - y \pm \frac{\omega''}{\omega'}\sqrt{\left(1 - \frac{y}{y_-}\right)\left(1 - \frac{y}{y_+}\right)}\right]\omega'^2,$$

$$y_\pm = \frac{\varrho \pm \sqrt{\varrho^2 - \omega'^2\omega''^2}}{\omega'^2}.$$

Some fanatic might think this should be multiplied by 2 since $\mathrm{d}u'' = 2\omega''\mathrm{d}\omega''$ but this 2 disappears because of our need really to have integrated over ω'' from $-\infty$ to ∞ in order to pick up both values of Σ. The formula for B may be reduced to manageable form by a series of transformations which can surely be done more elegantly and directly:

$$B = \frac{1}{(2\pi)^6}(s')^{\alpha(t)}\left(\frac{s}{s'}\right)^{\gamma}\int_{-\infty}^{\infty} d\beta'' \int_0^{\infty} d\omega'' K(t'_{\pm}, t''_{\pm}, t)(y_-)^{\alpha(t)+1}[2\omega'^2\sqrt{y_+ y_-}]^{\gamma} I\,,$$

where

$$I = \frac{1}{2}\int_0^1 dx x^{\alpha(t)}\Big\{\big[\cosh\psi - \exp[-\psi]x + [(1-x)(1-x\exp[-2\psi])]^{\frac{1}{2}}\big]^{\gamma} +$$

$$+ \big[\cosh\psi - \exp[-\psi]x - [(1-x)(1-x\exp[-2\psi])]^{\frac{1}{2}}\big]^{\gamma}\Big\}\cdot$$

$$\cdot[(1-x)(1-x\exp[-2\psi])]^{-\frac{1}{2}}\,,$$

where

$$\exp[-\psi] = \sqrt{\frac{y_-}{y_+}}\,, \qquad \cosh\psi = \frac{\varrho}{(\omega'\omega'')} = \frac{\cosh q_2 - z'z''}{\sqrt{(1-z'^2)(1-z''^2)}}\sqrt{y_+ y_-} = \omega''/\omega'\,.$$

This form follows simply by scaling, $y = xy_-$. Then let

$$x = \frac{\cosh\psi - \sinh\psi\cosh\theta}{\exp[-\psi]}\,,$$

and find

$$I = \exp[[\alpha(t)+1]\psi][\sinh\psi]^{\gamma+\alpha(t)}\int_0^{\cosh^{-1}(\mathrm{ctgh}\,\psi)} d\theta[\mathrm{ctgh}\,\psi - \cosh\theta]^{\alpha(t)}\cosh\gamma\theta =$$

$$= \exp[[\alpha(t)+1]\psi][\sinh\psi]^{\gamma-\frac{1}{2}}\sqrt{\frac{\pi}{2}}\,\Gamma(1+\alpha(t))\,P_{\gamma-\frac{1}{2}}^{-\alpha(t)-\frac{1}{2}}\,\mathrm{ctgh}\,\psi =$$

$$= \exp[[\alpha(t)+1]\psi][\sinh\psi]^{\gamma}\frac{\Gamma(1+\alpha(t))\exp[i\pi\gamma]}{\Gamma(\alpha(t)-\gamma+1)}Q_{\alpha}^{-\gamma}\cosh\psi\,,$$

where we have used standard formulae for Legendre functions. So far then we have for B

$$B = \frac{(2\omega'^2)^{\gamma}}{(2\pi)^6}(s')^{\alpha(t)}\left(\frac{s}{s'}\right)^{\gamma}\int_{-\infty}^{\infty} d\beta''\int_0^{\infty} d\omega'' K(t'_{\pm}, t''_{\pm}, t)\left(\frac{\omega''}{\omega'}\right)^{\alpha(t)+\gamma+1}\cdot$$

$$\cdot(\sinh\psi)^{\gamma}\frac{\Gamma(1+\alpha(t))}{\Gamma(\alpha(t)+1-\gamma)}\exp[i\pi\gamma]Q_{\alpha}^{-\gamma}\cosh\psi\,.$$

This is, in fact, as far as one can go without explicit assumptions about K.

We shall close with a demonstration that our result agrees with that of Sect. **3**.2. in the limit $t\to 0$. To do this we remark that as $t\to 0$, $K(t'_\pm, t''_\pm, t)$ becomes independent of β'' or, if you prefer, z''. This follows from $t''_\pm = u'' \pm \pm\sqrt{-t}\,\beta'' + t/4$. We also put, by hand, $t'_+ = t'_- = t'$. In this limit we have $\omega'' = \sqrt{-u''}\sqrt{1-z''^2}$, $\omega' = \sqrt{-u'}\sqrt{-t'}$; the external z' should be set $=0$. We also use $\mathrm{d}\omega''\,\mathrm{d}\beta'' = \frac{1}{2}\mathrm{d}u''\,\mathrm{d}z''(1-z''^2)^{-\frac{1}{2}}$ and find

$$B = \frac{2^\gamma(\sqrt{-t'})^{\gamma-\alpha(0)-1}}{(2\pi)^6}\,\frac{\Gamma(1+\alpha(0))}{\Gamma(\alpha(0)+1-\gamma)}\,(s')^{\alpha(0)}\left(\frac{s}{s'}\right)^\gamma \frac{1}{2}\int_{-\infty}^{0}\mathrm{d}u''\,(\sqrt{-u''})^{\alpha(0)+\gamma+1}K(t=0)\cdot$$
$$\cdot\int_{-1}^{+1}\mathrm{d}z''(\sqrt{1-z''^2})^{\gamma+\alpha}\left[\frac{\cosh q}{1-z^2}-1\right]^{\gamma/2}\exp[i\pi\gamma]Q^{-\gamma}_{\alpha(0)}\left(\frac{\cosh q}{\sqrt{1-z^2}}\right),$$

where, as before

$$\cosh q = \frac{m^2-u''-t'}{2\sqrt{u''t'}}\,.$$

The integration over z may be done by changing variables to $x=(\sqrt{1-z^2})^{-1}$, replacing the Q-function by its representation as a hypergeometric function, expanding the latter in a power series, integrating and resumming. The result is

$$\int_{-1}^{+1}\mathrm{d}z = \pi[\sinh q]^\gamma\,\Gamma(\alpha-\gamma+1)P_\gamma^{-\alpha-1}(\operatorname{ctgh}\psi)\,.$$

Recalling that $\gamma=\alpha(t'_+)+\alpha(t'_-)=2\alpha(t')$ we have as a final result

$$B = \frac{\pi(s')^{\alpha(0)}(s/s')^{2\alpha(t')}}{(2\pi)^3 16\pi^3}\,\Gamma(\alpha(0)+1)\int_{-\infty}^{0}\mathrm{d}u'\,K(t=0)[2\sqrt{u't'}\sinh q]^{2\alpha(t')}\cdot$$
$$\cdot\left(\sqrt{\frac{u'}{t'}}\right)^{\alpha(0)+1}P^{-\alpha(0)-1}_{2\alpha(t')}(\operatorname{ctgh} q)\,,$$

which agrees precisely with the result of Sect. **2**.2!

APPENDIX

In connection with a study of the multiperipheral model taking into account the high-energy pomeron tail of the potential one encounters in a certain approximation an integral of the form

$$J_l(t) = \frac{1}{\pi}\int\mathrm{d}t_1\int\mathrm{d}t_2\,\frac{\Theta(-\Delta)}{\sqrt{-\Delta(t_1,t_2,t)}}\,\frac{f_1(t_1)f_2(t_2)}{l+1-\alpha(t_1)-\alpha(t_2)}\,,$$

and it is of interest to study the t-dependence of $J_l(t)$. Now we know that

$$\frac{\Theta(-\Delta)}{\sqrt{-\Delta(t_1, t_2, t)}} \to \pi\delta(t_1 - t_2) \quad \text{as} \quad t \to 0\,,$$

but a straightforward expansion in powers of t is not easy if one makes no assumption about the various functions in the integrand.

By writing

$$(l+1) - (\alpha(t_1) - \alpha(t_2))^{-1} = \int_0^\infty \mathrm{d}s \exp\left[-s(l+1) - \alpha(t_1) - \alpha(t_2)\right],$$

we see that we can confine attention to a slightly simpler integral, namely

$$K(t) = \frac{1}{\pi}\int \mathrm{d}t_1 \int \mathrm{d}t_2 \frac{\Theta(-\Delta)}{\sqrt{-\Delta(t_1, t_2, t)}}\, g_1(t_1) g_2(t_2)\,.$$

To discuss this we remark that if

$$g_i(t_i) = \exp\left[\gamma_i t_i\right], \qquad i = 1, 2\,,$$

the integral can be readily evaluated. Writing

$$\frac{t_1 + t_2}{2} - \frac{t}{4} = u\,, \qquad t_1 = u + \frac{t}{4} + \sqrt{ut}\, z\,,$$

$$\frac{t_1 - t_2}{2\sqrt{ut}} = z\,, \qquad t_2 = u + \frac{t}{4} - \sqrt{ut}\, z\,,$$

we find that

$$\int \mathrm{d}t_1 \int \mathrm{d}t_2 \frac{\Theta(-\Delta)}{\sqrt{-\Delta}} \ldots = \int_{-\infty}^{0} \mathrm{d}u \int_{-1}^{+1} \frac{\mathrm{d}z}{\sqrt{1 - z^2}} \ldots,$$

so that with the above g_i's

$$K(t) = \frac{1}{\pi}\int_{-\infty}^{0} \mathrm{d}u \int_{-1}^{+1} \frac{\mathrm{d}z}{\sqrt{1-z^2}} \exp\left[(\gamma_1 + \gamma_2)(u + (t/4)) + \sqrt{ut}(\gamma_1 - \gamma_2) z\right] =$$

$$= \int_{-\infty}^{0} \mathrm{d}u \exp\left[(\gamma_1 + \gamma_2)(u + (t/4))\right] \frac{1}{\pi} \int_0^\pi \mathrm{d}\varphi \exp\left[\sqrt{ut}(\gamma_1 - \gamma_2)\cos\varphi\right] =$$

$$= \int_0^\infty \mathrm{d}v \exp\left[-(\gamma_1 + \gamma_2)(v - (t/4))\right] I_0\left[(\gamma_1 - \gamma_2)\sqrt{-vt}\right] =$$

$$= \frac{\exp\left[(\gamma_1 + \gamma_2)t/4\right]}{\gamma_1 + \gamma_2} \exp\left[-\frac{(\gamma_1 - \gamma_2)^2 t}{4(\gamma_1 + \gamma_2)}\right] = \frac{\exp\left[(\gamma_1\gamma_2/(\gamma_1 + \gamma_2))t\right]}{\gamma_1 + \gamma_2}\,.$$

To evaluate $K(t)$ in general we write a double Laplace representation

$$g_1(t_1\, g_2(t_2) = \frac{1}{(2\pi i)^2} \int_{c_1-i\infty}^{c_1+i\infty} \mathrm{d}s_1 \int_{c_2-i\infty}^{c_2+i\infty} \mathrm{d}s_2\, G_1(s_1)\, G_2(s_2) \exp[s_1 t_1 + s_2 t_2]\,,$$

where

$$G_i(s_i) = \int_{-\infty}^{0} \mathrm{d}t_i \exp[-s_i t_i] g_i(t_i)\,.$$

The contours run to the left of all singularities in the s_i-planes. The procedure then is to carry out the t_1, t_2 integrations to obtain

$$K(t) = \frac{1}{(2\pi i)^2} \int \mathrm{d}s_1 \int \mathrm{d}s_2 \frac{\exp[(s_1 s_2/(s_1+s_2))\, t]}{s_1+s_2} G_1(s_1)\, G_2(s_2)\,,$$

expand the exponential in powers of t and use the fact that

$$s^n G(s) = \int_{-\infty}^{0} \mathrm{d}t (-1)^n g(t) \frac{\mathrm{d}^n}{\mathrm{d}t^n} \exp[-st]$$

and

$$\frac{\mathrm{d}^n}{\mathrm{d}s^n} s^n G(s) = \int_{-\infty}^{0} \mathrm{d}t \frac{\mathrm{d}^n}{\mathrm{d}t^n} [t^n \exp[-st]] \cdot g(t) = \int_{-\infty}^{0} \mathrm{d}t (-1)^n t^n \exp[-st] g^{(n)}(t)\,, \qquad n > 0\,,$$

to evaluate the multiple pole at $s_2 = -s_1$, say, assuming $c_2 > c_1$. One finds then, using

$$\delta(t_1 - t_2) = \int \frac{\mathrm{d}s}{2\pi i} \exp[s(t_1 - t_2)]\,,$$

$$K(t) = \sum_{n=0}^{\infty} \frac{(-1)^n t^n}{(n!)^2} \int_{-\infty}^{0} \mathrm{d}u\, u^n g_1^{(n)}(u)\, g_2^{(n)}(u)\,.$$

This may be written as

$$K(t) = \int_{-\infty}^{0} \mathrm{d}u\, J_0 \left[2\sqrt{ut} \sqrt{\frac{\partial}{\partial x_1} \frac{\partial}{\partial x_2}} \right] g_1(x_1)\, g_2(x_2) \bigg|_{x_1=x_2=u}\,.$$

It should be possible to manipulate this expression using integral representations for J_0 to get a result for $K(t)$ for large t if that were desirable.

Dual Models: Their Group-Theoretic Structure.

V. ALESSANDRINI (*).

Institut de Physique Théorique, Université de Genève - Genève

D. AMATI

CERN - Geneva

Introduction.

The dual-resonance models developed in the last three years represent, in our opinion, an important step in the direction of obtaining a theory of elementary processes. Many aspects of these models are not yet fully understood, and, worst of all, they are not yet at the stage where they can be confronted directly with experiment. However, one may try to use them as a laboratory to understand some gross features of strong interactions, and in this sense it is clear that even the simplest dual model—the original one due to VENEZIANO [1]—sheds some light onto the mechanisms that control high-energy reactions. For example, one starts with models made out of resonances, and consequently, Regge poles only and finds that when the theory is suitably unitarized, diffraction effects—related to the Pomeranchuk singularity—begin to creep in [2] in the way one naturally expects it to happen from general dualiy considerations [3].

Dual models were originated by the observation that Regge trajectories seem to be linearly rising and that resonances seem to be pretty narrow. They were found as a solution of the problem of constructing meromorphic amplitudes with linearly rising trajectories and zero-width resonances, having Regge asymptotic behaviour [1]. Since it turned out that the residues of the poles factorize in the way demanded by S-matrix theory [4], and since they satisfy Lorentz invariance and crossing symmetry, we then see that dual models satisfy all our basic principles and theoretical prejudices with the exception of unitarity.

These lectures deal with models that exhibit *planar* duality only. More general models can be constructed, exhibiting nonplanar duality, but they

(*) On leave of absence from University of La Plata, La Plata. Member of the National Research Council of Argentina.

are not compatible with the experimentally observed fact that exchange degeneracy seems to be approximately valid for mesonic Regge trajectories. Nonplanar duality is certainly needed for amplitudes with baryons and/or currents, but their inclusion in dual theories is far from being well understood. Reasonable models exist for amplitudes with only one external baryon-antibaryon pair [5], or only one external current [6], but this situation for more complicated systems is still rather unclear, though we expect it to be better understood in a foreseable future.

The main purpose of these lectures is to discuss the basic underlying simplicity of dual models due to the existence of a symmetry group—the group of real projective transformations isomorphic to $SU_{1,1}$ or $SL_{2,R}$—that guarantees the duality and factorization properties of the amplitudes given by the model [7]. Indeed, one of the most important theoretical ideas in this field was the observation that the problem of the construction of a dual model was largely equivalent to the problem of finding suitable unitary representations of $SU_{1,1}$ [8]. This indicates a well-defined algebraic way of generalizing the original Veneziano model in order to construct new amplitudes that may hopefully compare more favourably with experiment, in the sense that they may possess a more realistic spectrum of states. The method also allows to solve the problem of the inclusion of intrinsic spin in dual models [9], though the solution is not yet complete because, as we said before, one does not know how to introduce more than one fermion line per amplitude. However, any number of particles of intrinsic integer spins can be dealt with at the present time.

In these lectures we then concentrate on the problem of the group-theoretic approach to generalized dual models. They are organized as follows: in Sect. **1** we start by discussing the conventional Veneziano model in the Koba-Nielsen formulation [7], and consider the operator formalism originally developed by FUBINI and VENEZIANO, in which the projective invariance of the Koba-Nielsen integrand is explicitly exhibited [8]. This leads to a simple way of analysing the spectrum of states of the model.

Section **2** consists of a group-theoretic interlude where the infinite-dimensional unitary representations of $SU_{1,1}$ of relevance for dual models are introduced. The connection with the operatorial formalism of Sect. **2** is discussed and it is shown that the conventional Veneziano model corresponds to the choice of a particular unitary irreducible representation characterized by a Casimir operator with eigenvalue $J \to 0$.

Section **3** contains a discussion of the N-Reggeon vertex, namely dual amplitudes where the external lines are excited states that belong to the spectrum of states predicted by the model. Here again the basic simplicity—due to a group-theoretical interpretation of the integrand of the N-Reggeon vertex—is emphasized [10, 11]. Moreover it becomes clear that the couplings among Reggeons are uniquely determined once a particular unitary representation of $SU_{1,1}$ is chosen, and that consequently they need not be introduced as an independent dynamical postulate. The conclusion is that the choice of a representation of $SU_{1,1}$ completely determines the dual model.

Since unitary representations of $SU_{1,1}$ are labelled by a continuous parameter J, there is, in principle, an infinity of dual models that can be constructed. However, another criterion helps to narrow the choice down to a few possible cases. The spectrum of states of dual models is in general plagued with ghosts, as a consequence of the Lorentz-invariance of the model. However, when the intercept of the leading Regge trajectory is 1 ($\alpha_0 = 1$), one finds a ghost killing mechanism in the form of gauge conditions [12] which state that many *a priori* different states have the same couplings and very likely cancel the ghosts away, in the same way as the conventional gauge invariance of quantum electrodynamics decouples the timelike and the longitudinal photons.

In Sect. **4** we discuss how the requirement of having these gauge conditions at work leaves a few values of J (in fact, $J = 0, -\frac{1}{2}, -1$) to construct acceptable dual models [11, 13]. However, in the conventional model $\alpha_0 = 1$ implies that the ground state is a tachyon of mass squared $\mu^2 = -1$. We end the Section by discussing the very interesting model suggested by NEVEU and SCHWARZ [14], where the tachyon problem is solved (the tachyon decouples from all physical states) by introducing spin into the model.

As regards the notations we shall use the metric $(+---)$ and unit slope for Regge trajectories.

1. – The dual-resonance model.

1'1. *The Koba-Nielsen form.* – The starting point of dual-resonance models has been the Veneziano proposal for the four-point function [1]. The Veneziano amplitude for the scattering of scalar spinless particles with momenta k_i is given by

$$T_4(s, t, u) = B(-\alpha(s), -\alpha(t)) + B(-\alpha(t), -\alpha(u)) + B(-\alpha(u), -\alpha(s)) \tag{1.1}$$

where

$$s = (k_1 + k_2)^2, \qquad t = (k_2 + k_3)^2, \qquad u = (k_1 + k_3)^2,$$

and $\alpha(s) = \alpha_0 + s$ is the Regge trajectory exchanged in all three channels and $B(x, y)$ is the beta-function:

$$B(-\alpha(s), -\alpha(t)) = \frac{\Gamma(-\alpha(s))\Gamma(-\alpha(t))}{\Gamma(-\alpha(s) - \alpha(t))} = \int_0^1 \mathrm{d}x\, x^{-\alpha(s)-1}(1-x)^{-\alpha(t)-1}. \tag{1.2}$$

The basic duality property is expressed by the fact that the beta-function can be written [15] either as a sum of poles in the s-channel or as a sum of poles in the t-channel:

$$B(-\alpha(s),-\alpha(t))=\sum_{n=0}^{\infty}\frac{C_n(t)}{\alpha(s)-n}=\sum_{n=0}^{\infty}\frac{C_n(s)}{\alpha(t)-n}, \tag{1.3}$$

where $C_n(t)$ are n-th degree polynomials.

We can think of eq. (1.1) as being obtained first by choosing a cyclic ordering for the external momenta k_i, as, for example, the one exhibited in Fig. 1, and then by associating with this cyclic ordering an amplitude—$A_4(k_1k_2k_3k_4)=$ $=B(-\alpha(s),-\alpha(t))$ for the case of Fig. 1—which is invariant under cyclic or anticyclic permutations of the external momenta.

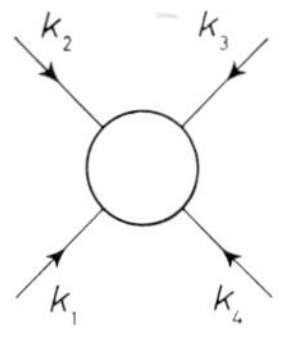

Fig. 1. – The four-point function.

Finally, the whole amplitude T_4 is given by the sum of the contributions over all noncyclic or anticyclic ordering of external particles, *i.e.*

$$T_4=A_4(k_1k_2k_3k_4)+A_4(k_1k_3k_2k_4)+A_4(k_1k_2k_4k_3)\,.$$

This is essentially the content of planar duality. The same procedure is followed for the N-point function. One writes an amplitude $A_N(k_1, k_2, \dots, k_n)$ for a particular cyclic ordering of external momenta and then constructs the full amplitude T_N by adding all noncyclic or anticyclic permutations, *i.e.*

$$T_N=\sum_{\substack{\text{noncyclic or anticyclic}\\ \text{permutations of } k_i}} A_N(k_1\dots k_N)\,. \tag{1.4}$$

A form for the amplitude $A_N(k_1, \dots, k_N)$ satisfying the cyclic symmetry and with the linearly rising pole structure was first given by CHAN, GOEBEL and SAKITA and by BARDAKÇI and RUEGG [16]. When written in the Koba-Nielsen representation [7], it reads

$$A_N=\int\frac{1}{\mathrm{d}V_{abc}}\prod_{i=1}^{N}\mathrm{d}z_i|z_{i+1}-z_i|^{\alpha_0-1}\prod_{i<j}^{N}|z_i-z_j|^{-2k_i\cdot k_j}\,. \tag{1.5}$$

Here we have associated with each external particle of momentum k_i an integration variable z_i on the real axis of the projective plane (the point at $\pm\infty$

is a single point, so the real axis should be regarded as the limit of a circle when the radius tends to infinity). These variables are ordered cyclically

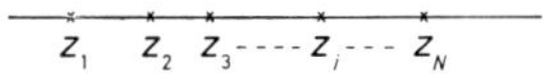

Fig. 2. – Koba-Nielsen variables for the N-point function.

in the same way as the external momenta, and the integration is carried out over the real axis, keeping their cyclic ordering. The coincidence of two or more variables corresponds to end-points of the region of integration. The singularity of the integrand at the end-points $(|z_i - z_j|^{-2k_ik_j})$ generates the poles in A_N in the same way as the singularities of the integrand of the beta-function $[x^{-\alpha(s)-1}$ and $(1-x)^{-\alpha(t)-1}]$ generate the s- and t-channel poles for the four-point function.

If we forget for the moment about $1/\mathrm{d}V_{abc}$, we can see that the formula exhibits planar duality, because it is invariant under the cyclic permutation $k_i \to k_{i+1}$ $(k_{N+1} \equiv k_1)$. This is easily shown by the change of variables $z_i \to z_{i+1}$ $(z_{N+1} \equiv z_1)$. However, since there are N variables of integration, we can have N simultaneous end-points and therefore N simultaneous poles, and this is clearly wrong because there can be at most $(N-3)$ simultaneous poles in the N-point function.

We must therefore get rid of three variables *without spoiling the cyclic invariance of* A_N. At first sight this would look impossible because the integral will depend on which are the three variables that have been fixed and on the values assigned to them. Fortunately, the integrand of A_N (still forgetting about $1/\mathrm{d}V_{abc}$) is projective invariant, *i.e.* it is invariant under a three-parameter real Möbius transformation of the form

$$z_i \to z'_i = \frac{az_i + b}{cz_i + d}\,, \qquad ad - bc = 1\,, \tag{1.6}$$

which maps three arbitrary points into other three arbitrary points by a suitable choice of the three parameters. Therefore, we can assign arbitrary values to three arbitrary variables without changing the integrand. The most elegant way to get rid of three variables is to divide by a volume element which is also projective invariant namely

$$\mathrm{d}V_{abc} = \frac{\mathrm{d}z_a\,\mathrm{d}z_b\,\mathrm{d}z_c}{|z_a - z_b||z_b - z_c||z_c - z_a|}\,, \tag{1.7}$$

where z_a, z_b, z_c are the arbitrarily chosen fixed variables. A simple way to check these statements is to notice that any projective transformation is a combination of

i) scale transformations $z' = \lambda z$,

ii) translations $z' = z + a$,

iii) inversions $z' = 1/z$,

and use momentum conservation, $\sum_{i=1}^{N} k_i = 0$, and the mass-shell condition $\alpha_0 = -\mu^2 = -k_i^2$, because $\alpha(s) = \alpha_0 + s$ and $\alpha(\mu^2) = 0$.

1·2. *The operator formalism.* – The dual-resonance model is the one for which N-particle scattering amplitudes T_N are given by eqs. (1.4) and (1.5). We could use that form to investigate its properties, but this turns out to be a complicated task [4]. It has been revealed much more enlightening to rewrite the same amplitude of eq. (1.5) in an operator form which exhibits explicitly the projective invariance of the integrand [8].

In order to do this, let us consider the factor

$$\prod_{i<j} |z_i - z_j|^{-2k_i\cdot k_j} = \prod_{i<j} z_i^{-2k_i\cdot k_j} \cdot \prod_{i<j} \left|1 - \frac{z_j}{z_i}\right|^{-2k_i\cdot k_j}, \tag{1.8}$$

appearing in the integrand of eq. (1.5).

If we can find two operators $Q_\mu^+(z)$ and $Q_\mu^-(z)$ such that

$$[Q_\mu^-(z), Q_\nu^-(z')] = [Q_\mu^+(z), Q_\nu^+(z')] = 0\,, \quad [Q_\mu^-(z), Q_\nu^+(z')] = g_{\mu\nu} \log\left(1 - \frac{z'}{z}\right), \tag{1.9a}$$

$$Q_\mu^-(z)|0\rangle = 0\,, \tag{1.9b}$$

then (*) we can write

$$\begin{aligned}\exp[i\sqrt{2}k_i\cdot Q^-(z_i)]\exp[i\sqrt{2}k_j\cdot Q^+(z_j)] =\\ = \exp[i\sqrt{2}k_j\cdot Q^+(z_j)]\exp[i\sqrt{2}k_i\cdot Q^-(z_i)]\exp\left[-2k_i\cdot k_j \log\left(1 - \frac{z_j}{z_i}\right)\right].\end{aligned} \tag{1.10}$$

It is then easy to prove [8], by commuting all Q^- to the right and all Q^+ to the left, that

$$\begin{aligned}\langle 0|\exp[i\sqrt{2}k_1\cdot Q^+(z_1)]\exp[i\sqrt{2}k_1\cdot Q^-(z_1)]\ldots\\ \ldots\exp[i\sqrt{2}k_N\cdot Q^+(z_N)]\exp[i\sqrt{2}k_N\cdot Q^-(z_N)]|0\rangle =\\ = \prod_{i<j}\exp\left[-2k_i\cdot k_j\log\left(1 - \frac{z_j}{z_i}\right)\right] = \prod_{i<j}\left(1 - \frac{z_j}{z_i}\right)^{-2k_i\cdot k_j}.\end{aligned} \tag{1.11}$$

(*) We use here and in the following the identities $\exp[A]\exp[B] = \exp[B]\exp[A]\cdot\exp[A,B]$, $\exp[A+B] = \exp[A]\exp[B]\exp\left[-\frac{1}{2}[A,B]\right]$, which are valid when $[A,B]$ commutes with A and B, as in our case.

A representation of $Q^+(z)$ and $Q^-(z)$ satisfying (1.9) can be found in the Fock space defined by creation and annihilation operators $a_{n,\mu}$, $a^+_{n,\mu}$, $n = 1, 2, \ldots$, μ being a Lorentz index ($\mu = 0, 1, 2, 3$) such that

$$[a_{n,\mu}, a_{m,\nu}] = [a^+_{n,\mu}, a^+_{m,\nu}] = 0\,, \qquad [a_{n,\mu}, a^+_{m,\nu}] = -g_{\mu\nu}\delta_{nm}\,; \qquad a_{n,\mu}|0\rangle = 0\,. \tag{1.12}$$

Then it is simple to verify that $Q^+(z)$ and $Q^-(z)$ defined by

$$Q^-_\mu(z) = \sum_{n=1}^{\infty} \frac{a_{n,\mu} z^{-n}}{\sqrt{n}}\,, \qquad Q^+_\mu(z) = \sum_{n=1}^{\infty} \frac{a^+_{n,\mu} z^{n}}{\sqrt{n}}\,, \tag{1.13}$$

satisfy eq. (1.9), and therefore, eq. (1.11).

Equation (1.11) provides us an operatorial representation of the last factor of the right-hand side of eq. (1.8). We still miss the first factor. A representation for it can be found by introducing one extra vector mode $a_{0,\mu}$ satisfying the usual commutation relations (1.12). Let us define [8]

$$Q^0_\mu(z) = q_\mu + ip_\mu \log z\,, \tag{1.14}$$

where

$$q_\mu = \frac{1}{\sqrt{\varepsilon}}(a_{0,\mu} + a^+_{0,\mu})\,, \quad p_\mu = \frac{i\sqrt{\varepsilon}}{2}(a_{0,\mu} - a^+_{0,k}) \rightarrow [p_\mu, q_\nu] = -ig_{\mu\nu}\,, \tag{1.15}$$

ε being a constant, arbitrary for the time being.

The vacuum of the zero mode is defined by

$$p_\mu|0\rangle = 0\,. \tag{1.16}$$

It is then easy to prove

$$\exp[i\sqrt{2}\,k_1\cdot Q^0(z_1)]\exp[i\sqrt{2}\,k_2\cdot Q^0(z_2)] =$$
$$= \exp[i\sqrt{2}\,(k_1+k_2)\cdot q]\exp[-\sqrt{2}\,p\cdot(k_1\log z_1 + k_2\log z_2)]\,z_1^{-k_1^2} z_2^{-k_2^2} z_1^{-2k_1\cdot k_2}\,.$$

Using $\sum_{i=1}^{N} k_i = 0$ and eq. (1.16) we find that

$$\langle 0|\prod_{i=1}^{N}\exp[i\sqrt{2}\,k_i\cdot Q^0(z_i)]|0\rangle = \prod_{i=1}^{N} z^{-\mu^2}\prod_{i<j} z^{-2k_i\cdot k_j}\,. \tag{1.17}$$

Let us now define [6]

$$V(k, z) = \tag{1.18}$$
$$= \exp[i\sqrt{2}\,k\cdot Q^+(z)]\exp[i\sqrt{2}\,k\cdot Q^0(z)]\exp[i\sqrt{2}\,k\cdot Q^-(z)] = :\exp[i\sqrt{2}\,k\cdot Q(z)]:$$

where the ordered product refers only to the nonzero modes, and

$$Q_\mu(z) = Q^+_\mu(z) + Q^0_\mu(z) + Q^-_\mu(z) \,. \tag{1.19}$$

Then, using eqs. (1.11) and (1.17), we can write the N-point amplitude A_N given by (1.5) as

$$A_N = \int \frac{1}{\mathrm{d}V_{abc}} \prod_{i=1}^{N} \mathrm{d}z_i\, z_i^{-\alpha_0}(z_{i+1}-z_i)^{\alpha_0-1} \langle 0|V(k_1, z_1) \dots V(k_N, z_N)|0\rangle \,. \tag{1.20}$$

We shall use later on this form for the amplitude. In order to discuss the spectrum it is, however, better to rewrite the amplitude in a different form. One can define an operator

$$L_0 = -\frac{p^2}{2} - \sum_{n=1}^{\infty} n a_n^+ \cdot a_n \,, \tag{1.21}$$

which satisfies the following relation:

$$z^{L_0} V(k, 1) z^{-L_0} = V(k, z) \,. \tag{1.22}$$

The meaning of L_0 and of eq. (1.22) will become more transparent when we shall discuss the projective invariance—duality—in the operatorial formalism.

Let us now choose z_1, z_2 and z_N to be the variables z_a, z_b and z_c, which are not integrated in eq. (1.20), and let us assign to them fixed values $z_1 = -\infty$, $z_2 = -1$, $z_N = 0$. We must be careful in approaching this configuration with a limiting procedure. By using the explicit expressions given by eqs. (1.18), (1.14) and (1.13), it is easy to check that

$$V(k_N, z_N)|0\rangle = |\sqrt{2}\, k_N, 0\rangle z_N^{\alpha_0} \,,$$

$$\langle 0|V(k_1, z_1) = z_1^{-\alpha_0} \langle -\sqrt{2}\, k_1, 0| \,,$$

where

$$|k, 0\rangle = \exp[ik \cdot q]|0\rangle \,, \tag{1.23}$$

represents a state with momentum k for the zero mode and the vacuum of modes $n = 1, 2, \dots$, *i.e.* it satisfies

$$p_\mu|k, 0\rangle = k_\mu|k, 0\rangle \,, \qquad a_{n,\mu}|k, 0\rangle = 0 \,, \qquad n = 1, 2 \dots .$$

Inserting now eq. (1.22) into eq. (1.18) for $V(k_i, z_i)$, $i = 2, \dots, N-1$, it is easy to show that

$$(1.24)\qquad A_N = \int \prod_{i=2}^{N-2} \mathrm{d}x_i x_i^{-\alpha_0-1}(1-x_i)^{\alpha_0-1}\langle -\sqrt{2}\,k_1,\, 0|V(k_2,\,1)x_2^{L_0}V(k_3,\,1)\,\dots$$
$$\dots\, V(k_{N-2},\,1)x_{N-2}^{L_0}V(k_{N-1},\,1)|\sqrt{2}\,k_N,\,0\rangle\,,$$

where $x_i = z_{i+1}/z_i$.

By explicitly performing the matrix element over the zero mode $a_{0,\mu}$, eq. (1.24) can be rewritten in the form

$$(1.25)\qquad A_N = \langle 0|V(k_2)D(s_2)V(k_3)\,\dots\,D(s_{N-2})V(k_{N-2})|0\rangle\,,$$

$$(1.26)\qquad D(s) = \int_0^1 \mathrm{d}x(1-x)^{\alpha_0-1}x^{-\sum\limits_{n=1}^{\infty} a_n^+\cdot a_n - 1 - \alpha(s)}\,,\qquad s_i = \left(\sum_{j=1}^{i} k_j\right)^2,$$

$$(1.27)\qquad V(k) = \exp\left[i\sqrt{2}\,k\cdot\sum_{n=1}^{\infty}\frac{a_n^+}{\sqrt{n}}\right]\exp\left[i\sqrt{2}\,k\cdot\sum_{n=1}^{\infty}\frac{a_n}{\sqrt{n}}\right].$$

We can interpret eq. (1.25) as the expression for the multiperipheral amplitude of Fig. 3 with vertices $V(k)$ and propagators D given by eqs. (1.27) and (1.26),

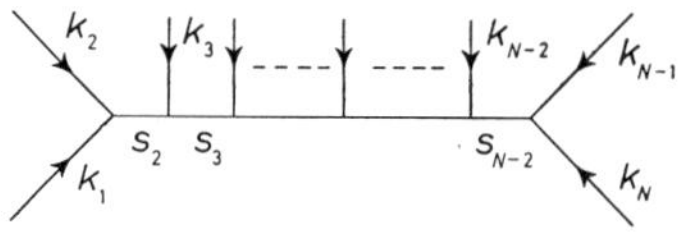

Fig. 3. – Multiperipheral representation of the N-point function.

respectively. The form of eq. (1.25), being explicitly factorized, is well suited to study the spectrum of states. In order to do that let us expands the factor $(1-x)^{\alpha_0-1}$ in the integrand of eq. (1.26), so as to write

$$(1.28)\qquad D(s) = \sum_{l_0=0}^{\infty}\binom{l_0-\alpha_0}{l_0}\frac{1}{H-\alpha(s)}\,,$$

$$(1.29)\qquad H = l_0 + \sum_{n-1}^{\infty} n a_n^+\cdot a_n\,.$$

The complete set of states $|\{l\},\, l_0\rangle$ which factorizes the amplitude and diagonalizes the propagator is characterized by an integer l_0 and a set of integers l_i, $i = 1, \dots, n$ which represent the occupation number of mode a_i [4]. Clearly

$$(1.30)\qquad \begin{cases} H|\{l\},\, l_0\rangle = (M^2_{\{l\},l_0} + \alpha_0)|\{l\},\, l_0\rangle\,, \\[2ex] M^2_{\{l\},l_0} = l_0 + \sum\limits_{n=1}^{\infty} n l_n - \alpha_0\,, \end{cases}$$

The factorization of the amplitude A_N into M and $N-M$ is evident by inserting

$$1=\sum |\{l\}\rangle\langle\{l\}|\,,$$

in between $V(k_M)$ and $D(s_M)$ in eq. (1.25). By using eqs. (1.28) and (1.30), it follows that

$$(1.31)\qquad A_N=\sum_{l_i}\sum_{l_0}\sqrt{\binom{l_0-\alpha_0}{l_0}}\,\langle 0|V(k_2)D(s_2)\ldots$$
$$\ldots V(k_M)|\{l\}\rangle\frac{1}{M^2_{\{l\},l_0}-s_M}\langle\{l\}|V(k_{M+1})D(s_{M+1})\ldots V(k_{N-1})|0\rangle\sqrt{\binom{l_0-\alpha_0}{l_0}}\,.$$

The states characterized by the set of integers $\{l\}$ and l_0 are therefore excited states of the theory and have square mass $M^2_{\{l\},l_0}$ given by eq. (1.30). The spin content of every such state can be easily analysed [4]. It turns out that it contains a superposition of spins ranging from 0 up to $\sum_{i=1} l_i$. The degeneracy of states is very high. Indeed, it increases as the number of partitions of integers satisfying eq. (1.30). This gives, for a given mass M, a number of states which increases as [4]

$$(1.32)\qquad N_M\sim\exp\left[2\pi\sqrt{2/3}\,M\right],$$

for $M\to\infty$.

An amplitude containing two excited states $|\{l\},l_0\rangle$, $|\{l'\},l'_0\rangle$ and an arbitrary number of scalars, as those of Fig. 4, can be obtained by a double factorization of an arbitrary tree. It is given by

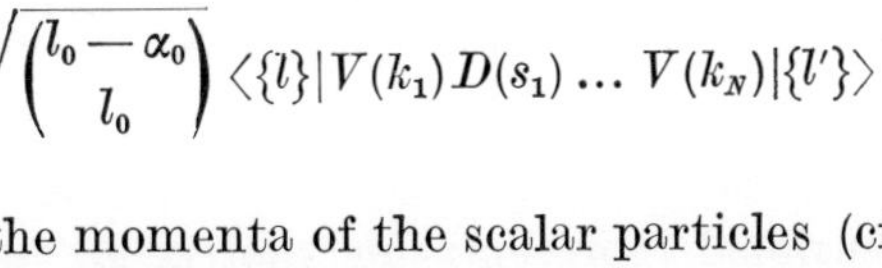

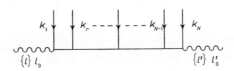

Fig. 4. – A $(N+2)$-point function with two external excited states obtained by a double factorization of a multiperipheral tree.

$$(1.33)\qquad A_{\substack{\{l\},l_0\\ \{l'\},l'_0}}=\sqrt{\binom{l_0-\alpha_0}{l_0}}\,\langle\{l\}|V(k_1)D(s_1)\ldots V(k_N)|\{l'\}\rangle\sqrt{\binom{l'_0-\alpha_0}{l'_0}}\,,$$

where $k_1,\ldots,k_N$ are the momenta of the scalar particles (cf., Fig. 4). However, in order to obtain an amplitude with more than two excited states, we need to further factorize the amplitude of eq. (1.33) in channels which are dual to the ones having poles explicitly represented in the propagators $D(s_i)$ of eq. (1.33). By writing the amplitude A_N in the factorized form of eq. (1.25) we have obscured its duality properties. In order to tackle this problem we need a better understanding of duality in the operatorial formalism. We then discuss the group of duality transformations in the next Section to prepare the ground for the introduction of the dual amplitude for an arbitrary number of excited states.

2. – The projective group.

To every real projective transformation of the form

$$z' = T(z) = \frac{az+b}{cz+d}\,, \tag{2.1}$$

normalized in such a way that $ad - bc = 1$, one associates the 2×2 matrix

$$\Lambda = \begin{pmatrix} a & b \\ c & d \end{pmatrix}, \tag{11.4}$$

which acts on a two-component object $\xi = (\xi_1, \xi_2)$. The components of ξ are the « homogeneous co-ordinates » of the variable z. The set of matrices defines the group $SL_{2,R}$, isomorphic to $SU_{1,1}$. The unitary irreducible representations are obtained in the standard way by considering the transformation properties of the monomials of the components of the spinor ξ, namely $\xi_1^a \xi_2^b$, which provides a basis for them. The standard procedure is to arrange the powers a and b in such a way that the basic states are written as [17]

$$|Jkm\rangle = N(J, k, m)\, (\xi_1\, \xi_2)^J \left(\frac{\xi_1}{\xi_2}\right)^{k+m}, \tag{2.3}$$

where N is a normalization coefficient. Irreducible representations are labelled by the quantum numbers J and k, which correspond to the eigenvalues $J(J+1)$ of the Casimir operator and the fractional part of $\frac{1}{2}(a-b)$.

Under a transformation Λ of the form given in eq. (2.2), the spinor transforms like

$$\begin{pmatrix} \xi'_1 \\ \xi'_2 \end{pmatrix} = \begin{pmatrix} a & b \\ c & d \end{pmatrix} \begin{pmatrix} \xi_1 \\ \xi_2 \end{pmatrix}, \tag{2.4}$$

and the infinite-dimensional representation matrix $D^{(J,k)}(\Lambda)$ is defined by expanding the new basis $\xi_1'^a \xi_2'^b$ in terms of the old one

$$|Jkm\rangle' = N(Jkm)(\xi'_1 \xi'_2)^J \left(\frac{\xi'_1}{\xi'_2}\right)^{k+m} = \sum_{n=0}^{\infty} D^{(J,k)}_{mn}(\Lambda)\, |Jkn\rangle\,. \tag{2.5}$$

We shall only be concerned here with representations for which $0 > J = \pm k$. Then eq. (2.5) can be written more explicitly by introducing the variable $\xi_1/\xi_2 = z$ for the case $k = -J$

$$N(J, -J, m)(cz + d)^{2J}\left(\frac{az + b}{cz + d}\right)^m = \sum_{n=0}^{\infty} D_{mn}^{(J,+)}(\Lambda)N(J, -J, n)z^n, \tag{2.6}$$

and in terms of the variable $x = z^{-1}$ for the case $k = J$

$$N(J, J, m)(a + bx)^{2J}\left(\frac{c + dx}{a + bx}\right)^m = \sum_{n=0}^{\infty} D_{mn}^{(J,-)}(\Lambda)N(J, J, n)x^n. \tag{2.7}$$

The generators of $SU_{1,1}$ will be denoted by L_0, $L_{\pm 1}$ and satisfy the commutation relations

$$[L_0, L_{\pm 1}] = \mp L_{\pm 1}, \tag{2.8}$$

$$[L_{+1}, L_{-1}] = 2L_0. \tag{2.9}$$

The representation of these generators in terms of 2×2 matrices is

$$\Lambda_0 = \tfrac{1}{2}\begin{pmatrix} 1 & 0 \\ 0 & -1 \end{pmatrix}, \quad \Lambda_+ = \begin{pmatrix} 0 & 0 \\ -1 & 0 \end{pmatrix}, \quad \Lambda_- = \begin{pmatrix} 0 & 1 \\ 0 & 0 \end{pmatrix}. \tag{2.10}$$

Then their matrix elements in the (J, k) irreducible representations follow at once from eq. (2.5):

$$D_{mn}^{(J,k)}(\Lambda_0) = (k + m)\,\delta_{m,n}, \tag{2.11}$$

$$D_{mn}^{(J,k)}(\Lambda_+) = -\frac{N(J, k, m)}{N(J, k, n)}(k + m - J)\,\delta_{n,m+1}, \tag{2.12}$$

$$D_{mn}^{(J,k)}(\Lambda_-) = \frac{N(J, k, m)}{N(J, k, n)}(k + m + J)\,\delta_{n,m-1}. \tag{2.13}$$

Finally, the normalization coefficient $N(J, k, m)$ is determined up to a phase by requiring [as in the case of SU_2] that $L_{-1} = L_1^+$. Fixing the arbitrary phase equal to 1, one obtains for the case $k = \pm J$

$$N(J, \pm J, n) = \sqrt{\frac{\Gamma(n - 2J)}{\Gamma(n + 1)}}. \tag{2.14}$$

We are interested in the relation between the representation $k = -J$ and what is usually called its « conjugate representation », $k = +J$. From eqs. (2.11) and (2.13) it easily follows that for the generators both representations are related by

$$D^{(J,+)}(\Lambda_i)^T = - D^{(J,-)}(\Lambda_i) , \tag{2.15}$$

so for finite transformations, obtained by exponentiating generators, it follows that

$$D^{(J,+)}(\Lambda)^T = D^{(J,-)}(\Lambda^{-1}) . \tag{2.16}$$

It is useful to introduce the 2×2 matrix [10]

$$\Gamma = \begin{pmatrix} 0 & 1 \\ 1 & 0 \end{pmatrix} , \tag{2.17}$$

which satisfies $\Gamma^2 = 1$ and corresponds to the transformation $z' = 1/z$. This matrix will play the role of a metric tensor in relating the $D^{(J,+)}$ and $D^{(J,-)}$ representations. Indeed, from eqs. (2.6) and (2.7), it follows at once that

$$D^{(J,-)}(\Lambda)^T = D^{(J,+)}(\Gamma \Lambda \Gamma) , \tag{2.18}$$

so that eq. (2.16) can be rewritten as

$$D^{(J,+)}(\Lambda)^T = D^{(J,+)}(\Gamma \Lambda^{-1} \Gamma) . \tag{2.19}$$

This equation just reflects the noncompactness of the $SU_{1,1}$ group, because the 2×2 matrices, being finite-dimensional, are not unitary, so that $\Lambda^T \neq \Lambda^{-1}$.

So much for the representations of $SU_{1,1}$. Let us now discuss the realization of the group in a Fock space generated by creation operators $a^+_{n,\mu}$ satisfying the canonical commutation relations

$$[a_{n,\mu}\, a^+_{m,\nu}] = - g_{\mu\nu} \delta_{\mu m} . \tag{2.20}$$

In this Fock space it is easily shown by explicit commutation that the operators

$$L_i = \sum_{\mu,\nu=0}^{3} - g_{\mu\nu} \sum_{\mu,m=0}^{\infty} a^+_{n,\mu} D^{(J,k)}_{nm}(\Lambda_i) a_{m,\nu} . \tag{2.21}$$

satisfy the commutation relations of the $SU_{1,1}$ algebra given by eqs. (2.8) and (2.9). They are therefore the operator representations of the $SU_{1,1}$ generators. Lorentz indices are rather irrelevant here because in these bilinear forms in $a^+_{n,\mu}$ and $a_{m,\nu}$ they are always contracted in a trivial way since the $D^{(J,k)}_{nm}$ matrices only act on the index m, and we shall consistently ignore them. This means that there is no coupling between the Lorentz symmetry associated with space-time and the «internal» symmetry group $SU_{1,1}$.

An arbitrary element of the group can be expressed in the form

$$U = \exp[\alpha_+ L_+] \exp[\alpha_0 L_0] \exp[\alpha_- L_-] . \tag{2.22}$$

The exponentiation of an operator L_i given by eq. (2.21) is simply obtained by noting that [10]

$$\exp\left[\sum_{nm} a_n^+ D_{nm}(\Lambda) a_m\right] = :\exp\left[\sum_{nm} a_n^+ (\exp[D(\Lambda)] - 1)_{nm} a_m\right]: = \\ = :\exp\left[\sum_{nm} a_n^+ (D_{nm}(\exp[\Lambda]) - \delta_{nm}) a_m\right]: . \tag{2.23}$$

The last form of eq. (2.22) is called a canonical form [10]. It is easy to check, from the commutation relations of a and a^+ that canonical forms satisfy [10] the group multiplication properties, *i.e.*

$$:\exp[a^+(D(\Lambda_1) - 1)a]: \times :\exp[a^+(D(\Lambda_2) - 1)a]: = \\ = :\exp[a^+(D(\Lambda_1)D(\Lambda_2) - 1)a]: = :\exp[a^+(D(\Lambda_1\Lambda_2) - 1)a]: . \tag{2.24}$$

From eqs. (2.23) and (2.24) we see that the arbitrary group element given by eq. (2.22) can be written in the canonical form [10]

$$U(\Lambda) = :\exp[-g_{\mu\nu} a_{n,\mu}^+ (D_{nm}(\Lambda) - \delta_{nm}) a_{m,\nu}]: , \tag{2.25}$$

where the 2×2 matrix Λ associated with U is

$$\Lambda = \exp[\alpha_+ \Lambda_+] \exp[\alpha_0 \Lambda_0] \exp[\alpha_- \Lambda_-] , \tag{2.26}$$

the Λ_i being the 2×2 representations of the L_i given in eq. (2.10). By explicit computation the matrix Λ of eq. (2.26) reads

$$\Lambda = \begin{pmatrix} \exp[\sqrt{\alpha_0}] & \alpha_- \exp[-\sqrt{\alpha_0}] \\ -\alpha_+ \exp[\sqrt{\alpha_0}] & \exp[-\sqrt{\alpha_0}](1 - \alpha_- \alpha_+) \end{pmatrix} . \tag{2.27}$$

The canonical form (2.25) of an arbitrary operator (group element) is particularly convenient for computing the action of the operator $U(\Lambda)$ on coherent states $|\alpha\rangle$ defined as

$$|\alpha\rangle \equiv \prod_{n=0}^{\infty} \prod_{\mu=0}^{3} |\alpha_{n,\mu}\rangle , \qquad |\alpha_{n,\mu}\rangle = \exp[\alpha_{n,\mu} a_{n,\mu}^+] |0\rangle . \tag{2.28}$$

Since they are eigenstates of the annihilation operators $a_{n,\mu}$, it follows at once from eq. (2.23) that one obtains a new coherent state given by

$$U(\Lambda)|\alpha\rangle = |D^{(J,k)}_{(\Lambda)}\alpha\rangle \equiv \prod_{m=0}^{\infty}\prod_{\mu=0}^{3}|\sum_{n=0}^{\infty} D^{(J,k)}_{mn}(\Lambda)\alpha_{n,\mu}\rangle\,.$$

We shall next introduce operators in our Fock space that transform like irreducibile tensor operators under $SU_{1,1}$. Let us define operator-valued functions $F^J_{n,\mu}(A)$ and $\tilde{F}^J_{n,\mu}(A)$, A being an $SU_{1,1}$ element, by

$$F^J_n(A) = \sum_{m=0}^{\infty} a^+_m D^{(J,+)}_{mn}(A)\,, \tag{2.29 a}$$

$$\tilde{F}_n(A) = \sum_{m=0}^{\infty} a_m D^{(J,-)}_{mn}(A\Gamma) = \sum_{m=0}^{\infty} a_m D^{(J,+)}_{mn}(\Gamma A)\,, \tag{2.29 b}$$

where again a Lorentz index in the F and a operators is understood. The motivation for introducing such operators is that, as we shall see later on, these « field operators » will allow us to factorize the N-Reggeon vertex in the same way as the field $Q(z)$—which will turn out to be a particular case of $(F^J_m + \tilde{F}^J_m)$ when $m = 0$ and $J \to 0$—allowed us to factorize the Koba-Nielsen formula for scalars [11,18]. The transformation properties of these field operators are obtained by using the commutation relations

$$:\exp\Big[\sum_{nm} a^+_n(D_{nm}(\Lambda)-\delta_{nm})a_m\Big]:a^+_r = \Big(\sum_{j=0}^{\infty} a^+_j D_{jr}(\Lambda)\Big):\exp\Big[\sum_{nm} a^+_n(D_{nm}(\Lambda)-\delta_{nm})a_m\Big],$$

$$a_r:\exp\Big[\sum_{nm} a^+_n(D_{nm}(\Lambda)-\delta_{nm})a_m\Big]: = :\exp\Big[\sum_{nm} a^+_n(D_{nm}(\Lambda)-\delta_{nm})a_n\Big]:\Big(\sum_{j=0}^{\infty} D_{rj}(\Lambda)a_j\Big),$$

which are in turn easily deduced by acting with these operators on coherent states. Indeed, writing a group element in Fock space $U(\Lambda)$ in canonical form, it follows that

$$U(\Lambda)F^J_n(A)U^{-1}(\Lambda) = \sum_{j,r} a^+_j D^{(J+)}_{jr}(\Lambda)D^{(J)}_{rn}(A) = \sum_j a^+_j D^{(J+)}_{jn}(\Lambda A) = F^J_n(\Lambda A)\,, \tag{2.30 a}$$

$$U(\Lambda)\tilde{F}^J_n(A)U^{-1}(\Lambda) = \sum_{j,r} a_j D^{(J+)}_{rj}(\Lambda^{-1})D^{(J-)}_{rn}(A\Gamma) = \tag{2.30 b}$$

$$= \sum_{j,r} a_j D^{(J+)}_{rj}(\Lambda^{-1})D^{(J+)}_{nr}(\Gamma A^{-1}) = \sum_j a_j D^{J+}_{nj}(\Gamma A^{-1}\Lambda^{-1}) = \tilde{F}^J_n(\Lambda A)\,.$$

We shall show now [19] that by choosing the representation $J \to 0$, one finds the expression of the operators which appeared in Sect. **1**, when the conventional dual-resonance model was factorized. We must be careful in performing the $J \to 0$ limit because of the singularity of the normalization coefficient $N(J, \pm J, 0)$ corresponding to the $n = 0$ mode [cf., eq. (2.14)].

Let us indeed define $\varepsilon \equiv -2J$ and identify it with the arbitrary constant appearing in the definition (1.15) of p and q from a_0 and a_0^+. By developing in ε and then taking $\varepsilon \to 0$ we obtain, from eqs. (2.21) and (2.11)-(2.13), for $J \to 0$

$$(2.31)\qquad \begin{cases} L_0 = -\dfrac{p^2}{2} - \sum\limits_{m=1}^{\infty} m a_m^+ \cdot a_m\,, \\ L_{-1} = L_1^+ = ipa_1^+ - \sum\limits_{m-1}^{\infty} \sqrt{m(m+1)}\, a_{m+1}^+ \cdot a_m\,, \end{cases}$$

which are the usual expression for the Gliozzi operators. The form for L_0 was already introduced in eq. (2.21).

We will now establish a connection between the irreducible tensors $F_n^J(A)$ and $\tilde{F}_n^J(A)$ and the operator $Q(z)$ of eq. (1.19). If

$$(2.32)\qquad A = \begin{pmatrix} a & b \\ c & d \end{pmatrix}, \qquad ad - bc = 1\,,$$

it is easy to see that at lowest order in $\varepsilon = -2J$

$$(2.33)\qquad \begin{cases} D_{m0}^{(J+)}(A) = \dfrac{\sqrt{\varepsilon}}{\sqrt{m}} \left(\dfrac{b}{d}\right)^m, \qquad D_{m0}^{(J+)}(\Gamma A) = \dfrac{\sqrt{\varepsilon}}{\sqrt{m}} \left(\dfrac{b}{d}\right)^{-m}, \qquad m = 1, 2, \ldots, \\ D_{00}^{(J+)}(A) = 1 - \varepsilon \log d\,, \quad D_{00}^{(J+)}(\Gamma A) = 1 - \varepsilon \log b\,. \end{cases}$$

By introducing eqs. (2.3), (2.26) and (2.27), it can be shown that $(F_0^J(A) + \tilde{F}_0^J(A))/\sqrt{\varepsilon}$ has a definite limit for $\varepsilon \to 0$ which happens to depend on only one of the three parameters appearing in A, *i.e.* b/d. Indeed

$$(2.34)\qquad \lim_{\varepsilon = 0} \frac{F_0^{2\varepsilon}(A) + \tilde{F}_0^{2\varepsilon}(A)}{\sqrt{\varepsilon}} = \sum_{m=1}^{\infty} \left(\frac{a_m^+}{\sqrt{m}} z^m + \frac{a_m}{\sqrt{m}} z^{-m} \right) + q + ip \log z = Q(z)\,,$$

where $z = b/d$ and $Q(z)$ is the operator defined by eq. (1.19).

Let us define the invariant quantity

$$(2.35)\qquad \mathscr{V}^J(k, A) = \exp\left[i\sqrt{2} \frac{kF_0^J(A)}{\sqrt{-2J}} \right] \exp\left[i\sqrt{2} \frac{k\tilde{F}_0^J(A)}{\sqrt{-2J}} \right].$$

For $-2J = \varepsilon \to 0$ it follows that

$$(2.36)\qquad \mathscr{V}^{-2\varepsilon}(k, A) = \exp\left[-\frac{k^2}{\varepsilon} \right] \left(\frac{b}{d}\right)^{k^2} d^{2k^2} V\left(k, \frac{b}{d}\right),$$

where $V\ (k, z = b/d)$ is the Fubini-Veneziano vertex [8] given in eq. (1.18). If $U(\Lambda)$ is an arbitrary group element corresponding to a 2×2 matrix $\Lambda = \begin{pmatrix}\alpha & \beta\\ \gamma & \delta\end{pmatrix}$, we have from eqs. (2.30) and (2.35) that

$$U(\Lambda)\mathscr{V}^J(k, A)\,U^{-1}(\Lambda) = \mathscr{V}^J(k, \Lambda A)\,. \tag{2.37}$$

Equations (2.36) and (2.37) imply

$$z^{k^2}U(\Lambda)\,V(k, z)\,U^{-1}(\Lambda) = z'^{k^2}(\gamma z + \delta)^{2k^2}V(k, z)\,, \tag{2.38}$$

where

$$z' = \frac{\alpha z + \beta}{\alpha z + \delta}\,. \tag{2.39}$$

By choosing U to be the infinitesimal transformation $\exp[\alpha_i L_i](\alpha_i \sim 0)$ and by using the explicit form of the corresponding 2×2 matrices, one obtains from eq. (2.38)

$$[L_n, V(k, z)] = z^n\left(z\frac{\partial}{\partial z} - k^2 n\right)V(k, z)\,, \qquad n = 0, \pm 1\,. \tag{2.40}$$

By making use of eq. (2.38) it is easy to prove directly the projective invariance of the integrand of the dual amplitude written in the operatorial form of eq. (1.20). Indeed, if z'_i and z_i are related by the projective transformation (2.39), it is easy to check that

$$\prod_i \mathrm{d}z'_i|z'_{i+i} - z'_i|^{\alpha_0 - 1} = \prod_i \frac{\mathrm{d}z_i|z_{i+1} - z_i|^{\alpha_0-1}}{(\gamma z_i + \delta)^{2\alpha_0}}\,. \tag{2.41}$$

Then

$$\begin{aligned}\prod_{i=1}^N \mathrm{d}z'_i z'^{-\alpha_0}_i|z'_{i+1} - z'_i|^{\alpha_0-1}\langle 0|\prod_{i=1}^N V(k_j, z'_i)|0\rangle &= \\ = \prod_{i=1}^N \mathrm{d}z_i|z_{i+1} - z_i|^{\alpha_0-1}\langle 0|\prod_{i=1}^N z'^{k^2}_i(\gamma z_i + \delta)^{2k^2}V(k_i, z'_i)|0\rangle &= \\ = \prod_{i=1}^N \mathrm{d}z_i|z_{i+1} - z_i|^{\alpha_0-1}\langle 0|\prod_{i=1}^N z_i^{k^2}\,U(\Lambda)V(k_i, z_i)U^{-1}(\Lambda)|0\rangle &= \\ = \prod_{i=1}^N \mathrm{d}z_i z_i^{-\alpha_0}|z_{i+1} - z_i|^{\alpha_0-1}\langle 0|\prod_{i=1}^N V(k_i, z_i)|0\rangle &,\end{aligned} \tag{2.42}$$

where we have used eq. (2.38) and that $U(\Lambda)$ leaves the vacuum invariant. Equation (2.42) shows explicitly the projective invariance of the integrand of A_N given by (1.20).

Finally, we shall show the connection between the canonical forms introduced before and the ones used in the conventional model [10]. Let us again look for the limit $J \equiv -2\varepsilon \to 0$ in eq. (2.25). The expressions for D_{m0} and D_{00} were given in eq. (2.33). On the other hand, from eqs. (2.18) and (2.33)

$$D_{0m}^{(J+)}(A) = \frac{\sqrt{\varepsilon}}{\sqrt{m}}\left(\frac{-c}{d}\right)^m. \tag{2.43}$$

Introducing eqs. (2.33) and (2.43) into eq. (2.25) for $J \to 0$, we find

$$U(A) = \exp\left[ip\sum_{m=1}^{\infty} a_m^+ \frac{1}{\sqrt{m}}\left(\frac{b}{d}\right)^m\right] : \exp\left[-\sum_{n,m=1}^{\infty} a_m^+ (D_{mn}(A) - \delta_{mn})\, a_n\right] : \cdot \\ \cdot \exp\left[-ip\sum_{m=1}^{\infty} a_m \frac{1}{\sqrt{m}}\left(\frac{-c}{d}\right)^m\right] \exp\left[-p^2 \log \frac{d}{\sqrt{|ad-bc|}}\right] \tag{2.44}$$

with

$$\sum_{n=a}^{\infty} D_{mn}(A)\frac{z^n}{\sqrt{n}} = \left(\frac{az+b}{cz+d}\right)^m \frac{1}{\sqrt{m}} - \left(\frac{b}{d}\right)^m \frac{1}{\sqrt{m}}, \qquad m = 1, 2, \ldots, \tag{2.45}$$

as given by eqs. (2.6) and (2.33) for $J = 0$. The canonical form, (2.44) and (2.45), coincides with the one introduced in ref. [10].

3. – The N-Reggeon vertex.

We start in this Section the discussion of dual amplitudes when the external lines are excited states, namely any one of the resonant states that factorize the Born approximation for scalar particles. These amplitudes are usually called N-Reggeon vertices. Their construction is an important step in order to obtain a true bootstrap theory. Since the N-Reggeon vertices are factorized again by the same basic set of states that occur in the external lines, one has a perfect « nuclear democracy » in the sense that the scalar particles we started with are not more fundamental than the excited ones.

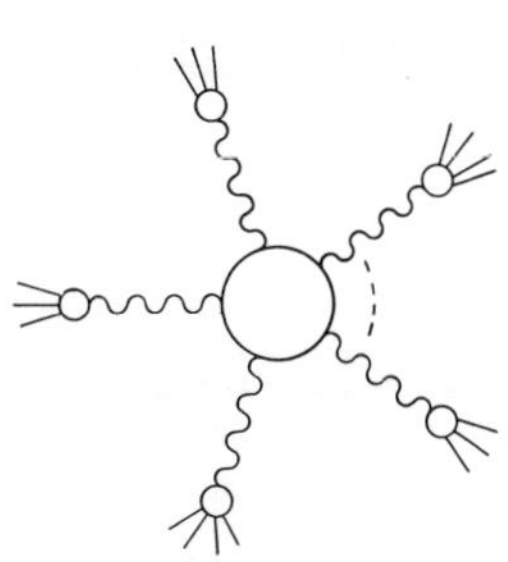

Fig. 5. – Multiple factorization leading to the N-Reggeon vertex.

We have strongly emphasized the underlying projective invariance of the Koba-Nielsen integrand because, as seen before, it is this invariance that guarantees the duality properties of the Veneziano amplitude. N-Reggeon vertices should enjoy similar duality properties and therefore the projective group should again be extremely relevant in their construction. In principle, the way to obtain the N-Reggeon vertex is perfectly straightforward. One starts from an amplitude with a large number of scalar particles and factorizes N excited states by calculating the residue of a multiple pole in the invariant masses of bunches of adjacent particles, as exhibited in Fig. 5. The calculations involved are quite messy, but the N-Reggeon vertex was historically first calculated in this way [20, 21]. The result, however, is particularly simple [18, 22] and has a nice group-theoretic interpretation. We shall see that the N-Reggeon vertex is uniquely determined in terms of unitary infinite-dimensional representations of $SU_{1,1}$ [11].

A slight change in notation will be necessary in what follows due to the fact that, in order to describe external excited particles, we need some way of labelling their state of internal excitation. In other words, we must give a set of quantum numbers like occupation numbers, or coherent-state indices, which will specify for each external line the particular resonant state we want to consider. The best way to proceed is to associate to an N-legged object as shown in Fig. 6, where all legs are pointing to the left, an operator in a

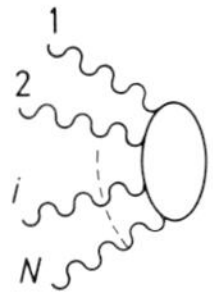

Fig. 6. – The N-Reggeon vertex.

bigger Fock space which is N times the tensor product of the ordinary Fock space we have been using up to now. In other words, we shall add to the operators $a_{n,\mu}$ yet another index i ($i = 1, \ldots, N$) which is a sort of particle index that will distinguish between different Reggeons. Operators with different particle index i are completely independent, that is to say, they always commute with one another. In order to simplify notations, we shall skip Lorentz indices from now on and write the operators as $a_n^{(i)}$ (n = oscillator index) or simply $a^{(i)}$. We then associate with the object of Fig. 6 the operator

$$\exp\left[\frac{1}{2}\sum_{i\neq j=1}^{N} a_n^{(i)+} D_{nm}^J(A_{ij}) a_m^{(j)+}\right]|0_1, \ldots 0_N\rangle \,, \tag{3.1}$$

where $|0_1, \ldots, 0_N\rangle$ is the vacuum state of all oscillators and $D_{nm}^J(A_{ij})$ the infinite-dimensional representation of some projective transformation A_{ij} to be defined later on. To be more specific, the 2×2 projective transformation will be determined in terms of the Koba-Nielsen variables of the Reggeons [18, 22].

The final step necessary before we give the N-Reggeon vertex is to learn to multiply two objects of the type given by eq. (3.1) or, in other words, to sew together an $(N+1)$-legged object and a $(N'+1)$-legged object in order to obtain a $(N+N')$-legged object, as indicated in Fig. 7. From now on we shall call these objects generalized canonical forms, for reasons that will soon become clear.

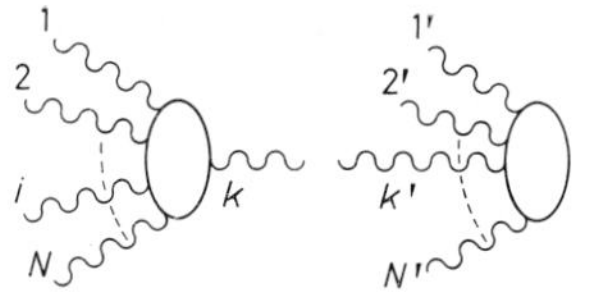

Fig. 7. – Sewing together two vertices along the lines $k'=k$.

Let us denote by i ($i=1, \ldots, N+1$) the particle indices of the first generalized canonical form and by i' ($i'=1, \ldots, N'+1$) those of the second one, and let us decide to sew the lines k and k', as indicated in Fig. 7. The sewing is done first by passing the k-th line to the right, which amounts to replacing $a^{(k)^+}$ by $a^{(k)}$ and shifting the vacuum state $|0_k\rangle$ from the right to the left ($|0_k\rangle \to \langle 0_k|$), and secondly by identifying $k=k'$ in such a way that the oscillators $a^{(k)}$, $a^{(k')}=a^{(k)}$ are the same [10].

Consider now a typical term, product of a term connecting lines i and k and a term connecting lines k' and l'. The sewing procedure we have just described gives the following vacuum expectation value with respect to the oscillators $a^{(k)}$:

$$(3.2) \qquad \langle 0_k| \exp[-a_n^{(i)^+} D^J_{nn'}(A_{ik})a_{n'}^{(k)}]\cdot \\ \cdot\exp[a_{m'}^{(k)^+} D^J_{m'm}(A_{kl'})a_m^{(l')^+}]\,|0_k\rangle|0_1 \ldots 0_N 0_{1'} \ldots, 0_{N'}\rangle,$$

which we can easily work out, using the techniques explained in Sect. **2**, because when computing the vacuum expectation value with respect to k the remaining oscillators are effectively c numbers. The result is easily seen to be

$$(3.3) \qquad \exp[a_n^{(i)^+}[D^J(A_{ik})D^J(A_{kl'})]_{nm}a_m^{(l')^+}]\,|0\ldots\rangle = \\ = \exp[a_n^{(i)^+} D^J_{nm}(A_{ik}A_{kl'})a_m^{(l')^+}]\,|0\ldots\rangle,$$

so we see that, after the sewing, the matrix connecting external lines associated to different generalized canonical forms is simply the product of the matrices $D^J(A_{ik})$ and $D^J(A_{kl'})$, where k is the line used to do the sewing. In the right-hand side of eq. (3.3) we have used the fact that D^J-matrices are infinite-dimensional representations of some—yet to be specified—projective transformations A_{ij} so that we can first multiply the 2×2 matrices in computing the product of two infinite matrices.

The final result for the sewing of two generalized canonical forms is the following [10]: one must write again a form in which the D^J-matrix connecting two pairs of lines that originally belonged to the same form remains unchanged, and the D^J-matrix connecting two external lines coming from different forms are computed according to the multiplication law explained above.

The generalized canonical forms are equivalent to the canonical form of operators discussed in Sect. **2** for the case of only two external lines. In this case the matrix elements of the generalized form between states $|\lambda_2\rangle$ and $|\lambda_1'\rangle$ in the Fock space of oscillators $a^{(2)}$ and $a^{(1)}$, respectively,

$$\langle\lambda_1'|\langle 0_2|\exp[a_n^{(1)+}D^J_{nm}a_m^{(2)}]|0_1\rangle|\lambda_2\rangle\,, \tag{3.4}$$

are identical to the matrix element of the following canonical form:

$$\langle\lambda'|:\exp[a_n^+(D^J_{nm}-\delta_{nm})a_m]:|\lambda\rangle\,, \tag{3.5}$$

defined in terms of only one set of operators, the D^J-matrix in both equations being, of course, the same. The relationship is further clarified by the multiplication law: both the sewing of objects defined by eq. (3.4) and the multiplication of canonical forms (3.5) are given in terms of D^J-matrix products.

After all these preliminaries on notations and formalism we are ready to introduce the N-Reggeon vertex. We would like to associate an amplitude to the duality diagram of Fig. 8, but in such a way that we can also describe

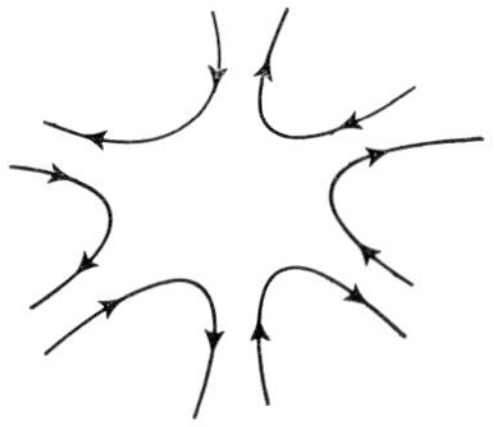

Fig. 8. – Duality diagram corresponding to the N-Reggeon vertex.

excitations in external lines represented by quark-antiquark pairs. One way of doing this is to take Fig. 8 literally and introduce for each external line two Koba-Nielsen variables which we shall label z_i, $\bar{z}_i$ and imagine that they are variables associated with the quark and the antiquark. We are following here an approach introduced by Olive [22]. A simple form for the N-Reggeon vertex was first introduced by Lovelace [18] without doubling the number of Koba-

Nielsen variables; but they are essentially equivalent in the sense that they differ by a gauge transformation on the external lines. However, the Olive form of the N-Reggeon vertex [22] is slightly more convenient to describe the sewing and bootstrap properties of the vertex.

The situation is now pictured in Fig. 9. Following OLIVE, one can define a Chan variable x_i for the external lines, as the anharmonic ratio

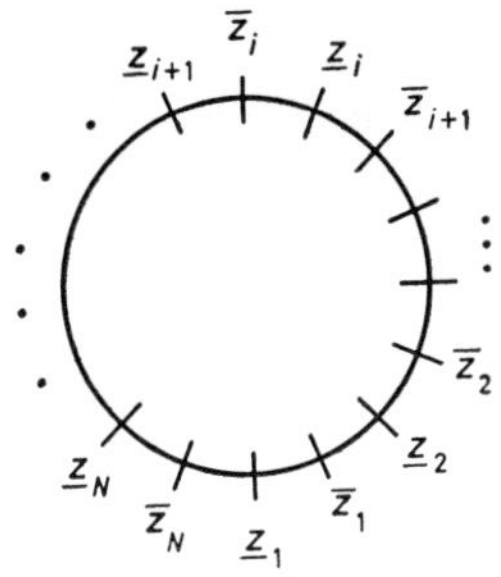

Fig. 9. – Koba-Nielsen variables for the Olive form of the N-Reggeon vertex. Ths $\underline{z}$'s and $\bar{z}$'s are Koba-Nielsen variables of the quark and the antiquark, respectively.

$$x_i = (\bar{z}_i, \underline{z}_i, \underline{z}_{i+1}, \bar{z}_{i-1}) \equiv \frac{(\bar{z}_i - \underline{z}_i)(\underline{z}_{i+1} - \bar{z}_{i-1})}{(\bar{z}_i - \underline{z}_{i+1})(\underline{z}_i - \bar{z}_{i-1})}\,. \tag{3.6}$$

This is the variable that one would naturally use to write an integral representation for a propagator attached to an external line. The poles of such a propagator occur when $x_i = 0$, that is to say when $\underline{z}_i = \bar{z}_i$. The definition of x_i is identical to the usual definition of Chan variables in the conventional Koba-Nielsen formalism for scalar particles [7]. All these remarks imply that x_i is a parameter that describes the structure of the external Reggeon, and should therefore not be integrated over. We must then integrate $\underline{z}_i$ and $\bar{z}_i$ keeping x_i constant, and this is done by changing variables $\mathrm{d}\underline{z}_i \mathrm{d}\bar{z}_i \to \mathrm{d}\underline{z}_i \mathrm{d}x_i$ and dividing out the $\mathrm{d}x_i$'s [22]. In other words, we integrate keeping the x_i's constant, freezing in that way the poles on the external lines. It is important to notice that the x_i's, being anharmonic ratios of four variables on the unit circle, are projective invariants. The Olive form of the N-Reggeon vertex is then [22]

$$V_N = \int \frac{1}{\mathrm{d}V_{abc}} \mathrm{d}\mu_{2N} \prod_{i=1}^{N} \left(\frac{1-x_i}{\mathrm{d}x_i}\right) \left(\frac{\bar{z}_i - \underline{z}_{i+2}}{\bar{z}_i - \underline{z}_{i+1}}\right)^{\alpha_0 - 1} \cdot \tag{3.7}$$

$$\cdot \exp\left[\frac{1}{2}\sum_{i\neq j} a_n^{(i)+} D^J_{nm}(U_i V_j) a_m^{(j)+}\right] |0_1, \dots, 0_N\rangle\,.$$

Let us discuss carefully the meaning of this formula. Consider first the measure of integration. The factor $d\mu_{2N}$ is the usual Koba-Nielsen measure for $2N$ variables, ordered as shown in Fig. 9. The remaining factor containing the dx_i, just cancels away N integration variables as explained before. How would we recover the usual Koba-Nielsen formula for scalars? Neglecting for the moment the operatorial part, one can work out explicitly the change of variables $\bar{z}^i \to x_i$ $(i = 1, \ldots, N)$ and check that the measure of integration collapses to the usual one for N particles in terms of the z_i-variables when the x_i are set equal to zero. The factor dV_{abc} is the usual projective invariant volume element for three particles that further reduces the number of independent variables from N to $(N-3)$. Notice also that apart from the factor raised to the power $(\alpha_0 - 1)$, the measure of integration is projective invariant.

Let us now consider the operatorial part of the N-Reggeon vertex. The matrix $D^J_{nm}(U_i V_j)$ is, as usual, an infinite-dimensional representation of the projective transformation $U_i V_j$ and, as explained before, the limit $J \to 0$ should be taken in order to define the vertex for the conventional model [11]. However, since duality is only a consequence of the algebraic properties of the vertex, any representation defines a dual model. We shall see later on that only when one insists in having a theory in which the infinite set of gauge conditions found by VIRASORO are at work, J becomes quantized and takes the values 0, $-\frac{1}{2}$ and -1 [13].

The important point to remark is that the projective transformation A_{ij} connecting the particles labelled i and j factorizes into the product of two projective transformations, U_i and V_j, depending only on the Koba-Nielsen variables related, respectively, to particles i and j, and their respective neighbourhoods which are defined by

$$V_i = \begin{pmatrix} \bar{z}_{i-1}(z_{i+1} - z_i) & z_i(\bar{z}_{i-1} - z_{i+1}) \\ z_{i+1} - z_i & \bar{z}_{i-1} - z_{i+1} \end{pmatrix}, \tag{3.8}$$

$$U_i = \Gamma V_i^{-1}. \tag{3.9}$$

This form of the U and V matrices is not very illuminating. We could also define them, using an alternative notation introduced by LOVELACE [18]. Any projective transformation is uniquely determined once one specifies how three given points transform. Using the notation

$$\begin{pmatrix} x & y & z \\ x' & y' & z' \end{pmatrix}$$

to represent the projective transformation that maps $(x, y, z) \to (x', y', z')$, we have

$$U_i = \begin{pmatrix} \bar{z}_{i-1} & \underline{z}_i & \underline{z}_{i+1} \\ 0 & \infty & 1 \end{pmatrix}, \tag{3.10}$$

$$V_i = \begin{pmatrix} \infty & 0 & 1 \\ \bar{z}_{i-1} & \underline{z}_i & \underline{z}_{i+1} \end{pmatrix}. \tag{3.11}$$

Let us now consider the duality properties of V_N as given by (3.7). First of all, it is quite straightforward to show that the integrand is invariant under projective transformations. Let us consider, for example, the operatorial part of the vertex, and let us imagine that we perform an over-all projective transformation A that maps all the variables $\underline{z}_i$, $\bar{z}^i$ ($i = 1, \dots, N$) to the new values $\underline{z}'_i$, $\bar{z}'_i$. Since A is uniquely defined by specifying the mapping of three points, we can write

$$A = \begin{pmatrix} \bar{z}_{i-1} & \underline{z}_i & \underline{z}_{i+1} \\ \bar{z}'_{i-1} & \underline{z}'_i & \underline{z}'_{i+1} \end{pmatrix}, \tag{3.12 a}$$

$$A^{-1} = \begin{pmatrix} \bar{z}'_{i-1} & \underline{z}'_i & \underline{z}'_{i+1} \\ \bar{z}_{i-1} & \underline{z}_i & \underline{z}_{i+1} \end{pmatrix}, \tag{3.12 b}$$

but notice that the index i is arbitrary, in the sense that replacing i by $j \neq i$ in eq. (3.12 *a*) defines the *same* A since this is an over-all transformation that acts in the same way on all the variables.

The new infinite matrix that occurs in the vertex will be now an infinite-dimensional representation of the projective transformation $U'_i V'_j$, since U'_i and V'_j are the same 2×2 matrices but as functions of the transformed variables $\underline{z}'_i$, $\bar{z}'_i$. However, we can write both U'_i and V'_j as follows:

$$V'_j = \begin{pmatrix} \infty & 0 & 1 \\ \bar{z}'_{j-1} & \underline{z}'_j & \underline{z}'_{j+1} \end{pmatrix} = \begin{pmatrix} \bar{z}_{j-1} & \underline{z}_j & \underline{z}_{j+1} \\ \bar{z}'_{j-1} & \underline{z}'_j & \underline{z}'_{j+1} \end{pmatrix} \begin{pmatrix} \infty & 0 & 1 \\ \bar{z}_{j-1} & \underline{z}_j & \underline{z}_{j+1} \end{pmatrix} = A V_j, \tag{3.13}$$

$$U'_i = \begin{pmatrix} \bar{z}'_{i-1} & \underline{z}'_i & \underline{z}'_{i+1} \\ 0 & \infty & 1 \end{pmatrix} = \begin{pmatrix} \bar{z}_{i-1} & \underline{z}_i & \underline{z}_{i+1} \\ 0 & \infty & 1 \end{pmatrix} \begin{pmatrix} \bar{z}'_{i-1} & \underline{z}'_i & \underline{z}'_{i+1} \\ \bar{z}_{i-1} & \underline{z}_i & \underline{z}_{i+1} \end{pmatrix} = U_i A^{-1}, \tag{3.14}$$

so we conclude that $U'_i V'_j = U_i V_j$, and therefore it follows that each individual D^J-matrix element in the operatorial part of the N-Reggeon vertex is projective invariant:

$$D^J_{nm}(U'_i\,V'_j) = D^J_{nm}(U_i\,V_j)\,. \tag{3.15}$$

This result neatly shows that the state of internal excitation of the external line—or, in other words, the operators $a_n^{(i)+}$—are irrelevant as far as the projective invariance of the vertex is concerned. And this, in turn, means that there is a complete decoupling between the spin properties of the external particles and the duality properties of the amplitude. Of course, the external « spins » generated in this way are just orbital excitations of the ground-state scalar particle. We will see later on how an « intrinsic » spin can be introduced into the formalism [9].

The N-Reggeon vertex in the conventional model is the limit $J\to 0$ of the operator defined by eq. (3.7). When the limit is taken, eq. (2.6) is, strictly speaking, no longer true for $n=m=0$ as explained in detail in the previous Section. In other words, one must be very careful with the « zero modes », but after the discussion given before, the interested reader will be able to reproduce the argument without difficulty.

The form of the N-Reggeon vertex we have been considering up to now is a generalization of the original Koba-Nielsen formula (1.5). Indeed, it reduces to that more familiar amplitude when all external particles are taken on the ground state. If only zero modes are kept for all external particles, the only matrix elements needed are [cf., eq. (2.33)] $D_{00}^{J\to 0}(U_i V_j)\sim 1 + (J/2)\log(z_i - z_j)$ and this gives back the Koba-Nielsen form of the N-point function. The proof of projective invariance we have given is a refined version of the simple proof one uses for the N-point function, namely just checking the invariance of the integrand.

One can ask oneself whether it is possible to write the N-Reggeon vertex as the vacuum expectation value of a product of vertex operators, as was done in Sect. **1**.2. for the corresponding formula for scalars. The answer to this question is that it is, provided one introduces yet another set of operators $c_{n,\mu}$ [18]

$$[c_{n,\mu},\, c^+_{m,\nu}] = -\,g_{\mu\nu}\,\delta_{nm}\,. \tag{3.16}$$

These c-operators should not be confused with the $a^{(i)}$'s. The latter ones describe the excitation of external particles, and the former ones will describe states that propagate « inside » the vertex. These c's will play here the same role as the a's of Sect. **2**. Introducing the operators

$$Q^J_{n,\mu}(V_i) = \sum_{n=0}^{\infty}\{c^+_{m,\mu}D^{(J,+)}_{mn}(V_i) + c_{m,\mu}D^{(J,+)}_{mn}(\Gamma V_i)\}\,, \tag{3.17}$$

analogous to $[F^J_n(V_i) + \tilde{F}^J_n(V_i)]$ of Sect. **2**, the operatorial part of the vertex can be written as a vacuum expectation value with respect to c [21, 11]:

$$\exp\left[\tfrac{1}{2}\sum_{i\neq j} a_n^{(i)^+} D^J_{nm}(U_i V_j) a_m^{(i)^+}\right]|0_1, \ldots, 0_N\rangle = \tag{3.18}$$

$$= \langle 0_c| \prod_{i=1}^{N} :\exp\,[i a_n^{(i)} Q_n^J(V_i)]:|0_c\rangle|0_1 \ldots, 0_N\rangle\,,$$

and the projective invariance of the vacuum expectation value over c follows from the covariance properties of the operator $Q_n^J(V_i)$, as explained in the previous Section. Notice that the factorization of the matrix $A_{ij} = U_i V_j$ into a projective transformation U_i depending on the variables associated to particle i and its neighbourgs and a transformation $V_j = U_j^{-1}\Gamma$ is crucial in order to factorize the vertex as in eq. (3.17).

Finally, we shall show that this factorization of the matrix A_{ij} plays also an important role in the discussion of the bootstrap properties of the N-Reggeon vertex. Strictly speaking, we have not yet proved that the amplitude defined by eq. (3.7) *is* the N-Reggeon vertex. The best way to do so is to prove that they reproduce themselves under the sewing procedure or, in other words, check that if two vertices are sewn together as explained at the beginning of this section, the resulting vertex has again the same form. Here again we refer to ref. [10] for full details of the sewing procedure; we shall concentrate here only on the operatorial part of the N-Reggeon vertex.

Suppose we want to sew two vertices with N_1 and N_2 external lines to form a vertex with $(N_1 + N_2 - 2)$ external lines, in the form suggested by the duality diagram of Fig. 10. This means that the sewing will be made using the

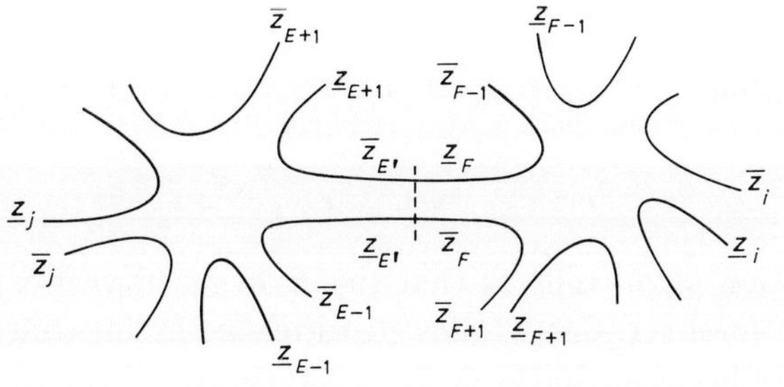

Fig. 10. – The sewing procedure applied to two vertices in terms of duality diagrams.

line E of vertex 1 and the line F of vertex 2. In sewing the two vertices one has to introduce a propagator for the intermediate Reggeon used in the sewing, and in this case a *twisted* propagator must be used. This is a consequence of the fact that Reggeons in this theory have an « orientation » due to the fact that their couplings to a given final state are not invariant if an anticyclic permutation of the particles in the final state is made. They must consequently remember the way they are to be coupled. Since the N-Reggeon vertex we are discussing is a symmetric one (all external Reggeons have the same orientation), a twist is necessary when one is sewing two of them. This is easily understood in the duality diagram of Fig. 10; the quark with the Koba-Nielsen variable z_E is facing the antiquark with variable $\bar{z}_F$, $\bar{z}_E$ and z_F are also facing each other. This mismatch between z and $\bar{z}$'s must be compensated by a quark-antiquark exchange on one side, that is to say, by a twist. The twisted propagator $P(x)$ is given by a canonical form corresponding to the group element [10]

$$P(x) = \begin{pmatrix} 0 & \infty & 1 \\ x & 0 & 1 \end{pmatrix}, \tag{3.19}$$

x being the Chan variable associated with the propagator.

Let us then sew the two vertices we are considering by introducing a twisted propagator with variable x_F, where x_F is the Chan variable of the sewn Reggeon as calculated from the second vertex. As explained at the beginning of this Section, the operatorial part will have exactly the same form, except that the matrix D^J connecting $a^{(i)+}$ and $a^{(j)+}$—where i belongs to the first and j to the second vertex—will be an infinite-dimensional representation of the 2×2 matrix obtained by multiplying first $U_i V_E$, then the matrix associated with the propagator, and finally $U_F V_j$. Using the explicit form of x_F, and the $P(x)$, V_E and U_F matrices, one obtains

$$U^{(i)} V_E P(x_F) U_F V^{(j)} = U^{(i)} \begin{pmatrix} z_{F+1} & z_F & \bar{z}_F \\ z_E & z_{E+1} & \bar{z}_{E-1} \end{pmatrix} V^{(j)} = U^{(i)} V^{(j)} \tag{3.20}$$

if the identification

$$z_E = z_{F+1}, \qquad z_{E+1} = z_F, \qquad \bar{z}_{E-1} = \bar{z}_F, \tag{3.21}$$

is made. Equation (3.20) tells us that the N-Reggeon vertex reproduces itself under sewing. Moreover, one is allowed to make the identification defined by eq. (3.21) simply because the projective invariance of vertices 1 and 2 allows us to fix arbitrarily three variables in each one of them. Moreover, a consequence of this identification is that the Chan variable of the sewn Reggeon calculated from the first vertex, which we shall denote by x_E, turns out to be equal to x_F, making the sewing procedure entirely consistent. This identification can be understood from the point of view of duality diagrams, see Fig. 10, as if the quark-antiquark lines of the two Reggeons that were used in the sewing had been made to overlap.

We have used up to now operators $a_{n,\mu}$ that satisfy commutation relations. However, due to the fact that all operators are bilinear in a and a^+, all the results are still valid if they are made to satisfy anticommutation relations of the form

$$\{a_{n,\mu}^{(i)}, a_{m,\nu}^{(j)+}\}_+ = -g_{\mu\nu}\delta_{nm}\delta_{ij}\,.$$

One must only be careful with the i, j summation in the integrand of eq. (3.7). Clearly, sums of the form

$$\sum_{i\neq j} a_n^{(i)+} D^J_{nm}(A_{ij}) a_m^{(j)+}\,.$$

are identically zero because of the anticommutation relations of the a's, and must be replaced by

$$2\sum_{i<j} a_n^{(i)+} D^J_{nm}(A_{ij}) a_m^{(j)+}\,.$$

4. – Ghosts and Gauge conditions.

The states that factorize the dual amplitude cannot be all identified with bona fide particles. Indeed, many of them are ghosts, *i.e.* states with negative norm. Or, in other words, states that are coupled with imaginary coupling constants to stable particles [4]. These ghosts are generated by the operators $a^+_{n,0}$ (time components) and their perversity stems from the minus sign of the time-component commutator in

$$[a_{n,\mu}, a_{m,\nu}] = -\delta_{nm} g_{\mu\nu}\,, \tag{4.1}$$

This ghost problem is common to all relativistic theories with spinning particles. Also the photon—or the ϱ—have a ghost component which, however, is harmless due to gauge conditions of the form

$$k_\mu T_\mu = 0\,, \tag{4.2}$$

$T_\mu \varepsilon_\mu$ being the amplitude for producing a photon with momentum k and polarization ε. Due to eq. (4.2), the time component T_0 is guaranteed to appear in a definite proportion together with the good components T_k which always cancel the ghost contribution.

We need also here gauge conditions to kill ghosts, and indeed, as soon as factorization was understood, it was recognized that conditions of this kind were at work [12]. We will show that

$$W_1 = L_0 - L_1 - \alpha_0\,, \tag{4.3}$$

is an operator which annihilates every state $\|k_i\rangle$ made up by scalars through a multiperipheral chain

$$\|k_i\rangle = DV(k_1, 1)DV(k_2, 1) \dots DV(k_{N-1}, 1)|\sqrt{2}\, k_N, 0\rangle ,$$

as that appearing when an arbitrary tree is factorized (*). In other words, we must prove that

$$W_1\|k_i\rangle = 0 . \tag{4.4}$$

In order to see how W_1 propagates through the multiperipheral chain of eq. (4.4), we notice that, using the following commutation relations:

$$[L_1, L_0] = L_1 \Rightarrow L_1 x^{L_0} = x^{L_0+1} L_1 , \tag{4.5}$$

$$[L_0, V(k, 1)] = 0 , \qquad [L_1, V(k, 1)] = \alpha_0 V(k, 1) ,$$

one easily obtains

$$W_1 x^{L_0} V(k, 1) = x^{L_0+1} V(k, 1) W_1 + V(k, 1) x^{L_0}[(1-x)(L_0-\alpha_0)-\alpha_0 x] . \tag{4.6}$$

Multiplying therefore by the integration measure $x^{-\alpha_0-1}(1-x)^{\alpha_0-1}$ appearing in the expression (1.27) for D, it follows that

$$W_1 x^{L_0-\alpha_0-1}(1-x)^{\alpha_0-1} V(k, 1) = x^{L_0-\alpha_0}(1-x)^{\alpha_0-1} V(k, 1) W_1 + \\ + V(k, 1)\frac{\mathrm{d}}{\mathrm{d}x}[x^{L_0-\alpha_0}(1-x)^{\alpha_0}] . \tag{4.7}$$

The total derivative gives a vanishing contribution when (4.7) is integrated between 0 and 1, so that

$$[W_1, DV(k, 1)] = 0 . \tag{4.8}$$

Then W_1 commutes with all DV's in the multiperipheral chain of eq. (4.4) until it arrives to the vacuum. But then, due to

$$L_1|\sqrt{2}\, k_N, 0\rangle = 0 , \qquad L_0|\sqrt{2}\, k_N, 0\rangle = -k_N^2|\sqrt{2}\, k_N, 0\rangle = \alpha_0|\sqrt{2}\, k_N, 0\rangle ,$$

$$\therefore W_1|\sqrt{2}\, k_N, 0\rangle = 0 ,$$

(*) We prefer to keep the zero mode $(a_{0,\mu})$ and therefore work with the chain of eq. (1.24) instead of the one of eq. (1.25), in order to use the full operatorial expression for L_0 and $L_{\pm 1}$, given by eq. (2.31).

we immediately see that eq. (4.4) holds. Then, if $\langle\lambda|$ is an arbitrary state, $\langle\lambda|W_1$ is a spurious state that decouples from physical ones. Let us show in detail how W_1 eliminates the first ghost. Choosing

$$\langle\lambda|\equiv 0\,,$$

then

$$\langle 0|W_1 = i\langle 0|a_{1\mu}p_\mu\,,$$

is indeed a spurious state, which implies that in the rest frame the first ghost $\langle 0|a_{1,0}$ is uncoupled from physical states.

Even if the first ghost is cancelled, it is easy to see that other ghosts are still there, generated, for instance, by the time components of $a^+_{n,0}$, $(n>1)$. Clearly, to eliminate them we would need gauge operators W_n containing a term of the form $k\cdot a_n$ that couples these modes to the momentum. Operators of this form cannot be constructed from the $SU_{1,1}$ generators already considered. It appears therefore that one needs a higher symmetry in the theory to find more gauge conditions. This is indeed what happens (as first found by VIRASORO) for the $\alpha_0=1$ case [12]. One can then define an infinite set of operators W_n which have commutation relations with vertices and propagators, such that they go through the multiperipheral tree and annihilate the vacuum as W_1 does.

The Virasoro operators L_n were given by

$$(4.9)\qquad L_n = L^+_{-n} = -i\sqrt{n}\,pa_n - \tfrac{1}{2}\sum_{m=1}^{n-1}\sqrt{m(n-m)}\,a_m a_{n-m} - \sum_{m=1}^{\infty}\sqrt{m(n+m)}\,a^+_m a_{n+m}\,,$$

and satisfy the commutation relations

$$(4.10)\qquad [L_n, L_m] = (n-m)L_{n+m} + \delta_{n,-m}\frac{n(n^2-1)}{3}\,.$$

Let us define

$$(4.11)\qquad W_n = L_0 - L_n - a_0 \qquad W_n|\sqrt{2}\,k, 0\rangle = 0\,.$$

With the same method used to derive (4.7) it is easy to find

$$(4.12)\qquad W_n x^{L_0-\alpha_0}(1-x)^{\alpha_0-1}V(k,1) = x^{L_0-\alpha_0}(1-x)^{\alpha_0-1}V(k,1)\,W_n + \frac{\mathrm{d}}{\mathrm{d}x}B(x) + (\alpha_0-1)\,C(x)\,,$$

where B and C are operators we do not need to specify, apart from the fact that B vanishes at $x=0$ and $x=1$. If $\alpha_0=1$, the last term on the right-hand side of eq. (4.11) vanishes. Integrating eq. (4.12), the total derivative does not contribute either. Then for $\alpha_0 = 1$ one has

$$[W_n, DV(k, 1)] = 0 . \tag{4.13}$$

Therefore, for $\alpha_0 = 1$, W_n annihilates all physical states and defines an infinite set of gauge operators.

As discussed in detail by FUBINI and VENEZIANO [8] and GALLI [23], the L_n are the generators of arbitrary conformal transformations characterized—of course—by an infinite number of parameters. It can be proved that the commutation relations between L_n and the vertex $V(k, z)$ given by eq. (2.60) are true for all n. The fact that for $\alpha_0 = 1$ the theory has this extra conformal invariance, besides the projective invariance discussed before, can now be easily visualized. Indeed, for $\alpha_0 = 1$ the commutation relations (2.40) can be written as

$$\left[L_n, \frac{V(k, z)}{z}\right] = \frac{\mathrm{d}}{\mathrm{d}z}[z^n V(k, z)] . \tag{4.14}$$

On the other hand, the amplitude given by eq. (1.20) can be written as

$$A_N = \int \prod_{i=1}^{N} \mathrm{d}z_i \langle 0| \prod_{i=1}^{N} \frac{V(k_i, z_i)}{z_i} |0\rangle \qquad (\alpha_0 = 1) . \tag{4.15}$$

It is clear from eq. (4.14) that if we perform in the V an infinitesimal transformation generated by L_n, the integrand of eq. (4.15) remains invariant apart from a total derivative which vanishes after integration.

Going back to our ghost problem, we have seen that every W_n given by eq. (4.12) annihilates physical states and defines, therefore, spurious states of the form

$$\langle \lambda | W_n .$$

The hope, of course, is that all ghosts are spurious states so that they actually never appear as intermediate states in amplitudes defined from physical particles. The counting of the gauge conditions agrees with the one necessary to eliminate all ghosts but, for the time being, no rigorous proof that nonspurious states have positive norm has been provided (*). It has been proved numerically up to the 10th daugheter [24] and on an infinite but not complete set by FUBINI, DEL GIUDICE and DI VECCHIA [25]. GODDARD will discuss this problem further in his seminar [26]. Hopefully, this failure of ghost finding is not an

(*) *Footnote added in proofs.* – R. BROWER (MIT preprint, to be published) has recently succeeded in proving that the nonspurious states have positive norm; so that there are no ghosts in the Veneziano model with $\alpha_0 = 1$.

accident; let us work for the moment on this assumption. But even so we have no reason to be happy. Indeed, $\alpha_0 = 1$ implies $\mu^2 = -1$. Therefore, we have traded an infinite number of ghosts by one tachyon and to live with it is certainly not a pleasure. We must therefore get rid of it. The oldest among you will remember that we had a similar problem when in 1962 we realized that Regge poles seem to have positive intercept and would therefore provide a tachyon when they cross zero at negative values of the square mass. But it was soon understood how to handle it, by providing reasonable arguments to justify the addition of a ghost killing factor $\alpha(t)$. The arguments were related with spin. Regge trajectories interpolate bound states and if these are $\mathcal{N}\overline{\mathcal{N}}(q\bar{q})$ we could have the situation in which the bound states we interpolate are triplets, *i.e.* ^3L_J. The lowest in the family would be $^3S_1$; no one appearing with $J = 0$ because this implies a nonsense value for the orbital angular momentum ($L = -1$). So clearly, we must introduce spin in order to get rid of the tachyon. The problem here is that, dealing with a dynamical theory (dual, crossing symmetric, etc.) we cannot just add an *ad hoc* factor. This factor should be provided by the theory itself if it can incorporate spin excitations. This is what has been obtained recently, for the first time in a convincing way, by NEVEU and SCHWARZ [14].

5. – The introduction of spin.

In Sect. **3** we have emphasized the connection between the dual properties of the N-Reggeon vertex V_N and the group-theoretical structure underlying the theory. Indeed, the duality properties of V_N, as given by eq. (3.7), are a consequence of the fact that the infinite matrices D^J_{nm} are unitary representations of $SU_{1,1}$, corresponding to the 2×2 matrix given by eq. (3.8). For every representation we shall therefore find a specific form of the D^J_{nm}-matrix elements in terms of Koba-Nielsen variables which must then be integrated, as shown in (3.7). We remarked that the irreducible $J\to 0$ representation gave us the N-Reggeon vertex corresponding to the conventional model.

The preceding discussion suggests a simple way to look for new dual models. It is sufficient to find other representations of $SU_{1,1}$ which would meet our physical requirements.

Firts of all, the dual amplitude we are looking for must have the characteristic pole structure. It is easy to realize that no irreducible representation other than $J\to 0$ is able to satisfy that requirement. Indeed, by analysing the expressions (2.6) and (2.7), the matrix elements $D^J_{nm}(U_i V_j)$ depend on powers of differences of Koba-Nielsen parameters. It is therefore only for $J\to 0$, and thanks to the singularity of such a limit, that the power law gives rise to the logarithmic behaviour of $D^{J\to 0}_{00}$ [cf., eq. (2.33)] and then to the pole structure of the amplitude.

The only possibility left to generalize the conventional model is to look for reducible representations of $SU_{1,1}$ which contain the $J\to 0$ one in its reduction. This is equivalent to introduce, besides the $a^{(i)}_{n,\mu}$ operators of the conventional model, other operators $b^{(i)}_{n,\mu}$ and rewrite eq. (2.7) in the form

$$(5.1)\qquad V_N=\int\frac{1}{\partial V_{abc}}\,\mathrm{d}\mu_{2N}\prod_{i=1}^{N}\left(\frac{1-x_i}{\mathrm{d}x_i}\right)\left(\frac{\bar{z}_i-z_{i+2}}{\bar{z}_i-z_{i+1}}\right)\exp\left[\frac{1}{2}\sum_{i\neq j}a^{(i)+}_{n}D^{J\to 0}_{nm}(U_iV_j)\,a^{(j)+}_{m}\right]\cdot$$
$$\cdot\exp\left[\sum_{i<j}b^{(i)+}_{n}D^{J}_{nm}(U_iV_j)\,b^{(i)+}_{m}\right]|O_a,O_b\rangle\,.$$

The 2×2 matrix U_iV_j being again given by eq. (3.8).

The a_n-operators will then be responsible for the pole structure of V_N and will be referred to as orbital excitations. The b_n-operators will be called spin excitations. We now must decide which are the J-representations and the commuting or anticommuting character of the b-operators which will meet our needs. Let us remember that we are introducing the spin excitations in order to eliminate the tachyon that appeared in the conventional model for $\alpha_0=1$. But, of course, all our efforts would be vain if the spin excitations would introduce new ghosts. We must therefore provide to the spin modes the same Virasoro algebra [12] we invoked for the orbital modes, in order to eliminate ghosts.

Starting therefore from the operators L_0, L_1 and L_{-1} which were given in eqs. (2.11)-(2.13) and (2.21), we want to construct other operators L_n ($n=\pm 2,\ldots$) which satisfy the algebra, described by eq. (4.10). We decomposed our reducible representation into the $J\to 0$, assigned to a operators, and another $J\neq 0$ assigned to b operators. Also for the L_n we will write

$$(5.2)\qquad L_n=L^{J\to 0}_{n}+L^{J}_{n}\,,$$

where the $L^{J\to 0}_n$ are the Virasoro operators constructed with a modes of eq. (4.9). The L^J_n must therefore satisfy the algebra (4.10) themselves.

We will not enter into the details of the construction of the L^J_n, but by using the forms of L^J_0, L^J_1 and L^J_{-1} and the commutation relations of the b operators, it is possible to show [13] that the solutions of (4.10) with $L^+_n=L_{-n}$ can only be found for $J=0$ and -1 for commuting b's and for $J=-\frac{1}{2}$ for anticommuting b's. The solutions is ($n\geqslant 2$)

$$L^J_n=-\sum_{m=n}^{\infty}c_{n,m}\,b^{+}_{m-n}b_n+\sum_{m=0}^{\infty}d_{n,m}\,b_m b_{n+2J-m}\,,$$

where the coefficients $c_{n,m}$ and $d_{n,m}$ are given in the following Table:

TABLE I.

	J	$c_{n,m}$	$d_{n,m}$
Commuting b's	$0,\,-1$	$\sqrt{(m-J)(m-n-J)}$	$-\frac{1}{2}\sqrt{(m-J)(n-m+J)}$
Anticommuting b's	$-\frac{1}{2}$	$m-(n-1)/2$	$\frac{1}{2}(m-(n-1)/2)$

The $J=0$ case is the conventional model discussed previously. The $J=-1$ case is a relabelling of the same model (it corresponds to the same value of the Casimir operator $J(J+1)$) [13]. The $J=-\frac{1}{2}$ case remains as the only interesting one for spin excitations in terms of the new set of anticommuting b_n operators [27]. The model is, however, not completely specified. Indeed, we must still decide which one is the b content of the ground state, which we shall call π for reasons that will become clear later. As discussed before, the a content of the ground state with momentum k was $\exp[i\sqrt{2}\,k\cdot q]|0_a\rangle$, where q_μ was the operator defined in eq. (1.15) in terms of the zero mode $a_{0,\mu}$. If we would associate to the π the vacuum of the b modes, the N-pion amplitude would again be given by the same Koba-Nielsen formula as before. This follows from the fact that the vacuum expectation value of the b part of the N-Reggeon vertex of eq. (5.1) is 1.

The Neveu-Schwarz choice, instead, is to associate a pion with momentum k with the state [28]

$$\sqrt{2}\,k\cdot b_0^+ \exp[i\sqrt{2}\,k\cdot q]\,|0_a, 0_b\rangle\,, \tag{5.4}$$

By saturating the N-Reggeon vertex of eq. (5.3) with N-pion states given by eq. (5.4), we obtain the usual expression (1.20) from a modes, while the b modes give

$$\langle 0_b|\prod_{i=1}^{N}\sqrt{2}\,k_i\cdot Q_0^{-\frac{1}{2}}(V_i)|0_b\rangle\,, \tag{5.5}$$

where $Q_0^{-\frac{1}{2}}(V_i)$ is the $J=-\frac{1}{2}$ representation of the tensor defined by eq. (3.17) for the group element V_i given by eq. (3.8) in which all $z_i=\bar{z}_i\equiv z_i$. By introducing the explicit expression (3.8) for V_i in the form of $D_{n0}^{(-\frac{1}{2},+)}$ that can be obtained by making $m=0$ in eq. (2.6), we find

$$\begin{cases} D_{n0}^{(-\frac{1}{2},+)}(V_i)=\dfrac{z_i^n}{z_{i-1}-z_{i+1}}\,,\\[2ex] D_{n0}^{(-\frac{1}{2},+)}(\Gamma V_i)=\dfrac{z_i^{-n-1}}{z_{i-1}-z_{i+1}}\,. \end{cases} \tag{5.6}$$

Inserting eq. (5.6) into (3.17) we obtain

$$Q_{0,\mu}^{-\frac{1}{2}}(V_i)=i\sqrt{\frac{z_i(z_{i-1}-z_{i+1})}{(z_i-z_{i+1})(z_i-z_{i-1})}}\,H_\mu(z_i)\,, \tag{5.7}$$

$$H_\mu(z)=\sum_{m=0}^{\infty}(z^{m+\frac{1}{2}}b_{m,\mu}^+ + z^{-m-\frac{1}{2}}b_{m,\mu})\,. \tag{5.8}$$

Collecting everything, the N-pion scattering amplitude is given by

$$(5.9)\qquad A_N = \int \prod_{i=1}^{N} \mathrm{d}z_i \left(\frac{z_i - z_{i+2}}{z_i - z_{i+1}}\right)^{\alpha_0 + k^2 - \frac{1}{2}} \frac{z^{k^2-\frac{1}{2}}}{(z_i - z_{i+1})^{\frac{1}{2}+k^2}} \cdot \langle O_a O_b | \prod_{i=1}^{N} V(k_i, z_i)\sqrt{2} k_i \cdot H(z_i) | O_a O_b \rangle ,$$

which is the Neveu-Schwarz amplitude [14]. The analysis of such an amplitude has been carried out carefully in ref. [14]. We state without details some of the results.

First we have chosen the $J = -\frac{1}{2}$ representation to have the Virasoro algebra. Let us check that this gives indeed rise to gauge conditions.

We remember that $Q_0(V_i)$ is an invariant under $SU_{1,1}$ transformations, cf., eq. (2.30). Therefore we can easily compute from eq. (5.8) how $H_\mu(z)$ transforms. It turns out that

$$(5.10)\qquad z_i^{-\frac{1}{2}} U(\Lambda) H_\mu(z_i) U^{-1}(\Lambda) = z_i'^{-\frac{1}{2}} (\gamma z_i + \delta) H_\mu(z_i') ,$$

where

$$\Lambda = \begin{pmatrix} \alpha & \beta \\ \gamma & \delta \end{pmatrix}, \qquad z_i' = \frac{\alpha z_i + \beta}{\gamma z_i + \delta} .$$

Exactly as we obtained eq. (2.40) from eq. (2.38), eq. (5.10) implies for $n = \pm 1$

$$(5.11)\qquad [L_n^{-\frac{1}{2}}, H_\mu(z)] = z^n \left(z \frac{\mathrm{d}}{\mathrm{d}z} + \frac{n}{2}\right) H_\mu(z) .$$

By explicit computation it is possible to check that eq. (5.11) is valid for all the L_n given by eq. (5.3). Then, from eqs. (5.2) and (1.40), which is valid for all n and (5.11), we obtain

$$(5.12)\qquad [L_n, V(k, z) k \cdot H(z)] = z^n \left(z \frac{\mathrm{d}}{\mathrm{d}z} - n(k^2 - \tfrac{1}{2})\right) V(k, z) k \cdot H(z) ,$$

so that, if $\alpha_0 = 1$ and $k^2 = -\frac{1}{2}$, it follows that

$$(5.13)\qquad \left[L_n, \frac{V(k, z) k \cdot H(z)}{z}\right] = \frac{\mathrm{d}}{\mathrm{d}z} \left(z^n V(k, z) k \cdot H(z)\right) .$$

On the other hand, for $\alpha_0 = 1$ and $k^2 = -\frac{1}{2}$ we have from eq. (5.9)

$$(5.14)\qquad A_N = \int \frac{\prod_{i=1}^{N} \mathrm{d}z_i}{\mathrm{d}V_{abc}} \langle O_a, O_b | \prod_{i=1}^{N} \frac{V(k_i, z_i)\sqrt{2} k H(z_i)}{z_i} | O_a, O_b \rangle .$$

The last two equations show that the L_n commute through the multiperipheral chain. Indeed, their commutators give total derivatives which vanish after integration.

The factorization of eq. (5.14) can be done exactly as we did before for the conventional model. With the method used to obtain (1.25) from (1.20)

$$A_N = \langle 0_a 0_b | k\cdot b_0 \sqrt{2}\, DW(k_2) D \dots DW(k_{N-1}) \sqrt{2}\, k_N \cdot b_0^+ | 0_a 0_b \rangle , \tag{5.15}$$

where

$$D = \left[-\frac{p^2}{2} - \sum_{n=1}^{\infty} n a_n^+ a_n + \sum_{n=0}^{\infty} \left(n + \frac{1}{2} \right) b_n^+ b_n \right]^{-1} \tag{5.16}$$

and

$$W(k) = \sqrt{2}\, k \cdot H(1) V(k) , \tag{5.17}$$

where $V(k_i)$ is the vertex of the a modes given in (1.27). One property which is clearly exhibited by eq. (5.15) is that A_N vanishes unless N is even, *i.e.* the total number of pions (incoming + outgoing) must be even. This is simply due to the fact that the vacuum expectation value of an odd number of b operators is zero. Therefore, there is a G-parity operator defined as

$$G = (-1)^{\sum\limits_{n=1}^{\infty} b_n^+ \cdot b_n} ,$$

and the pion has a negative G-parity.

From eq. (5.15) it is easy to analyse the spectrum. We have for occupation number 0 of b-modes the same states as before, including the zero-mass ρ and the tachyon of square mass (-1). Then, with occupation number 1 for b_0^+, we have the pion at square mass $k^2 = -\frac{1}{2}$ (still a tachyon!) and its recurrences. With occupation number 2 of b_0^+, we have an ugly ancestor, a massless spin-two particle and its recurrences. At first sight it would seem that not only have we not eliminated the old tachyon, but that we have introduced a new one (calling it π) and a whole unpleasant ancestor trajectory.

Let us first shift the pion to a physical mass. This can easily be done with a trick first introduced by FUBINI and VENEZIANO [8] and used by HALPERN-THORN [29] in this context. Everything said up to now never used the four dimensionality of space-time. One can therefore introduce a fifth spacelike dimension and assign a value $+c$ or $-c$ to the fifth component of all momenta, k_5. Consecutive pions in the Koba-Nielsen circle have opposite value of k_5 and—therefore— $\sum_{i=1}^{N} k_{5i} = 0$ due to the fact that we have an even number of pions. But now

$$k^2 = -\tfrac{1}{2} = k_0^2 - \boldsymbol{k}^2 - c ,$$

i.e.

$$\mu^2 = -\tfrac{1}{2} + c\,, \tag{5.18}$$

so μ^2 is arbitrary. Therefore the pion trajectory can be shifted to an arbitrary physical intercept [29].

We have introduced the spin to kill the $m^2 = -1$ tachyon and it is rewarding that this goal has been achieved. Indeed, the ghost at $m^2 = -1$ and the whole ancestor trajectory (intercept 2) discussed before, decouple from physical states composed with an arbitrary number of pions [14]. An elegant way to prove this fact has been proposed by NEVEU, SCHWARZ and THORN [14]. We will sketch the idea referring to the original paper for the details.

The basic observation is that besides the Virasoro L_n algebra there are other operators G_n which form—together with the L_n—a larger algebra. While the L_n are bilinear combinations of a_n operators on one side and b_m operators on the other, the G_n are bilinear in a_n and b_m and mix, therefore, the orbital excitations with spin excitations. The G_n-operators satisfy the commutation relations

$$\{\mathscr{G}_n, \mathscr{G}_m\} = 2L_{m+n} \tag{5.19}$$

[m and n are half-integers in the notations of ref. [14]] and

$$\sqrt{2}\, k\cdot b_0^+|0\rangle = \mathscr{G}_{-\frac{1}{2}}|0\rangle\,. \tag{5.20}$$

By commuting conveniently the G's, the N-pion amplitude can also be written as

$$\begin{aligned} A_N = \langle 0|\mathscr{G}_{\frac{1}{2}} W(k_1)\frac{1}{L_0-1} W(k_2)\,\dots\, W(k)\,\mathscr{G}_{-\frac{1}{2}}|0\rangle = \\ = \langle 0|W(k_1)\frac{1}{L_0-\frac{1}{2}} W(k_2)\dots\frac{1}{L_0-\frac{1}{2}} W(k_N)|0\rangle\,. \end{aligned} \tag{5.21}$$

The last expression for A_N allows a reinterpretation of the model in a new Fock space in which the pion with $k^2 = -\frac{1}{2}$ ($\mu^2 = -\frac{1}{2} + c$, cf., eq. [5.18]) is the ground state, *i.e.* the lowest pole of the propagator. In this form it appears clearly that the $k^2 = -1$ tachyon, as well as the ancestor trajectories, do not appear as intermediate states of pion-amplitudes, and are therefore spurious states uncoupled to physical ones. The disadvantage of the last form for A_N in eq. (5.21) is that the duality properties are not explicit any more. One must go back to the first form of eq. (5.21)—and make use therefore of the whole Fock space found previously—in order to see the equivalence of different multiperipheral configurations.

Anyhow, we have finally constructed a bootstrap model with only poles, Regge behaviour and with no tachyon and (hopefully) no ghosts. This, in itself, is quite an achievement.

After the introduction of isospin with Chan-Paton factors [30], the spectrum is that of Fig. 11. The pion trajectory (with arbitrary intercept) has zero residue at $J = 1$. The φ, f′ and μ trajectories are absent. The σ trajectory (with the ρ' degenerate with f^0) appears due to the fifth component introduced to avoid the pion tachyon. The ω-A_2 trajectory is degenerate with the pion trajectory. At the μ' level, there are two trajectories with opposite parities (μ' and a scalar for $J = 0$).

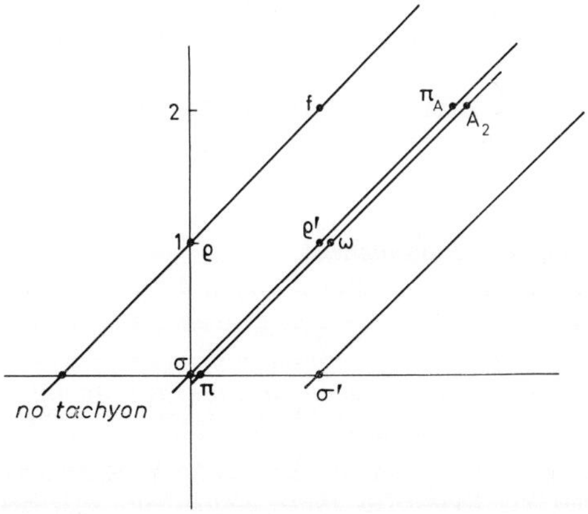

Fig. 11. – Low-lying states in the Neveu-Schwarz model.

To conclude, we can remark that the basic difficulties with the spectrum predicted by the model are the vanishing of the ρ mass and the absence of a ρ-ω degeneracy, due to the degeneracy of the π and ω trajectories.

We notice that the spin excitations that have been introduced so far correspond to integer spin. To introduce half-integer spin in a dual theoretical form is by far nontrivial. A first step in this direction has been recently achieved by Neveu and Schwarz and Bardakçi and Halpern, following a method first proposed by Ramond [5]. A single-fermion line, emitting an arbitrary number of mesons is introduced by means of another set of operators $d_{n,\mu}$ with spinor indices. The $d_{0,\mu}$ is associated with the γ_μ matrices similarly as the $a_{0,\mu}$ was to momentum. A rewarding property is that the poles in the fermion-antifermion channel (boson states) give residues which are indeed the multipion amplitudes of the Neveu-Schwarz model [14] discussed before. For the time being, the dual treatment of arbitrary half-integer spin excitations, *i.e.* creation of fermion pairs or amplitudes with more than a fermion line, has not been found. To do it in a way similar to that introduced for the mesons meets the difficulty of exponentiating operators with spinorial indices.

REFERENCES

[1] G. VENEZIANO: *Nuovo Cimento*, **57** A, 190 (1968).

[2] G. FRYE and L. SUSSKIND: *Phys. Lett.*, B **31**, 537 (1970); D. J. GROSS, A. NEVEU, J. SCHERK and J. SCHWARZ: *Phys. Rev. D*, **2**, 697 (1970); C. S. HSUE, B. SAKITA and M. A. VIRASORO: *Phys. Rev. D*, **2**, 2857 (1970).

[3] H. HARARI: *Phys. Rev. Lett.*, **20**, 1395 (1968); P. G. O. FREUND: *Phys. Rev. Lett.*, **20**, 235 (1968).

[4] S. FUBINI and G. VENEZIANO: *Nuovo Cimento*, **64** A, 811 (1969); K. BARDAKÇI and S. MANDELSTAM: *Phys. Rev.*, **184**, 1640 (1969).

[5] P. RAMOND: *Phys. Rev. D*, **3**, 2415 (1971); A. NEVEU and J. SCHWARZ: *Quark model of dual pions*, Princeton preprint, to be published; K. BARDAKÇI: *New Gauge identities in dual-quark models*, Berkeley preprint, to be published.

[6] In our opinion, the most promising theoretical possibility is the one suggested by I. DRUMMOND: CERN preprint TH. 1301, to be published; C. REBBI: Trieste preprint, to be published.

[7] Z. KOBA and H. B. NIELSEN: *Nucl. Phys.*, **10** B, 633 (1969).

[8] S. FUBINI and G. VENEZIANO: *Nuovo Cimento*, **67** A, 29 (1970).

[9] See, for example, P. RAMOND: *Group-theoretical properties of dual-resonance models* NAL preprint THY-15 (*Lectures at the 1971 Boulder Summer School*) and references contained therein.

[10] V. ALESSANDRINI, D. AMATI, M. LE BELLAC and D. OLIVE: *The operator approach to dual-multiparticle theory*, in *Phys. Rep.*, Vol. **1** C, Sect. **6**, July 1971.

[11] E. CORRIGAN and C. MONTONEN: *General dual operatorial vertices*, Cambridge preprint DAMTP 71/32, August 1971.

[12] M. A. VIRASORO: *Phys. Rev. D*, **1**, 2933 (1970).

[13] J. L. GERVAIS and B. SAKITA: Orsay preprint LPTHE 71/23, to be published.

[14] A. NEVEU and J. SCHWARZ: *Nucl. Phys.*, **31** B, 86 (1971); A. NEVEU, J. SCHWARZ and C. B. THORN: *Reformulation of the dual-pion model*, Princeton preprint, to be published; see also, K. BARDAKÇI: *New gauge identities in dual-quark models*, Berkeley preprint, to be published.

[15] *Bateman Manuscript Project*, edited by A. ERDÉLYI (New York, 1955).

[16] CHAN H.-M.: *Phys. Lett.*, **28** B, 425 (1968); C. G. GOEBEL and B. SAKITA: *Phys. Rev. Lett.*, **22**, 257 (1969); K. BARDAKÇI and H. RUEGG: *Phys. Rev.*, **182**, 1884 (1969).

[17] See, for example, A. O. BARUT and C. FRONSDAL: *Proc. Roy. Soc.*, A **287**, 532 (1965). We use here the same conventions as the authors of ref. [11].

[18] C. LOVELACE: *Phys. Lett.*, **32** B, 703 (1970).

[19] This was first realized by L. CLAVELLI and P. RAMOND: *Phys. Rev. D*, **2**, 973 (1970); **3**, 988 (1971); S. SCIUTO: *Lett. Nuovo Cimento*, **2**, 411 (1969).

[20] I. DRUMMOND: *Nuovo Cimento*, **67** A, 71 (1970); G. CARBONE and S. SCIUTO: *Lett. Nuovo Cimento*, **3**, 246 (1970); J. KOSTERLITZ and D. WRAY: *Lett. Nuovo Cimento*, **3**, 491 (1970).

[21] D. COLLOP: *Nuovo Cimento*, **1** A, 217 (1971); L. P. YU: *Phys. Rev. D*, **2**, 1010, 2256 (1970).

[22] D. I. OLIVE: *Nuovo Cimento*, **3** A, 399 (1971).

[23] A. GALLI: *Nuovo Cimento*, **69** A, 275 (1970).

[24] R. C. BROWER and C. B. THORN: CERN preprint TH. 1293 (1971), to be published in *Nucl. Phys.*

[25] E. DEL GIUDICE, P. DI VECCHIA and S. FUBINI: *General properties of the dual-resonance model*, MIT preprint LTP 209, June 1971, to be published in *Ann. of Phys.*

[26] R. C. BROWER and P. GODDARD: this Volume p. 98.

[27] There are other possible generalizations using $SU_{1,1}$ representations with $k \neq \pm J$, by using two sets of operators, say, $b^{(n)}$ and $c^{(n)}$ for spin excitations. D. OLIVE, private communication.

[28] The discussion of the Neveu-Schwarz model closely parallels the one given in ref. [11].

[29] M. B. HALPERN and C. THORN: *Dual model of pions with no tachyon*, Berkeley preprint (1971).

[30] CHAN H.-M. and J. PATON: *Nucl. Phys.*, **10** B, 519 (1969).

Physical States in the Dual-Resonance Model (*).

R. C. Brower

Massachusetts Institute of Technology - Cambridge, Mass.

P. Goddard (**)

Trinity College, University of Cambridge - Cambridge

1. – Introduction.

Historically, the dual-resonance model (DRM) has progressed from the fundamental suggestions of Veneziano [1] of a dual-resonance amplitude for four spinless particles to amplitudes for an arbitrary number of spinless particles. From this generalized Veneziano formula, amplitudes for spinning particles can be obtained by factorization. This process satisfies a bootstrap condition, in that all these amplitudes factorize on the same spectrum of intermediate resonance states in terms of DRM amplitudes [2].

Here we shall briefly review the formalism of the DRM, in the case of unit intercept, to establish notation [3]. Then we shall describe a recent extension of this formalism by Del Giudice, Di Vecchia and Fubini [4, 5] to vertices for the coupling of physical excited states. Using this formalism they were able to produce an orthogonal basis for a large subspace of the physical states and verify the absence of ghosts. Finally, we will give an account of a very recent attempt [6] to extend this into a spectrum generating algebra which will produce all the physical states for $\alpha(0) = 1$.

2. – Formalism of the dual-resonance model.

In the DRM an amplitude for N spinless particles $A^{(N)}$ is defined as the integral of a certain function of the momenta k_i of the particles and a variable z_i which describes internal excitations

$$A^{(N)}(k_i) = \int F(z_i, k_i)\, \mathrm{d}\Omega_z \,. \tag{2.1}$$

(*) Based on part of Lectures given by Prof. S. Fubini and a Lecture by P. Goddard.
(**) Present address: CERN, Geneva.

To define the fundamental integrand F we need the annihilation and creation operators $a_n^{\mu\dagger}$, a_n^{μ}, $n = 1, 2, \ldots$, which determine the spectrum

$$[a_n^{\mu}, a_n^{\nu\dagger}] = g^{\mu\nu}\delta_{mn} , \tag{2.2}$$

and positron and momentum operators q_0, p_0 such that

$$[q_0^{\mu}, p_0^{\nu}] = ig_{\mu\nu} . \tag{2.3}$$

Then

$$F(z_i, k_i) = \langle 0| \frac{V(z_1, k_1)}{z_1^{\alpha(0)}} \frac{V(z_2, k_2)}{z_2^{\alpha(0)}} \ldots \frac{V(z_N, k_N)}{z_N^{\alpha(0)}} |0\rangle \prod_{i=1}^{N} (z_{i+1} - z_i)^{\alpha(0)} , \tag{2.4}$$

where the vertices $V(z, k)$ are constructed in terms of the field

$$Q^{\mu}(z) \quad = Q^{(-)}(z) + Q^{(0)\mu}(z) + Q^{(+)\mu}(z) , \tag{2.5}$$

$$Q^{(\pm)\mu}(z) = \sum_{n=1}^{\infty} \frac{a_{\pm n}^{\mu}}{\sqrt{n}} z^{\pm n} , \qquad a_{-n}^{\mu} = a_n^{\mu\dagger} , \tag{2.6}$$

$$Q^{(0)\mu}(z) = q_0^{\mu} + ip_0^{\mu} \log z , \tag{2.7}$$

$$V(z, k) = :\exp[i\sqrt{2}\, k, Q(z)]: . \tag{2.8}$$

In eq. (2.1) $d\Omega_z$ is an integration measure

$$d\Omega z = \prod_{i=1}^{N} \frac{dz_i}{(z_{i+1} - z_i)} \Big/ \left\{ \frac{dz_a dz_b dz_c}{(z_b - z_a)(z_c - z_b)(z_a - z_c)} \right\} . \tag{3.9}$$

Rather than integrating over all N of the z_i's only $N-3$ of them are integrated over, three of them, z_a, z_b and z_c, being held fixed.

An important point, elaborated elsewhere [3], is that the result is independent of which three z_i's are fixed, or which values these variables take. Particular values which play a special role are $z_i \to 0$, $z_N \to \infty$:

$$\lim_{z_1 \to 0} \langle 0| \frac{V(z_1, k_1)}{z_1^{\alpha(0)}} = \langle 0| \exp[i\sqrt{2}k_1 q_0] \equiv \langle k_1| , \tag{2.10}$$

$$\lim_{z_N \to \infty} \frac{V(z_N, k_N)}{z_N^{-\alpha(0)}} |0\rangle = \exp[i\sqrt{2}k_N q_0]|0\rangle \equiv |-k_N\rangle , \tag{2.11}$$

and using these particular values we can rewrite eq. (2.4) for the case of unit intercept

$$F(z_i, k_i)\,\mathrm{d}\Omega_z = \langle k_1|\,\frac{V(z_2, k_2)}{z_2^{\alpha(0)}}\dots\frac{V(z_{N-1}, k_{N-1})}{z_{N-1}{}^{\alpha(0)}}\,|-k_N\rangle\cdot\prod_{i=2}^{N-1}\mathrm{d}z_i\bigg/\left\{\frac{\mathrm{d}z_b}{z_b}\right\}. \tag{2.12}$$

The creation operators $a_n^{0\dagger}$ associated with the time dimension create states of negative norm which are called ghosts. These states are a bad feature of the DRM. However, a similar feature appears in quantum electrodynamics if care is not taken. It is attractive to quantize the electromagnetic field covariantly, and then there are apparent ghosts for the same reason. But in electrodynamics these states do not couple; the space of « physical » states which couple is only a subspace of the full Fock space made by the creation operators of the field. The decoupling of the undesirable states is expressed in terms of gauge conditions.

The Veneziano model also allows for this possibility. The space of physical states is again smaller than the full space created by the oscillators. In general, there is one gauge condition, but in the special case $\alpha(0) = 1$, when the ground state particle is a tachyon (*i.e.* has negative mass squared) and there is a massless spin-one particle (« photon »), there is an infinity of gauge conditions. Such an infinity is needed because of the infinity of operators creating ghosts. As a result the question of the absence of ghosts in the case of unit intercept has received much attention. This spectrum was first investigated by DEL GIUDICE and DI VECCHIA [7] and then explored further by BROWER and THORN [8] using a computer. From this it is known that there are no ghosts on the first nine trajectories and it seems very unlikely that there are any ghosts in this special case. We shall now discuss the physical states when $\alpha(0) = 1$, first constructing the gauge operators.

Introduce a field $P^\mu(z)$

$$P^\mu(z) = -iz\,\frac{\mathrm{d}Q^\mu}{\mathrm{d}z}\,. \tag{2.13}$$

Then

$$[Q^\mu(z_1), P^\nu(z_2)] = ig^{\mu\nu}\delta(\theta_1 - \theta_2)\,, \qquad z_j = \exp[2\pi i\theta_j]\,. \tag{2.14}$$

It is helpful to define the bilinear operators

$$L_f = \frac{1}{2\pi i}\oint :P^2(z):zf(z)\,\mathrm{d}z\,, \tag{2.15}$$

where the integral encircles $z = 0$. From this is follows

$$\left[L_f, \frac{V(z, k)}{z}\right] = -\frac{\mathrm{d}}{\mathrm{d}z}\{f(z)\,V(z, k)\} \quad \text{if} \quad k^2 = 0\,. \tag{2.16}$$

To obtain this result care must be taken over the normal ordering. In particular we consider the operators

$$L_{z^{-n}} \equiv L_n \,. \tag{2.17}$$

In particular

$$L_0 = p_0^2 + \sum_{n=1}^{\infty} a_n^{\dagger} a_n \,. \tag{2.18}$$

It is easy to check that the one-particle states $|-k_N\rangle$ satisfy

$$(L_0 - 1)|-k_N\rangle = 0 \,, \qquad L_n|-k_N\rangle = 0 \,, \qquad n > 0 \,, \tag{2.19}$$

which are the mass shell and gauge conditions, respectively. To show that all excited states satisfy these conditions, consider

$$V(k_{N-1}, z_{N-1})|-k_N\rangle \,. \tag{2.20}$$

From this we can extract the pole at $(k_N + k_{N-1})^2$ equal to an integer by constructing

$$I = \oint \frac{\mathrm{d}z_{N-1}}{z_{N-1}} V(z_{N-1}, k_{N-1}) \,. \tag{2.21}$$

Then it is easy to deduce $[L_n, I] = 0$ which extends eq. (2.19) to two-particle states. But at this point it should be emphasized that eq. (2.21) is only well defined when the integrand is single-valued, and this happens exactly when $(k_{N-1} + k_N)^2$ is at one of the energy levels. We can construct an arbitrary physical state by repeating this process any number of times, always restricting the momenta so that the mass-shell conditions are satisfied. Thus, if $|\varphi\rangle$ is an arbitrary physical state

$$(L_0 - 1)|\varphi\rangle = 0 \,, \qquad L_n|\varphi\rangle = 0 \,. \tag{2.22}$$

3. – Vertices for excited physical states.

We now seek vertices which describe the coupling of excited states. From (2.10), (2.20) and (2.21) we see that a physical state coupling to two ground-state particles may be written in the form

$$\langle k_1|\oint \frac{\mathrm{d}z_1}{z_2}(z_2, k_2) = \lim_{z_1\to 0} \langle 0|\frac{V\alpha_1(k_1, z_1)}{z_1}\oint \frac{\mathrm{d}z_2}{z_2} V(z_2, k_2) = \lim_{z_1\to 0} \langle 0|\frac{V\alpha_1(z_1, k_1)}{z_1} \,, \tag{3.1}$$

where we have defined

$$V_{\alpha_1}(z_1, k_1) = V(z_1, k_1) \oint_{z_1} \frac{dz_2}{z_2} V(z_2, k_2) , \tag{3.2}$$

α_1 describes additional states labels (polarization, etc.) resulting from the relative orientation of k_1 and k_2. Again, we should stress that eqs. (3.1 and (3.2) have an unambiguous meaning only when the integrands are single-valued. This means that $k_1^2 = k_2^2 = 1$ and $k, = k_1 t k_2$, must correspond to some definite energy level. Equation (3.2) then gives the vertex for the coupling of the excited state of eq. (3.1). Of course, not all physical states couple to two ground-state particles. But it is clear that the process just described can be extended [5] by combining any number of vertices. In this way the general vertex is obtained and it clearly satisfies.

$$\left[L_n, \oint \frac{V_\alpha(k, z)}{z} dz\right] = 0 , \tag{3.3}$$

(but again only when the integral is unambiguously defined).

The vertices $V_\alpha(k, z)$ are really on the same footing as $V(k, z)$ and we can write an amplitude for N excited particles as

$$A^N_{\alpha_i}(k_i) = \int \langle 0| \frac{V_{\alpha_1}(k_1 z_1)}{z_1} \frac{V_{\alpha_2}(z_2 k_2)}{z_2} \dots \frac{V_{\alpha_N}(z_N, k_N)}{z_N} |0\rangle \cdot \left\{\prod_{i=1}^{N} (z_{i+1} - z_i)\right\} d\Omega_z . \tag{3.4}$$

This equation makes clear, in a beautiful way, the sense in which the DRM realizes a bootstrap.

Using these techniques it is possible, for example, to construct all the operators associated with all the states on the leading trajectory [9]. These are given by the simple formula

$$V_J(z, k) = :[\varepsilon, P(z)]^J V(z, k): , \tag{3.5}$$

where ε is a complex vector satisfying

$$\varepsilon \cdot k = 0 . \tag{3.6}$$

Of particular interest to us will be the case $J=1$, which gives the zero-mass vector state and, to show that the above analysis is useful calculationally as well as elegant, we will derive (3.5) for this case. Figure 1 illustrates the way we are using duality in eq. (3.2) to obtain the « photon » coupling

$$k = k_1 - k_2 , \qquad k^2 = 0 , \tag{3.7}$$

$$k_1^2 = k_2^2 = 1 , \qquad 2k_1 \cdot k_2 = -2 . \tag{3.8}$$

To evaluate the right-hand side of eq. (3.2) we need to determine the singularities of the integrand when $z_2 = z_1$. When one looks at the definition (2.8) no singularities are evident. To produce the singularities which are really there we have to normal-order all the operators in the expression. It is easy to

Fig. 1.

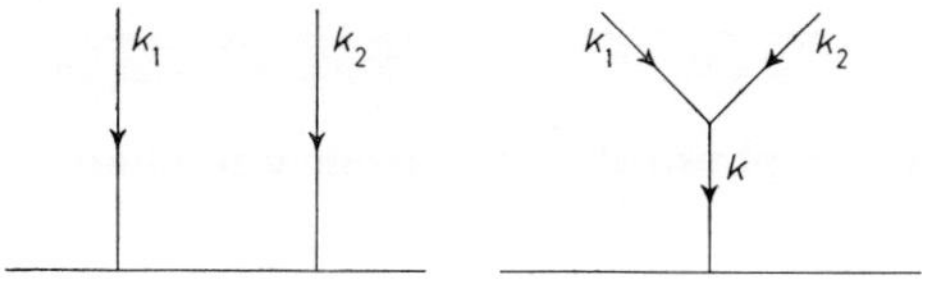

understand that this is precisely the step needed for the rigorous evaluation of (3.2). For the basic states on which our operators must be well behaved are those of an occupation number basis (*i.e.* those created with a finite number of finite powers of the operators $a_n^{\mu\dagger}$). Once we have normal-ordered, taking matrix elements of an operator between such states will introduce no new divergences. This will be exemplified by the case we are considering. To normal-order consider the vertex

$$V(z_i, k_i) = \exp\left[i\sqrt{2}\, k_i Q^{(-)}(z_i)\right] \exp\left[i\sqrt{2}\, k_i Q^{(0)}(z_i)\right] \exp\left[i\sqrt{2}\, k_i Q^{(+)}(z_i)\right], \tag{3.9}$$

where

$$\exp\left[i\sqrt{2}\, k_i Q^{(0)}(z_i)\right] = \exp\left[i\sqrt{2}\, k_i q^{(0)}\right] z_i^{-\sqrt{2}k_1\cdot p_0} z_i^{k_i^2}, \tag{3.10}$$

we must interchange the order of the terms involving $Q^{(-)}(z_i)$ and $Q^{(+)}(z_2)$, respectively, in eq. (3.2). This can be achieved using

$$\begin{aligned}\exp[i\sqrt{2}k_1 Q^{(+)}(z_1)] \exp[i\sqrt{2}k_2 Q^{(-)}(z_2)] = \\ = \exp[i\sqrt{2}k_2 Q^{(-)}(z_2)] \exp[i\sqrt{2}k_1 Q^{(+)}(z_1)] \left(1 - \frac{z_1}{z_2}\right)^{2k_1\cdot k_2}.\end{aligned} \tag{3.11}$$

Thus normal-ordering does produce a singularity, and we see from (3.8) that this singularity is a double pole. This, of course, acts like the derivative of a delta function, introducing

$$\left.\frac{\partial}{\partial z_2} V(z_2, k_2)\right|_{z_1=z_2} = i\sqrt{2}k_2 \frac{dQ^{(-)}}{dz_1}(z_1) V(z_1, k_1). \tag{3.12}$$

Carefully collecting the various factors we obtain, apart from an irrelevant overall constant

$$V_\varepsilon(z, k) = z\varepsilon \frac{dQ}{dz} V(z, k). \tag{3.13}$$

Here $k_2 \equiv \varepsilon$, the polarization, plays the rôle α_1 did in eq. (3.2). ε satisfies the familiar subsidiary condition (3.6), which immediately follows from eqs. (3.7) and (3.8). Because of our derivation we should consider (3.13) acting only on states on which the integrand of eq. (3.2) is single-valued.

4. – The construction of physical states from « photons ».

As we remarked eq. (3.4) makes it clear that the ground state and the excited physical states very much have equal status in the DRM. This being so, there is no need to consider a general physical state exclusively as built up from vertices describing the coupling of the ground state. In particular, in the basic form of the DRM we are discussing the ground state is a tachyon, an undesirable feature of the model. The massless vector meson, or « photon », state is more palatable theoretically if not phenomenologically. Indeed, even in a realistic theory the operator describing its vertex may be associated with the electromagnetic coupling through some sort of vector meson dominance.

Also photon vertices are technically easier to handle, resulting in a striking simplification in the algebra. From eq. (2.2) we see

$$[k_1\, a_n, k'_1, a^\dagger_m] = k \cdot k' \delta_{mn} . \tag{4.1}$$

Thus, if $k = k'$ and k is the lightlike vector of Sect. **3**, all the operators in the exponential in $V(k, z)$ commute. As a result of this and the supplementary condition (3.6), there is no need to normal-order in (3.13). The exponential parts of different « photon » vertices will commute only if the photon momenta are parallel. Thus we are led to consider a tachyon coming in and interacting with a succession of « photons » with parallel momenta to produce a sequence of resonance states.

The restriction to parallel momenta at first sight seems a severe one, but we shall see that many of the physical states are produced thus, and that, with a certain natural extension, we obtain an algebra which describes all the states (not just the physical ones).

For definiteness we will work in a particular frame in which the incoming tachyon has momentum p

$$p = (0, 0, 0, 1) , \qquad p^2 = 1 , \tag{4.2}$$

and we will choose a lightlike vector k

$$k = (\tfrac{1}{2}, 0, 0, -\tfrac{1}{2}) , \qquad k^2 = 0 . \tag{4.3}$$

It will also be convenient to define

$$k_L = (\tfrac{1}{2}, 0, 0, \tfrac{1}{2}) . \tag{4.4}$$

A « photon » of momentum λk will interact with a ground state of momentum p to produce a resonance of mass squared $n-1$ $(n = 0, 1, 2, \ldots)$ if

$$(p + \lambda k)^2 = 1 - n , \tag{4.5}$$

that is if λ is a positive integer. So we consider the situation illustrated in Fig. 2.

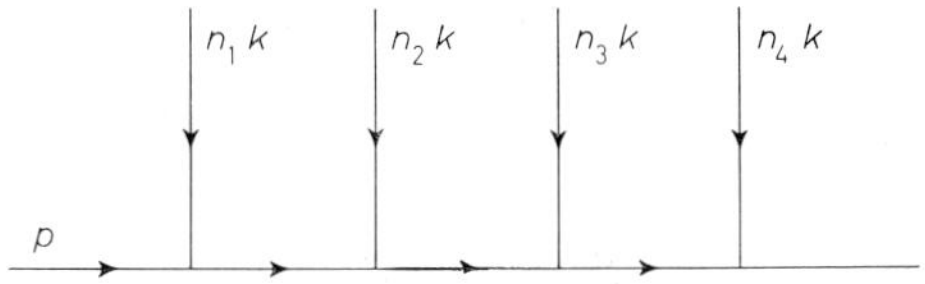

Fig. 2.

We want the operators which build up the successive resonances; these are obtained from the vertices by a loop integral as eq. (2.21)

$$\oint \frac{V_\varepsilon(z, nk)}{z} \, dz . \tag{4.6}$$

Now the subsidiary condition (3.6) means we can write

$$\varepsilon = \alpha \hat{l}_1 + \beta \hat{l}_2 + \gamma k , \tag{4.7}$$

where $\hat{l}_1$, $\hat{l}_2$ are unit vectors in the x and y directions, respectively.

α, β correspond to the two familiar polarizations of the photon and γ to the third transverse-polarization state which decouples. We shall see this decoupling explicitly shortly. Corresponding to the two photon polarization states we introduce operators

$$A_n^i = -\frac{1}{\sqrt{2\pi}} \oint \frac{dQ^i(z)}{dz} V(z, nk) \, dz , \tag{4.8}$$

where $i = 1, 2$; n is any integer and we have chosen a convenient normalization factor.

If we consider the third polarization state

$$k A_n = -\frac{1}{\sqrt{2\pi}} \oint k \frac{dQ}{dz} V(z, nk) \, dz , \tag{4.9}$$

$$= -\frac{1}{2\pi i n} \oint \frac{d}{dz} \{V(z, nk)\} \, dz , \quad n \neq 0 , \tag{4.10}$$

which is zero when single-valued. In fact, we can write

$$k \cdot A_n = \delta_{n0}\,, \tag{4.11}$$

when acting on states of momentum $p + \lambda k$. It is easy to show directly that

$$A^i_n = A^{i\dagger}_{-n}\,, \tag{4.12}$$

$$\langle p|A^i_n = 0\,, \qquad A^{i\dagger}_n|p\rangle = 0 \qquad \text{if } n<0\,. \tag{4.13}$$

Thus we have obtained an infinity by physical states

$$\langle p|A^{i_1}_{n_1} A^{i_2}_{n_2} \dots A^{i_r}_{n_r}\,, \qquad n_j > 0\,,\; i_j = 1, 2\,, \tag{4.14}$$

or, if we prefer to think of the operators acting to the right

$$A^{i_1\dagger}_{n_1} A^{i_2\dagger}_{n_2} \dots A^{i_r\dagger}_{n_r}|p\rangle\,. \tag{4.15}$$

These are physical states by construction but it is easy and illuminating to check this explicitly. We need the basic commutation relations [3]

$$[L_n, V(z, k)] = -\,z^{-n+1}\frac{\mathrm{d}V}{\mathrm{d}z} \qquad \text{if } k^2 = 0\,, \tag{4.16}$$

$$[L_n, Q^\mu(z)] = -\,z^{-n+1}\frac{\mathrm{d}Q^\mu}{\mathrm{d}z}\,. \tag{4.17}$$

From the last we can deduce, by differentiation with respect to z

$$\left[L_n, \frac{\mathrm{d}Q^\mu}{\mathrm{d}z}\right] = -\frac{\mathrm{d}}{\mathrm{d}z}\left\{z^{-n+1}\frac{\mathrm{d}Q^\mu}{\mathrm{d}z}\right\}, \tag{4.18}$$

and so

$$\left[L_n, \frac{\mathrm{d}Q^\mu}{\mathrm{d}z}V(z, k)\right] = -\frac{\mathrm{d}}{\mathrm{d}z}\left\{z^{-n+1}\frac{\mathrm{d}Q^\mu}{\mathrm{d}z}V\right\}. \tag{4.19}$$

Thus it follows that

$$[L_n, A^i_m] = 0 \qquad \text{for all } n,\, m\,, \tag{4.20}$$

and so we obtain the conditions (2.22).

In fact, we can say much more about the states (4.15) than just that they are physical states. Two such states are orthogonal if the number of time any given $A^{i\dagger}_n$ occurs in each is not the same for both, and the norm of each state is positive. Thus we have an infinity of independent physical states corresponding to two dimensions of harmonic oscillators like the $a^{i\dagger}_n$ which is ghost-free. To prove this important result we will not follow the original method of DI VECCHIA, DEL GIUDICE and FUBINI [5] but adopt a more algebraic approach [6] in line with the sort of further developments we will make.

The inner product of any two states formed with the $a_n^{i\dagger}$ follows from the basic commutation relations (2.2). Thus we calculate the commutator $[A_m^i, A_n^j]$; this will illustrate the basic method used to find other commutators needed later, though the details there are somewhat more subtle:

$$[A_m^j, A_n^i] = \frac{1}{2\pi^2}\oint \mathrm{d}z_1 \oint \mathrm{d}z_2 \left[\frac{\mathrm{d}Q^i}{\mathrm{d}z_1}, \frac{\mathrm{d}Q^j}{\mathrm{d}z_2}\right] V(z_1, mk) V(z_2, nk)\,, \tag{4.21}$$

exploiting the commutativity of the exponentials. The commutator of the derivatives of Q fields follows from eq. (2.14)

$$\left[z_1 \frac{\mathrm{d}Q^i}{\mathrm{d}z_1}(z_1), z_2 \frac{\mathrm{d}Q^j}{\mathrm{d}z_2}(z_2)\right] = \frac{g^{\mu\nu}}{2\pi i}\frac{\partial}{\partial\theta_2}\delta(\theta_1-\theta_2)\delta_{ij}\,. \tag{4.21}$$

From this we obtain

$$[A_m^i, A_n^j] = -2nk\cdot A_{m+n}\delta_{ij} = 2m\delta_{m+n,0}\delta_{ij}\,, \tag{4.22}$$

using eq. (4.11). Thus the operators A_m^i behave exactly like annihilation and creation operators, and the assertion of orthogonality and positivity follows.

5. – Colinear algebra for the dual-resonance model.

At this point we can easily understand what has been achieved so far. We have physical states corresponding to a two-dimensional infinite set of harmonic oscillators, in a sense associated with oscillations at right angles to the plane of p and k.

But we know from the work of BROWER and THORN that the space of all physical states corresponds to three dimensions of oscillators, rougly speaking [10]. If we were working in n-dimensional space-time we would have physical states for $(n-2)$ dimensions, always being one dimension short. We lack, each time, one dimension of states which clearly should be associated with oscillations just in the plane of p and k.

To rectify this deficiency we will enlarge the algebra of the A_n^i. A natural suggestion [6] is to try to extend A_n^i into a four-vector A_n^μ. Since $k\cdot A_n = 0$, $n \neq 0$ this would give the three independent components needed. The obvious first choice for A_n^μ is perhaps

$$-\frac{1}{\sqrt{2}\pi}\oint \frac{\mathrm{d}Q^\mu}{\mathrm{d}z} V(k, nz)\,\mathrm{d}z\,. \tag{5.1}$$

This choice is reinforced by the fact that the direct proof we gave that $[L_n, A^i_m]=0$ did not depend on the subsidiary condition. So formally we would have $[L_n, A^\mu_m]=0$ and $k\cdot A_m=0$, $m\neq 0$, with this definition. The third independent operator would be defined as A^L_n, where this is obtained by taking the product of k^μ_L (see eq. (4.4)) with (5.1). If this were correct the problem of constructing the physical states would be solved. But really the longitudinal photon state associated with A^L_n is not physical and there must be an error in the argument.

The catch is that we now need to normal-order. Whereas the product of $\hat{l}_1$, $\hat{l}_2$ or k with (5.1) is trivially normal-ordered, the product of k_L with it is not. Such an operator is easy seen to give infinities in the basic occupation number states. If we normal-order (5.1),

$$-\frac{1}{\sqrt{2\pi}}\oint :k_L\frac{\mathrm{d}Q}{\mathrm{d}z}V(z, nk):\mathrm{d}z\,, \tag{5.2}$$

it will no longer commute with L_m, but one may hope that, as in other cases, the additional terms resulting from normal-ordering do not present an insuperable problem. Thus we shall consider the algebra of operators like (5.2).

As $k_L\cdot\mathrm{d}Q/\mathrm{d}z$ does not commute with $V(z, nk)$, (5.2) is not a very symmetric definition for A^L_n. We might equally have written

$$-\frac{1}{\sqrt{2\pi}}\oint :V(z, nk)k_L\frac{\mathrm{d}Q}{\mathrm{d}z}:\mathrm{d}z\,. \tag{5.3}$$

In fact, we should compromise by taking the average

$$A^L_n=-\frac{1}{\sqrt{2\pi}}\oint\frac{1}{2}:\left\{V(z, nk), k_L\frac{\mathrm{d}Q}{\mathrm{d}z}\right\}_+:\mathrm{d}z\,, \tag{5.4}$$

where $\{\,,\,\}_+$ denotes the anticommutator. This definition ensures the hermiticity property

$$A^{L\dagger}_n=A^L_{-n}\,. \tag{5.5}$$

We must now check that the algebra still closes by calculating the following commutators by methods we have indicated:

$$[A^L_m, A^j_n]=-nA^j_{m+n}\,, \tag{5.6}$$

$$[A^L_m, A^L_n]=(m-n)A^L_{m+n}+m^3\delta_{m+n,0}\,. \tag{5.7}$$

This last result should be compared with the algebra of the gauge operators

$$[L_m, L_n]=(m-n)L_{m+n}+\frac{m(m^2-1)}{2}D\delta_{m+n,0}\,. \tag{5.8}$$

The algebras differ only in the « c-number » terms (note that the c-number in (5.8) depends on D, the dimension of space-time, but that in (5.7) it does not), and this encourages the hope that these operators are significant for creating the physical states.

However, at present we have too few operators because we have only three dimensions of operators, but they do not all give physical states. The further extension of the algebra can be motivated by the discrepancy between A_n^L and either (5.2) or (5.3). These can be written as $A_n^L \pm (n/2)\Phi_m^n$ where

$$\Phi_m^n = -\frac{1}{2\pi i}\oint z^{n-m}\frac{V(z, nk)}{z}\,\mathrm{d}z\,. \tag{5.9}$$

Clearly, not all the operators defined by eq. (5.9) can be independent. But they can all be related to the infinite set $\{\Phi_n'\}$, for example, using the Fourier or Laurent expansions

$$V(z, nk) = \{V(z, k)\}^n = -\sum_{m=-\infty}^{\infty} \Phi_{m+n}^n z\,. \tag{5.10}$$

The Φ_m^n-operators are remarkable in that they commute.

We can now see how the whole algebra closes:

$$[\Phi_m^r, \Phi_n^s] = 0\,, \qquad [A_m^i, \Phi_n^r] = 0\,, \tag{5.11}$$

$$[A_m^L, \Phi_n^r] = -r\Phi_{m+n}^{m+r}\,, \tag{5.12}$$

$$A_n^{L\dagger} = A_{-n}^L\,, \qquad \Phi_s^{r\dagger} = \Phi_{-s}^{-r}\,, \tag{5.13}$$

$$A_n^\mu|0\rangle = \Phi_n^m|0\rangle = 0 \qquad \text{if } n>0\,. \tag{5.14}$$

Further, the commutation relations with L_n are

$$[L_m, A_n^L] = -\frac{m^2 n}{2}\Phi_{m+n}^n\,, \tag{5.15}$$

$$[L_m, \Phi_n^r] = (m+n-r)\Phi_{m+n}^r\,. \tag{5.16}$$

Now it can be shown that together A_n^i, A_n^L, Φ_n^m generate all their states formed by the original $a_n^{\mu\dagger}$. Thus we can replace the original harmonic oscillators by this algebra. The advantage of the new algebra is that the Φ-operators commute and they are the operators produced by most the of commutators. Further, instead of A_n^L we may use

$$\hat{A}_n^L = A_n^L + \frac{n}{2}\Phi_n^n\,, \tag{5.17}$$

which commutes with L_1 and L_0. Thus, because only two of the L_n are independent it only remains to modify the states created by $\hat{A}_n^{L\dagger}$, so that they satisfy the second gauge condition. In doing this it is useful to note that

$$[L_1, \Phi_n^{n+1}] = 0 . \tag{5.18}$$

At present work on constructing all the physical states by these techniques is at a preliminary stage but already we have expressions for infinities of physical states containing just one of the $\hat{A}_n^{L\dagger}$ operators many times. This is given in a sort of coherent state form by the expression

$$\exp [z\hat{A}_1^{L\dagger}]\left\{\Phi_0^{-1} \sum_r \Phi_{-1-r}^r z^r\right\}|p\rangle . \tag{5.19}$$

The proof of this results depends on relations between the Fourier coefficients of functions. We are optimistic that this can be extended to give a linear independent basis for the physical states in a compact form, and hope that it will reveal more about the algebraic structure of the spectrum. The problem of their checking the norm of all these physical states to *prove* there are ghosts would still seem likely to be a very difficult one.

* * *

We thank Prof. S. FUBINI for encouragement and Profs. S. FUBINI and R. GATTO for the opportunity to present our work at the Varenna Summer School.

Note added in proofs.

Since these lectures the absence of ghosts in the DRM has been proved independently by R. C. BROWER (MIT preprint) and by P. GODDARD and C. B. THORN (CERN preprint).

REFERENCES

[1] G. VENEZIANO: *Nuovo Cimento*, **57** A, 190 (1968).
[2] S. FUBINI and G. VENEZIANO: *Nuovo Cimento*, **64** A, 811 (1969); K. BARDAKÇI and S. MANDELSTAM: *Phys. Rev.*, **184**, 1640 (1969).
[3] For a further account of the DRM formation see V. ALESSANDRINI and D. AMATI: this volume p. 58; S. FUBINI: *Duality in strong interaction physics*, in *Lectures at Ecole d'Eté de Physique Théorique, Les Houches, 1971.*
[4] P. DI VECCHIA and S. FUBINI: *Lett. Nuovo Cimento*, **1**, 823 (1971).
[5] E. DEL GIUDICE, P. DI VECCHIA and S. FUBINI: *Ann. of Phys.*, **70**, 378 (1972).
[6] R. C. BROWER and P. GODDARD: *Nucl. Phys.*, **40** B, 437 (1972).
[7] E. DEL GIUDICE and P. DI VECCHIA: *Nuovo Cimento*, **70** A, 579 (1970).
[8] R. C. BROWER and C. B. THORN: *Nucl. Phys.*, **31** B, 163 (1971).
[9] C. ROSENZWEIG and V. P. SUKHATME: *Nuovo Cimento*, to be published.
[10] For the sake of simplicity we will not discuss the subtelties in the counting associated with zero-norm states, referring to ref. [8] on this point.

Veneziano Amplitudes, Angular-Momentum Coefficients and Related Symmetries.

T. REGGE

Istituto di Fisica dell'Università - Torino
Institute for Advanced Study - Princeton, N. J.

I wish to illustrate in this brief lecture some interesting properties of the Veneziano amplitudes and their connection with the well-known Clebsh-Gordan and Racah coefficients, so useful in the theory of angular momentum. These properties have been discovered and are being developed by a group from Torino [1, 2].

The starting point is the well-known 4-point Veneziano amplitude [3] with unit intercept

$$B(1234) = \int_0^1 \mathrm{d}x\, x^{-\alpha_{12}-1}(1-x)^{-\alpha_{13}-1} = \frac{\Gamma(-\alpha_{12})\Gamma(-\alpha_{13})}{\Gamma(-\alpha_{12}-\alpha_{13})}, \tag{1}$$

where $p_1^2 = 1$, $\alpha_{ij} = 1-(p_i+p_j)^2$, i and $j = 1, ..., 4$; the constraint $\sum_{i=1}^{4} p_i = 0$ implies: $-\alpha_{12}-\alpha_{13} = 1+\alpha_{23}$. This formula clearly shows both the structures of poles and zeros of the amplitude; we notice that the latter—which appear when $\alpha_{12}+\alpha_{13}$ is positive integer—just prevent from having simultaneous poles in the 12 and 13 channels.

In the generalized N-point Veneziano amplitude [4] the structure of the zero varieties is much more complicated and worthy of a detailed analysis. Starting from the Koba-Nielsen representation

$$\begin{cases} B(123 \dots N) = \int_0^\infty \dots \int_0^\infty \left(\prod_{i=2}^{N-2} \mathrm{d}x_i\right) \left\{\prod_{\substack{i,j=1 \\ j>i}}^{N-1} (x_j - x_i)^{-\alpha_{ij}-1}\right\} \left\{\prod_{i=1}^{N-2} \theta(x_{i+1}-x_i)\right\}, \\ x_1 = 0, \qquad x_{N-1} = 1, \qquad x_N = \infty. \end{cases} \tag{2}$$

PLAHTE [5] was able to deduce linear relations of the type

$$(3)\qquad \begin{aligned} &B(123\,\ldots\,N) + \exp\,[2\pi i p_1\cdot p_2]B(213\,\ldots\,N) + \\ &+ \exp\,[2\pi i(p_1\cdot p_2 + p_1\cdot p_3)]B(231\,\ldots\,N) + \ldots + \\ &+ \exp\,[2\pi i(p_1\cdot p_2 + p_1\cdot p_3 + \ldots + p_1\cdot p_{N-1})]B(23\,\ldots\,N-1, 1, N) = 0\,. \end{aligned}$$

As the B-functions are all real, they may be regarded as the modulus of a complex number whose phase is the multiplying exponential in eq. (3), so that this equation appears as a closed polygon in the complex plane (see Fig. 1). Choosing

$$(4)\qquad 2p_1\cdot p_j = -\alpha_{1j} - 1 = n_{1j}, \qquad j = 3, \ldots, N-1,$$

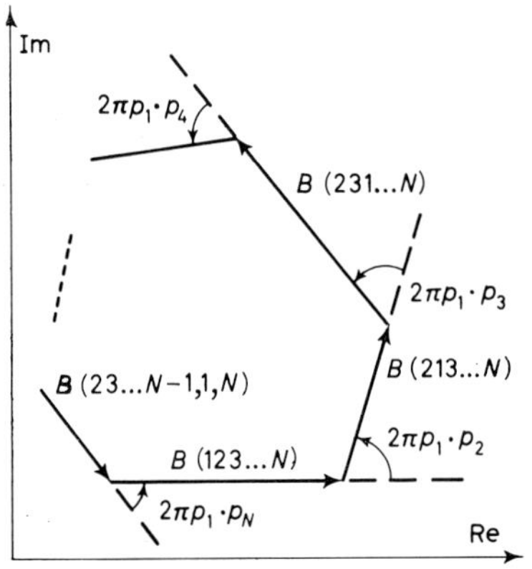

Fig. 1.

where n_{1j} are all nonnegative integers, we see that all vectors but $B(123\,\ldots\,N)$ have the same directions as $\exp\,(2\pi i p_1\cdot p_2)B(213\,\ldots\,N)$, so that—since $p_1\cdot p_2$ is arbitrary—the only way for the polygon to be closed is to have $B(123\,\ldots\,N) = 0$. The restriction to nonnegative values of n_{1j} guarantees that we do not run into poles of some amplitudes in eq. (3), which would destroy the argument. Clearly, when $N = 4$, the polygon of Fig. 1 becomes a triangle and the zeros prescribed by eq. (4) coincide with the ones of the Veneziano amplitude.

In ref. [1] more complicated classes of zeros are obtained by joining two or more polygons together, but I shall skip this point; rather I want to sketch how eq. (4) can be deduced for the case $N = 5$ in an entirely different way. DIXON [6] in 1905 discovered that

$$(5)\qquad \begin{aligned} B(12\,345) = \Gamma(-\alpha_{12})\Gamma(-\alpha_{23})\Gamma(-\alpha_{34})\Gamma(-\alpha_{45})\Gamma(-\alpha_{51})\,\times \\ \times F(\alpha_{12}+\alpha_{23},\, \alpha_{23}+\alpha_{34},\, \alpha_{34}+\alpha_{45},\, \alpha_{45}+\alpha_{51},\, \alpha_{51}+\alpha_{12})\,, \end{aligned}$$

where F turns out to be an entire function invariant under any permutation of its arguments. We know that B satisfies the duality constraint so that it does not have simultaneous poles in overlapping channels, for instance, 12, 23. Therefore, at positive integer values of α_{12}, α_{23}, F must vanish to compensate one of the two exploding Γ's. If we call the arguments of F: $\varrho_1, \ldots, \varrho_5$, the above condition to have a zero of F reads

$$(6)\qquad \begin{cases} \varrho_1 = n_1\,, & n_1 \geqslant |n_2|\,,\\ \varrho_2 - \varrho_3 + \varrho_4 - \varrho_5 = n_2\,, \end{cases}$$

where n_1, n_2 are integers. Due to the symmetry of F, any permutation of ϱ's in eq. (6) still gives zeros of F; for instance, if we exchange ϱ_3, ϱ_4, we get a zero of F which—in terms of α's—reads: $\alpha_{24} = -n_1'$, $\alpha_{25} = -n_2'$, with n_1', n_2 positive integers. As no Γ-function in eq. (5) explodes, these conditions yield a zero of B in agreement with eq. (4).

Dixon's symmetry is relevant not only in the context of dual amplitudes but has also a deep connection with the symmetries of the $3j$-coefficient of SU_2. In fact, it is not difficult to see that $B\,(12345)$ is proportional to ${}_3F_2(1 + \alpha_{24}, -\alpha_{12}, -\alpha_{45}; -\alpha_{12} - \alpha_{23}, -\alpha_{34} - \alpha_{45}; 1)$; since also the $3j$-coefficient can be written in terms of a ${}_3F_2$ of unit argument with 5 independent parameters, it turns out that

$$(7)\quad \begin{pmatrix} j_1 & j_2 & j_3 \\ \mu_1 & \mu_2 & \mu_3 \end{pmatrix} = \begin{pmatrix} (r_5 - r_4 - 1)/2 & (r_1 - r_2 - 1)/2 & (r_3 - r_0 - 1)/2 \\ (r_1 + r_2 - r_0 - r_3)/2 & (r_0 + r_3 - r_4 - r_5)/2 & (r_4 + r_5 - r_1 - r_2)/2 \end{pmatrix} =$$

$$= \begin{bmatrix} (\;\;r_0 - r_1 - r_2 + r_3 - r_4 + r_5 - 1)/2 & (-r_0 + r_1 - r_2 - r_3 + r_4 + r_5 - 1)/2 & (-r_0 + r_1 + r_2 + r_3 - r_4 - r_5 - 1)/2 \\ (-r_0 + r_1 + r_2 - r_3 - r_4 + r_5 - 1)/2 & (\;\;r_0 + r_1 - r_2 + r_3 - r_4 - r_5 - 1)/2 & (-r_0 - r_1 - r_2 + r_3 + r_4 + r_5 - 1)/2 \\ (-r_0 + r_1 - r_2 + r_3 + r_4 - r_5 - 1)/2 & (-r_0 - r_1 + r_2 + r_3 - r_4 + r_5 - 1)/2 & (\;\;r_0 + r_1 - r_2 - r_3 - r_4 + r_5 - 1)/2 \end{bmatrix} =$$

$$= (-1)^{r_2 - r_4}\{\Gamma(\alpha_{123}, \alpha_{134}, \alpha_{125}, \alpha_{145}, \alpha_{235}, \alpha_{345}, \alpha_{135})/\Gamma(\alpha_{013}, \alpha_{015}, \alpha_{035})\}^{\frac{1}{2}} \times$$
$$\times F(r_0 - r_1 - 1,\; r_0 - r_2 - 1,\; r_0 - r_3 - 1,\; r_0 - r_4 - 1,\; r_0 - r_5 - 1)\,,$$

where

$$(8)\qquad \begin{cases} \displaystyle\sum_{i=0}^{5} r_i = 0\,, \\ \alpha_{ijh} = r_i + r_j + r_h + \tfrac{1}{2}\,, \end{cases}$$

and $\Gamma(x_1, x_2, \ldots) = \prod_i \Gamma(x_i)$. Clearly, in terms of r's, the 72 symmetries of the

$3j$ consist of: permutations of r_0, r_2, r_4; permutations of r_1, r_3, r_5; exchange of r_0, r_2, r_4 with $-r_1, -r_3, -r_5$. These operations leave invariant the sum of rows and columns of the square symbol [7]: $j_1+j_2+j_3=(r_1+r_3+r_5-\\ -r_0-r_2-r_4-3)/2$.

On the other hand, Dixon's function is obviously invariant under any permutation of r's which keeps r_0 fixed. This group shares a subgroup of 12 elements (permutations of 0, 2 and of 1, 3, 5) with the symmetry group of the $3j$; the remaining elements are not strictly speaking symmetries of the whole r.h.s. of eq. (7) because several arguments of the Γ-functions are changed and sent into negative integers. However, one can easily check that, by adopting the convention [8] (which amounts to taking the residue of Γ's)

$$\Gamma(-m)/\Gamma(-n)=(-1)^{m-n}\Gamma(n+1)/\Gamma(m+1)\,, \tag{9}$$

with m, n positive integers, the initial Γ-functions are restored. Therefore one is led to consider the larger group of order $6!\otimes 2$ realized by: permutations of $r_0, \ldots, r_5$; sign inversion of all r's. This group makes it possible to generalize the square symbol to negative integer values of its arguments; part of the new simbols correspond to $3j$-coefficients with negative j's (see also ref. [8]), the others are $3j$-coefficients of $SU_{1,1}$ discrete unitary representations [9] and involve antitriangular relations among j's.

A similar extension can be realized also in the case of $6j$-coefficients. To this end let us adopt the following notation:

$$\begin{bmatrix} p_1 & p_2 & p_3 \\ q_1 & q_2 & q_3 \end{bmatrix} = \begin{Bmatrix} a & b & e \\ d & c & f \end{Bmatrix}, \tag{10}$$

where

$$\left\{\begin{array}{ll} p_1=a+d+1\,, & q_1=a-d\,, \\ p_2=b+c+1\,, & q_2=b-c\,, \\ p_3=e+f+1\,, & q_3=e-f\,. \end{array}\right. \tag{11}$$

Clearly the $144=3!\otimes 3!\otimes\left(1+\binom{3}{2}\right)$ symmetries of the $6j$ are realized by: permutations of p_1, p_2, p_3; permutations of q_1, q_2, q_3; sign inversion of any pair of q's. Recently a new symmetry has been claimed to hold for the $6j$-coefficient [10]; actually it leads to nonphysical values of the parameters [11] and requires the same convention of eq. (9). In terms of our parameters this transformation can be expressed as the product of the usual symmetry $q_1\to -q_1$,

$q_2 \to -q_2$ and of

$$\begin{bmatrix} p_1 & p_2 & p_3 \\ q_1 & q_2 & q_3 \end{bmatrix} = (-1)^{p_2 - q_3 + 1} \begin{bmatrix} p_1 & q_3 & p_3 \\ q_1 & p_2 & q_2 \end{bmatrix} . \tag{12}$$

If we combine eq. (12) with the usual symmetries—and this is possible in the same sense as for the $3j$, by expressing the $6j$ in terms of a well-poised ${}_7F_6$ series—we are led to the group of: permutations of p_i, q_i, $i = 1, 2, 3$; change of sign of any pair of p's and q's. Its order amounts to

$$6! \otimes \left\{ \binom{6}{0} + \binom{6}{2} + \binom{6}{4} + \binom{6}{6} \right\} = 720 \times 32 .$$

Finally we may tentatively apply to the extended $6j$-coefficient the asymptotic result of ref. [12], where d, c, f are made arbitrarily large. With our present notations it reads

$$\lim_{R \to \infty} \begin{bmatrix} p_1 + R & p_2 + R & p_3 + R \\ q_1 - R & q_2 - R & q_3 - R \end{bmatrix} = (2R)^{-\frac{1}{2}} (-1)^{\left\{ \sum_{i=1}^{3} (3p_i - q_i) - 1 \right\} / 2} . \tag{13}$$

$$\cdot \begin{pmatrix} (r_5 - r_4 - 1)/2 & (r_1 - r_2 - 1)/2 & (r_3 - r_0 - 1)/2 \\ (r_1 + r_2 - r_0 - r_3)/2 & (r_0 + r_3 - r_4 - r_5)/2 & (r_4 + r_5 - r \ - r_2)/2 \end{pmatrix},$$

where $r_0 = -q_3 + c$, $r_2 = -q_2 + c$, $r_4 = -q_1 + c$, $r_1 = p_2 + c$, $r_3 = p_3 + c$, $r_5 = p_1 + c$ and c, being arbitrary, is chosen in such a way to have $\sum_{i=0}^{5} r_i = 0$. We notice that the asymmetry introduced into the $6j$ of eq. (13) by R lets only 720 elements of the $6j$ symmetry group survive, that is, all possible permutations of p_1, p_2, p_3 , $-q_1$, $-q_2$, $-q_3$. They correspond to permutations of $r_0, \ldots, r_5$, namely to one-half of the symmetries enjoyed by the extended $3j$-coefficent.

REFERENCES

[1] A. D'ADDA, R. D'AURIA, F. GLIOZZI and S. SCIUTO: *Zeros of dual resonant amplitudes*, Università di Torino preprint, February 1971.

[2] A. D'ADDA, R. D'AURIA and G. PONZANO: *On generalized Wigner and Racah coefficients*, internal report, September 1971.

[3] G. VENEZIANO: *Nuovo Cimento*, **57** A, 190 (1968).

[4] K. BARDAKÇI and H. RUEGG: *Phys. Lett.*, **28** B, 242 (1968); Z. KOBA and H. B. NIELSEN: *Nucl. Phys.*, **10** B, 633 (1969).

[5] E. PLAHTE: *Nuovo Cimento*, **66** A, 713 (1970).
[6] A. C. DIXON: *Proc. London Math. Soc.*, **2**, 8 (1905).
[7] T. REGGE: *Nuovo Cimento*, **11**, 116 (1959).
[8] This convention was introduced by A. A. BANDZAITIS, A. V. KAROSENE, A. YU. SAVUKINAS and A. P. YUTSIS: *Sov. Phys. Doklady*, **9**, 139 (1964), to extend the $3j$-coefficient to negative values of j's.
[9] W. HOLMAN and L. C. BIEDENHARN: *Ann. of Phys.*, **39**, 1 (1966); **47**, 205 (1968).
[10] B. M. MINTON: *Journ. Math. Phys.*, **11**, 3061 (1970).
[11] V. JOSHI: *A remark on Minton's symmetry for Racah functions*, Duke University preprint (1971).
[12] P. PONZANO and T. REGGE: *Semiclassical limit of Racah coefficients*, in *Spectroscopic and Group-Theoretical Methods in Physics* (*Racah Memorial Volume*) (Amsterdam, 1968).

Conservation Laws in Inclusive Reactions (*).

G. Veneziano (**)

Laboratory for Nuclear Science and Department of Physics
Massachusetts Institute of Technology - Cambridge, Mass.

I shall discuss today some of the consequences of conservation of energy momentum (or of any other additive quantum number) on inclusive cross-sections. I remind first that an inclusive cross-section is, by definition, the cross-section for a process (Fig. 1)

$$a + b \to c_1 + c_2 + \dots c_k + X\,, \tag{1}$$

where X can be anything, while particles c_1 through c_k have well-defined quantum numbers and momenta [1].

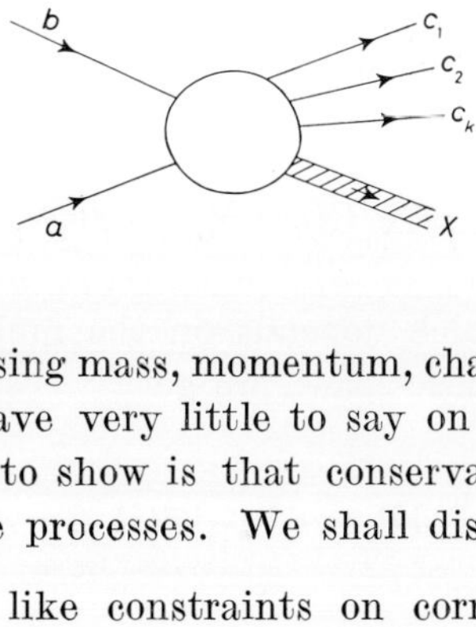

Fig. 1.

Since X can eat up missing mass, momentum, charge, etc. one would imagine that conservation laws have very little to say on inclusive reactions. Nevertheless, what I shall try to show is that conservation laws yield very non-trivial results in inclusive processes. We shall discuss two types of results:

a) *phenomenological*, like constraints on correlation functions, on the Feynman gas analog model, and on the question of exotics and early scaling;

b) *theoretical*, like determination of Regge parameters, connections with bootstrap programs and a new approach to unitarity.

(*) This work was supported in part through funds provided by the U.S. Atomic Energy Commission under grant AT (30-1) 2098.
(**) On leave of absence from the Weizmann Institute of Science, Rehovot.

We define $P = P_a + P_b$, while P_i will represent the 4-momentum of particle c_i. Also we shall denote the invariant phase-space volume element by

$$\mathrm{d}P_i \equiv \frac{\mathrm{d}^3\boldsymbol{P}_i}{2E_i} = \mathrm{d}^4P_i\,\delta(P_i^2-\mu^2)\theta(E_i)\,. \tag{2}$$

We shall denote by σ_{ab} the total cross-section

$$\sigma_{ab} = \sum_X \sigma(a+b\to X)\,, \tag{3}$$

and by $\mathrm{d}\sigma_{ab}/\mathrm{d}P_1 \dots \mathrm{d}P_k$ the k-particle inclusive cross-section

$$\frac{\mathrm{d}\sigma_{ab}}{\mathrm{d}P_1 \dots \mathrm{d}P_k} = \sum_X \sigma(a+b\to c_1+c_2+\dots c_k+X)\,. \tag{4}$$

Setting equal to one an incident flux factor which appears all the time on both sides of our final equations, we have, in a world with one type of stable particle,

$$\begin{cases} \sigma_{ab} = \sum\limits_{n=2}^{\infty}\dfrac{1}{n!}\displaystyle\int \mathrm{d}P_1 \dots \mathrm{d}P_n |T(ab\to n)|^2\delta^{(4)}\Big(P-\sum\limits_1^n P_i\Big)\,,\\[2ex] \dfrac{\mathrm{d}\sigma_{ab}}{\mathrm{d}P_1} = \sum\limits_{n=1}^{\infty}\dfrac{1}{n!}\displaystyle\int \mathrm{d}P_2 \dots \mathrm{d}P_{n+1} |T(ab\to c_1+n)|^2\delta^{(4)}\Big(P-\sum\limits_1^{n+1} P_i\Big)\,,\\ \vdots \\ \dfrac{\mathrm{d}\sigma_{ab}}{\mathrm{d}P_1 \dots \mathrm{d}P_k} = \sum\limits_{n=0}^{\infty}\dfrac{1}{n!}\displaystyle\int \mathrm{d}P_{k+1}\mathrm{d}P_{k+n} |T(ab\to c_1+\dots c_k+n)|^2\delta^{(4)}\Big(P-\sum\limits_1^{n+k} P_i\Big)\,, \end{cases} \tag{5}$$

where T is the usual T-matrix ($S_{fi} = \delta_{fi} + i\delta^{(4)}(P_i - P_f)T_{fi}$). Although they will not be our main concern, we will recall first some general sum rules relating cross-sections in a way which depends on the multiplicity. If we integrate $\mathrm{d}\sigma_{ab}/\mathrm{d}P_1$ over the whole phase-space, we get

$$\begin{aligned} \int \mathrm{d}P_1 \frac{\mathrm{d}\sigma}{\mathrm{d}P_1} &= \sum_{n=1}^{\infty}\frac{1}{n!}\int \mathrm{d}P_1\mathrm{d}P_2 \dots \mathrm{d}P_{n+1}|T(ab\to n+1)|^2\delta^{(4)}\Big(P-\sum_1^{n+1}P_i\Big) = \\ &= \sum_{m=2}^{\infty} m\frac{1}{m!}\int \mathrm{d}P_1 \dots \mathrm{d}P_m |T(ab\to m)|^2\delta^{(4)}\Big(P-\sum_1^m P_i\Big) = \sum_{m=2}^{\infty} m\sigma_m^{ab} \equiv \bar{n}\sigma_{ab}\,, \end{aligned} \tag{6}$$

where σ_m^{ab} is the cross-section σ ($ab\to m$ particles) and $\bar{n}$ (the average multiplicity) is defined by the last eq. in (6). Equation (6) gives

$$\bar{n} = \int \mathrm{d}P_1 \frac{1}{\sigma}\frac{\mathrm{d}\sigma}{\mathrm{d}P_1}\,. \tag{7}$$

If we now assume scaling, *i.e.*

$$\frac{1}{\sigma}\frac{\mathrm{d}\sigma}{\mathrm{d}P_1}\xrightarrow[s\to\infty]{} F(P_1^\perp, x = 2P_1^L/\sqrt{s}), \tag{8}$$

where $P_1^\perp$, P_1^L are the c.m. transverse and longitudinal momenta of particle c_1, we have

$$\bar{n}\xrightarrow[s\to\infty]{}\int \mathrm{d}P_1^\perp \frac{\mathrm{d}P_1^L}{2E_1}F(P_1^\perp, x) = \frac{1}{2}\int \mathrm{d}P_1^\perp \frac{\mathrm{d}x}{2E_1/\sqrt{s}}F(x, P_1^\perp), \tag{9}$$

where $x = 2P_1^L\sqrt{s}$. Definining $\bar{x} = 2E_1/\sqrt{s}$, one finds

$$\bar{x} = \sqrt{x^2 + \frac{4m_T^2}{s}}, \qquad m_T^2 = m^2 + P^{\perp 2}. \tag{10}$$

Therefore

$$\bar{n}\to\frac{1}{2}\int \mathrm{d}P_\perp \int_{-1}^{+1}\frac{\mathrm{d}x}{\sqrt{x^2 + 4m_T^2/s}}F(P_\perp, x)\xrightarrow[s\to\infty]{}\frac{1}{2}\log s\int \mathrm{d}P_\perp F(P^\perp, 0) \tag{11}$$

provided $F(P_\perp, 0)\neq 0$. This is the well-known result, first obtained by A.F.S. (*), that the average multiplicity $\bar{n}$ grows logarithmically with energy. Similarly one has

$$\int \mathrm{d}P_1\mathrm{d}P_2 \frac{\mathrm{d}\sigma}{\mathrm{d}P_1\mathrm{d}P_2} = \sum_n n(n-1)\sigma_n = \langle n(n-1)\rangle\sigma\,. \tag{12}$$

In general, knowledge of $\mathrm{d}\sigma/(\mathrm{d}P_1 \ldots \mathrm{d}P_k)$ gives information on the behavior of $\psi_n(s) = \sigma_n/\sigma$ as a function of n (distribution of multiplicities) as recently pointed out by MUELLER (2).

We now come to derive sum rules which do not depend upon $\bar{n}$ [3]. The simplest example is obtained by the following algebraic manipulations on the definitions, eq. (5):

$$P_\mu\sigma_{ab} = \sum_{n=2}^{\infty}\frac{1}{n!}\int \mathrm{d}P_1 \ldots \mathrm{d}P_n|T(ab\to n)|^2\delta^{(4)}\Big(P - \sum_1^n P_i\Big)\Big(\sum_1^n P_i\Big)_\mu. \tag{13}$$

Now, because of the symmetry in the P_i, we can replace $\sum_1^n P_i$ by nP_1 and get

$$P_\mu\sigma_{ab} = \int \mathrm{d}P_1 P_{1,\mu}\sum_{n=2}^{\infty}\frac{1}{(n-1)!}\int \mathrm{d}P_2 \ldots \mathrm{d}P_n|T|^2\delta^{(4)}(P - \sum P_i) = \\ = \int \mathrm{d}P_1 P_{1,\mu}\frac{\mathrm{d}\sigma_{ab}}{\mathrm{d}P_1}. \tag{14}$$

(*) AMATI, FUBINI and STANGHELLINI.

It is amusing to compare eqs. (6) and (14). Take $\mu = 0$ in (14) in the c.m.s. Then we have

$$E\sigma_{ab} = \int \mathrm{d}P_1 E_1 \frac{\mathrm{d}\sigma_{ab}}{\mathrm{d}P_1}, \tag{15}$$

this can be rewritten as

$$1 = \int \mathrm{d}P_1^{\perp} \frac{\mathrm{d}P_1^L}{2E} \frac{\mathrm{d}\sigma}{\sigma \mathrm{d}P_1} = \frac{1}{4}\int \mathrm{d}P_1^{\perp} \mathrm{d}x \frac{1}{\sigma}\frac{\mathrm{d}\sigma}{\mathrm{d}P_1}. \tag{16}$$

We seen now that if $(1/\sigma)(\mathrm{d}\sigma/\mathrm{d}P_1) \to F(x, P_\perp)$, we get no s-dependence on the r.h.s. of eq. (16) compared to the log s-dependence in eq. (7).

If there are many types of final particles, eq. (14) is simply modified into

$$P_\mu \sigma_{ab} = \sum_c \int \mathrm{d}P_c P_{c,\mu} \frac{\mathrm{d}\sigma_{ab}}{\mathrm{d}P_c}. \tag{17}$$

In exactly the same way charge conservation gives the sum rule

$$(Q_a + Q_b)\sigma_{ab} = \sum_c \int \mathrm{d}P_c Q_c \frac{\mathrm{d}\sigma_{ab}}{\mathrm{d}P_c}, \tag{18}$$

where Q_i is the charge of the i-th particle.

We now turn to the derivation of the most general sum rule (for a fixed initial state $a+b$). Define a vector $\hat{P} = (P_\mu, Q, Y, B, \dots)$, which has as many components as there are additive conserved quantum numbers. As we would do in statistical mechanics, let us introduce a «multiplier» for each component, $\hat{\beta} = (\beta_\mu, \beta_Q, \beta_Y, \beta_B, \dots)$ and consider the quantity

$$(\exp[-\hat{\beta}\cdot\hat{P}] - 1)\sigma_{ab} = \sum_{n=2}^{\infty} \frac{1}{n!} \int \mathrm{d}P_1 \dots \mathrm{d}P_n |T|^2 \delta^{(4)}\Big(\hat{P} - \sum_1^n \hat{P}_i\Big)\Big(\exp\Big[-\hat{\beta}\sum_1^n \hat{P}_i\Big] - 1\Big). \tag{19}$$

Defining $y_i = \exp[-\hat{\beta}\hat{P}_i]$, we notice the identities

$$\begin{cases} y_1 y_2 - 1 = (y_1 - 1)(y_2 - 1) + (y_1 - 1) + (y_2 - 1), \\ \vdots \\ y_1 y_2 \dots y_n - 1 = \sum_{i-1}^{n} (y_i - 1) + \sum_{i<j} (y_i - 1)(y_j - 1) + \\ \quad + \sum_{i<j<k} (y_i - 1)(y_i - 1)(y_k - 1) + \dots (y_1 - 1)(y_2 - 1)\dots(y_n - 1). \end{cases} \tag{20}$$

Using again the symmetry in the n-final momenta as well as the fact that the $\sum$ with k factors in eq. (20) has $n!/k!(n-k)!$ terms, we find

$$(21)\qquad (\exp[-\beta P]-1)\,\sigma_{ab}=\sum_{n=2}^{\infty}\sum_{k=1}^{n}\frac{1}{k!(n-k)!}\cdot$$
$$\cdot\int dP_1\dots dP_n|T|^2\delta(P-\sum P_i)(y_1-1)(y_2-1)\dots(y_k-1)=\sum_{k=1}^{\infty}\frac{1}{k!}\int dP_1\dots dP_k\cdot$$
$$\cdot\frac{d\sigma_{ab}}{dP_1\dots dP_k}(\exp[-\beta P_1]-1)(\exp[-\beta P_2]-1)\dots(\exp[-\beta P_k]-1)\,,$$

where the superscript ^ has been dropped for simplicity. This is our main result. Before going into applications of eq. (21) we check it in a few cases:

1) $\beta\to\infty$ gives

$$(22)\qquad -\sigma_{ab}=\sum_{k=1}^{\infty}\frac{1}{k!}\int dP_1\dots dP_k\frac{d\sigma_{ab}}{dP_1\dots dP_k}=$$
$$=\sum_{k=1}^{\infty}\frac{1}{k!}\sum_{n=k}^{\infty}n(n-1)\dots(n-k+1)\sigma_n^{ab}=\sum_{n=2}^{\infty}\sigma_n\sum_{k=1}^{n}(-1)^k\binom{n}{k}=-\sum_{n=2}^{\infty}\sigma_n^{ab}\,.$$

2) $\beta=0$ gives $0=0$. The first order in β gives

$$(23)\qquad -\hat{P}\sigma_{ab}=-\int dP_1\,\hat{P}_1\,d\sigma/dP_1\,,$$

which is just eq. (14).

We now turn to some applications of our result:

a) Predictions on correlation functions

Taking the 2*nd* order in $\hat{\beta}$ of eq. (21) one gets

$$(24)\qquad \hat{P}_\mu\hat{P}_\nu\sigma_{ab}=\int dP_1\hat{P}_{1,\mu}\hat{P}_{1,\nu}\frac{d\sigma_{ab}}{dP_1}+\int dP_1 d\hat{P}_2\hat{P}_{1,\mu}\hat{P}_{2,\nu}\frac{d\sigma_{ab}}{dP_1dP_2}\,,$$

which we want to use together with eq. (23). We now define the 2-particle correlation function $\varrho^{(2)}(P_1, P_2)$ by

$$(25)\qquad \frac{1}{\sigma_{ab}}\frac{d\sigma_{ab}}{dP_1dP_2}=\frac{1}{\sigma_{ab}}\frac{d\sigma_{ab}}{dP_1}\cdot\frac{1}{\sigma_{ab}}\frac{d\sigma_{ab}}{dP_2}+\varrho^{(2)}(P_1,P_2)\,.$$

Dividing both sides of eq. (24) by σ_{ab}, using eqs. (25) and (23), one gets

$$(26)\qquad 0=\int dP_1\hat{P}_{1,\mu}\hat{P}_{1,\nu}\frac{1}{\sigma}\frac{d\sigma}{dP_1}+\int dP_1dP_2\hat{P}_{1,\mu}\hat{P}_{2,\nu}\,\varrho^{(2)}(P_1,P_2)\,.$$

Equation (26) connects the 2-particle correlation and the single-particle spectrum. It can be used to prove that $\varrho^{(2)}$ has to be different from zero. Indeed, taking $\mu=\nu$ in eq. (26)

$$0=\int dP(\hat{P}_\mu)^2\frac{1}{\sigma}\frac{d\sigma}{dP}+\int dP_1 dP_2 P_{1,\mu}P_{2,\mu}\varrho^{(2)}(P_1,P_2)\,. \tag{27}$$

Since the first term in the r.h.s. of eq. (27) is positive definite the equation is inconsistent with $\varrho^{(2)}=0$. In order to gain further insight we take in eq. (27) $\mu=0$ first and $\mu=3$ second. The two resulting equations are

$$\begin{cases}\mu=0\Rightarrow\dfrac{1}{4}\displaystyle\int dP_1^\perp dP_2^\perp\int dP_1^L dP_2^L\varrho^{(2)}=-\int dP_1 E_1^2\frac{1}{\sigma}\frac{d\sigma}{dP_1}\,,\\[2ex]\mu=3\Rightarrow\dfrac{1}{4}\displaystyle\int dP_1^\perp dP_2^\perp\int dP_1^L dP_2^L\frac{P_1^L P_2^L}{E_1E_2}\varrho^{(2)}=-\int dP_1 P_1^{L^2}\frac{1}{\sigma}\frac{d\sigma}{dP_1}\,.\end{cases} \tag{28}$$

Under the assumption of scaling the r.h.s. of both equations are the same while $(P_1^L P_2^L)/(E_1E_2)\simeq\pm1$ according to whether P_1 and P_2 are parallel or antiparallel. One then gets

$$\int\limits_{P_1^L\cdot P_2^L>0} dP_1^L dP_2^L\int dP_1^\perp dP_2^\perp\varrho^{(2)}=-4\int dP_1 E_1^2\frac{1}{\sigma}\frac{d\sigma}{dP_1}<0\,, \tag{29}$$

while

$$\int\limits_{P_3^L\cdot P_2^L<0} dP_1^L dP_2^L\int dP_1^\perp dP_2^\perp\varrho^{(2)}=0\,. \tag{30}$$

Equations (29), (30) are consistent with the results recently obtained in various models like the multi-Regge model [4] or the dual-resonance model [5].

It is interesting to write down eq. (24) for $\mu=\nu=0$ in terms of the variable $\bar{x}$ of eq. (10). One has

$$1=\int dP(\bar{x}/2)^2\frac{d\sigma}{\sigma dP}+\int dP_1 dP_2\frac{\bar{x}_1\bar{x}_2}{4}\frac{d\sigma}{\sigma dP_1 dP_2}\,. \tag{31}$$

If there is no correlation, eq. (31) is inconsistent, *i.e.*

$$1=k^2+1\,. \tag{32}$$

It is interesting to compare this with the elastic scattering case where eq. (31)

reads (there is a δ-function–type correlation and $\bar{x}=1$)

$$1=\tfrac{1}{2}+\tfrac{1}{2}\,. \tag{33}$$

Probably high-energy physics is somehow in between eqs. (32) and (33). But where?

Next we consider the case $\mu=1,2$ in eq. (26) and we sum over μ. One gets

$$0=\int \mathrm{d}P_1(P_1^\perp)^2\frac{\mathrm{d}\sigma}{\sigma\,\mathrm{d}P_1}+\int \mathrm{d}P_1\mathrm{d}P_2 P_1^\perp\cdot P_2^\perp\,\varrho^{(2)}(P_1,P_2)\,. \tag{34}$$

As noticed already by Bassetto and Toller [6], eq. (34) demands a dependence of $\mathrm{d}\sigma/\mathrm{d}P_1\mathrm{d}P_2$ on φ, the angle between $P_1^\perp$ and $P_2^\perp$, in the pionization region ($x\sim 0$). As a last example take $\mu=\nu=Q$. One has then

$$0=\sum_c\int \mathrm{d}P_c Q_c^2\frac{1}{\sigma}\frac{\mathrm{d}\sigma}{\mathrm{d}P_c}+\sum_{c,d}\int \mathrm{d}P_c\mathrm{d}P_d Q_c Q_d\frac{\mathrm{d}\sigma}{\sigma\,\mathrm{d}P_c\mathrm{d}P_d}\,. \tag{35}$$

Correlations are again necessary.

We now rewrite our sum rule in terms of correlation functions directly. In order to do that, one performs a cluster decomposition of $(1/\sigma)(\mathrm{d}\sigma/\mathrm{d}P_1\ldots\mathrm{d}P_k)$, *i.e.*

$$\begin{aligned}\frac{1}{\sigma}\frac{\mathrm{d}\sigma}{\mathrm{d}P_1\ldots\mathrm{d}P_k}&=\frac{1}{\sigma}\frac{\mathrm{d}\sigma}{\mathrm{d}P_1}\cdot\frac{1}{\sigma}\frac{\mathrm{d}\sigma}{\mathrm{d}P_2}\cdots\frac{1}{\sigma}\frac{\mathrm{d}\sigma}{\mathrm{d}P_k}+\varrho^{(2)}(P_1,P_2)\cdot\frac{1}{\sigma}\frac{\mathrm{d}\sigma}{\mathrm{d}P_3}\cdots\frac{1}{\sigma}\frac{\mathrm{d}\sigma}{\mathrm{d}P_k}+\\&+(\text{permutations})+\varrho^{(3)}(P_1,P_2,P_3)\frac{\mathrm{d}\sigma}{\mathrm{d}P_4}\cdots+\varrho^{(k)}(P_1,P_2\ldots P_k)\,.\end{aligned} \tag{36}$$

Inserting eq. (36) in the sum rule eq. (21) one finds that, as in a similar calculation in statistical mechanics [7], all the combinatorial algebra works out to give simply [8]

$$-P\cdot\beta=\sum_{i=1}^{\infty}\frac{1}{i!}\tilde{\varrho}^{(i)}(\beta)\,, \tag{37}$$

where $\varrho^{(1)}(P)\equiv(1/\sigma)(\mathrm{d}\sigma/\mathrm{d}P)$ and

$$\tilde{\varrho}^{(i)}(\beta)=\int \mathrm{d}P_1\ldots\mathrm{d}P_i\,\varrho^{(i)}(P_1\ldots P_i)(\exp[-\beta P_1]-1)\ldots(\exp[-\beta P_i]-1)\,. \tag{38}$$

One can see very easily from eq. (37) that it is inconsistent to have $\varrho^{(i)}=0$ for $i>1$.

Our result on the correlation functions can be of interest for the study of the distribution of multiplicity $\psi_n(s)=\sigma_n/\sigma$ as a function of n. Indeed,

MUELLER [2] has shown that all the information on $\psi_n(s)$ is contained in a kind of generating function

$$F(h, s) = \sum_k \frac{1}{k!} \int \mathrm{d}P_1 \ldots \mathrm{d}P_k \frac{\mathrm{d}\sigma}{\mathrm{d}P_1 \ldots \mathrm{d}P_k} h^k . \tag{39}$$

Again, by using eq. (36) one finds that, in terms of correlation functions, $F(h, s)$ can be expressed as

$$F(h, s) = \exp\left[\sum_{i=1}^{\infty} \frac{h^i}{i!} \int \mathrm{d}P_1 \ldots \mathrm{d}P_i \varrho^{(i)}(P_1 \ldots P_i)\right] - 1 . \tag{40}$$

If all $\varrho^{(i)} = 0$ for $i \geqslant 2$, one would get

$$F(h, s) = \exp[h\bar{n}] - 1 , \tag{41}$$

which can be seen to generate a Poisson distribution

$$\psi_n(s) = \frac{(\bar{n})^n}{n!} \exp[-\bar{n}] . \tag{42}$$

Therefore our result on $\varrho^{(2)}$ seems to imply (in accord with ref. [2]) that an exact Poisson distribution is not to be expected.

b) Scaling.

As a second application we want to show that the sum rules are not only consistent with scaling, but can even suggest which variables could scale. Since β is arbitrary, we choose $\beta_\mu = \beta P_\mu / P^2$. Then:

$$\begin{cases} \exp[-\beta_\mu P_\mu] \to \exp[-\beta] , \\ \exp[-\beta_\mu P_{i,\mu}] \to \exp[-\beta P_i P / P^2] = \exp[-\beta\omega_i/2] , \end{cases} \tag{43}$$

where we have defined in analogy with the famous ω-variable of deep inelastic electron scattering

$$\omega_i = \frac{2P_i \cdot P}{P^2} . \tag{44}$$

Sum rule (21) can then be brought into the form

$$(\exp[-\beta] - 1) = \tag{45}$$

$$= \sum_k \frac{1}{k!} \int \mathrm{d}P_1 \ldots \mathrm{d}P_k \frac{1}{\sigma} \frac{\mathrm{d}\sigma}{\mathrm{d}P_1 \ldots \mathrm{d}P_k} \left(\exp\left[-\frac{\beta\omega_1}{2}\right] - 1\right) \ldots \left(\exp\left[-\frac{\beta\omega_k}{2}\right] - 1\right) .$$

The l. h. s. of eq. (45) is independent of $s = P^2$. So is the r. h. s. if $(1/\sigma)(\mathrm{d}\sigma/\mathrm{d}P_1 \dots \mathrm{d}P_k)$ depends on $P_i^\perp$ and ω_i. As a practical application consider $e^+e^- \to P_1 + X$. Equation (45) suggests then that

$$\frac{1}{\sigma^{\text{tot}}_{e^+e^-}} \frac{\mathrm{d}\sigma_{e^+e^-}}{\mathrm{d}P_1} \xrightarrow[s\to\infty]{} (\text{angular factors}) \times f(\omega = 2P_1 \cdot P/P^2)\,. \tag{46}$$

If we say that $\sigma^{\text{tot}}_{e^+e^-} \to 1/s$, one gets scaling for νW_2 in $e^+e^- \to P + X$.

c) Exotics and early scaling.

We now combine sum rules and duality concepts in order to obtain information on which inclusive cross-sections should be expected to exhibit scaling at subasymptotic energies. We assume the usual connection of HARARI and FREUND [9] between background and pomeron which yields for the total cross-section σ_{ab}:

$$\sigma_{ab} = \sigma^R_{ab} + \sigma^B_{ab}\,, \tag{47}$$

where σ^B_{ab} is the diffractive (background) cross-section ($\sigma^B_{ab} \to \text{const}$) with vacuum quantum numbers in the t-channel, and σ^R_{ab} is the resonant cross-section $\sigma^R_{ab} \to s^{\alpha_0 - 1} \sim s^{-\frac{1}{2}}$.

If ab is exotic $\sigma^R_{ab} = 0$ and therefore we have a fast approach to constant cross-sections. In the naive unitarization approach which I have recently discussed [10] the two components of the total cross-section are given (if one neglects absorption) by the two dual graphs of Fig. 2. A similar analysis has

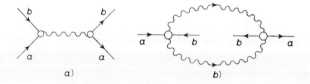

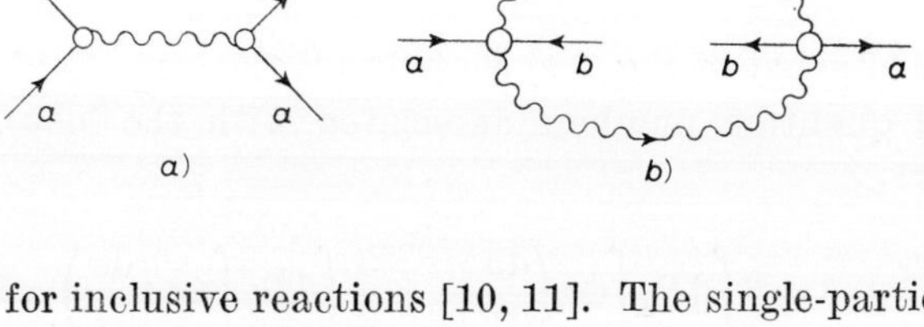

Fig. 2.

been performed for inclusive reactions [10, 11]. The single-particle distribution can also be written (still neglecting absorption and also diffractive dissociation) as

$$\frac{\mathrm{d}\sigma_{ab}}{\mathrm{d}P_c} = \left(\frac{\mathrm{d}\sigma_{ab}}{\mathrm{d}P_c}\right)_R + \left(\frac{\mathrm{d}\sigma_{ab}}{\mathrm{d}P_c}\right)_B\,, \tag{48}$$

where the subscripts R and B refer to the presence of resonances or background in the ab channel.

One can see clearly that the sum rules must hold separately for the resonant and background-type contributions. As an example of the conditions one can get from the sum rules consider the resonant part only in a world of pions for $ab = \pi^+\pi^0$. Then energy and charge conservation give, respectively

$$(49)\qquad 4\sigma^R_{\pi^+\pi^0} = \sum_i \int \mathrm{d}P^{\perp}_{\pi_i} \int \mathrm{d}x_i \left(\frac{\mathrm{d}\sigma_{\pi^+\pi^0}}{\mathrm{d}P_{\pi_i}}\right)_R ,$$

$$(50)\qquad \sigma^R_{\pi^+\pi^0} = \int \mathrm{d}P_{\pi^+} \left(\frac{\mathrm{d}\sigma_{\pi^+\pi^0}}{\mathrm{d}P_{\pi^+}}\right)_R - \int \mathrm{d}P_{\pi^-} \left(\frac{\mathrm{d}\sigma_{\pi^+\pi^0}}{\mathrm{d}P_{\pi^-}}\right)_R$$

is given by Fig. 2 *a*). There are three contributions to $\mathrm{d}\sigma(\pi^+\pi^0)/\mathrm{d}P_{\pi^i}$ which are indicated in Fig. 3. Those of Figs. 3 *a*), 3 *b*) contribute only to π^0 and π^+ produc-

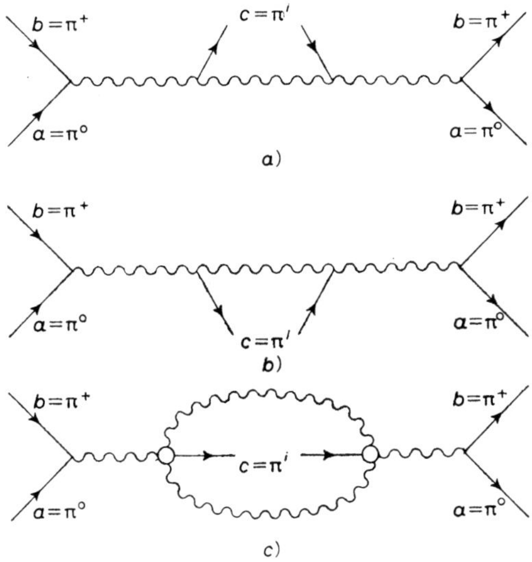

Fig. 3.

tion in a ratio 1 : 2; that of Fig. 3 *c*) contributes the same to π^0, π^+, π^- (because of the internal quantum numbers associated with the loop). The two sum rules then give

$$(51)\qquad 4\sigma^R = 3\int \mathrm{d}P_\perp \int \mathrm{d}x \left(\frac{\mathrm{d}\sigma}{\mathrm{d}P}\right)_2 + \frac{3}{2}\int \mathrm{d}P_\perp \int \mathrm{d}x \left(\frac{\mathrm{d}\sigma}{\mathrm{d}P}\right)_1 ,$$

$$(52)\qquad \sigma^R = \int \mathrm{d}P \left(\frac{\mathrm{d}\sigma}{\mathrm{d}P}\right)_1 ,$$

where $(\mathrm{d}\sigma/\mathrm{d}P)_{1,2}$ are the contributions of Figs. 3 *a*), *b*) and Fig. 3 *c*), respectively. Eliminating $(\mathrm{d}\sigma/\mathrm{d}P)_{2,2}$, one gets

$$(53)\qquad \int \mathrm{d}P_\perp \mathrm{d}x \frac{\mathrm{d}\sigma_{\pi^+\pi^0}}{\mathrm{d}P}\begin{pmatrix}\pi^+\\ \pi^0\\ \pi^-\end{pmatrix} = \frac{2}{3}\sigma^R_{\pi^+\pi^0}\left(1 + \begin{matrix}\frac{3}{4}\langle\bar{x}\rangle\\ 0\\ -\frac{3}{4}\langle\bar{x}\rangle\end{matrix}\right),$$

where

$$\langle \bar{x} \rangle = \int \mathrm{d}P_\perp \int \mathrm{d}x \bar{x} \left(\frac{\mathrm{d}\sigma}{\mathrm{d}P}\right)_1 \Big/ \int \mathrm{d}P_\perp \mathrm{d}x \left(\frac{\mathrm{d}\sigma}{\mathrm{d}P}\right)_1, \qquad 0 < \langle \bar{x} \rangle < 1 . \tag{54}$$

This seems to teach us that even production of π^- (in which case $ab\bar{c}$ is exotic) has some nonscaling piece (remember $\sigma^R_{\pi^+\pi^0} \sim s^{-\frac{1}{2}}$) although this is quite smaller than for π^0 and π^+.

A more complete analysis of the consistency between duality and conservation sum rules is certainly needed (*). However, it is almost certain that this will shed some light on the controversy in the problem of subasymptotic scaling [12].

We now pass to more theoretical applications of the sum rules. These come when we combine the sum rules with optical-type theorems such as Mueller's analysis [13] of $\mathrm{d}\sigma/\mathrm{d}P$.

d) *Bounds on cross-sections and coupling constants.*

From eq. (16) one has

$$\int \mathrm{d}P_\perp \int_k^1 \mathrm{d}x \frac{\mathrm{d}\sigma}{\mathrm{d}P} < 2\sigma_{\text{tot}} < \text{const} (\log s)^2 , \tag{55}$$

where the last inequality is, of course, the Froissart bound. Also by taking $1-k$ small one knows that $\mathrm{d}\sigma/\mathrm{d}P$ can be approximated by the triple-Regge formula [14]:

$$\frac{1}{\sigma} \frac{\mathrm{d}\sigma}{\mathrm{d}P} \xrightarrow[x\sim 1]{} g_{TR}(t)(1-x)^{-2\alpha_t + \alpha_0^V} \xrightarrow[t\to 0]{} g_{TR}(0)(1-x)^{-2\alpha_0 + \alpha_0^V - 2\alpha' t} , \tag{56}$$

where α_t is the trajectory of the $b\bar{c}$ channel and α^V that of the $a\bar{a}$ channel. If we are considering the case of the triple-pomeron vertex, $\alpha = a^V$ and then (because of eq. (55))

$$\int \mathrm{d}P_\perp \int_k^1 \mathrm{d}x g_{TR}(0)(1-x)^{-\alpha_0} \sim \frac{g_{TR}(0)}{-\log(1-\alpha_0)} < \text{const} . \tag{57}$$

As $\alpha_0 \to 1$ one must have [15] $g^{TR}(0) \to 0$. Amazingly the dual-resonance model does possess this zero automatically [16].

(*) *Note added in Proofs.* – Such more detailed analysis (S. H. H. TYE and G. VENEZIANO: *Phys. Lett.*, **38** B, 30 (1972)) confirms these results although eq. (53) needs a slight modification.

e) Sum rules and unitarity.

We now turn to a very general question, that is a possible approach to unitarity from the point of view of the sum rules.

We first generalize our results to nonforward matrix elements of TT^+ (*i.e.* quantities which are not connected to cross-sections). The most comprehensive generalization of eq. (14) that can be proven along the same line is the following:

$$(58)\qquad (P_\alpha - P_\gamma)_\mu \sum_n \langle\alpha|T|\gamma+n\rangle\langle n+\beta|T^+|\delta\rangle = \\ = \int \mathrm{d}P_1\, P_{1,\mu} \sum_n \langle\alpha|T|\gamma+n+1\rangle\langle n+1+\beta|T^+|\delta\rangle\,,$$

where α, β, γ, δ represent any state of free particles. We now invoke unitarity (in the spirit of Mueller's observation) to state that, in general [17]

$$(59)\qquad \operatorname*{disc}_{(P_\alpha-P_\gamma)^2} \langle\alpha+\beta|T|\gamma+\delta\rangle = i\sum_n \langle\alpha|T|\gamma+n\rangle\langle n+\beta|T^+|\delta\rangle\,.$$

Then eq. (58) can be rewritten in terms of discontinuities

$$(60)\qquad (P_\alpha - P_\gamma)_\mu \operatorname*{disc}_{(P_\alpha-P_\gamma)^2} T(\alpha+\beta\to\gamma+\delta) = \int \mathrm{d}P_1\, P_{1,\mu} \operatorname*{disc}_{(P_\alpha-P_\gamma-P_1)^2} T(\alpha+\beta+1\to\gamma\delta+1)\,.$$

To summarize, unitarity (eq. 59) plus conservations laws (eq. 58) allow us to write down sum rules, for discontinuities (eq. 60). Is it possible to reverse the argument and derive unitarity from the sum rules written in terms of discontinuities?

A positive answer to this question appears hard to believe because eq. (60) is linear in T, while unitarity is not. However, amazingly, the answer is almost yes [18]. All we have to add to eq. (60) in order to derive eq. (59) is the assumption of isolated factorized pole-type singularities, namely the statement that

$$(61)\qquad \operatorname*{disc}_{0<(P_\alpha-P_\gamma)^2<4\mu^2} T(\alpha+\beta\to\gamma+\delta) \\ = i\delta((P_\alpha-P_\gamma)^2-\mu^2)\,T(\alpha\to\gamma+1)T^+(1+\beta\to\delta)\,.$$

Then from eqs. (60) and (61) it is possible to derive eq. (59) by induction. For the explicit proof see ref. [18].

The most apparent advantage in replacing eq. (59) with eqs. (60), (61), is, as we said, the linear character of (60). It seems that we have been able to shift all the nonlinearity of (59) in the pole-term assumption, which however is not so hard to satisfy (the dual model is an example in which eq. (61) is true). Equation (60) could then provide a new (and hopefully simpler) way of carrying out bootstrap type programs.

REFERENCES

[1] A more complete discussion can be found in M. L. GOLDBERGER: This Volume p. 1.
[2] A. H. MUELLER: *Phys. Rev. D*, **4**, 150 (1971).
[3] T. T. CHOU and C. N. YANG: *Phys. Rev. Lett.*, **25**, 1072 (1971); C. E. DETAR, D. Z. FREEDMAN and G. VENEZIANO: *Phys. Rev. D*, **4**, 906 (1971): F. PREDAZZI and G. VENEZIANO: *Lett. Nuovo Cim.*, **2**, 749 (1971).
[4] D. Z. FREEDMAN, C. E. JONES, F. E. LOW and J. E. YOUNG: *Phys. Rev. Lett.*, **26**, 1197 (1971).
[5] See, *e.g.*, B. HASSLACHER, C. S. HSUE and D. K. SINCLAIR: SUNY preprint (1971); C. JEN, K. KANG, P. SHEN and C. I. TAN: Brown University preprint (1971).
[6] A. BASSETTO and M. TOLLER: CERN preprint TH 1337 (1971).
[7] B. KAHN and G. E. UHLENBECK: *Physica*, **5**, 399 (1938); see also, K. HUANG: *Statistical Mechanics* (New York, 1963), p. 303.
[8] This is the analog of the virial expansion in ref. [7].
[9] H. HARARI: *Phys. Rev. Lett.*, **20**, 1396 (1968); P. G. O. FREUND: *Phys. Rev. Lett.*, **20**, 235 (1968).
[10] G. VENEZIANO: *Proceedings of the International Conference on Duality and Symmetry in Hadron Physics* (Tel Aviv, 1971).
[11] G. VENEZIANO: *Lett. Nuovo Cimento*, **1**, 681 (1971); R. C. BROWER and R. E. WALTZ: CERN preprint TH 1335 (1971).
[12] H. M. CHAN, C. W. HSUE, C. QUIGG and J. M. WANG: *Phys. Rev. Lett.*, **26**, 672 (1971); J. ELLIS, J. FINKELSTEIN, P. H. FRAMPTON and M. JACOB: *Phys. Lett.*, B **35**, 227 (1971).
[13] A. H. MUELLER: *Phys. Rev. D*, **2**, 296 (1971).
[14] See, *e.g.*, C. E. DETAR, C. E. JONES, F. E. LOW, J. H. WEIS, J. E. YOUNG and CHUNG-I TAN: *Phys. Rev. Lett.*, **26**, 675 (1971).
[15] This result was first derived by H. D. ABARBANEL, G. F. CHEW, M. L. GOLDBERGER and L. M. SAUNDERS: *Phys. Rev. Lett.*, **26**, 537 (1971).
[16] D. GORDON and G. VENEZIANO: *Phys. Rev. D*, **3**, 2116 (1971); M. A. VIRASORO: *Phys. Rev. D*, **3**, 2843 (1971); C. E. DETAR, K. HUANG, C. I. TAN and J. H. WEIS: *Phys. Rev. D*, **4**, 425 (1971). More recent studies have been done by C. E. DETAR and J. H. WEIS: MIT preprint CTP No. 218 (1971); SHAU-J. CHANG, D. GORDON, F. E. LOW and S. B. TREIMAN: *Phys. Rev. D*, **4**, 3055 (1971).
[17] For a justification of eq. (59) see H. P. STAPP: *Phys. Rev. D*, **3**, 3177 (1971); C. I. TAN: *Phys. Rev. D*, **4**, 2412 (1971).
[18] G. VENEZIANO: *Phys. Lett.*, **36** B, 397 (1971).

High-Energy Collisions of Hadrons.

L. Van Hove
CERN - Geneva

1. – Introduction.

The present text summarizes a lecture series given at the 1971 E. Fermi International School of Physics at Varenna, and contains a number of references and figures illustrating the main theme of the lectures. The aim is to show by examples that high-energy collision data contain a number of interesting features which have hardly been exploited until now, partly because they do not directly relate to predictions or tests of current theoretical models, partly because they are only uncovered by making the unfamiliar effort of analysing the data in multidimensional phase space.

The existence of these unexploited features of high-energy hadron collisions is welcome and should be of interest to theorists. Indeed, the current theoretical models have strayed very far from the ideals of conceptual simplicity and genuine power which satisfactory models should fulfil. In this situation, a careful examination of new features in the experimental data may be the best way for theorists to get fresh inspiration.

It is in collisions with three or more (meta)stable particles in the final state that new aspects are still to be found for $p_{lab} \leqslant 30$ GeV/c at the level of reasonably large cross-sections. Indeed, for $p_{lab} \leqslant 30$ GeV/c, we are already familiar with the main properties of collisions which have only two (meta)stable particles in the final state, and there is no indication of profound changes as one goes to Serpukhov or ISR energies (we do not regard the slight increase of the K^+p total cross-section as a particularly significant development, and indeed most current models adjust without difficulty to slowly increasing cross-sections).

The features illustrated below are essentially of two types:

i) For general collisions of about average multiplicity, in particular in inclusive experiments, single-particle distributions and multiparticle correlations have a number of detailed structure properties which may give us useful hints toward systematic descriptions of main production mechanisms.

ii) In few-body collisions (multiplicity 3, 4, 5) the distribution in complete phase-space separates very clearly at high energy into parts belonging to different collision types with different energy dependences (diffraction dissociation, charged meson exchange, baryon exchange, etc.). This provides the basis for an empirical classification of such collisions, which may possibly be generalized to all common collisions at sufficiently high energy.

The text is only a summary of the lectures. A detailed introduction is found in a recent review article on the same subject [1]. Considerable data are available in the review papers given by W. KITTEL [2] and the present author [3] at the Colloquium on Multiparticle Dynamics (Helsinki, 25-28 May 1971), as well as in the rapporteur paper of DEUTSCHMANN at the International Conference on High-Energy Physics (Amsterdam, 30 June - 6 July, 1971).

2. – Single-particle distributions.

The old question of single-particle distributions attracted recently a new wave of interest (under the name of inclusive experiments), firstly because simple asymptotic conjectures were formulated (scaling, limiting fragmentation), secondly because MUELLER related it to a «3 into 3 particle» forward amplitude to which Regge and dual models can be applied [4]. While these theoretical considerations concern mainly the energy dependence of the distributions, it is good for theorists also to get familiar with their detailed shapes which are far from trivial.

We discuss the shape properties for the following pion production experiments:

$$\mathrm{pp} \to \pi^- + \text{anything}, \tag{1}$$

$$\pi^-\mathrm{p} \to \pi^\pm + \text{anything}, \tag{2}$$

for which detailed studies were presented at the Helsinki Colloquium, for (1) at 19 GeV/c from the Scandinavian Collaboration [5] and for (2) at 16 GeV/c from the Aachen-Berlin-Bonn-CERN-Cracow-Heidelberg-Warsaw Collaboration [6]. We begin by recalling which variables seem appropriate for the analysis of single-particle distributions.

The differential production cross-section can be written

$$\mathrm{d}^3\sigma/\mathrm{d}^3\boldsymbol{p} = f/E\,, \qquad E = \sqrt{m_\pi^2 + p^2}\,, \tag{3}$$

where $\boldsymbol{p}$ and E are the c.m. momentum and energy of the pion produced, and f is the Lorentz-invariant distribution function depending on $\boldsymbol{p}$ and on the incident energy. The dependence on $\boldsymbol{p}$ is very different in the longitudinal and transverse directions, so that it is useful to separate $\boldsymbol{p}$ into its longitudinal part p_L (which we shall also denote by q) and its 2-dimensional transverse part $\boldsymbol{p}_T$ (which we shall also denote by $\boldsymbol{r}$). With unpolarized beams and targets, we have

$$f = f(q, r)\,, \qquad r = |\boldsymbol{r}|\,, \tag{4}$$

and one may integrate (3) over the azimuthal angle obtaining

$$\mathrm{d}^2\sigma/\mathrm{d}q\,\mathrm{d}r^2 = \pi f/E\,. \tag{5}$$

The transverse momenta being small at all energies, one usually plots transverse momentum distributions in terms of r or r^2. The former choice has the disadvantage that equal bins in r get very low statistics at small r. This difficulty is avoided by plotting in terms of r^2. Experimentally, however, it is found that f has a sharp peak at $r^2 = 0$ when plotted in r^2 (Fig. 1), and it

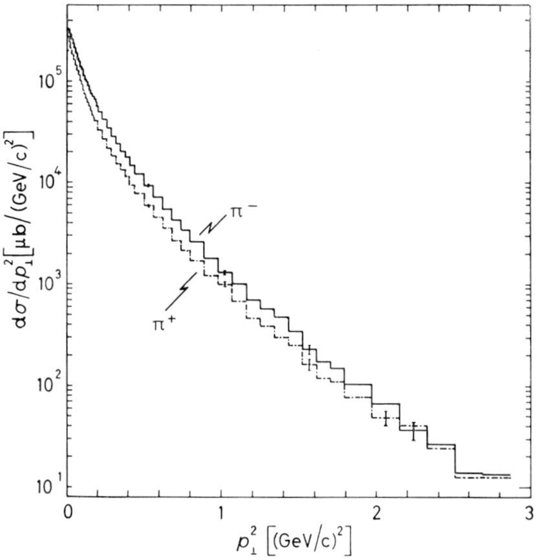

Fig. 1. – p_T^2-distribution for π^+ and π^- produced in π^-p collisions at 16 GeV/c, from ref. [6]: 102 779 π^+-particles, 140 770 π^--particles.

may therefore also be useful to examine plots of f *vs.* r. The more traditional plots of rf in terms of r are less informative concerning the region $r \simeq 0$ where f is maximum.

Regarding the longitudinal motion, in addition to the momentum variable $q = p_L$, we shall see that it is also desirable to consider the *longitudinal rapidity* y

defined by

(6) $$\sinh y = \frac{q}{(m^2 + r^2)^{\frac{1}{2}}} \qquad \text{or} \qquad y = \ln \frac{q+E}{(m^2+r^2)^{\frac{1}{2}}}\,.$$

This is a dimensionless velocity parameter which has the advantage of transforming additively ($y \to y +$ constant) under longitudinal Lorentz transformations. Since experimental distributions in q and y are found to extend over most of the kinematically accessible range of these variables, it is often also useful to consider reduced variables ranging over a fixed interval at all energies. One can replace q by the *reduced longitudinal momentum*

(7) $$x = q/p_{\text{c.m.}}\,, \qquad p_{\text{c.m.}} = \text{incident c.m. momentum}\,.$$

Similarly, one can use a *reduced longitudinal rapidity*

(8) $$\xi = (y - y_B)/(y_A - y_B)\,,$$

(9) $$y_A = \sinh^{-1}(p_{\text{c.m.}}/m_A)\,, \qquad y_B = -\sinh^{-1}(p_{\text{c.m.}}/m_B)\,,$$

A, B being the incident particles moving forward and backward respectively in the c.m. system.

It does not always seem to be realized that the variables q and y (or their reduced forms x and ξ) become complementary for $p_{\text{c.m.}} \to \infty$ and that it is therefore instructive to plot high-energy data in both of them. This fact is explained in detail in ref. [1]. We here recall briefly that it results from the following properties. Consider production of a particle C of finite transverse momentum in the limiting case $p_{\text{c.m.}} \to \infty$. The following asymptotic cases can be distinguished:

i) x tends to a positive value < 1. Then $\xi \to 1$, the momentum $\boldsymbol{P}_A$ of C in the rest system of the forward incident particle A tends to a finite limit and its momentum $\boldsymbol{P}_B$ in the rest system of the backward incident particle B tends to infinity. C can then be called a *forward fragment* or fragment of A.

ii) x tends to a negative value > -1. Then $\xi \to 0$, $\boldsymbol{P}_A \to \infty$, $\boldsymbol{P}_B$ tends to a finite limit, and C can be called a *backward fragment* or fragment of B.

iii) x tends to zero. Then both $\boldsymbol{P}_A$ and $\boldsymbol{P}_B \to \infty$, whereas ξ can tend to any value in the closed interval $(0, 1)$. Conversely, if ξ tends to any value ξ_0 in $0 < \xi_0 < 1$, one has $x \to 0$, $\boldsymbol{P}_A$ and $\boldsymbol{P}_B \to \infty$. For such values of ξ, C can be called an *intermediate* or *central particle*.

It must be noted that in cases i) and ii) the convergence of ξ to 0 or 1 is only logarithmic and in practice very slow indeed. The data reveal that pions are produced with practically all accessible x-values ($-1 < x < 1$) with a concentration at $x \simeq 0$. One should therefore use x or $p_L = q$ which are differential for the fragments ($|x| > 0$) but lump the many intermediate particles together, as well as ξ or y which are differential for the latter but lump

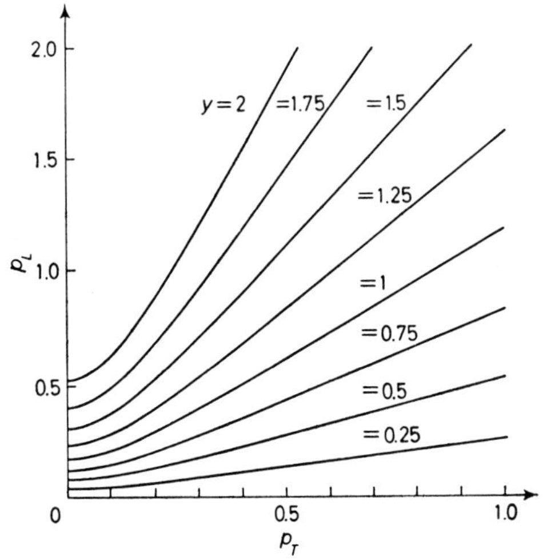

Fig. 2. – Lines of constant longitudinal rapidity y in the p_T-, p_L-plane for pions.

fragments together. A disadvantage of the rapidity for pions which are fragments is that y varies too rapidly with p_T when $|p_L| \gtrsim \langle p_T \rangle$. This is seen in Fig. 2 and is due to the fact that $m_\pi \lesssim 0.5 \langle p_T \rangle$. No such strong variation is found for protons (Fig. 3). It would of course be nice to find a single longitudinal variable combining all good properties of x and ξ, but none is known.

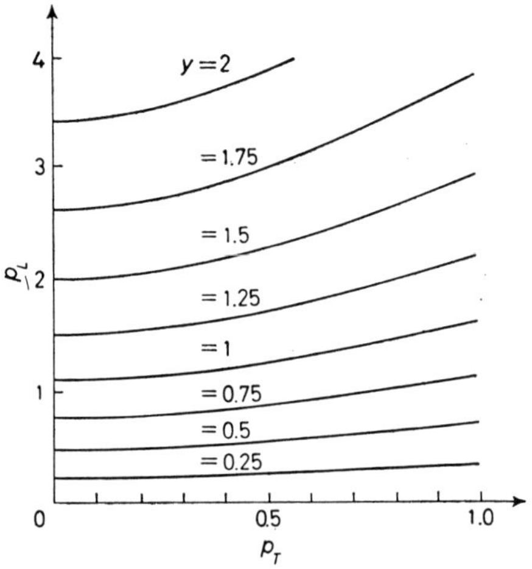

Fig. 3. – Lines of constant longitudinal rapidity y in the p_T-, p_L-plane for protons.

The π^- distribution in 19 GeV/c pp collisions has been carefully analysed by BØGGILD *et al.* [7]. The distribution in x

$$\text{(10)} \qquad \mathrm{d}\sigma/\mathrm{d}x = \pi\int(\mathrm{d}^3\sigma/\mathrm{d}^3\boldsymbol{p})\,\mathrm{d}r^2 = \pi\int(f/E)\,\mathrm{d}r^2\,,$$

is well represented by an exponential

$$\text{(11)} \qquad \mathrm{d}\sigma/\mathrm{d}x \propto \exp\left[-9|x|\right].$$

The very sharp peak at $x \approx 0$ is partly due to the phase-space denominator E of eq. (3). Indeed, the weighted distribution (weight E being given to each pion)

$$\text{(12)} \qquad \pi\int f\,\mathrm{d}r^2 = \pi\int E(\mathrm{d}^3\sigma/\mathrm{d}^3\boldsymbol{p})\,\mathrm{d}r^2\,,$$

is well fitted by en exponential

$$\text{(13)} \qquad \pi\int f\,\mathrm{d}r^2 \propto \exp\left[-5|x|\right],$$

for $|x| \leqslant 0.5$, the distribution dropping faster at larger $|x|$. An earlier fit of (13) by $\exp[-10x^2]$, proposed by BALI *et al.* [8] on the basis of incomplete counter data, must now be regarded as unsatisfactory. The logarithmic slope in (13) is seen to be about half the one in (11). The peak is still very sharp, however, and half of the weighted distribution, *i.e.* half of the total energy of all π^-, is contained in the region $|x| < 0.14$. This small interval has more than 70 % of the total number of π^-.

Obviously, a more differential analysis of the $x \simeq 0$ region is needed. It is provided by the rapidity distribution in Fig. 4, contributed by BØGGILD *et al.* to the Helsinki Colloquium [5]. A rounding-off is now visible for $|y| \leqslant 1$, but the distribution is nevertheless still a very peaked one. Nothing is yet seen of the flattening off at $y \simeq 0$ which is predicted by most current models for the high-energy behaviour of the y-distribution. We must therefore try to get a more quantitative feeling for what should be regarded as a narrow or a broad y-distribution. This we have attempted as follows. We chose a c.m. momentum distribution $\mathrm{d}^3\sigma/\mathrm{d}^3\boldsymbol{p}$ which is isotropic, *i.e.* depends only on $|\boldsymbol{p}|$, and gives rise to a transverse momentum distribution similar to the observed one. In fact, we adopted (p in GeV/c)

$$\text{(14)} \qquad \mathrm{d}^3\sigma/\mathrm{d}^3\boldsymbol{p} \propto \exp\left[-13p^2\right] + c\exp\left[-3.5p^2\right],$$

c being so determined that $\langle p_T\rangle \simeq 0.35$ GeV/c. The form (14) makes the algebra particularly simple. We then calculated the corresponding y-distribution $I_0(y)$. It corresponds to the curve marked $a = 0$ in Fig. 5. This distribution

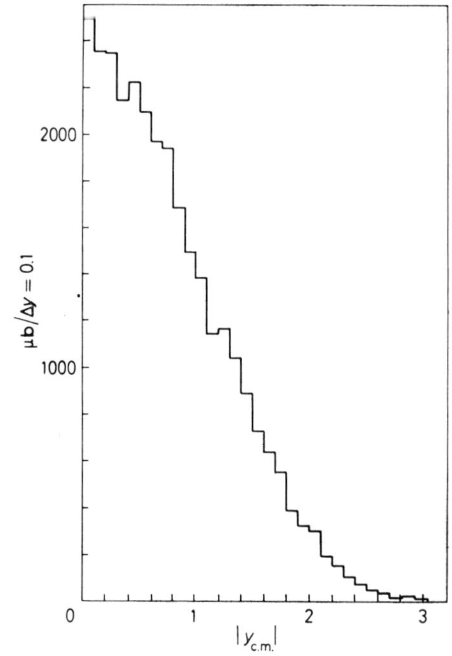

Fig. 4. – Longitudinal rapidity distribution of π^- produced in pp collisions at 19 GeV/c, from ref. [5].

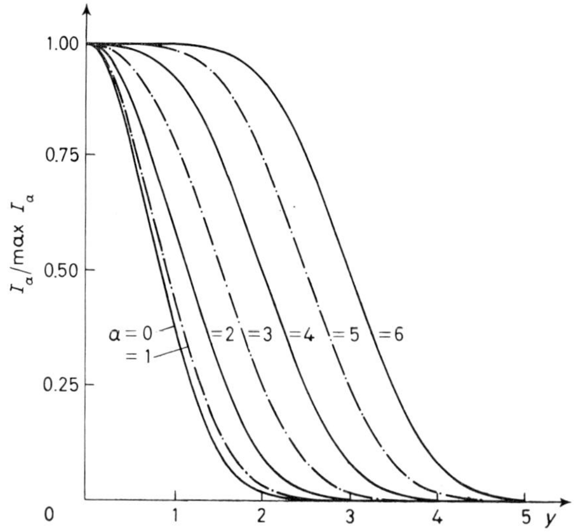

Fig. 5. – The functions $I_0(y)$ and $I_a(y)$ for various values of a, as they are defined in the text, eq. (15) and preceding lines. The Figure gives the functions normalized to value 1 at $y = 0$.

was then uniformly smeared out in rapidity, according to the formula

$$I_a(y) = a^{-1}\int_{-a/2}^{a/2} I_0(y - y')\,\mathrm{d}y' \,. \tag{15}$$

This function is shown in Fig. 5 for various values of a. The smearing out has a simple physical meaning because the transformation $y \to y + \text{constant}$ corresponds to a Lorentz transformation in longitudinal direction.

Fig. 5 allows us now to judge better whether an experimental rapidity distribution is narrow or broad. No experimental distribution is of course expected to be narrower than I_0, and this is certainly not the case for the distribution of Fig. 4. The latter has a half-width at half-height of $\delta y \simeq 1.2$, and Fig. 5 shows that the same value of δy is obtained for I_a with $a \simeq 2$, not an unreasonable value. One will also note from Fig. 5 that I_a only develops a flat plateau for $a \geqslant 4$, corresponding to $\delta y \geqslant 2$. But Fig. 2 indicates that, for $p_T \simeq \langle p_T \rangle \simeq 0.35$ GeV/c, a pion of rapidity 2 has a value p_L as large as 1.4 GeV/c; if such a pion has to have reasonably small x, say $x = 0.2$, one needs $p_{\text{c.m.}} \simeq$ $\simeq 7$ GeV/c, *i.e.* $p_{\text{lab}} \simeq 100$ GeV/c. It is therefore only above the latter incident momentum that a flattening off at $y \simeq 0$ in the y-distribution of pions can reasonably be expected.

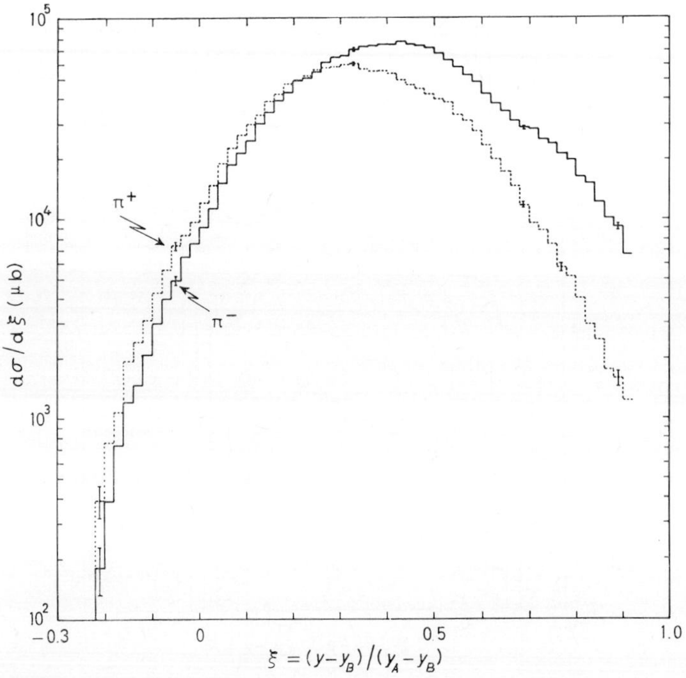

Fig. 6. – Reduced rapidity distribution for π^+ and π^- produced in π^-p collisions at 16 GeV/c, from ref. [6].

The general trend observed for reaction (1) by the Scandinavian Collaboration is also found for π^+ production in π^-p collisions at 16 GeV/c [6]. For π^- production in π^-p, the main difference is that more π^- are produced forward, *i.e.* for $x > 0$, which is indeed to be expected as a leading particle effect. We illustrate these facts with rapidity distributions in Fig. 6 where it should be noted that $y = 0$ corresponds to $\xi = 0.329$ and that $\xi_B - \xi_A = 5.49$. It is remarkable that in $\pi^-p \to \pi^+ + \dots$ at 16 GeV/c the half-width δy at half-height is found to be $\delta y \simeq 1.3$, close to the value 1.2 for $pp \to \pi^- + \dots$ at 19 GeV/c.

Close examination of the distributions of Fig. 6 shows that they contain fine structure, by which we mean small but sudden changes of slope. The π^-

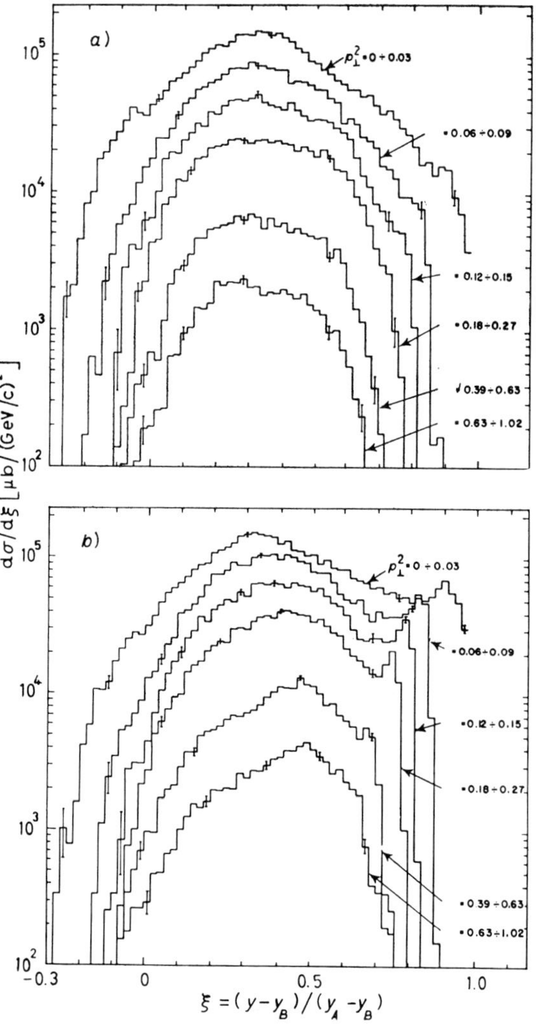

Fig. 7. – Reduced rapidity distribution for *a*) π^+ and *b*) π^- produced in π^-p collisions at 16 GeV/c for various intervals of transverse momentum, from ref. [6].

distribution has such structure at $\xi \simeq 0.7$. Less visible is a weak but undeniable structure of both distributions at $\xi \simeq 0$. A more differential analysis is needed to find out about possible origins of such structure effects. Large statistics bubble chamber experiments like the one on which Fig. 6 is based [6] offer many possibilities for this type of work. One example is given here in Fig. 7, which contains the ξ-distributions for successive intervals of square transverse momentum. The $\xi \simeq 0.7$ structure in the π^- distribution clearly originates from the sharp peaks visible in the lower part of Fig. 7, which are due to the leading π^-. It would be of interest to analyse this further and look at all properties of the collisions having a π^- in those peaks, contrasting them with the properties of the remaining collisions.

The weak structure of the overall distributions of Fig. 6 at $\xi \simeq 0$ are also found back more clearly in Fig. 7. They here show up as marked shoulders at $\xi \leqslant 0$ in the $0 \leqslant r^2 \leqslant 0.03$ (GeV/c)^2 interval. It is remarkable that they occur at $\xi = 0$, which is the minimum ξ-value reachable for a nucleon. It is very likely that some sort of nucleon isobar effect is involved, and much can again be done to study the matter in greater detail.

3. – Correlations.

Beyond effective-mass distributions, the most natural correlation effects to study are those between longitudinal momenta, because they are known to be very strong in the case of few-body collisions. An obvious thing to try is to start from the 2-particle distribution

$$\mathrm{d}^6\sigma/\mathrm{d}^3\boldsymbol{p}_1\mathrm{d}^3\boldsymbol{p}_2 = f^{(2)}/(E_1E_2)\,, \tag{16}$$

in a notation analogous to eq. (3) above. The function $f^{(2)}$ is the Lorentz-invariant 2-particle distribution. Using $\boldsymbol{p}_1$, $\boldsymbol{p}_2$ as co-ordinates, one obtains it from the data by giving weight E_1 E_2 to each event. It can also be obtained by using the longitudinal rapidities, because (16) can be rewritten as

$$\mathrm{d}^6\sigma/\mathrm{d}y_1\mathrm{d}^2\boldsymbol{r}_1\mathrm{d}y_2\mathrm{d}^2\boldsymbol{r}_2 = f^{(2)}\,, \tag{17}$$

where the $\boldsymbol{r}_i$ denote the transverse momenta. For the longitudinal variable distributions, one can take either of the two functions

$$F(q_1, q_2) = \int f^{(2)}\mathrm{d}^2\boldsymbol{r}_1\mathrm{d}^2\boldsymbol{r}_2\,, \tag{18}$$

$$\tilde{F}(y_1, y_2) = \int f^{(2)}\mathrm{d}^2\boldsymbol{r}_1\mathrm{d}^2\boldsymbol{r}_2\,, \tag{19}$$

(18) is obtained by integrating at constant longitudinal momenta q_1, q_2, or at

constant reduced variables $x_i = q_i/p_{\text{c.m.}}$ see eq. (7). The distribution $\tilde{F}$ of (19) is obtained by integrating at constant longitudinal rapidities y_1, y_2, or at constant reduced variables ξ_1, ξ_2, see eq. (8). As explained earlier for single-particle distributions, these are not only distinct but also complementary ways of examining the data.

The ABBCCHW Collaboration [6] presented at the Helsinki Colloquium the data of Figs. 8 and 9 concerning 16 GeV/c π^-p collisions which have a proton

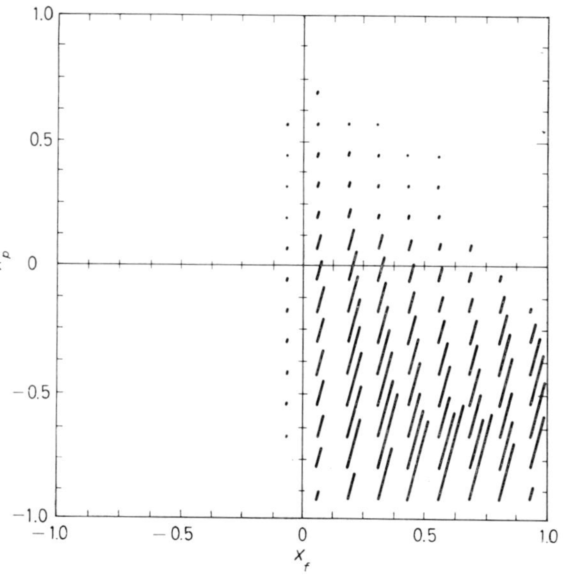

Fig. 8. – Weighted distribution of proton and most forward π^- and π^-p collisions at 16 GeV/c, from ref. [6].

in the final state. They refer to the proton and the most forward π^-, *i.e.* the π^- of largest q. The plot uses the variables x_i, x_f referring to the most forward π^-. Fig. 8 gives the function F of eq. (18). Correlations are immediately visible. For $x_p \simeq 0$, the x_f-distribution has one maximum at small x_f. It has two maxima for x_p between $-\frac{1}{8}$ and $-\frac{7}{8}$, one being at small x_f, the other at $x_f \simeq 0.8$. Finally, as x_p is in the last bin $(-1 < x_p < -\frac{7}{8})$, the two maxima of the x_f-distribution merge into a single one at $x_p \simeq 0.6$. Figure 9 gives the same distribution for some more specific channels. Z_0 denotes a system of 2 or more neutral particles. Correlations are strong at low multiplicity and the overall modification of the x_f, x_p-distribution with increasing multiplicity is pronounced.

We turn now to other correlation effects pointed ont in a comparative study of $\pi^\pm p \to 6$ prongs from 5 to 16 GeV/c made by a « collaboration of collaborations » [9]. This is the first instance of an extensive and thorough investi-

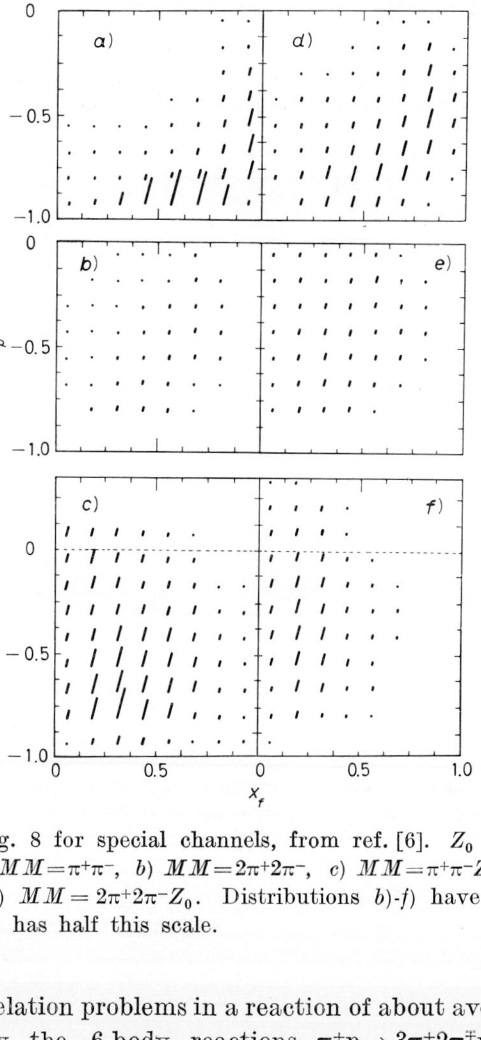

Fig. 9. – Same as Fig. 8 for special channels, from ref. [6]. Z_0 denotes 2 or more neutral particles: *a*) $MM=\pi^+\pi^-$, *b*) $MM=2\pi^+2\pi^-$, *c*) $MM=\pi^+\pi^-Z_0$, *d*) $MM=\pi^+\pi^-\pi^0$, *e*) $MM=2\pi^+2\pi^-\pi^0$, *f*) $MM=2\pi^+2\pi^-Z_0$. Distributions *b*)-*f*) have the same scale as Fig. 8, distribution *a*) has half this scale.

gation of many correlation problems in a reaction of about average multiplicity. It concerns mainly the 6-body reactions $\pi^\pm p \to 3\pi^\pm 2\pi^\mp p$, with occasional inclusion of an additional π^0. Many empirical questions are treated in the general field of correlations between longitudinal momenta, nature and electric charge of particles, resonance production, etc. Here we select a few of the many interesting results presented and advise the interested reader to study the original work in detail.

Figure 10 concerns the collisions where the proton is very backward and gives the distribution of the pions of $\pi^- p \to 3\pi^- 2\pi^+ p$ at 16 GeV/c in the longitudinal component of their momentum calculated in the rest frame of the $3\pi^- 2\pi^+$ system. The selection criterion on the proton corresponds to $x_p < -0.72$ (one has $p^{\text{c.m.}} = 2.7$). Both the identity in shape of the two distributions and their forward-backward symmetry are quite remarkable. The angular distributions show a certain amount of forward and backward peaking.

Largely complementary to this class of final states is the one where the proton has small backward momentum, say $-0.3 \leqslant x_p \leqslant 0$. The pion distributions in p_L in the overall c.m. frame are rather similar for $p_L < 1$ GeV/c, but there is a large surplus of π^- for $p_L > 1$ GeV/c. This suggests to take the most for-

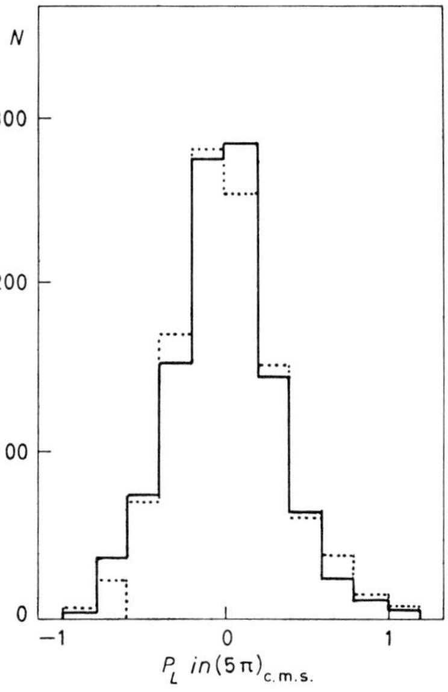

Fig. 10. – Distribution of π^+ and π^- produced in a subset of collisions $\pi^-p \to 3\pi^-2\pi^+p$ at 16 GeV/c, in the longitudinal momentum as defined in the 5π rest frame, from ref. [9]. The subset is defined by $p_L < -1.944$ GeV/c for the proton (p_L = longitudinal momentum in overall c.m. system): ---- π^-, —— $\pi^+ \times \frac{3}{2}$; π^+: $\langle p_L^{\text{c.m.s.}} \rangle = -0.009$, $\langle p_\perp^{\text{c.m.s.}} \rangle = 0.293$; π^-: $\langle p_L^{\text{c.m.s.}} \rangle = 0.006$, $\langle p_\perp^{\text{c.m.s.}} \rangle = 0.304$.

ward π^- in the interval $p_L > 1$ GeV/c (corresponding to $x_f > 0.37$ in a notation we used earlier) while limiting x_p as indicated above. The remaining pions and the proton, examined in their rest frame, show again a strong forward-backward symmetry, as seen in Fig. 11. The angular distributions have forward and backward peaking for $\pi^\pm$, not for p.

This type of longitudinal phase-space (LPS) analysis gives also remarkable results on resonance production. Figure 12 shows the effective-mass distributions of $\pi^+\pi^-$ in the complete reactions $\pi^-p \to 2\pi^+3\pi^-p$ and $\to 2\pi^+3\pi^-\pi_0 p$. Despite the high statistics, it seems difficult to claim that they are informative. In contrast, by limiting oneself to the sector of LPS where a π^+ and a π^- are the only particles going forward, one obtains the solid histogram in Fig. 13, while the sector where $2\pi^-$ are the only forward particles gives the dashed histogram for their effective mass distribution. The difference is shown in Fig. 14. Finally, Fig. 15 gives the effective-mass distribution of the π^0 and one π^- in the LPS sector where they are the only forward particles in the reaction $\pi^-p \to 2\pi^+3\pi^-\pi^0p$. All these distributions are obviously interesting. Despite the high multiplicity, collisions with a dipion as forward moving system show impressive resonance

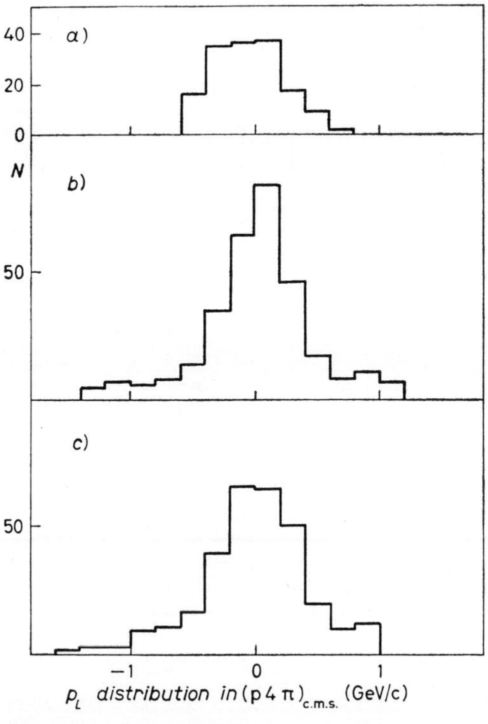

Fig. 11.

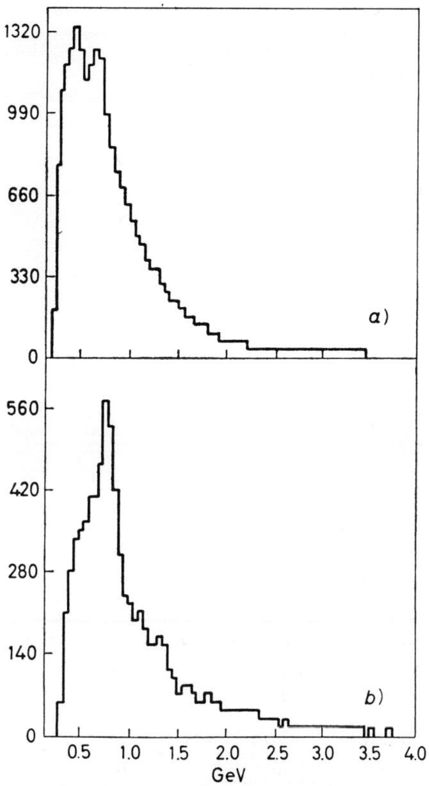

Fig. 12.

Fig. 11. – Distribution of p, π^+ and nonleading π^- produced in a subset of collisions $\pi^-p \to 3\pi^-2\pi^+p$ at 16 GeV/c, in the longitudinal momentum as defined in rest frame of $2\pi^-2\pi^+p$ (where the π^- are the nonleading ones), from ref. [9]. The subset is defined by -0.864 GeV/c $< p_L < 0$ for the proton and $p_L > 1$ GeV/c for the leading (*i.e.* most forward) negative pion (150 jets with leading π^-), where p_L denotes the longitudinal momentum in the overall c.m. system, from ref. [9]: *a*) p, *b*) π^+, *c*) π^-.

Fig. 12. – Effective-mass distributions of $\pi^+\pi^-$ in *a*) $\pi^-p \to 3\pi^-2\pi^+\pi^0p$ collisions (upper graph) and *b*) $\pi^-p \to 3\pi^-2\pi^+p$ collisions (lower graph), from ref. [9]: $M(\pi^+\pi^-)$.

dominance for nonexotic states, the background being smooth and comparable to the complete distribution for the exotic state $2\pi^-$. This can become a rich field for further work on the low-mass dipion system, which has attracted up to now an enormous effort in the special case of single-pion production ($\pi p \to 2\pi\mathcal{N}$). As is well known although rarely stressed, the latter studies suffer from the

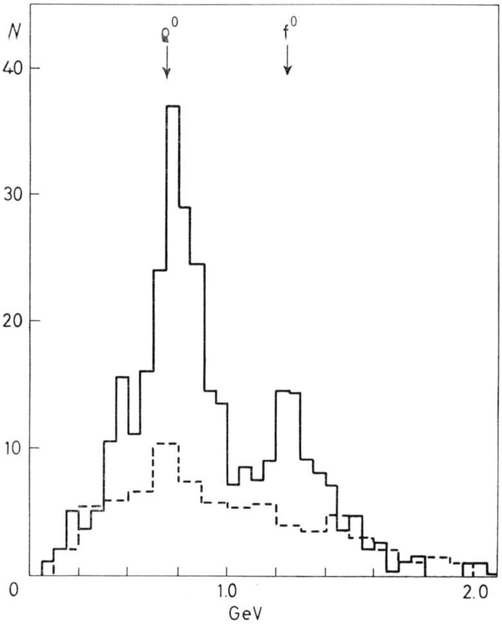

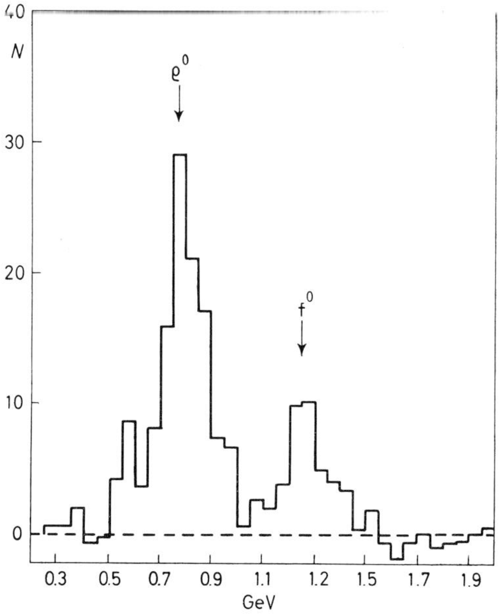

Fig. 13. Fig. 14.

Fig. 13. – Solid histogram: effective-mass distribution of forward pions in $\pi^-p \to \to (\pi^+\pi^-)(\pi^+2\pi^-p)$ and $\pi^-p \to (\pi^+\pi^-)(\pi^+2\pi^-\pi^0p)$, where the first bracket denotes the forward going particles and the second one the backward going particles in the overall c.m. system. Dashed histogram: same for $\pi^-p \to (2\pi^-)(2\pi^+\pi^-p)$ and $\pi^-p \to (2\pi^-)\cdot \cdot(2\pi^+\pi^-\pi^0p)$. All data at 16 GeV/c from ref. [9]: mass of $(\pi^+\pi^-)_B$ 1 combination per event (316 events), mass of $(\pi^-\pi^-)_F$ 1 combination per event (145 events).

Fig. 14. – Difference between solid an dashed histograms of Fig. 13, from ref. [9]: 16 GeV/c π^-p.

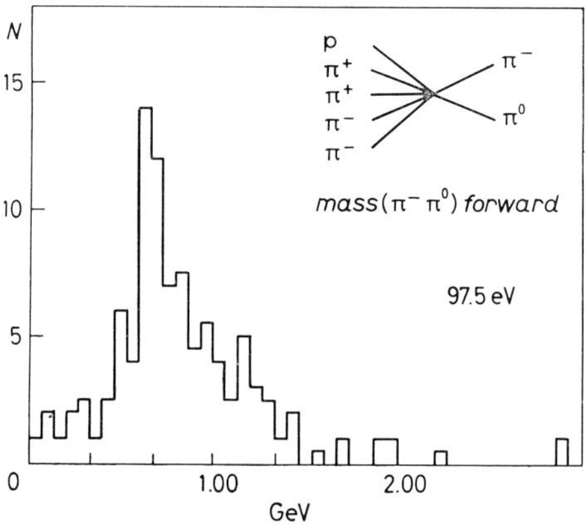

Fig. 15. – Effective-mass distribution of forward pions in $\pi^-p \to (\pi^-\pi^0)(2\pi^+2\pi^-p)$ at 16 GeV/c, from ref. [9]. Brackets have same meaning as in caption of Fig. 13.

difficulty to separate two dynamical effects, the production mechanism and the pion-pion interaction. One natural way to approach this question is to produce dipions in a variety of collisions, in the hope that significant similarities will be found in distinct production conditions. The results just reported give good hope in this direction.

4. – Exclusive experiments on few-body collisions.

Despite the fact that n-particle final states have the large value $3n-5$ for the dimensionality of their phase-space, it has been possible to obtain considerable insight in the overall phase-space distribution of high-energy collisions of multiplicity $n = 3, 4$ and to some extent, $n = 5$ [1, 2, 3, 10]. The reason is that all transverse momenta remain small while only the longitudinal momenta

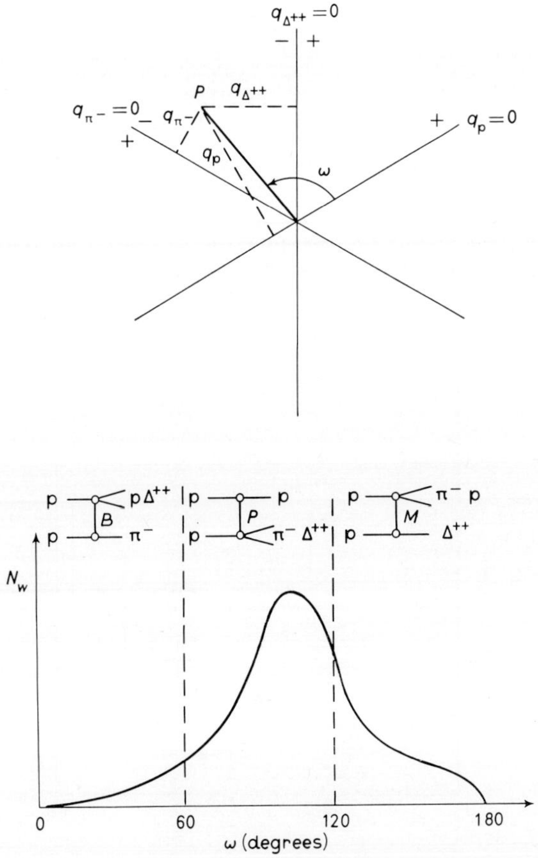

Fig. 16. – Definition of angle ω and characteristics of LPS plot for $pp \to p\Delta^{++}\pi^-$, from ref. [11]. The q denote the c.m. longitudinal momenta.

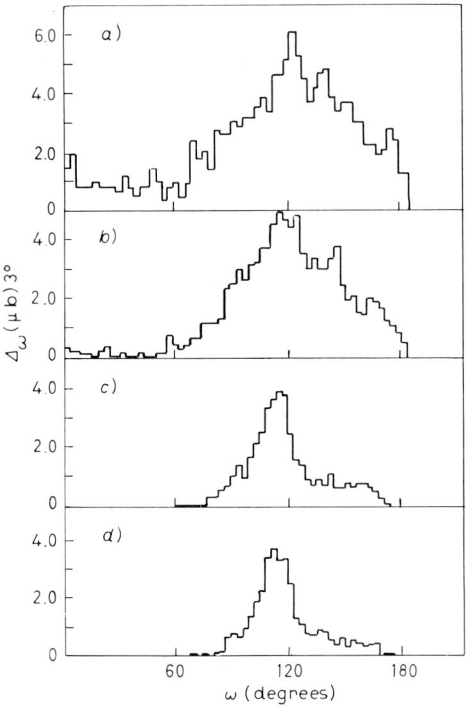

Fig. 17. – Weighted ω-distribution for $pp \to p\Delta^{++}\pi^-$ at 4 energies, from ref. [11]: *a*) 5 GeV/c, *b*) 8 GeV/c, *c*) 19 GeV/c, *d*) 25 GeV/c.

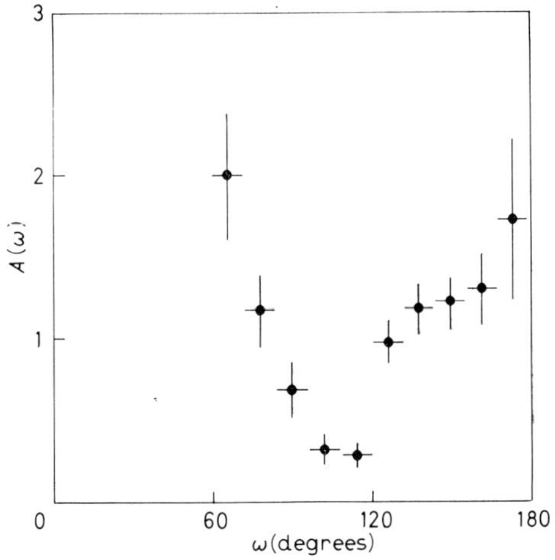

Fig. 18. – Exponent giving the $p_{lab}^{-A(\omega)}$ variation of the weighted distribution of Fig. 17, from ref. [11].

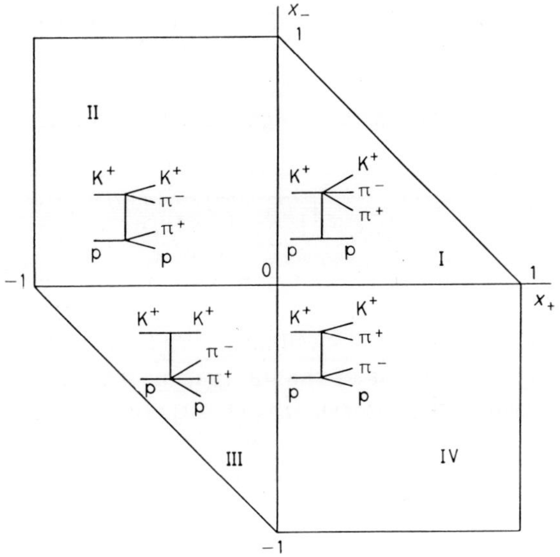

Fig. 19. – Most populated sectors of LPS for $K^+p \to K^+\pi^+\pi^-p$ and their interpretation in terms of kinematic diagrams (upper lines indicate forward particles, lower lines backward particles). The variables $x_\pm$ refer to $\pi^\pm$. From ref. [12].

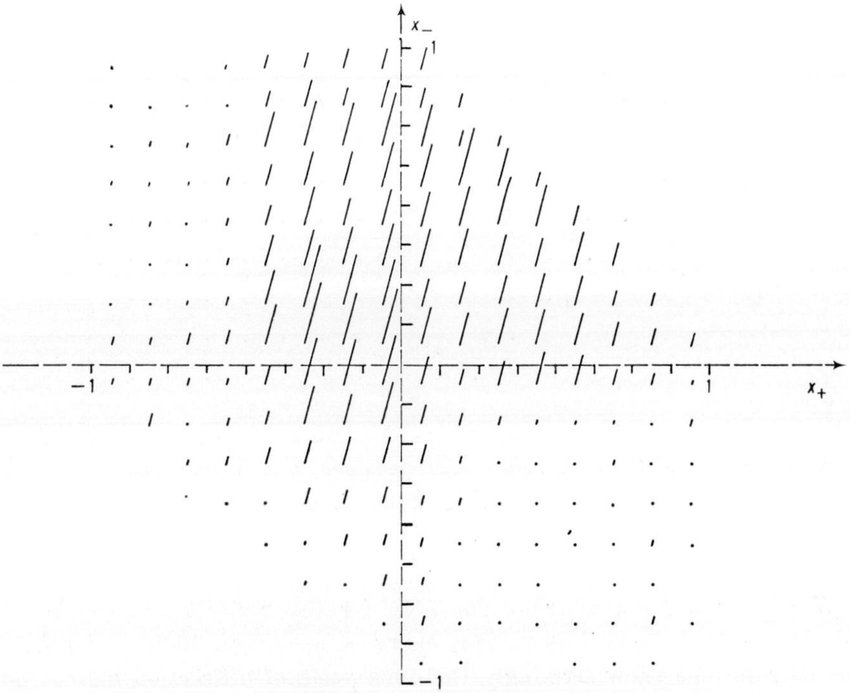

Fig. 20. – Weighted LPS distribution for $K^+p \to K^+\pi^+\pi^-p$ at 7.3 GeV/c, plotted with convention of Fig. 19, from ref. [12].

grow with energy, so that a good survey of the overall distribution is obtained from an analysis in longitudinal phase-space (LPS), the dimension of which reduces in effect to $n-2$ at high energy. As explained elsewhere [1, 10], one can extract from the data a so-called weighted LPS distribution which represents the quantity

$$p_{\text{c.m.}}^{-1}\, s^{-\frac{3}{2}} \int |M|^2 \prod_{1}^{n-1} d^2 \boldsymbol{r}_i\,,$$

where M is the invariant collision amplitude and the integral over all transverse momenta is carried out so that the reduced longitudinal momenta $x_i = q_i/p_{\text{c.m.}}$ remain approximately constant in the integration (the approximate constancy is up to terms of order $s^{-\frac{1}{2}}$). References [1, 10] contain examples discussed in detail.

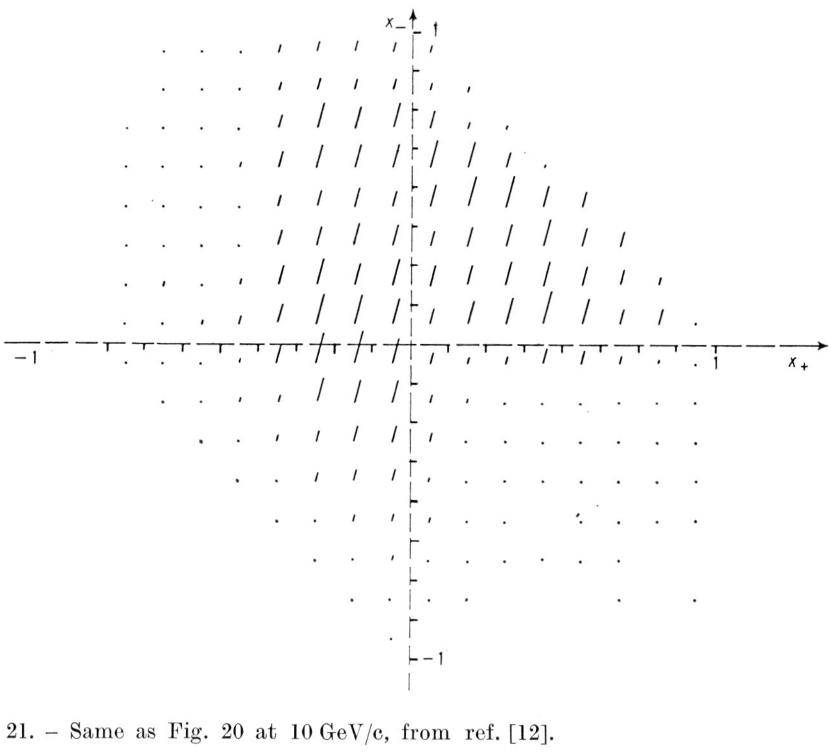

Fig. 21. – Same as Fig. 20 at 10 GeV/c, from ref. [12].

We illustrate such distributions here for the reactions $pp \to p\Delta^{++}\pi^-$ [11] and $K^+p \to K^+\pi^+\pi^-p$ [12] at various energies, see Figs. 16-22 and refs. [2, 3]. The distributions show strikingly the emergence of diffraction dissociation as dominant mechanism at high energy in well-defined LPS regions, the interval $60° \leqslant \omega \leqslant 120°$ in Figs. 16, 17 and the triangles I and III in Figs. 19-22. These are exactly the regions where vacuum quantum number exchange is allowed between (groups of) particles flying forward and backward in the c.m. system (see the «kinematic» diagrams in Figs. 16 and 19, where forward particles are drawn on top and backward ones below).

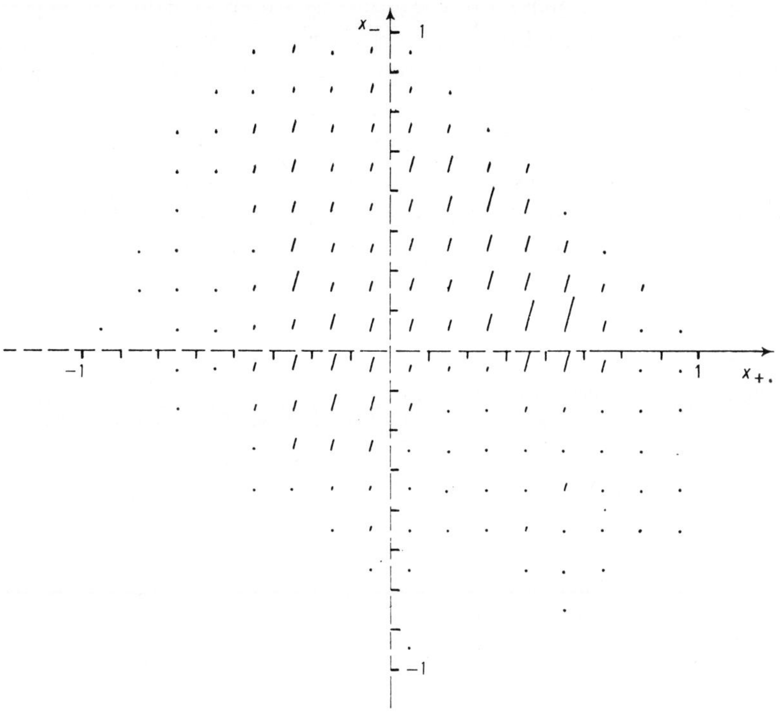

Fig. 22. – Same as Fig. 20 at 16 GeV/c, from ref. [12].

The region $120° \leqslant \omega \leqslant 180°$ of Figs. 16, 17 as well as regions II, IV in Figs, 19-22 correspond to exchange of a singly charged mesonic object. For $pp \to p\Delta^{++}\pi^-$, Figure 20 shows that charge exchange separates well from vacuum exchange in the LPS distribution at the highest energies, but not yet at $p_{lab} \simeq 8$ GeV/c. For $K^+p \to K^+\pi^+\pi^-p$, the two charge exchange regions II and IV behave very differently. In IV, the distribution is small already at $p_{lab} \simeq 7$ GeV/c, and there is a clear boundary separating it from the vacuum exchange regions I and III. In II, on the contrary, the distribution is very large at $p_{lab} \simeq 7$ GeV/c and one has to go to 16 GeV/c to find it smaller than in regions I, III and to distinguish a clear boundary. This difference between two cases of charge exchange must of course be due to their very different resonance contents (K^* and Δ^{++} in II, not in IV). A comparison study of single charge exchange in $Ap \to A\pi^+\pi^-p$ with $A = K^\pm$, $\pi^\pm$ suggests itself and is easily accessible experimentally.

We note finally the baryon exchange region $0° \leqslant \omega \leqslant 60°$ in Figs. 16, 17. The distribution of Fig. 17 is very flat in this region and disappears rapidly. The LPS regions corresponding to baryon exchange in $K^+p \to K^+\pi^+\pi^-p$ have very low statistics and are not plotted.

The above examples illustrate the clear separation of LPS distributions of few-body collisions into parts interpretable as different collision types corresponding to various internal quantum number exchanges. We should add that such separations have been found in complete phase space at even lower energies by PLESS and his collaborators at M.I.T. [13]. They carry out an ambitious programme of fully differential analysis for few-body reactions. They combine LPS variables with c.m. energies and use detailed inspection of plots by cathode ray tube displays to disentangle the mechanisms at work in the reaction. They find a remarkable degree of separation between various mechanisms even at low energy ($p_{lab} = 3.9$ GeV/c).

5. – A remark on energy dependence.

We end by illustrating a serious difficulty affecting all studies which concern the s-dependence of any physical cross-section in 2- or more-body processes.

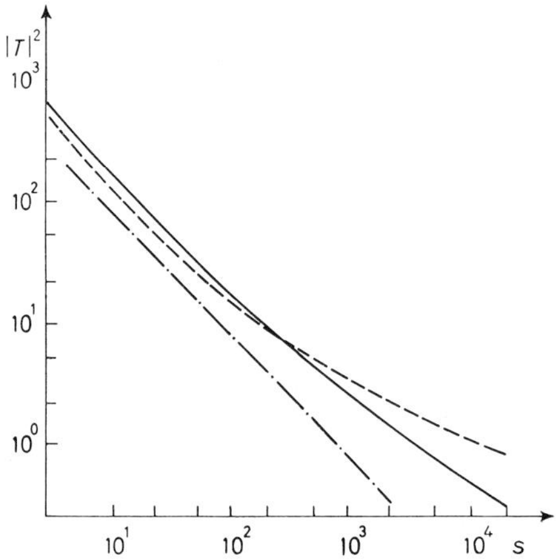

Fig. 23. – $|T|^2$ *vs.* s for the functions defined in eqs. (20), (21) and (22): — — — $0 \leqslant \alpha \leqslant 1$ ——— $0.18 \leqslant \alpha \leqslant 0.88$, —·—·— $\alpha = 0.5$.

The fact of finding experimentally an almost pure power dependence s^{-n} over even a large s-interval does not guarantee at all that the matrix element itself contains a single power of s. This is illustrated in Fig. 23 which gives in logarithmic scales $|T|^2$ *vs.* s for the following choices of T (arbitrary normalization):

$$(20)\qquad T\propto\int_0^1(1-\alpha)\,s^{\alpha-1}\exp\left[i\frac{\pi}{2}\alpha\right]d\alpha\,,$$

$$(21)\qquad T\propto\int_{0.18}^{0.88}(0.88-\alpha)\,s^{\alpha-1}\exp\left[i\frac{\pi}{2}\alpha\right]d\alpha\,,$$

$$(22)\qquad T\propto s^{\alpha-1}\exp\left[i\frac{2}{\pi}\alpha\right]\quad\text{with } \alpha=\tfrac{1}{2}\,.$$

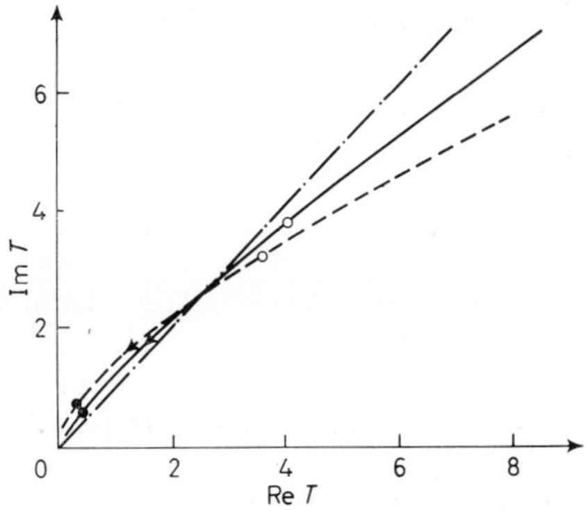

Fig. 24. – Imaginary *vs.* real parts of T as defined in eqs. (20), (21) and (22): – – – $0\leqslant\alpha\leqslant1$, ——— $0.18\leqslant\alpha\leqslant0.88$, —·—·— $\alpha=0.5$; ○ $s=10$, ▲ $s=50$, ● $s=500$.

It is striking to see the parallelism of these curves in a range $5\leqslant s\leqslant 50$ or even larger, covering the medium and high-energy domains at CERN, Brookhaven and even Serpukhov. The Argand diagrams of the three above amplitudes are given in Fig. 24 and show that even the phases are approximately the same over a wide energy range.

* * *

In addition to a general acknowledgment to the various collaborations whose data were used, the author wants to express his profound gratitude for detailed discussions, based on extensive work with the data, to the following physicists: Mrs. T. Besliu (CERN), Dr. H. Bøggild (Copenhagen), Dr. O. Czyzewski (Krakow, CERN), Mrs. P. Dalpiaz (CERN), Dr. K. Hansen (Copenhagen), Dr. W. Kittel (CERN), Miss L. Ljung (Helsinki), Dr. L. Michejda (Warsaw), Dr. S. Ratti (Milano), Dr. F. Verbeure (Brussels), Dr. A. Wroblewski (Warsaw), and Dr. N. Yamdagni (Stockholm).

REFERENCES

[1] L. Van Hove: *Phys. Lett. C*, **1**, 347 (1971).

[2] W. Kittel: *Experimental Results on Few-body reactions*, in *Rapporteur Lecture at Colloquium on Multiparticle Dynamics, Helsinki, 25-28 May 1971*, see Proceedings edited by J. Tuominiemi *et al.*, Research Institute for Theoretical Physics, Helsinki.

[3] L. Van Hove: *Longitudinal phase-space analysis and related questions*, in *Lecture at Helsinki Colloquium*, see Proceedings edited by J. Tuominiemi *et al.*, Research Institute for Theoretical Physics, Helsinki.

[4] A. H. Mueller: *Phys. Rev. D*, **2**, 2963 (1970).

[5] Scandinavian Collaboration: pp *at* 19 GeV/c, contributions to *Helsinki Colloquium.*

[6] Aachen-Berlin-Bonn-CERN-Cracow-Heidelberg-Warsaw Collaboration: π^-p *at* 16 GeV/c, contributions to *Helsinki Colloquium.*

[7] H. Bøggild, K. H. Hansen and M. Suk: *Nucl. Phys.*, **27** B, 1 (1971)

[8] N. F. Bali, L. S. Brown, R. D. Peccei and A. Pignotti: *Phys. Rev. Lett.*, **25**, 557 (1970).

[9] European Collaboration for Analysis of High-Multiplicity Events: *Comparative study of* πp $\rightarrow$ *six prongs*, paper submitted to the *Helsinki Colloquium and the Amsterdam Conference*, data from following Collaborations: π^+p, 5 GeV/c, Bonn-Durham-Nijmegen-Paris (E.P.)-Torino; π^+p, 8 GeV/c, Warsaw; π^+p, 16 GeV/c, Aachen-Bonn-CERN-Cracow-Warsaw; π^-p, 11 GeV/c, Saclay; π^-p, 16 GeV/c, Aachen-Bonn-Berlin (Zeuthen)-CERN-Cracow-Warsaw.

[10] W. Kittel, S. Ratti and L. Van Hove: *Nucl. Phys.*, **30** B, 333 (1971).

[11] The World Collaboration on the Reaction pp $\rightarrow$ pp$\pi^+\pi^-$. *Longitudinal phase-space analysis of the reaction* pp $\rightarrow$ pp$\pi^+\pi^-$ *between* 4 *and* 25 GeV/c, paper contributed to the *Helsinki Colloquium.*

[12] F. Verbeure: *A phenomenological analysis of the reactions* $K^+p \rightarrow K^{+}_{0}\pi^{-}_{0}\pi^+p$ *using the* LPS *method* (*Data from the International* K^+ *Collaboration*), paper contributed to the *Helsinki Colloquium.*

[13] J. E. Brau, F. T. Dao, M. F. Hodous, I. A. Pless and R. A. Singer: *Phys. Rev. Lett.*, **27**, 1481 (1971).

Elastic Scattering Experiments with High-Energy Protons.

K. SCHLÜPMANN (*)
CERN - Geneva

1. – Introductory remarks.

With reference to the title of this year's course and with a businessman's attitude that budgets reflect developments, no further justification is needed to present

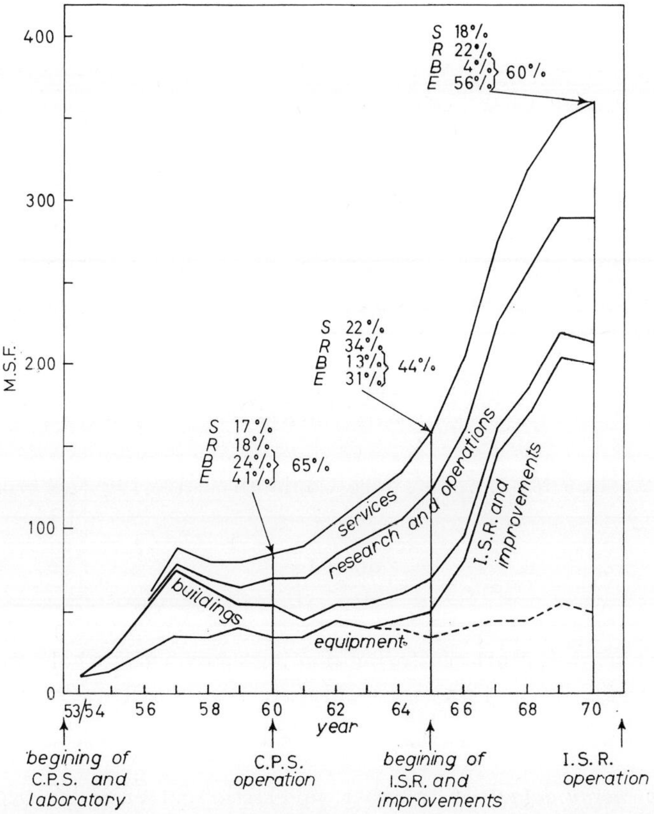

Fig. 1. – Annual expenditures of CERN, MEYRIN (from CERN ISR 210-425-4, CERN, Geneva, 1970).

(*) Present address: Institut für theoretische Physik, Arnimallee 3, 1 Berlin 33.

in Fig. 1 the curve of the annual expenditures of a big laboratory. The development over the past years has been rapid. Last year's budget amounted to 350 million Swiss francs.

With reference to the subject of this lecture it can be said that by the construction of the CERN intersecting storage rings (ISR) the effort and the interest in proton-scattering studies has been multiplied. Proton experiments at conventional accelerators count amongst the easiest experiments on strong interactions, since a p beam can be $10^6 \div 10^7$ times more intense than a π beam. With the ISR the additional attractive feature has appeared that p interactions can be studied by laboratory means up to energies corresponding to at least 1200 GeV for a conventional machine.

2. – Proton-proton elastic scattering.

2'1. *Definition of variables.* – A pp collision can be described by two variables, *e.g.*, the momentum k and the angle $\bar{\theta}$ in the centre-of-momentum system (c.m.s.):

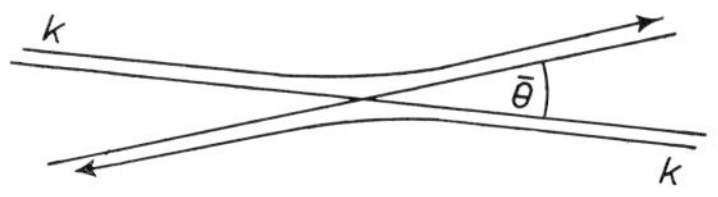

or, alternatively, by s, the square of the c.m.s. energy and the square of the momentum transfer: $t = 4k^2 \sin^2(\bar{\theta}/2)$. The interesting measurable quantity is the differential cross-section

$$\frac{d\sigma}{dt} = \frac{\pi}{k^2} \frac{d\sigma}{d\bar{\Omega}}$$

where $\bar{\Omega} = 2\pi \sin \bar{\theta}\, d\bar{\theta}$; *e.g.*, for pure Coulomb scattering this cross-section is given by

$$\frac{d\sigma}{dt} \approx \frac{4\pi}{t^2} \,. \tag{1}$$

At very low energies, Mott scattering dominates over the whole angular range; at high energies it is important only at very small scattering angle, for $t < 0.01\ (\mathrm{GeV})^2$.

2'2. *Low-energy data.* – In the past, pp elastic scattering has been a successful tool with which to study properties of the nucleon. WHITE, HAFSTAD, HEIDENBURG and TUVE [1] were the first to observe a deviation from Mott scattering and thus contributed to the discovery of strong interactions. Their results are plotted in Fig. 2. The observed number of scattered protons is lower than was predicted for pure-charge scattering at small angles and exceeds the prediction at scattering angles above 28°. s-wave nuclear scattering interferes with the Coulomb-scattering amplitude.

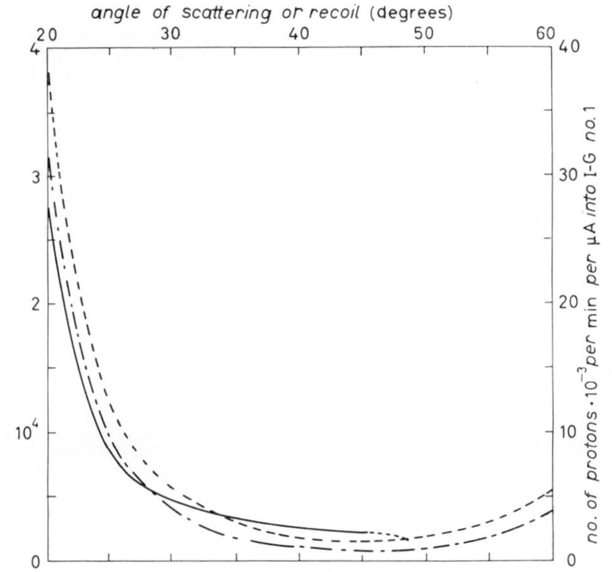

Fig. 2. – Scattering of protons by protons at 800 kV (from ref. [1]), actual numbers of scattered particles as a function of angle: – – –, classical, 155 000 at 15°; —·—·—, MOTT, 142 000 at 15°; ———, observed, 123 000 at 15°.

The incident proton momentum in this experiment was $p_0 = 0.038$ GeV/c. In the momentum range $p_0 = 0.1$ to $p_0 = 1$ GeV, the scattered wave is expected to consist of an increasing number of partial waves. The experimental data in this range were successfully described by partial-wave dispersion relations [2].

2·3. *The general features of the high-energy data.* – Above $p_0 = 1$ GeV/c the diffraction phenomenon, which is common to all elastic high-energy processes, starts to appear. The forward cross-section $d\sigma/d\Omega$ rises approximately linearly with s or k^2, *i.e.* $d\sigma/dt$ becomes nearly constant and most of the elastic scattering is concentrated in a sharp peak in $d\sigma/dt$ near $t = 0$, which apart from a slight narrowing (shrinkage) does not change between 10 and 70 GeV/c. Therefore, the total elastic cross-section is nearly constant in this momentum range and amounts to $\sigma_{el} = (8 \div 10)$ mb. Recently, this quantity was measured at corresponding laboratory momenta of 500 GeV/c in an experiment at the CERN ISR and found to be $\sigma_{el} = (6.8 \pm 0.6)$ mb. The results on σ_{el} are plotted in Fig. 3.

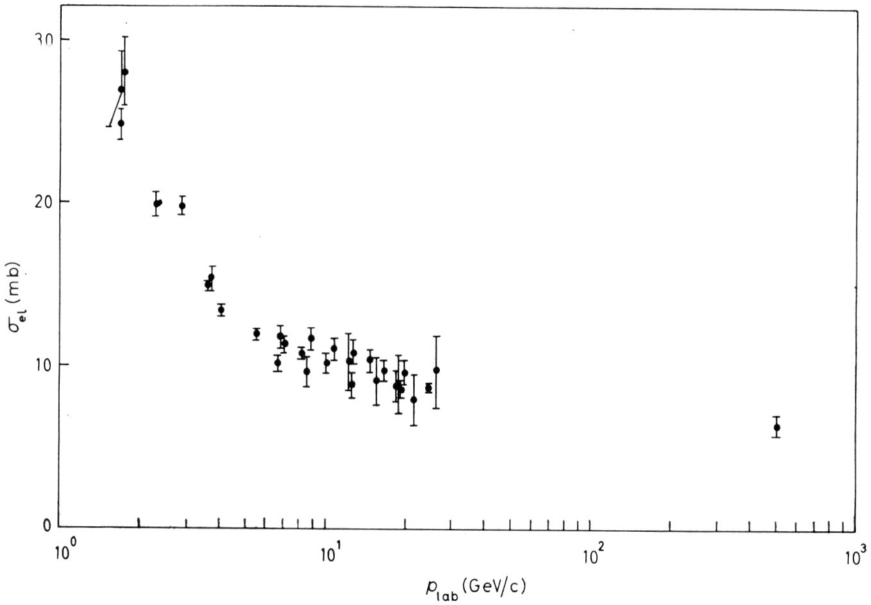

Fig. 3. – Elastic proton-proton cross-section σ_{el} (from ref. [3]).

The phase of the forward amplitude $A(t=0)$: $\alpha_{pp} = \text{Re}\, A/\text{Im}\, A$ has been deduced from observations of the Coulomb-nuclear interference [4]. It changes from $\alpha_{pp} = -0.3 \pm 0.08$ at $\beta_0 = 10$ GeV/c to $\alpha_{pp} = -0.12 \pm 0.06$ at $p_0 = 70$ GeV/c in agrement with forward dispersion-relation calculations [5]. α_{pp} is not known for $t \neq 0$.

Spin effects persist up to the highest momenta where polarization measurements [6] have been performed, $p_0 = 17.5$ GeV/c.

The polarization varies strongly with t in the range where it has been measured, $0 < t < 2.5\ (\text{GeV})^2$, it reaches values of $(15 \div 20)\%$ at 10 GeV/c and still is $(5 \div 10)\%$ at 17.5 GeV/c.

The differential cross-section at large angles is characterized by a gradual decrease of the t dependence and a strong s dependence. The experimental data can be approximated to within a factor of ~ 2 by Orear's empirical formula [7] $\text{d}\sigma/\text{d}\bar{\Omega} = 595/s \exp[-(P_t/0.158)]$ mb/sr, where P_t is the transverse momentum $P_t = k \sin \bar{\theta}$. However, at a higher degree of precision, systematic deviations [19] from this description exist.

2·4. *The optical model.* – The differential cross-section is often discussed in terms of an optical-model approximation:

$$\left.\frac{\text{d}\sigma}{\text{d}t}\right|_{pp} = |A(t)|^2 = \left|\int J_0(\sqrt{t}\cdot b)\cdot V(b)\cdot b\, \text{d}b\right|^2 = |\mathcal{F}(V(b))|^2. \tag{2}$$

The amplitude $A(t)$ at small t is written as a Fourier-Bessel transform of a profile function $V(b)$, where b is a transverse radial co-ordinate (impact parameter). If V(b) has a Gaussian shape

$$V(b) = V_0 \exp\left[-\left(\frac{b}{R_0}\right)^2\right],$$

$d\sigma/dt$ also has a Gaussian shape in $\sqrt{t}$

$$\frac{d\sigma}{dt} = \sigma_0 \exp\left[-\frac{R_0^2}{2}t\right].$$

With a value of $R_0 = 0.9$ fm for the effective radius of a proton, one obtains

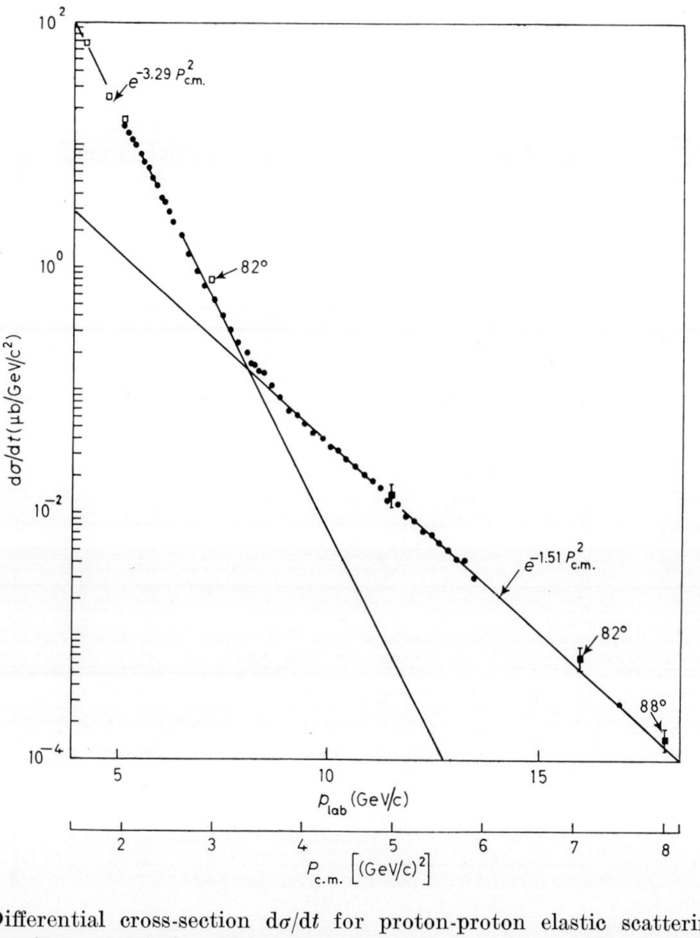

Fig. 4. – Differential cross-section $d\sigma/dt$ for proton-proton elastic scattering at 90° in the c.m.s. (from ref. [8]). Other data [2-4] are also plotted. The lines drawn are straight-line fits to the data: ▫ Berkeley, CLYDE *et al.*; ▪ Cornell BNL, COCCONI *et al.*; ◆ CERN, ALLABY *et al.*; • this experiment.

$d\sigma/dt = \sigma_0 \cdot \exp[-10t]$ in approximate agreement with experiment. Although eq. (3) is a small-angle approximation, it has stimulated the interpretation of observations at 90° in the c.m.s. [8] (Fig. 4). The data can be described by two exponential function in $k^2 = t(90°)/2$. It has therefore been speculated [8] that there was evidence of an « onion » structure of the proton.

2'5. *Comparison with electron-proton scattering.* – In a simple-minded way the laboratory cross-sections for electron-proton scattering can be compared to those for proton-proton scattering (Fig. 5). The pp data are experimental results [9] at $p_0 = 19.2$ GeV/c, the ep cross-section was calculated for the same momentum by using the known form factors [10]. The proton cross-section has structure whereas the ep distribution is smooth.

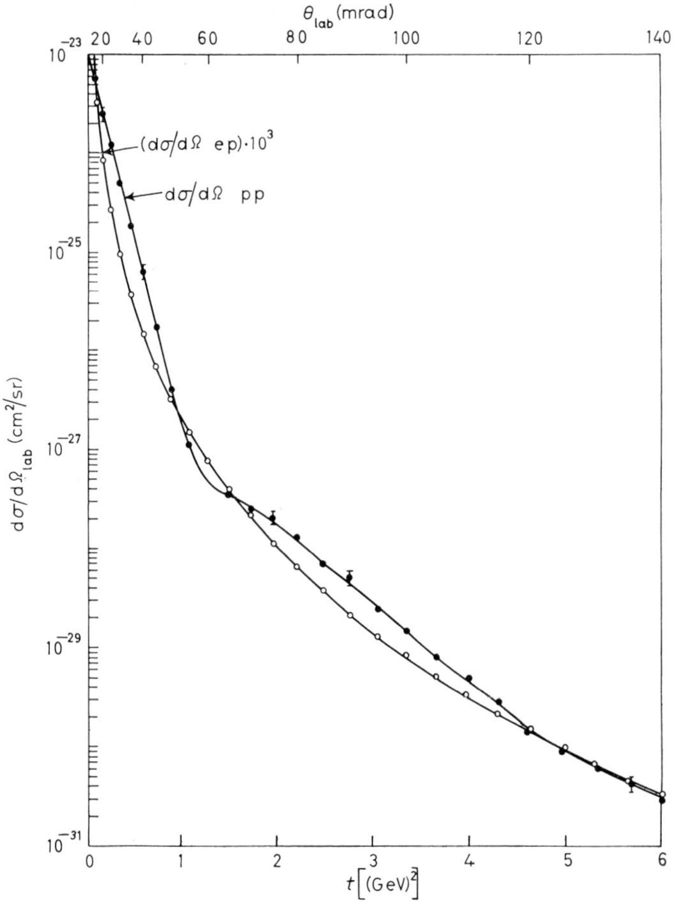

Fig. 5. – Comparison of the laboratory cross-sections $d\sigma/d\Omega$ for pp scattering at 19.2 GeV/c (ref. [9]) and ep scattering at the same momentum (calculated from ref. [10]): • pp, ○ ep·10^3.

In a casual manner the electron-proton cross-section can be written as

$$\left.\frac{d\sigma}{dt}\right|_{ep} \approx \sigma_{Mott} \cdot G^2(t) , \tag{3}$$

where $G(t)$ is the form factor or in a picturesque way of speaking, the persistence probability of a proton to survive a momentum transfer t without excitation or emission of pions. Since the persistence probability in this picture is expected to occur quadratically in the amplitude for a pp collision, WU and YANG [11] suggested the relation

$$\left.\frac{d\sigma}{dt}\right|_{pp} = |A(t)|^2 = |A(0)|^2 \cdot G^4(t) \tag{4}$$

as an asymptotic connection of ep and pp scattering for $s \to \infty$. VAN HOVE [12]

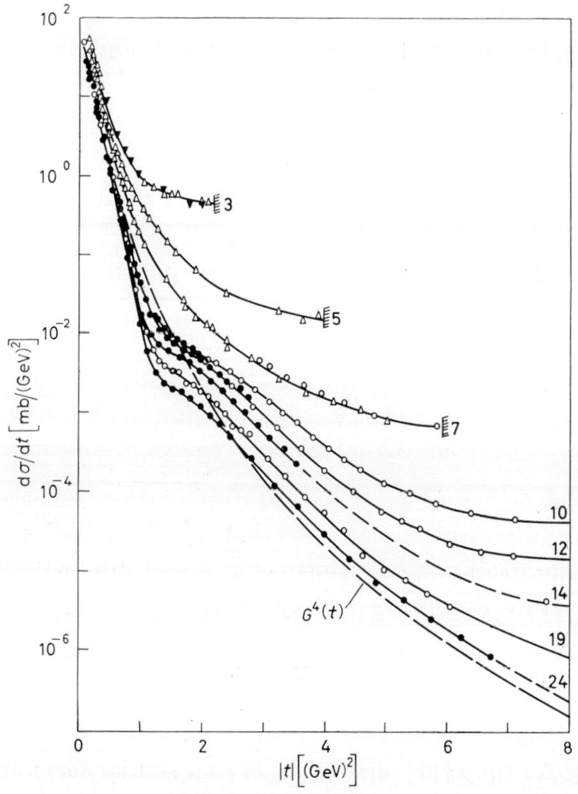

Fig. 6. – $d\sigma/dt$ for elastic pp scattering at different energies between $p_0 = 3$ GeV/c and $p_0 = 24$ GeV/c (from ref. [13]); momenta (GeV/c), ▵ CLYDE: 3.0, 5.0, 7.0, 7.1 [37]; ○ ALLABY *et al.*: 14.2 [19]; I 7.1, 10.1, 12.1 [9]; II 19.2 [9]; ▾ ANKENBRANDT *et al.*: 3.0 [38]; • ALLABY *et al.*: 10.0, 12.0, 14.2, 24.0 [25].

derived the same equation for small t in an additive quark model. Recent experimental results [13] together with earlier work allow a rather detailed comparison with the prediction of formula 4. The experimental data are plotted in Fig. 6 together with $G^4(t)$ for which the dipole form $G(t) = (1 + t/0.71)^{-2}$ was assumed. In the region of the diffraction peak for $t < 0.7\ (\text{GeV})^2$ the data roughly coincide with $G^4(t)$. So far, at large t the t-distribution does not approach an asymptotic curve, and by extrapolating the observed s dependence one can expect the experimental data at higher energy to fall below the $G^4(t)$ prediction.

2·6. *The « multiple-scattering » conjecture.* – The fact that the proton cross-section in Fig. 5 has structure, whereas the ep distribution is smooth, recalls nuclear scattering of electrons and protons. In a collision with a nucleus protons undergo multiple-scattering processes whereas electrons do not. This results in additional shoulders and minima in the t distributions of the scattered protons which, for example, in the case of pd scattering [14], resemble the structure at $t \approx 1.5\ (\text{GeV})^2$ in the pp cross-section in Fig. 5. Chou and Yang [15] derived a formula for elastic scattering of two extended bodies, which has the form of a multiple-scattering series [16]. The considerations are asymptotic; *i.e.* the calculated amplitude $A(t)$ is independent of s. To simplify one can write

$$A(t) = a_1 \exp[-b_1 t] - a_2 \exp[-b_2 t] + - \dots \tag{5}$$

with $a_2 \ll a_1$ and $b_2 \approx b_1/2$. The second term has a negative sign as in Glauber's theory [16] and the resulting cross-section can be sketched as follows:

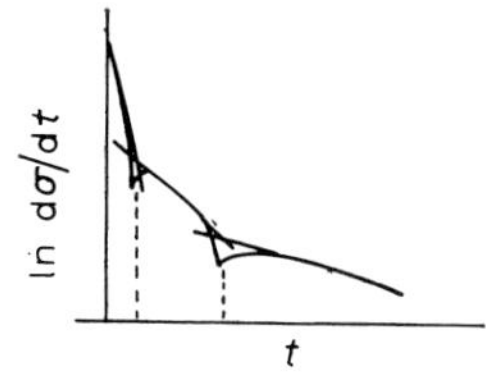

It has a gradually decreasing dependence on t and minima, with the position of the first structure depending mainly on b_1:

$$t_{\text{shoulder}} \approx \frac{2 \ln (a_1/a_2)}{b_1}\,.$$

Earlier work by Arnold [17] already leads to a similar form of the differential cross-section and, in addition, allows for an s-dependence in a phenomenological way by a Regge eikonal formalism. Many other papers [18] have been stimulated by the experiments on pp elastic scattering and in particular by the observations of structure in the angular distribution.

2·7. *Discussion of recent experimental data at intermediate momentum transfer.* – The experimental data plotted in Fig. 6 allow a fairly good analysis of the structure phenomenon. The fact that it begins to develop between $p_0 = 7$ GeV/c and $p_0 = 10$ GeV/c suggests a connection with the earlier observations of a change in the transverse-momentum dependence of large-angle scattering [19] and of a kink in the 90° scattering cross-section [8] in this momentum region. The latter phenomena have also been discussed as a reflection of baryon pair production [19, 20].

The energy-dependence of the structure is emphasized by plotting the data as in Fig. 7. Since this dependence is small compared to the variation with t in the range of the measurements, a representation of the data as relative cross-

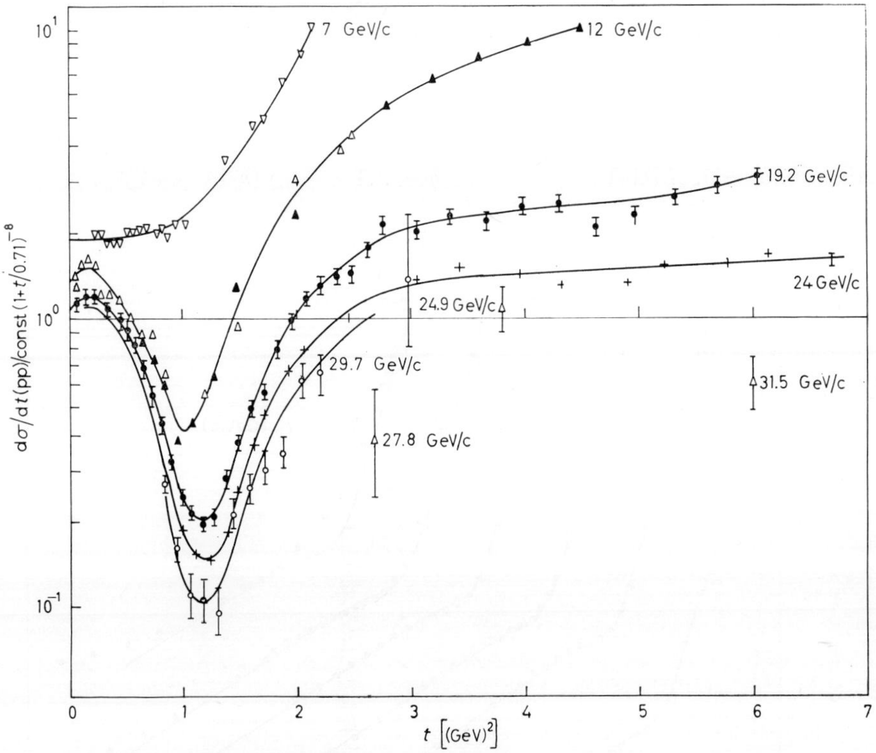

Fig. 7. – The ratio $(\mathrm{d}\sigma/\mathrm{d}t)(\mathrm{pp})/\mathrm{const}\,(1+t/0.71)^{-8}$ with const $= 80$ mb/(GeV)^2.

sections $(\mathrm{d}\sigma/\mathrm{d}t)/f(t)$ was chosen; $f(t)$ could be any function which approximates the t dependence, *e.g.*, $\mathrm{d}\sigma/\mathrm{d}t$ at a fixed p_0. The choice $f(t) = G^4(t)$ was prompted by the ideas discussed in Section **2·5**. From Fig. 7 it appears that the position of the structure in t does not change with energy in the range of the measurements. It also appears that the phenomenon becomes more pronounced at higher energies. This feature is characteristic, since in other reactions shoulders and dips tend to disappear with increasing energy with the exception, perhaps, of the minimum at $t = 2.8\ (\text{GeV})^2$ in π^-p elastic scattering [21].

The pp polarization distributions [6] also show structure at $t = (1 \div 2)(\text{GeV})^2$, which however exists as well at momenta below $p_0 = 7$ GeV/c. The polarization decreases with energy. Thus the polarization distribution and the shoulder in the differential cross-section have different behaviour in s. However, the coincidence of the structures in t can hardly be accidental.

An attempt was made [22] to parametrize the data plotted in Fig. 6 and 7 by a two-component amplitude:

$$A(t, s) = a_1 \exp[-b_1 t] + a_2 s^\alpha \exp[i\varphi] \exp[-(b_2 + b_3 \log s)t] .$$

This formula describes approximately the experimental results in the region $12 < p_0 < 24$ GeV/c and $0.2 < t < 4.5\ (\text{GeV})^2$. A best fit gave the following values for the constants a_2, α, φ, b_2, b_3.

$$a_2^2 \cdot s^{2\alpha} = 50 \cdot s^{2\alpha}\ \text{mb}/(\text{GeV})^2 , \qquad 2\alpha = -1.9 , \qquad 2b_2 = 1.1\ (\text{GeV})^{-2} ,$$

$$2b_3 = 0.17\ (\text{GeV})^{-2} , \qquad \varphi = 2.1 \qquad (s \text{ in } (\text{GeV})^2) .$$

Data, fit and an extrapolation to $p_0 = 70$ GeV/c are plotted in Fig. 8.

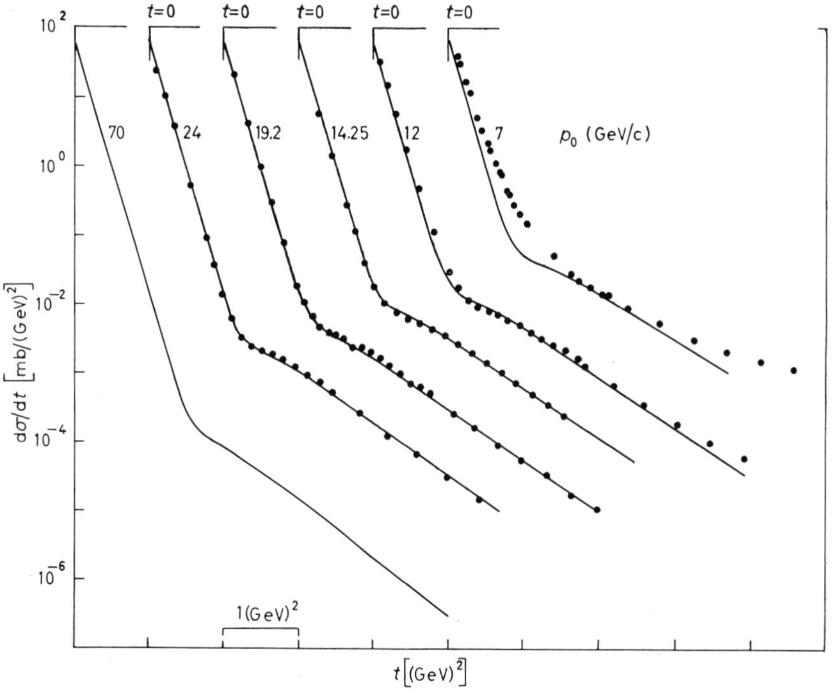

Fig. 8. – Experimental data of Fig. 6 together with a fit and an extrapolation of the fit to $p_0 = 70$ GeV/c, explanation in the text.

The discussion would be incomplete without a look at related reactions: pn elastic scattering was studied recently [23] by scattering protons at $p_0 = 24$ GeV/c on a deuterium target. The results are plotted in Fig. 9 together with the t distribution of the pp process. No difference was detected; it is interesting that the method allows a precise determination of the pn distribution at large t.

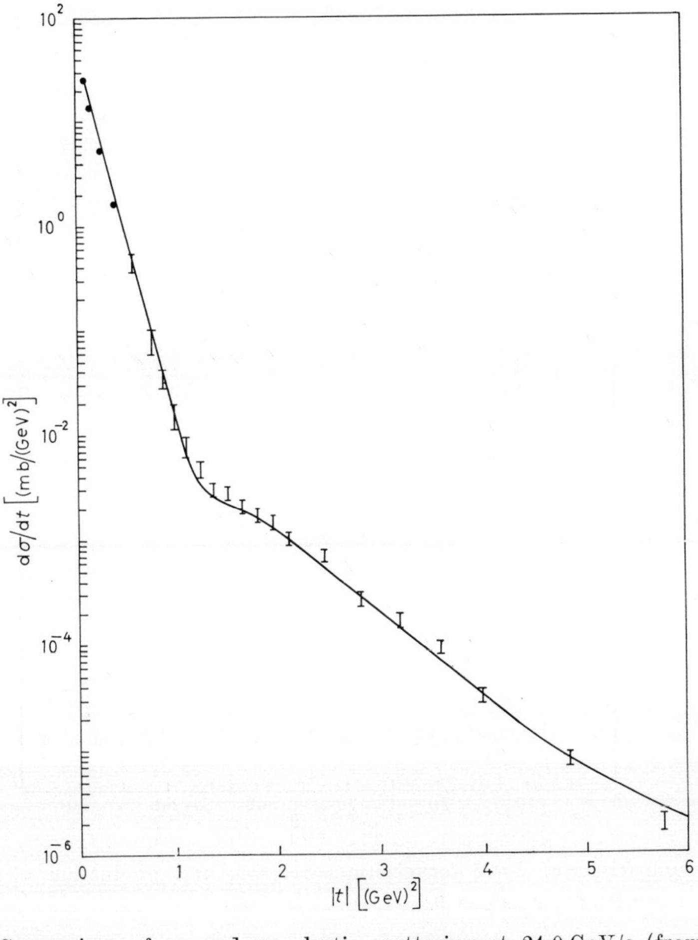

Fig. 9. – Comparison of pp and pn elastic scattering at 24.0 GeV/c (from ref. [23]): I p-n cross-sections from « single peak », this experiment; • p-n cross-sections from « single + double peak », this experiment; —— p-p cross-sections, ALLABY *et al.* [25].

Another related reaction is the production of nucleon resonances. The process pp → $\mathcal{N}^*$ (1690) + p is known [24] to have a diffractive nature. In comparison with the elastic case ($b = (8 \div 10)(\text{GeV})^{-2}$) the slope of the differential cross-section at small t is noticeably smaller. Since in models mentioned above the slope at large t is connected to the slope at small t, one would expect the differential cross-section to have a less strong t-dependence throughout the whole t range. Recently, this t distribution has been studied [25] up to $t = 6\ (\text{GeV})^2$ at $p_0 = 24\ \text{GeV}/c$. A change in slope was found in the t-region where the structure in the elastic reaction occurs. The two distributions are compared in Fig. 10. Above $t \approx 2\ (\text{GeV})^2$ the slopes are surprisingly similar, if not equal; there is therefore no simple connection between the distribution

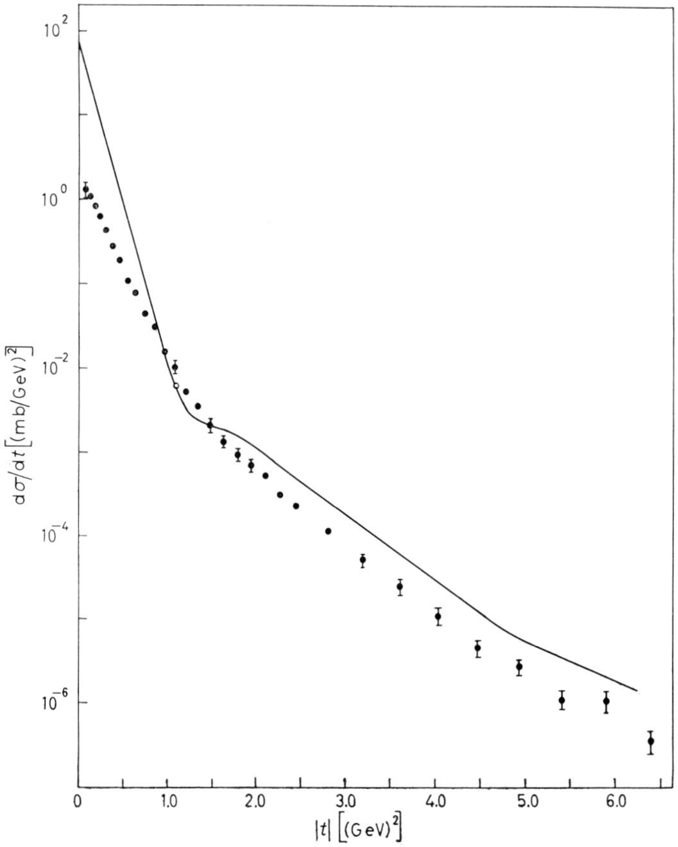

Fig. 10. – Comparison of $d\sigma/dt$ for elastic scattering and production of $\mathcal{N}^*(1690)$: 24 GeV/c; ——— $p+p \to p+p$, • $p+p \to p+\mathcal{N}^*(1690)$.

at small and at large t. The ratio $R_p = (d\sigma/dt)(p\mathcal{N}^*(1690))/(d\sigma/dt)(pp)$ has a value of $R \approx 0.4$ in the range $2 < t < 6\ (\text{GeV})^2$. This number seems to be independent of the nature of the incident particle, a similar value was observed for the ratio $R_e = (d\sigma/dt)(e\mathcal{N}^*(1690))/(d\sigma/dt)(ep)$ [26] around $t = (2 \div 3)\ (\text{GeV})^2$ and the results from pion experiments [27] on

$$R_\pi = \frac{(d\sigma/dt)(\pi\mathcal{N}^*(1690))}{(d\sigma/dt)(\pi p)}$$

at 16 GeV/c show the same trend as the electron data for $0.5 < t < 1.5$ (GeV)^2. R_p, R_e, R_π are plotted in Fig. 11. The characteristic behaviour of R_p in the range $0.5 < t < 2$ (GeV)^2 is due to the structure in the elastic reaction. No similar phenomenon occurs in any of the other reactions.

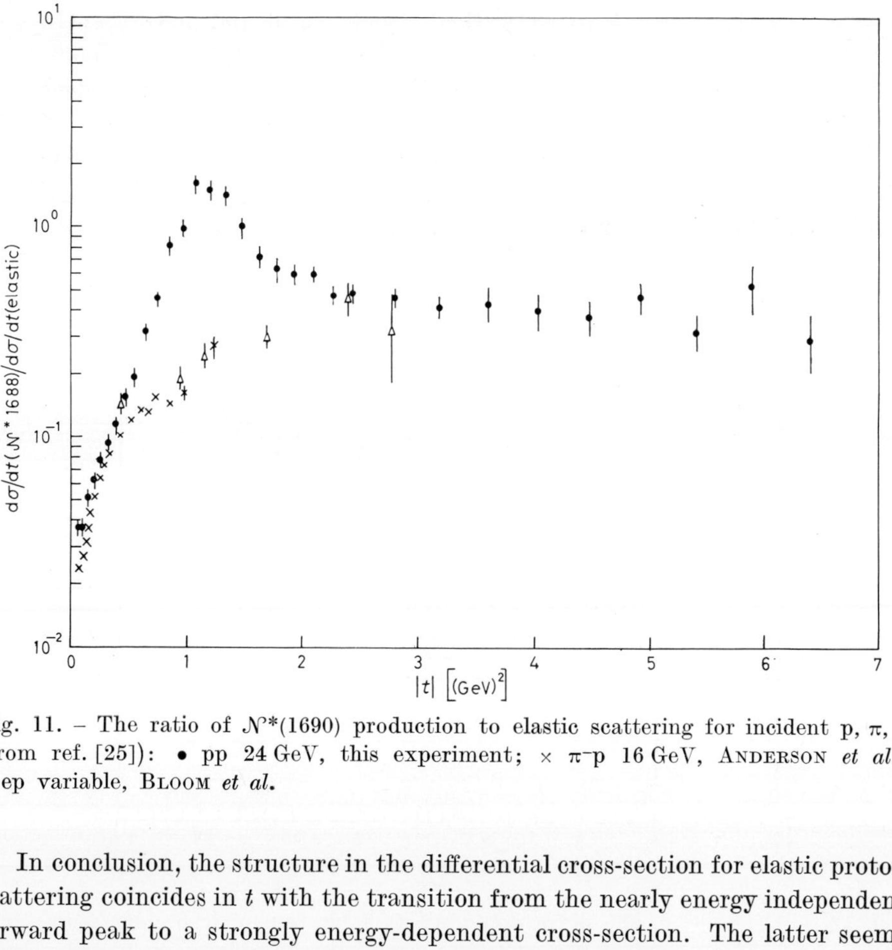

Fig. 11. – The ratio of $\mathcal{N}^*(1690)$ production to elastic scattering for incident p, π, e (from ref. [25]): • pp 24 GeV, this experiment; × π^-p 16 GeV, Anderson *et al.*; △ ep variable, Bloom *et al.*

In conclusion, the structure in the differential cross-section for elastic proton scattering coincides in t with the transition from the nearly energy independent forward peak to a strongly energy-dependent cross-section. The latter seems to be related to the distribution for $\mathcal{N}^*$ (1690) production. A Regge eikonal model cannot be excluded, it has to account for the very similar t-dependence on t of elastic scattering and resonance production above $t \approx 2$ (GeV)^2 and the pronounced shoulder which is typical for the elastic case only.

2·8. *Small-angle scattering.* – The scattering cross-sections of $\pi^\pm K^\pm$ and $p^\pm$ on protons at energies above the threshhold for π production exhibit a diffraction maximum at small scattering angles. The slope

$$b = \frac{d}{dt} \ln \frac{d\sigma}{dt}\bigg|_{t=0}$$

of the diffraction peak varies rapidly with momentum up to $p_0 = 3$ GeV/c. This variation is roughly a simple rise for K$^+$ and p. For π^+, $\bar{K}$ and p, b exhibits dips and bumps in the neighbourhood of s-channel resonances or maxima in the total cross-section. Above $p_0 = 3$ GeV/c the strongest variation is in K$^+$p from $b = 3$ (GeV)$^{-2}$ at $p_0 = 3$ GeV/c to $b = 7$ (GeV)$^{-2}$ at $p_0 \approx 16$ GeV/c; in π^-p b varies from 7 to ~ 10 (GeV)$^{-2}$ between 3 and 20 GeV/c; in π^+p from ~ 8 at 3 GeV/c to ~ 11 (GeV)$^{-2}$ at 25 GeV/c; in K$^-$p, b is about constant, $b \approx 7.5$ (GeV)$^{-2}$ from 3 to 16 GeV/c; in $\bar{\text{p}}$p b is constant or decreases slightly between $b = 14$ and $b = 12$ (GeV)$^{-2}$ from 3 GeV/c to 16 GeV/c. The pp slope increases from ~ 8 (GeV)$^{-2}$ at 3 GeV/c to ~ 11.5 (GeV)$^{-2}$ at 70 GeV/c. These latter results [28] are given in Fig. 12. Since the strong interaction is dominated by Coulomb forces near $t = 0$, all measurements of the slope para-

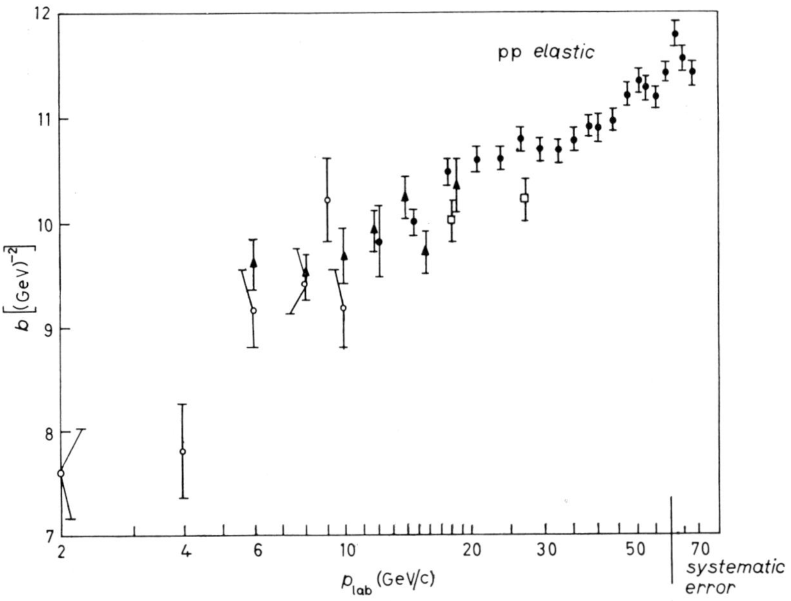

Fig. 12. – The slope of the diffraction peak as a function of laboratory momentum (data of ref. [28] and earlier work quoted therein).

meter b are performed at $t \neq 0$. The result of a precise experiment is likely to depend on the range in t of the data if the shape of the diffraction maximum is not exactly exponential. According to CARRIGAN [29], the results of experiments in different t-regions show systematic differences (Fig. 13). For $t < 0.11$ (GeV)^2, the measured slopes are around or above $b = 10$ (GeV)$^{-2}$. If the minimum t is > 0.13 GeV, the results scatter around $b \approx 8$ (GeV)$^{-2}$. Carrigan's observation seems to be confirmed by recent experiments at the ISR. In two independent measurements [30, 31] the slope parameter was

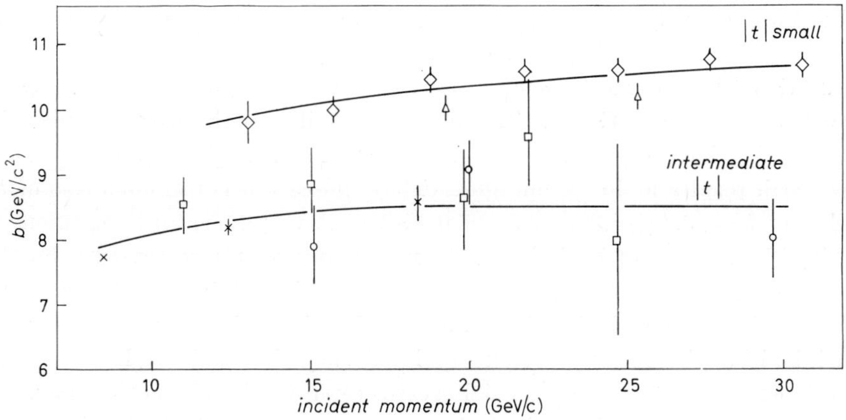

Fig. 13. – Comparison of measurements in different t intervals (from ref. [29]): ◇ SERPUKOFF, $|t| < 0.11$; ▵ BELLETTINI *et al.*, $|_1| < 0.11$; ▫ FOLEY *et al.*, $0.13 < |t| < 0.8$; ○ BNL-CMU, $0.13 < |t| < 0.8$; × HARTING *et al.*, $0.13 < |t| < 0.8$.

determined at corresponding laboratory momenta of $p_0 \approx 500$ GeV/c, 1000 GeV/c and 1400 GeV/c. Because of the geometry of the ISR vacuum chamber, the accessible regions in scattering angle for the two measurements were fixed to angles from $(7 \div 16)$ mrad [31] and $(18 \div 30)$ mrad [31] respec-

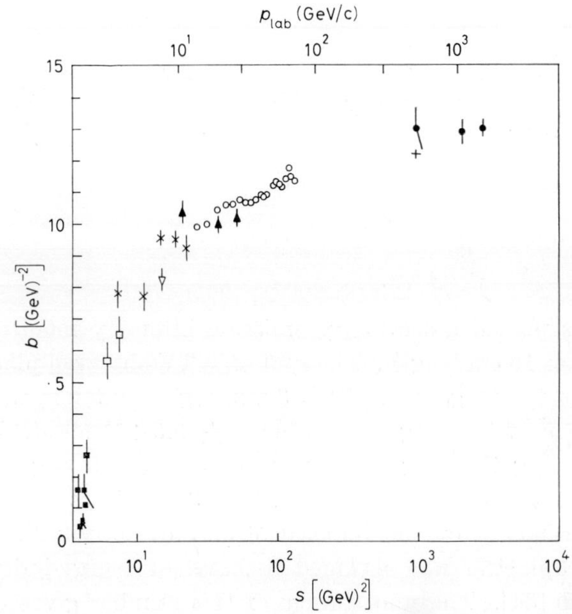

Fig. 14. – The slope for $0.01 < t < 0.16$ (GeV)^2 (from ref. [31]): ▪ compilation, LASINCKI *et al.*; ▫ FUJII *et al.*; ▿ CLYDE *et al.*; × AKIMOV *et al.*, KIRILLOVA *et al.*; ▴ BELLETTINI *et al.*; ○ BEZNOGIKH *et al.*; + HOLDER *et al.*; • this experiment.

tively. Consequently, the t-region of the two experiments was different and depended on energy. The results at larger angle [30] indicated a slope $b = 12.2\ (\text{GeV})^{-2}$ at $s \approx 950\ (\text{GeV})^2$, and lower values, around $b \sim 10.2\ (\text{GeV})^{-2}$ at $s = 2030$ and $2800\ (\text{GeV})^2$. The data at smaller angles [31] did not show this « antishrinkage » effect. These measurements are plotted in Fig. 14 together with results at lower energies. They indicate a continuous rise of b to a value of $b = 13.4 \pm 0.3$ at $s \approx 2800\ (\text{GeV})^2$. A very recent measurement [32] covering the t range of both the first experiments seems to reveal a break in the t-dependence at $t \approx 0.12$ in agreement with Carrigan's observation at lower energies. In conclusion, it seems that the diffraction maximum continues to shrink although not as fast as extrapolated from the results at lower energies. The t-dependence of the differential cross-section in the diffraction maximum appears to change rather abruptly around $t \approx 0.12\ (\text{GeV})^2$ in the 30 GeV range as well as at ISR energies. In a first interpretation, the effect was connected to the production of nucleon resonances [33] via an intermediate state involving $\mathcal{N}^*(1400)$.

2·9. *Experimental techniques.*

2·9.1. 27 GeV/c Single-arm spectrometer. – An apparatus [13] used to study elastic scattering and $\mathcal{N}^*$ production up to $p_0 = 24$ GeV/c is depicted in Fig. 15. The ejected proton beam of the CERN PS with 10^{12} protons in

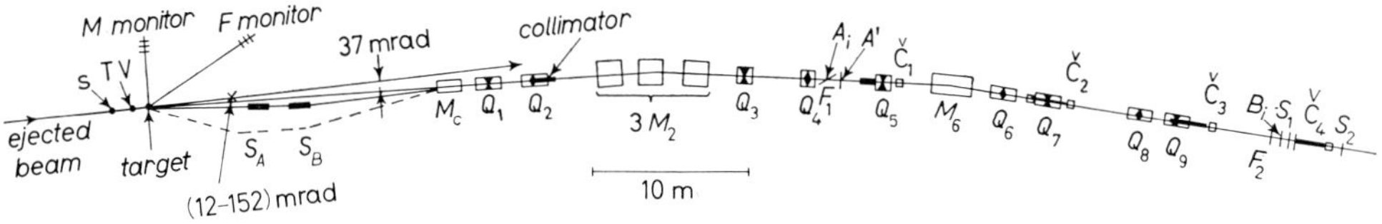

Fig. 15. – Single-arm spectrometer (from ref. [13]).

200 ms every 2 s passed a secondary emission intensity monitor and a liquid hydrogen target of 10 cm length. The spot on a TV observed plastic scintillator screen in front of the target was about 3 mm high and 5 mm wide. The divergence of the beam at the target was about $\pm(0.5 \div 1)$ mrad. For additional intensity monitoring the target was viewed by a small angle (F, 30°) and a large angle (M, 120°) counter telescope. The 3 monitors were calibrated by repeated measurements of the induced ^{24}Na radioactivity in Al foils. The reaction ^{27}Al p, 3pn ^{24}Na was assumed to have an energy-independent cross-section of 8.6 mb [34]. The uncertainty of this number gives rise to a systematic error of 8 % in all results obtained with this set up.

The scattered particles were observed by a single-arm spectrometer. The two movable septum-window-frame magnets S_A and S_B of $\int B\,\text{d}l \approx 4(\text{Wb/m})$

each, together with the C magnet M_c allowed the deflection of scattered beams from 12 to 152 mrad scattering angle into a fixed angle (37 mrad) spectrometer. This steering stage was followed by a momentum-analysing stage consisting of 4 quadrupole magnets Q_1 - Q_4 and three $2m$ deflection magnets $3M_2$. The dispersive image of the target at the focus F_1 showed a demagnification of 0.5 in the horizontal plane and 0.9 in the vertical plane. The magnets $3M_2$ deflected the beam by 120 mrad and produced a dispersion of 1.5 cm/%(Δp/p). Thus, the momentum resolution of the instrument was of the order of ± 0.3 %. Particles were detected at F_1 by a ladder of 9 scintillation counters 3 mm wide and 15 mm high, staggered over about 1.5 m downstream along the focal line. The total momentum bite accepted by these counters was about ± 1 %. The solid angle acceptance determined by the cylindrical collimator placed in Q_2 varied less than 1 % over the total momentum interval and amounted to $\cdot 10^{-5}$ sr for the largest collimator. The momentum-analysing stage was followed by a background rejection and particle identification stage, consisting of 5 quadrupole magnets Q_5 - Q_9, the dispersion-compensating 6 m long magnet M_6, the counters A', B_{1-5}, S_1, S_2 and the threshhold Čerenkov counters $\check{C}_1$ - $\check{C}_4$. The image of the target at the focus F_2 showed a magnification of 1:1 vertically and horizontally. The symmetric triplet $Q_3Q_4Q_5$ imaged the central plane of $3M_2$ into the central plane of M_6. Thereby, the dispersion compensation was improved so that by the use of 5 counters B_1 - B_5 at F_2 the origin of a particle in the target plane could be determined with a resolution of about 1 cm. This gave a continuous position control of the incident proton beam and allowed the use of a 20 cm long target for small cross-section work at large angles without excessive loss in momentum resolution. A parallel beam section between Q_7 and Q_8 was planned to allow for the eventual use of a differential Čerenkov counter. In this parallel section the emission angle of a particle in the target plane could be resolved to about 0.5 mrad. The total length of the instrument was 80 m. All magnets were set automatically to the appropriate currents for the desired momentum and scattering angle. The electronic signal for the detection of a particle was a coincidence between counters $A_iA'B_iS_1\,S_2$. $A'S$, S_2 were large scintillation counters providing the necessary redundancy. The Čerenkov counters were intended mainly for particle production studies and are of minor importance here. A typical momentum spectrum obtained with this apparatus at $p_0 = 24$ GeV/c and fixed scattering angle $\theta = 37$ mrad is plotted in Fig. 16. By integrating the elastic maximum appearing at $p_{lab} = 23.6$ GeV/c, the differential cross-section $d\sigma/dt$ at $t = 0.79$ (GeV)^2 was-obtained. The enhancements occuring at 22.9, 22.5 and 21.5 GeV/c are due to protons from the reactions $pp \to p\mathcal{N}^*$ for $\mathcal{N}^*$ (1520), $\mathcal{N}^*$ (1690), $\mathcal{N}^*$ (2190). A delicate background problem is involved in the evaluation of cross-sections for these processes.

Spectrometers have been built which are by an order of magnitude superior in performance [35]. The advantage of this design lies in the low cost. The

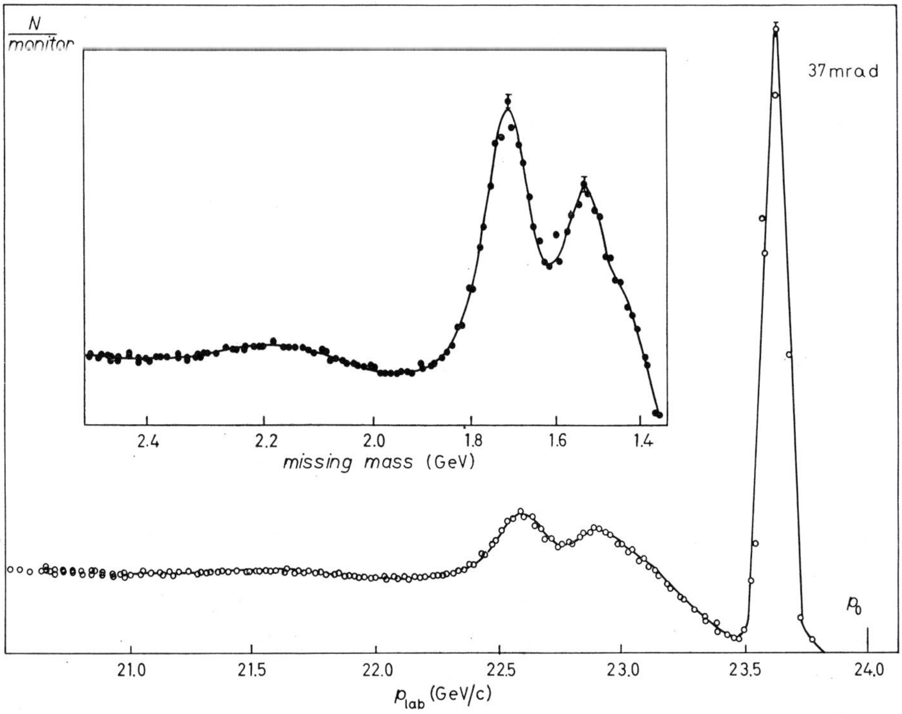

Fig. 16. – Momentum spectrum of scattered protons at fixed angle (data or ref. [13]).

disadvantages are a relatively low acceptance because of the space-consuming stearing stage in front of the actual spectrometer and a coupling in the observation of the angular and the momentum dependence of scattered particle spectra. This arises because the dispersion acts in the scattering plane. Finally, the t-range of the instrument was limited to $t_{MAX} = 6\ (GeV)^2$ at $p_0 = 24$ GeV/c by the bending power of the steering stage. It was not limited by background problems due to the single-arm technique. A movable arm spectrometer with vertical dispersion would not have these disadvantages.

2'9.2. Recoil-proton spectrometer. – Small-angle scattering can also be observed by detecting the low-enery recoil proton. This technique was developped since several years ago by the Dubna group [4, 28]. Since the energy and angle of the recoiling proton are related to t and in first approximation independent of p_0, the same apparatus is suitable to investigate a given t-region over the whole range in s. The apparatus which was used in experiments of diffraction scattering and Coulomb interference at Serpukhov [4, 28] is shown in Fig. 17. For the recent measurements at very small angles the CH_2 foil target was replaced by a hydrogen gas jet. Si detectors were used to determine the scattering angle and the energy of the recoiling proton. The ap-

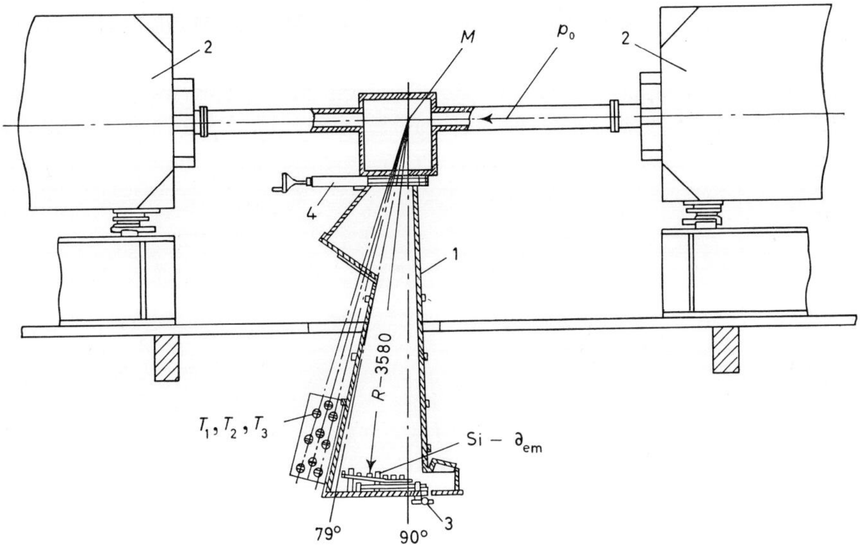

Fig. 17. – Recoil-proton spectrometer (from ref. [28]).

paratus was placed between two accelerator magnets (2) in the circulating proton beam.

2'9.3. Experimental techniques at the ISR. – Measurements with the CERN colliding proton-beam facility differ in several respects from conventional proton elastic-scattering experiments. The laboratory system coincides nearly with the c.m. system. The velocity β of the c.m.s. is small and directed radially. β is approximately independent of the energy; $\beta \approx \sin\theta/2 \approx 0.13$, where $\theta = 15°$ is the crossing angle of the two beams. However, compared to typical transverse velocity components of scattered or produced particles, β is not negligible. Second, the determination of a cross-section involves knowledge of the luminosity L. The number N_1 of interactions per s for a cross-section σ is $N_I = \sigma \cdot L$, where L stands for

$$L = \frac{I_1 \cdot I_2}{h_{\text{eff}} \cdot c \cdot e^2 \cdot \text{tg}\,(\theta/2)}\,.$$

This formula is valid for interacting relativistic proton currents I_1, I_2. h_{eff} is the effective height of the interacting beams:

$$\frac{1}{h_{\text{eff}}} = \frac{\int S_1(z) \cdot S_2(z)\,\text{d}z}{\int S_1(z)\,\text{d}z \cdot \int S_2(z)\,\text{d}z}\,,$$

$S_1(z)$, $S_2(z)$ are the two current-density distributions h_{eff} can be determined by observing the interaction rate when the two beams are displaced with respect to each other in vertical direction (van der Meer method [36]). The result of such a measurement [30] is shown in Fig. 18. The effective beam height was found to be $h_{eff} = 5.39$ mm. The design value for the luminosity is $L \approx 4.10^{30}\ cm^{-2}\ s^{-1}$ leading to an interaction rate of $N_I \approx 10^5/s$ for a cross-section of $\sigma = 40$ mb. Thus the ISR is a low-intensity machine, N_I being

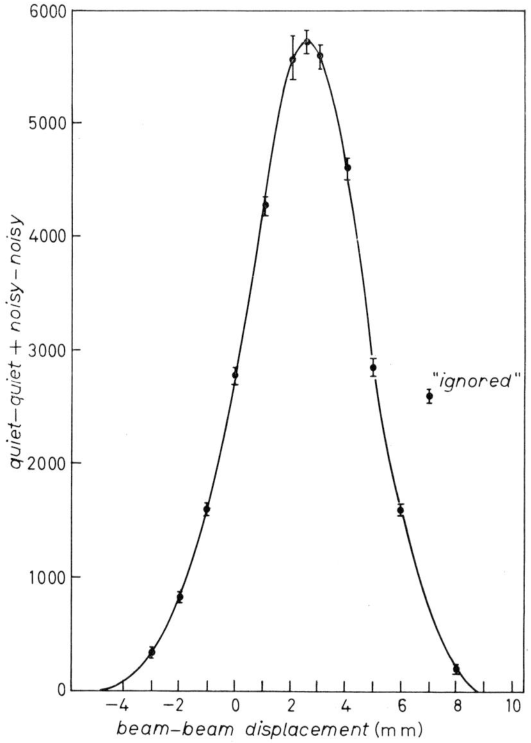

Fig. 18. – The counting rate as a function of relative beam-beam displacement (ref. [30], private communication P. STROLIN): (15.34 + 15.34) GeV/c, total area = = 30.870 counts/mm, $h_{eff} = 5.39$ mm, April 5, 1971.

lower by $10^5 \div 10^7$ compared to experiments with stationary targets. Third, the single-arm spectrometer missing-mass resolution Δm is of the order of several GeV in contrast to $\Delta m \approx 0.1$ GeV in the conventional experiment described above. This is due to the large momentum spread of the colliding beams (at least 100 MeV/c; ± 400 MeV/c for the full stack) and due to kinematics (laboratory frame $\approx$ c.m.s.). Thus, elastic scattering cannot be detected by single-arm spectrometer techniques.

The experimental set up used in one of the first experiments [31] is shown in Fig. 19. Four scintillation-counter hodoscopes *ABCD* were placed at about 5 m distance from the interaction region above and below the vacuum pipe,

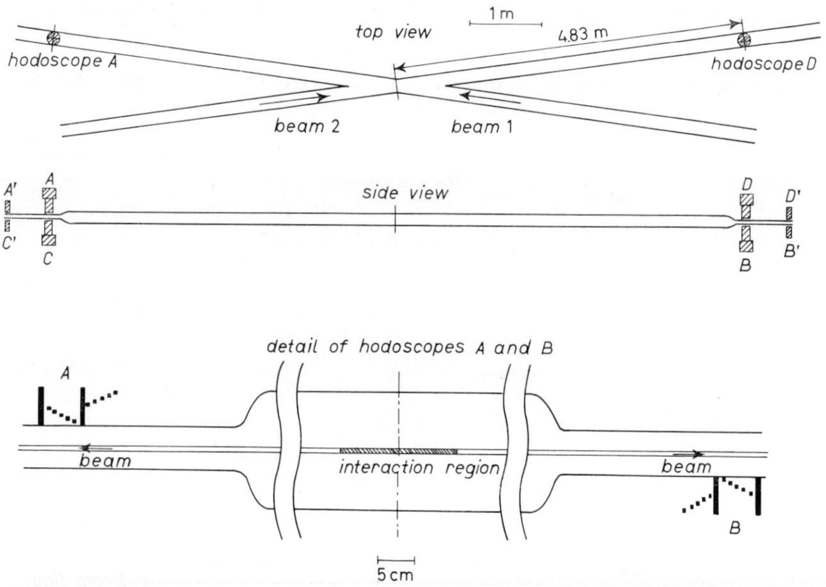

Fig. 19. – Apparatus to observe small-angle elastic scattering at the ISR (from ref. [31]).

2 on each side, each containing 10 scintillators 4.5 mm high and 10 mm wide in size and vertically staggered. An event was recorded whenever a left-right coincidence occured. Elastic events were selected only by the requirement of collinearity. This is illustrated by Fig. 20 *a*) where the two-dimensional distrib-

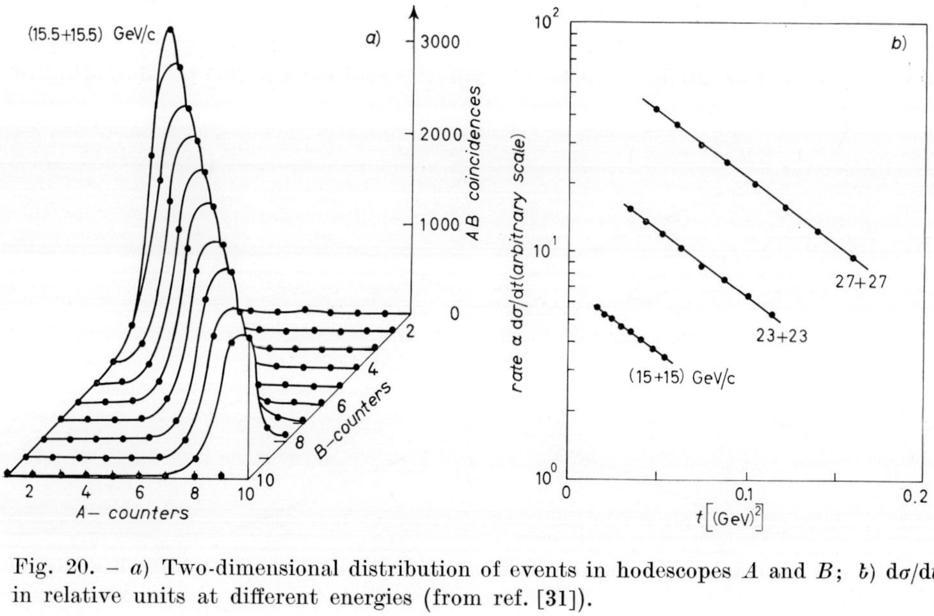

Fig. 20. – *a*) Two-dimensional distribution of events in hodescopes A and B; *b*) $d\sigma/dt$ in relative units at different energies (from ref. [31]).

ution of events in hodoscopes A and B is plotted. The majority of events are collinear; noncoplanar events contribute a few percent only. The resolution in collinearity angle was ± 1 mrad. The t-range of the hodoscopes was de-

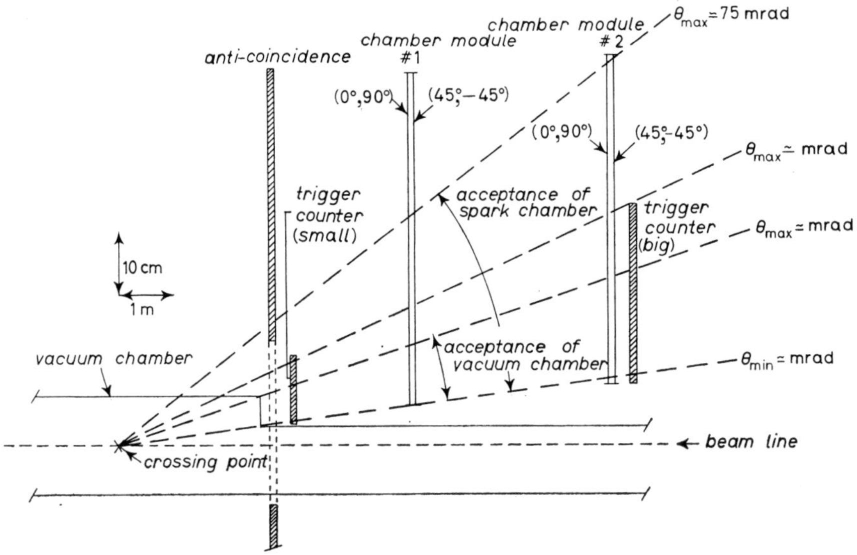

Fig. 21. – Spark-chamber set-up to observe small-angle scattering at the ISR (one arm only) (from ref. [30]): forward telescope.

termined from geometry. The counting rate was of the order of 1 event/s for $I_1 = I_2 = 1A$ ($2 \cdot 10^{13}$ p's stored in each ring). The decay rate of the stored beams was normally less than 10^{-3}/h. Resulting t-distributions for 3 different energies are plotted in Fig. 20 b). The other experimental set up [30] for detection of

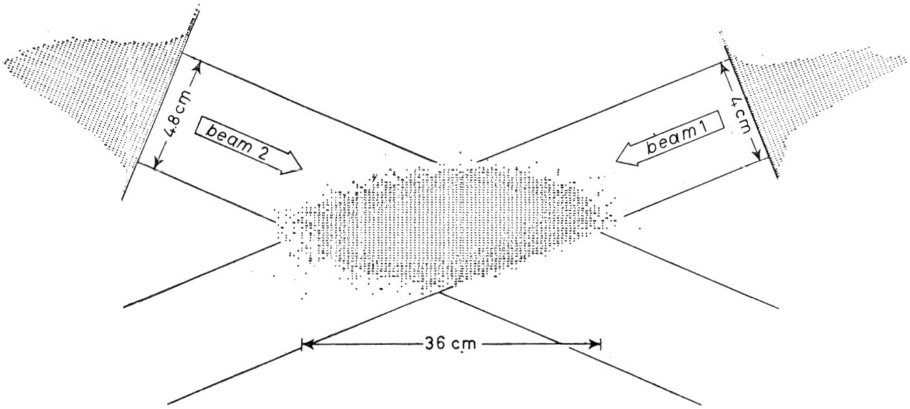

Fig. 22. – « Diamond » distribution of interaction points (from ref. [30]): beam-beam diamond.

small-angle scattering used magnetostrictive spark chambers instead of scintillation counter hodoscopes. One arm of this instrument is shown in Fig. 21. The limitation imposed by the ISR vacuum chamber on the angular acceptance is clearly visible. By reconstructing the trajectories of the scattered protons the « diamond » distribution of interaction points in Fig. 22 was obtained.

3. – Concluding remark.

High-energy nucleon scattering is an open field. The study of structure phenomena in the t-distributions is potentially of fundamental value, since information on basic properties of high-energy amplitudes and quarklike substructure is obtained. More data are forthcoming. The author hopes, that elastic-scattering experiments with protons may again contribute to physics as much as the low-energy studies mentioned at the beginning of these notes.

REFERENCES

[1] M. G. White: *Phys. Rev.*, **49**, 309 (1936); M. A. Tuve, N. P. Heydenburg and L. R. Hafstad: *Phys. Rev.*, **50**, 806 (1936).

[2] A. Scotti and D. Y. Wong: *Phys. Rev.*, **138** B, 145 (1965).

[3] H. Holder, E. Rademacher, A. Staude, G. Barbiellini, P. Darriulat, M. Hansroul, S. Orito, P. Palazzi, A. Santroni, P. Strolin, K. Tittel, J. Pilcher, C. Rubbia, G. De Zorzi, M. Macri, G. Sette, A. Fainberg, C. Grosso-Pilcher and G. Maderni: *Phys. Lett.*, **35** B, 361 (1971).

[4] V. D. Bartenev, G. G. Beznogikh, A. Buyak, K. I. Iovchev, L. F. Kirillova, P. K. Markov, B. A. Marozov, V. A. Nikitin, P. V. Nomokonov, U. K. Philipenko, A. Sandach, M. G. Shafranova, V. A. Sviridov, Truong Bien, V. I. Zayachki, N. K. Zhidkov and L. S. Zolin: *Cont. to International Conference on High-Energy Physics* (Kiev, 1970) and references quoted therein.

[5] P. Söding: *Phys. Lett.*, **8**, 285 (1964).

[6] M. Borghini, L. Dick, L. Di Lella, A. Navarro, J. C. Oliver, K. Reibel, C. Coignet, D. Kronenberger, G. Gregoire, K. Kuroda, A. Michalowicz, M. Poulet, D. Sillov, G. Bellettini, P. L. Braccini, T. Del Prete, L. Foa, G. Sanguinetti and M. Valdata: *Phys. Lett.*, **31** B, 405 (1970).

[7] J. Orear: *Phys. Lett.*, **13**, 190 (1964).

[8] C. W. Akerlof, R. H. Hieber, A. D. Krisch, K. W. Edwards, L. G. Ratner and K. Ruddick: *Phys. Rev. Lett.*, **17**, 1105 (1966).

[9] J. V. Allaby, F. Binon, A. N. Diddens, P. Duteil, A. Klowning, R. Meunier, J. P. Peigneux, E. J. Sacharidis, K. Schlüpmann, M. Spighel, J. P. Stroot, A. M. Thorndike and A. M. Wetherell: *Phys. Lett.*, **28** B, 67 (1968).

[10] W. Albrecht, H. J. Berend, F. W. Brasse, W. Flauger, H. Hultschig and K. G. Steffen: *Phys. Rev. Lett.*, **17**, 1192 (1966) and references quoted therein.

[11] T. T. Wu and C. N. Yang: *Phys. Rev.*, **137** B, 709 (1965).

[12] L. Van Hove: *Proceedings of the Conference on High-Energy Two-Body Reactions* (Stony Brook, 1966); L. Van Hove: in *Particle Interactions at High Energies,*

Scottish Universities Summer School, 1966, edited by T. W. PREIST and L. L. VICK (London, 1067), p. 63.
[13] U. AMALDI, R. BIANCASTELLI, C. BOSIO, G. MATTHIAE, J. V. ALLABY, A. N. DIDDENS, R. W. DOBINSON, A. KLOWNING, J. LITT, L. S. ROCHESTER, K. SCHLÜPMANN and A. M. WETHERELL: *Phys. Lett.*, **34** B, 431 (1971).
[14] F. BRADAMANTE, G. FIDECARO, M. FIDECARO, M. GIORGI, P. PALAZZI, A. PENZO, L. PIEMONTESE, F. SAULI, P. SCHIAVON and A. VASCOTTO: *Phys. Lett.*, **32** B, 303 (1970).
[15] T. T. CHOU and C. N. YANG: *Phys. Rev.*, **175**, 1832 (1968).
[16] R. J. GLAUBER: in *Lectures in Theoretical Physics*, edited by W. E. BRITTIN *et al.*, Vol. **1** (New York, 1959), p. 315; R. J. GLAUBER: *High-energy physics and nuclear structure*, in *Proceedings of the II International Conference on Rehovoth, 1967*, edited by G. ALEXANDER (Amsterdam, 1967), p. 311.
[17] R. C. ARNOLD: *Phys. Rev.*, **153**, 1523 (1967).
[18] A. P. CONTOGOURIS: *Phys. Lett.*, **23**, 698 (1966); K. HUANG, C. E. JONES and V. L. TEPLITZ: *Phys. Rev. Lett.*, **18**, 146 (1966); A. A. ANSELM and I. T. DYATLOV: *Phys. Lett.*, **24** B, 479 (1967); V. FRANCO: *Phys. Rev. Lett.*, **18**, 1159 (1967), D. R. HARRINGTON and A. PAGNAMENTA: *Phys. Rev. Lett.*, **18**, 1147 (1967), A. I. ABARBANEL, S. D. DRELL and F. J. GILMAN: *Phys. Rev. Lett.*, **20**, 280 (1968), L. BENOFY, D. W. CHO and E. SHRAUNER: *Phys. Rev. Lett.*, **20**, 1258 (1968); C. B. CHIU and J. FINKELSTEIN: *Nuovo Cimento*, **57** A, 649 (1968); **59** A, 92 (1969); L. DURAND III and R. LIPES: *Phys. Rev. Lett.*, **20**, 637 (1968); H. FLEMING, A. GIOVANNI and E. PREDAZZI: *Nuovo Cimento*, **56** A, 1131 (1968); S. FRAUTSCHI and B. MARGOLIS: *Nuovo Cimento*, **56** A, 1115 (1968); **57** A, 427 (1968); M. M. ISLAM and J. ROSEN: *Phys. Rev. Lett.*, **22**, 502 (1969); M. JACOB and S. POKORSKI: *Nuovo Cimento*, **61** A, 233 (1969); K. H. MÜTTER and E. TRÄNKLE: *Zeits. Phys.*, **239**, 211 (1970).
[19] J. V. ALLABY, G. COCCONI, A. N. DIDDENS, A. KLOWNING, G. MATTHIAE, E. J. SACHARIDIS and A. M. WETHERELL: *Phys. Lett.*, **25** B, 156 (1967).
[20] J. J. J. KOKKEDEE and L. VAN HOVE: *Phys. Lett.*, **25** B, 228 (1967).
[21] C. BAGLIN, P. BRIANDET, P. J. CARLSON, V. CHABAUD, M. DAVIER, A. EIDE, B. ENGLUND, P. FLEURY, V. GRACCO, E. JOHNNSON, P. LEHMANN, A. LUNDBY, R. MORAND, S. MUKHIN, J. MYRHEIM and D. TREILLE: *Contribution to the International Conference on Elementary Particles* (Amsterdam, 1971).
[22] L. ROCHESTER: private communication.
[23] U. AMALDI, R. BIANCASTELLI, C. BOSIO and G. MATTHIAE, J. V. ALLABY, A. N. DIDDENS, R. W. DOBINSON, A. KLOVNING, J. LITT, L. S. ROCHESTER. K. SCHLÜPMANN and A. M. WETHERELL: submitted to *Nucl. Phys.*, B.
[24] E. W. ANDERSON, E. J. BLESER, G. B. COLLINS, T. FUJII, J. MENES, F. TURKOT, R. A. CARRIGAN, R. M. EDELSTEIN, N. C. HIEN, T. J. MCMAHON and I. NADELHAFT: *Phys. Rev. Lett.*, **16**, 855 (1966).
[25] U. AMALDI, R. BIANCASTELLI, C. BOSIO, G. MATTHIAE, J. V. ALLABY, A. N. DIDDENS, R. W. DOBINSON, A. KLOWNING, J. LITT, L. S. ROCHESTER, K. SCHLÜPMANN and A. M. WETHERELL: *Phys. Lett.*, **34** B, 435 (1971).
[26] E. D. BLOOM, G. BUSCHHORN, R. L. COTTRELL, D. H. COWARD, H. DE STAEBLER, J. DREES, C. L. JORDAN, G. MILLER, L. MO, H. PIEL, R. E. TAYLOR, M. BREIDENBACH, W. R. DITZLER, J. I. FRIEDMANN, G. C. HARTMANN, H. W. KENDALL and J. S. POUCHER: paper presented at the *XV International Conference on High-Energy Physics* (Kiev, 1970).
[27] E. W. ANDERSON, E. J. BLESER, H. R. BLIEDEN, G. B. COLLINS, D. GARELICK, J. MENES, F. TURKOT, D. BIRNBAUM, R. M. EDELSTEIN, N. C. HIEN, J. J. MCMAHON, J. F. MUCCI and J. S. RUSS: *Phys. Rev. Lett.*, **25**, 699 (1970).

[28] G. G. Beznogikh, A. Buyak, K. I. Iovchev, L. F. Kirillova, P. K. Markov, B. A. Morozov, V. A. Nikitin, P. V. Nomokonov, M. G. Shafranova, S. B. Nurushev and V. L. Solovianov: *Phys. Lett.*, **30** B, 274 (1969).
[29] R. A. Carrigan: *Phys. Rev. Lett.*, **24**, 168 (1970).
[30] M. Holder, E. Radermacher, A. Staude, G. Barbiellini, P. Darriulat, M. Hansroul, S. Orito, P. Palazzi, A. Santroni, P. Strolin, K. Tittel, J. Pilcher, C. Rubbia, G. De Zorzi, M. Macri, G. Sette, A. Fainberg, C. Grosso-Pilcher and G. Maderni: *Phys. Lett.*, **35** B, 355 (1971); and submitted to *Phys. Lett.* (July 1971).
[31] U. Amaldi, R. Biancastelli, C. Bosio, G. Matthiae, J. V. Allaby, G. Cocconi, A. N. Diddens, R. W. Dobinson, V. Elings, J. Litt, L. S. Rochester and A. M. Wetherell: paper contributed to the *International Conference on Elementary Particles* (Amsterdam, 1971).
[32] P. Darriulat and K. Tittel: private communication.
[33] R. M. Edelstein: submitted to *Phys. Lett.* (Aug. 1971).
[34] J. B. Cumming: *Ann. Rev. Nucl. Sci.*, **13**, 261 (1963).
[35] L. Mo and C. Peck: SLAC report T.N.-65-29 (1965).
[36] S. Van der Meer: unpublished, CERN-ISR-PO/68-31 (June 1968).
[37] A. R. Clyde: UCRL-16275 (1966).
[38] C. M. Ankenbrandt, A. R. Clark, B. Lork, T. Elioff, L. T. Kerth and W. A. Wenzel: *Phys. Rev.*, **170**, 1223 (1968).

Experiments and Problems in High-Energy Neutrino Physics.

C. Franzinetti

Istituto di Fisica dell'Università - Torino

Introduction.

Weak interactions in low-energy processes so far observed, are well accounted for by a comparatively simple first-order perturbation theory and a number of « rules », some of which are supposed to be « strictly observed », while others have only a partial validity.

The study of weak interactions, in processes in which high-energy transfers are involved, has begun only in recent times ($\sim$10 years ago) and our knowledge of them is very scarce. It is based entirely on the study of neutrino reactions on nuclei and nucleons, observed in bubble chambers and multitonn optical spark chambers (see Table I). Up to the present, not more than 2000 events have been analysed in bubble chamber pictures and about ten thousand in spark chamber pictures.

Thus a comparison with theory has been carried out only to the extent of testing orders of magnitude rather than precise numbers. Nevertheless, some important discoveries and conclusions have been reached.

As for the future, new experiments with new powerful machines and detectors are about to begin or have just begun: at CERN, the Gargamelle heavy-liquid bubble chamber is now working with freon (CF_3Br) and is recording neutrino events at a rate of 1 ev./45 s; at Serpukhov, another large theory liquid bubble chamber (CKAT) should soon be ready and be used on a neutrino beam; at Batavia several experiments are planned with a variety of techniques for the same purpose. Thus there is hope that new light will be thrown soon on this interesting subject.

In these two lectures I propose to present a brief discussion of some particular problems on which—for different reasons—interest has been focussed recently.

TABLE I. – *Chronology of neutrino experiments.*

Laboratory	Detector	Period
Brookhaven	Spark chamber	1960
CERN	Spark chamber and freon-filled bubble chamber (1 m)	1963-65
CERN	Propane-filled bubble chamber (1 m)	1967
ANL (Argonne)	Spark chamber	1967
CERN	Freon-filled bubble chamber (Gargamelle)	1971
ANL (Argonne)	H_2-filled bubble chamber (12 ft)	1971
Planned experiments		(probable dates)
CERN	Propane-filled bubble chamber (Gargamelle)	1972
Brookhaven	H_2-filled bubble chamber (7 ft)	1971-72
Serpukhov	Heavy liquid bubble chamber (CKAT)	1972
NAL (Batavia)	Spark chamber Bubble chamber	1972 ?

I. – Neutrino Sources and Lepton Charge Conservation.

1. – Conservation of lepton charge in high-energy weak processes.

The principle of lepton charge conservation is formulated in a manner which is similar to that of the law of baryon conservation. To each lepton a lepton charge ℓ is assigned and, to the corresponding antilepton, $-\ell$. In this scheme neutrinos and antineutrinos are two distinct particles.

Historically, the first experimental evidence favouring the existence of a law of lepton charge conservation came from the study of low-energy processes such as β decay. The reaction

$$^{130}\text{Te} \rightarrow {}^{130}\text{Xe} + e^- \text{-} e^- ,$$

or

$$^{48}\text{Ca} \rightarrow {}^{48}\text{Ti} + e^- \text{-} e^-$$

should exist if neutrinos were Majorana particles. In this case the expected lifetimes would be of the order of 10^{16} years and the established upper limit, for example, for the second reaction is $\tau \geqslant 10^{21}$ years.

Moreover, antineutrinos, produced by the decay of reactor neutrons, can induce

$$\bar{\nu}+p\rightarrow e^{+}+n\,,$$

but cannot induce

$$\bar{\nu}+{}^{37}\mathrm{Cl}\nrightarrow{}^{37}\mathrm{Ar}+e^{-}\,.$$

Thus, insofar as processes involving electrons are concerned, the $\bar{\nu}$ is a different particle from ν. For interactions involving μ's, direct experimentation has shown that the transition [1]

$$\mu^{\pm}\rightarrow e^{\pm}+\gamma$$

occurs with a frequency of less than 10^{-7} with respect to the total μ-decay rate; and that the neutrinos created in the decay

$$\pi^{+}\rightarrow\mu^{+}+\nu_{\mu}$$

can produce reactions

(*a*) $$\nu_{\mu}+n\rightarrow\mu^{-}+p\,,$$

but not [2]

(*b*) $$\nu_{\mu}+p\nrightarrow\mu^{+}+n\,,$$

(*c*) $$\nu_{\mu}+n\nrightarrow e^{-}+p\,,$$

This selection rule was discovered at Brookhaven and put on firm grounds at CERN. In fact, from the Cern spark-chamber experiment, we obtain for the relative frequency of occurrence of relations (*b*) and (*c*) with respect to (*a*)

$$\frac{\mathscr{F}(\nu_{\mu}\rightarrow\mu^{+})}{\mathscr{F}(\nu_{\mu}\rightarrow\mu^{-})}\leqslant 0.5\,\% \qquad \text{with } 90\,\% \text{ confidence [2]}$$

and

$$\frac{\mathscr{F}(\nu_{\mu}\rightarrow e^{-})}{\mathscr{F}(\nu_{\mu}\rightarrow\mu^{-})}\leqslant 3\,\% \qquad \text{with } 90\,\% \text{ confidence [3].}$$

Moreover, from CERN experiments [2] one deduces also that neutrino reactions *not producing any visible lepton* are less than 20 % (with 90 % confidence) of the total rate of observed events with a charged final lepton.

All these experimental facts strongly suggest that some conservation laws must exist for leptons. However, the choice is not unique: in fact one can formulate it in different ways [4-7].

1) *There are two different leptonic charges, muonic* (ℓ_μ) *and electronic* (ℓ_e) *ones, which are both subject to additive conservation laws.* The following assignments are currently used:

	Leptons				Antileptons			
	ν_μ	ν_e	μ^-	e^-	$\bar{\nu}_\mu$	$\bar{\nu}_e$	μ^+	e^+
ℓ_μ	1	0	1	0	-1	0	-1	0
ℓ_e	0	1	0	1	0	-1	0	-1
$\ell=\ell_\mu+\ell_e$	1	1	1	1	-1	-1	-1	-1

Thus $\sum \ell_\mu$ and $\sum \ell_e$ are separately conserved quantities.

2) *There is only one additive conservation law for the lepton charge,* $e^-\,\mu^+$ *being leptons and* $e^+\,\mu^-$ *being antileptons; and one 4-component neutrino.* The leptonic charges ℓ and the helicities $\boldsymbol{H}$ of the neutrinos are as indicated in the following Table:

	Leptons				Antileptons			
	($\bar{\nu}_\mu$	ν_e	μ^+	e^-)	(ν_μ	$\bar{\nu}_e$	μ^-	e^+)
	ν_R	ν_L	μ	e	$\bar{\nu}_L$	$\bar{\nu}_R$	$\bar{\mu}$	$\bar{e}$
ℓ	$+1$	$+1$	$+1$	$+1$	-1	-1	-1	-1
H	$+1$	-1			-1	$+1$		

This assignment would allow the decays

$$\pi^+\rightarrow\mu^++\bar{\nu}_L\,,$$

$$\pi^-\rightarrow\mu^-+\nu_R\,,$$

$$\mu^+\rightarrow e^++\nu_R+\nu_L\,,$$

$$\mu^-\rightarrow e^-+\bar{\nu}_L+\nu_R\,,$$

and forbid unseen reactions like $\mu\rightarrow e+\gamma$.

For massless neutrinos this law is in fact identical with the previous one.

3) *There is only one additive conservation law, for the total lepton charge* ℓ, *and one multiplicative conservation law for the lepton parity* p. Using the assignments

	Leptons				Antileptons			
	ν_μ	ν_e	μ^-	e^-	$\bar{\nu}_\mu$	$\bar{\nu}_e$	μ^+	e^+
ℓ	$+1$	$+1$	$+1$	$+1$	-1	-1	-1	-1
p	-1	$+1$	-1	$+1$	-1	$+1$	-1	$+1$

and assuming that $\ell = \sum \ell_i$ and $p = \prod p_i$ are conserved quantities, the reactions

$$\mu \to e + \gamma ,$$

$$\mu \to 3e ,$$

would be equally forbidden, but the following ones would be allowed:

$$\mu^+ e^- \leftrightarrow \mu^- e^+ \tag{I.1}$$

and

$$e^- e^- \leftrightarrow \mu^- \mu^- . \tag{I.2}$$

The existence of this law should also show in a reaction

$$\nu_\mu + (A, Z) \to (A, Z) + \mu^- + e^+ + \nu_e \tag{I.3}$$

(not to be confused with $\nu_\mu + (A, Z) \to (A, Z) + \mu^- + e^+ + \bar{\nu}_e$).

Reaction (I.1) is the most difficult to detect. Reaction (I.2) should not present great difficulty to detect once intersecting electron beams will be available. At present (I.3) is the only one within the reach of experimental possibilities. It can be calculated exactly; in fact [8] one would predict for such a process in lead (averaged over the Cern neutrino spectrum)

$$\sigma_{\text{theor}}(\text{Pb}) \approx 0.2 \cdot 10^{-40} (G'/G)^2 \text{ cm}^2 ,$$

where G' is the coupling constant associated with the « multiplicative coupling », and G the universal Fermi constant.

From present experimental data one can already establish an upper limit for G'. In fact, one has (see ref. [3] and ref. [8]) $G' \leqslant 19G$.

2. – Two remarks on recent experimental results.

a) The discussion reported above is based on the assumption that the neutrino beams, on which experiments have been carried out, originate essentially from π and K decays.

There are, however, theoretical reasons [9, 10] to believe that other short-lived sources of neutrinos could exist. If they exist, they might produce effects which simulate violation of lepton-charge conservation laws. These effects might give the explanation of some results recently obtained at Stanford (see Sect. **3**).

b) The failure to observe solar neutrinos by DAVIS *et al.* [11] could be due to the breakdown of lepton charge conservation. The possible existence of such a breakdown was first pointed out by PONTECORVO [12] and later by others [13-15].

Although none of the quoted experiments *as yet* compels us to believe that new particles or new laws are being discovered, several authors have been trying to see how these facts—if proved true—could fit in some theoretical schemes. Today I shall briefly report on them.

3. – Other sources of neutrinos: a heavy lepton?

Experiments to detect new sources of neutrinos are being carried out at Stanford and CERN. Being both at very preliminary stages, no definite result can be quoted. I shall mainly discuss the experimental methods. At Stanford the electron beam was allowed to enter a dump of 200 ft of earth shielding. A hole, 35 ft deep was excavated behind the hill of the beam dump and a 20 tons spark chamber, equipped with a trigger system of counters was placed behind.

The chamber was triggered synchronously with the beam. The trigger required a coincidence across an 11-inch Al thickness and hence predisposed toward energetic events ($>$ 500 MeV).

Under these conditions neutrino events were observed at the expected rate. However, when the beam energy was raised to $E_b \geqslant 18$ GeV, a higher rate of events was observed, associated with large multiplicities and no obvious μ-candidate. No case of this kind was seen at $E \leqslant 15$ GeV. When a water target was placed 25 m upstream of the dump, no event of this kind was recorded either.

A similar (although not equivalent) experiment is being carried out at CERN. Here the beam is a proton beam instead of electrons; and the detector is the giant bubble chamber « Gargamelle » instead of a spark chamber.

The analysis of the first 33000 pictures has just been completed—only a small fraction of what is required for a meaningful experiment!

No event similar to those observed at Stanford has been observed at CERN. Subtracting known sources of backgrounds, the upper limit of the cross-section for the production of the Stanford hypothetical particle by the CERN 26 GeV/c proton beam has been found to be

$$\sigma \leqslant 10^{-29}\ \text{cm}^2/\text{nucleon}\,,$$

a value to be reduced by several orders of magnitudes before it becomes of interest.

However, suppose that the preliminary Stanford results are confirmed by the experiments which are being carried out at Stanford and Cern. Then we have to conclude that a weakly interacting neutral particle—capable of traversing 200 ft of earth shielding—is being produced when the available energy in the c.m. system is $\gtrsim 5$ GeV. This particle could interact with matter producing highly inelastic events, and no obvious μ-candidate.

The simplest way to explain it—and also the least damaging to the presently accepted scheme—consists in assuming [9, 10] the existence of another heavier lepton λ, associated with another neutrino field ν_λ, and another leptonic number ℓ_λ supposedly conserved. Such a lepton is assumed to be coupled to other leptons and hadrons in the normal way of weak interactions and by the same coupling constants.

Thus the lepton current is now written as

$$j_\sigma = i\bar{\psi}_{\nu_e}\gamma_\sigma(1+\gamma_5)\psi_e + i\bar{\psi}_{\nu_\mu}\gamma_\sigma(1+\gamma_5)\psi_\mu + i\bar{\psi}_{\nu_\lambda}\gamma_\sigma(1+\gamma_5)\psi_\lambda .$$

The λ would be subject to a decay analogous to that of the μ

$$\lambda^\pm \to e^\pm + \begin{pmatrix}\nu_e\\ \bar{\nu}_e\end{pmatrix} + \begin{pmatrix}\bar{\nu}_\lambda\\ \nu_\lambda\end{pmatrix}$$

with decay rate [9, 10]

$$\Gamma_{\lambda\text{-}e} = \frac{G^2 m_\lambda^5}{192\pi^3} \qquad (m_e = 0)\,,$$

G being the usual universal Fermi constant; or to the similar one

$$\lambda \to \mu + \nu_\mu + \bar{\nu}_\lambda$$

with decay rate [10]

$$\Gamma_{\lambda\to\mu} = \frac{G^2 m_\lambda^5}{192\pi^3}(1 - 8\delta + 8\delta^3 - \delta^4 - 12\delta^2 \ln \delta)\,,$$

where $\delta = (m_\mu/m_\lambda)^2$. (In the latter case the muon mass cannot be neglected: in fact, it reduces the decay rate by a factor $\sim 30\,\%$.)

A heavier lepton may decay also into other modes. The mode

$$\lambda \to \pi + \nu_\lambda$$

not allowed to muons by energy conservation, but allowed to the λ if

$m_\lambda > m_\pi$, should have the rate

$$\Gamma_{\lambda\to\pi} = \frac{G^2 f_\pi^2 m_\lambda^3 \cos^2\theta_A}{8\pi}\left(1-\frac{m_\pi^2}{m_\lambda^2}\right)^2,$$

where $Gf_\pi/\sqrt{2}$ is the coupling constant for the $\lambda\pi\nu$ vertex, taken to be equal to that of the $\pi\mu\nu$ vertex; θ_A is the axial-vector Cabibbo angle.

The rate of the two-pion decay mode

$$\lambda \to \pi + \pi' + \nu_\lambda,$$

could be calculated assuming conservation of the vector current (CVC). Then the $\lambda(2\pi)\nu_\lambda$ vertex is defined by the coupling

$$iGF_\pi(s)\,\bar\psi_{\nu_\lambda}\gamma_\sigma(1+\gamma_5)\psi_\lambda(p_\pi + p_{\pi'})_\sigma,$$

where

$$s = -(p_\pi + p_{\pi'})^2$$

and $F_\pi(s)$ is the pion form factor. Taking $F_\pi(s)$ to be represented by the measured electromagnetic form factor of the pion and assuming ρ dominance [10]

$$|F_\pi(s)|^2 = \frac{km_\rho^4}{(m_\rho^2 - s)^2 + n_\rho^2\Gamma_\rho^2}$$

($K = 1.26$; $m_\rho = 762\ (\text{MeV/c})^2$; $\Gamma_\rho = 117$ MeV). With this form factor, the $\sigma\pi$ mode is more frequent than the 1π mode when $m_\lambda \gtrsim 1.1\ (\text{GeV/c})^2$.

The three-pion mode

$$\lambda \to \pi + \pi' + \pi'' + \nu_\lambda$$

is more complicated to compute and not as obvious. One can assume that:

1) The decay is dominated by the A_1. In fact, it must be in a ($J=1$, $\varrho=+1$, $G=-1$) state or in a ($J=0$, $\varrho=-1$, $G=-1$) state. However, no $3\pi(0^{--})$ resonance has been observed, whereas the A_1 is known to be a 1^{+-} resonance.

2) The A_1 is coupled to the lepton current by a coupling constant [16]

$$g_A = \frac{\sqrt{2}\,m_\rho}{g_{\rho\gamma}},$$

where $g_{\rho\gamma}$ is the coupling constant of the $\rho\gamma$ vertex.

3) The $(A_1 3\pi)$ vertex is dominated by the ρ-meson, *i.e.* the three π production takes place according to the graph

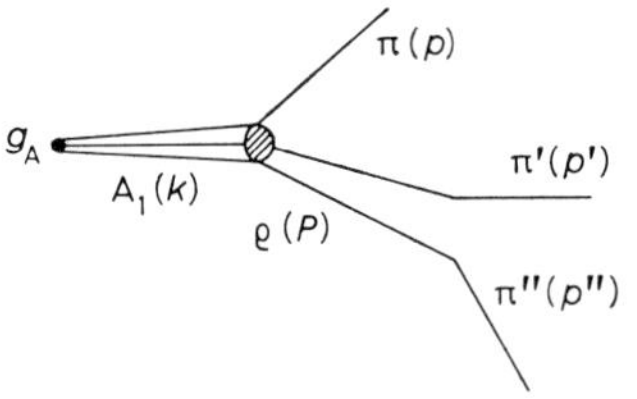

Fig. 1.

THACKER and SAKURAI [10] take a simple approach (neglecting the longitudinal coupling and treating the ρ as a stable particle); they in fact assume the $A_1\rho\pi$ coupling to be of the form

$$f_{A\rho\pi}\,\varepsilon^{(A)}\varepsilon^{(\rho)}\,,$$

and again determine the constants to agree with the observed A_1 width and Weinberg results (cit ref. [16]).

Finally, assuming the mass sufficiently large, it could decay into a kaon

$$\lambda \to K + \nu_\lambda\,,$$

the rate being the same as for the $\lambda \to \pi + \nu_\lambda$ apart from a factor of $\mathrm{tg}^2\,\theta_A$ and a different value of the mass of the final meson.

With increasing mass values of the λ, more and more channels open, characterized by increasing numbers of mesons in the final state. The estimate of the corresponding decay rates is in general difficult if not impossible. However, that adds interest to a serious experimental search of an effect of this kind.

A high value of m_λ, of course, is what is needed to explain events as those seen in the Stanford experiment. In such a case one should assume that λ's are created, presumably in $(\lambda\bar{\lambda})$ pairs by the electron beam; they decay into ν-'s and $\bar{\nu}$-'s (thus both the Stanford and CERN beams should have equal numbers of ν_λ's and $\bar{\nu}_\lambda$'s) which traverse the shielding and interact in the detectors, producing reactions of the type

$$\nu + \mathcal{N} \to \lambda + \beta$$

$$\hookrightarrow \nu_\lambda + X$$

or

$$\hookrightarrow \nu_\lambda + \mu + \bar{\nu}_\mu\,,$$

where $\mathcal{N}$ is the initial target nucleon β and X are hadronic systems of particles.

To begin with, it is worth remarking that a pionic decay of the λ would simulate a break-down of lepton charge conservation as the final ν_λ would escape undetected.

Moreover, the comparatively large number of secondaries associated with the events observed at Stanford could indicate that the λ decay produces several hadrons and—hence—it has indeed a large mass. In such a case its lifetime would be small: for example, for $m_\lambda \geqslant 1.2\ \text{GeV}/c(^2)$ THACKER and SAKURAI [10] estimate $\tau_\lambda < 2\cdot 10^{-12}$ s. Thus in the Stanford experiment, in which a 1 inch aluminum plate spark chamber was used, one could not distinguish in general the point of decay of the λ from its point of creation.

Let us see what one could hope to see in the giant CERN bubble chamber, Gargamelle. The production of a λ and its subsequent decay would appear as in Fig. 2.

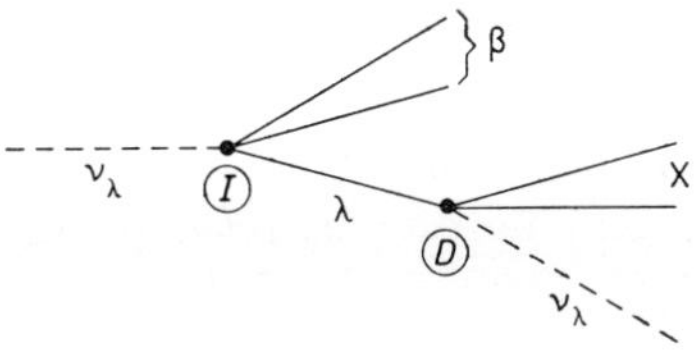

Fig. 2. – – – – unseen particles.

X represents here any number of mesons. To distinguish the vertex I from D in Gargamelle, the track λ must be at least 3 mm long. Assuming that the λ is ejected with a momentum of ~ 6 GeV/c, that it has a mass of 1.2 $(\text{GeV}/c)^2$ a track ~ 3 mm long would be produced. Smaller masses would produce longer tracks and—in general—fewer final particles. In conclusion, the detection of such a particle is not a simple experimental problem.

4. – Vacuum oscillations $\nu \rightleftarrows \bar{\nu}$ and $\nu_e \rightleftarrows \nu_\mu$.

It was pointed out by PONTECORVO [13, 15, 16], CABIBBO and GATTO [17] that a breakdown of lepton-charge conservation laws could lead to oscillations of the type $\nu \rightleftarrows \bar{\nu}$, $\nu_\mu \rightleftarrows \nu_e$.

From experiments one can fix upper limits for the magnitude of a hypothetical lepton charge, non conserving the coupling constant G'. In fact, from the upper limit on the rates

$$\begin{aligned} &\text{of} \quad {}^{48}\text{Ca} \to {}^{48}\text{Ti} + 2e^- && \text{one has} \quad G'/G < 0.02\,, \\ &\text{of} \quad \nu_\mu + p \to \mu^+ + n && \text{one has} \quad G'/G < 0.15\,, \\ &\text{of} \quad \mu^+ \to e^+ + \gamma && \text{one has} \quad G'/G < 10^{-3}\,. \end{aligned}$$

There is room for a sizeable effect.

If there are two 2-component neutrinos and two lepton charges, the transitions $\nu \rightleftarrows \bar{\nu}$ of an originally pure ν beam would produce «sterile» $\bar{\nu}$ because they would have the wrong helicity.

Vice versa, if one assumes only one (additive) conservation law, with different signs for μ^- and e^- and the helicity assignment given for the rule 2), Sect. **1**, then the oscillations would produce $\nu_L \rightleftarrows \bar{\nu}_L$ and $\nu_R \rightleftarrows \bar{\nu}_R$ which correspond to $\nu_e \rightleftarrows \nu_\mu$ oscillations. For sufficiently high energy, both types of oscillations would produce particles capable of interacting with matter.

GRIBOV and PONTECORVO [16] discuss the mechanism of the oscillation using a simple Lagrangian density

$$\mathscr{L}_{mf} = m_{ee}(\bar{\psi}'_{\nu_e}\psi_{\nu_e}) + m_{\mu\mu}(\bar{\psi}_{\nu_\mu}\psi_{\nu_\mu}) + m_{\mu e}(\bar{\psi}'_{\nu_e}\psi_{\nu_\mu}) + \text{h.c.}\,,$$

where ψ' is the charge conjugate field of ψ. (For the sake of simplicity they assume the m_{ij}'s to be real. In fact, the particular nature of the interaction chosen here does not affect the validity of this discussion.)

The above Lagrangian can be diagonalized by states of the type

$$\varphi_1 = \cos\xi(\nu_e + \nu'_e) + \sin\xi(\nu_\mu + \nu'_\mu)\,,$$

$$\varphi_2 = \sin\xi(\nu_e + \nu'_e) - \cos\xi(\nu_\mu + \nu'_\mu)\,.$$

In fact, putting

$$\langle\varphi_1|\mathscr{L}_{\text{int}}|\varphi_1\rangle = m_1\,,$$

$$\langle\varphi_2|\mathscr{L}_{\text{int}}|\varphi_2\rangle = m_2\,,$$

$$\langle\varphi_1|\mathscr{L}_{\text{int}}|\varphi_2\rangle = \langle\varphi_2|\mathscr{L}_{\text{int}}|\varphi_1\rangle = 0\,,$$

one has

$$m_{ee}\cos^2\xi + m_{\mu\mu}\sin^2\xi + m_{e\mu}\sin\xi\cos\xi = m_1\,,$$

$$m_{ee}\sin^2\xi + m_{\mu\mu}\cos^2\xi - m_{e\mu}\sin\xi\cos\xi = m_2\,,$$

$$(m_{ee} - m_{\mu\mu})\sin\xi\cos\xi - (\cos^2\xi - \sin^2\xi)\,m_{e\mu} = 0\,.$$

From the 3rd of these equations, one gets

$$\text{tg}\,2\xi = \frac{2m_{e\mu}}{m_{ee} - m_{\mu\mu}}\,,$$

and from the others

$$\left.\begin{matrix} m_1 \\ m_2 \end{matrix}\right\} = \frac{1}{2}\left[m_{ee} + m_{\mu\mu} \pm \sqrt{(m_{ee} - m_{\mu\mu})^2 + 4m_{e\mu}^2}\right].$$

Thus the two masses, m_1 and m_2 are not equal unless $m_{ee} - m_{\mu\mu} = m_{e\mu} = 0$. This mass difference leads to oscillations.

Considering a pure ν_e state at $t = 0$, at the subsequent time t one has

$$|\nu_e(t)|^2 = |\nu_e(0)|^2\{A + B\cos\delta t\},$$

where

$$\delta = \frac{m\,\delta m}{E}$$

with

$$m = \frac{1}{2}(m_1 + m_2),$$

$$\delta m = m_1 - m_2,$$

$$E = \text{neutrino energy},$$

$$A = \frac{(m_{ee} - m_{\mu\mu})^2 + 2m_{e\mu}^2}{(m_{ee} - m_{\mu\mu})^2 + 4m_{e\mu}^2}, \qquad B = \frac{2m_{e\mu}^2}{(m_{ee} - m_{\mu\mu})^2 + 4m_{e\mu}^2}.$$

Thus, for $\delta m = 0$, $\delta = 0$, $\cos\delta t = 1$ and

$$|\nu_e(t)|^2 = |\nu_e(0)|^2\{A + B\} = |\nu_e(0)|^2.$$

For $\delta \neq 0$ the period of oscillation is $1/\delta = E/(m\,\delta m)$.

If, for the sake of simplicity, we take $m \sim \delta m \sim 1$ eV, $E = 1$ GeV, then the space covered during one oscillation is

$$l_\delta = \frac{pc}{E}\frac{c}{8} \simeq 200m.$$

However, taking equally plausible values for $m\,\delta m$, one could get enormously different predictions for l_δ.

The above estimates indicate only that an experimental detection of this phenomenon although improbable and difficult, cannot be excluded « *a priori* ».

REFERENCES

[1] G. Conforto, M. Conversi, L. Di Lella, G. Penso, C. Rubbia and M. Toller: *Nuovo Cimento*, **26**, 261 (1962).

[2] M. Block, H. Burmeister, D. C. Cundy, B. Eiben, C. Franzinetti, J. Keren, R. Møllerud, G. Myatt, M. Nicolić, A. Orkin-Lecourtois, M. Paty,

D. H. Perkins, C. A. Ramm, K. Schultze, H. Sletten, K. Soop, R. Stump, W. Venus and H. Yashiki: *Phys. Lett.*, **12**, 281 (1964). See also: C. Franzinetti. CERN, 65-13 (1966); D. H. Perkins. *Topical Conference on Weak Interactions*, CERN-68-7 (1969). Previous Literature quoted there.
[3] K. Boren, B. Hahn, H. Hofer, H. Kasper, F. Krienen and P.-G. Seiler. *Phys. Lett.*, **29** B, 614 (1969).
[4] K. Nishijima: *Phys. Rev.*, **108**, 907 (1957).
[5] A. A. Sokolov: *Phys. Lett.*, **3**, 211 (1963).
[6] E. J. Konopiński and H. M. Mahmoud: *Phys. Rev.*, **92**, 1045 (1953).
[7] G. Feinberg and S. Weinberg: *Phys. Rev. Lett.*, **6**, 381 (1961).
[8] C. Y. Chang: *Phys. Rev. Lett.*, **24**, 79 (1970).
[9] S. S. Gershtein and V. N. Folomeshkin: *Sov. Journ. Nucl. Phys.*, **8**, 447 (1969).
[10] H. B. Thacker and J. J. Sakurai: preprint.
[11] R. Davis, D. Harmer and K. Hoffman: *Phys. Rev. Lett.*, **20**, 1205 (1968).
[12] B. Pontecorvo: *Sov. Phys. JETP*, **6**, 429 (1958); **7**, 172 (1959).
[13] B. Pontecorvo: *Sov. Phys. JETP*, **26**, 984 (1968).
[14] N. Cabibbo and R. Gatto: *Phys. Rev. Lett.*, **5**, 114 (1960).
[15] V. Gribov and B. Pontecorvo: *Phys. Lett.*, **28** B, 493 (1969).
[16] S. Weinberg: *Phys. Rev. Lett.*, **18**, 507 (1967).
[17] F. L. Gilman and H. Harari: SLAC preprint (8 Sept. 967).

II. – Problems in the Interpretation of ν Inelastic Reactions.

1. – General considerations on inelastic ν reactions.

There are several problems connected with the understanding of inelastic neutrino reactions. At comparatively low energy the inelastic production goes mainly via baryonic-resonance production [1], but no theoretical model made in analogy with photoproduction and electroproduction seems to be adequate to describe the phenomenon quantitatively.

The production of mesonic resonances (such as ρ, A_1, etc.) has not been observed yet. This is not in contradiction with theory: for example, assuming CVC, diffractive production of ρ^+ at high energy is expected to have a cross-section of $\sim 10^{-40}\,\text{cm}^2$ and with the experimental facilities which have been available in the past years it could have hardly been seen. Thus the problem here is an experimental rather than a theoretical one.

For highly-inelastic weak reactions, the cross-section—integrated over all the internal degrees of freedom of the hadronic final state—depends on three structure functions $W_i(Q^2, M_F^2)$ $(i = 1, 2, 3)$. The main interest is associated with testing the validity of certain « sum rules » which are predicted theoretically. The experimental study of these processes has just begun and I shall confine myself to a sketchy summary of the theory.

2. – Quasi-elastic production of baryonic resonances.

Experimental data tell us that the 3-3 resonance is the main contributor to neutrino single-pion production. The main features of this process are:

a value of the cross-section for

$$\nu + p \rightarrow \mu^- + \pi^+ + p\,, \tag{II.1}$$

which is of the order of $\sim 10^{-38}\ \text{cm}^2$, namely higher than any so far predicted by a factor ranging between 1.2 and 2;

a rather broad Q^2-distribution, not too dissimilar from that observed for elastic scattering of neutrinos on neutrons.

Let us consider the production of a 3-3 isobar as if it were a stable particle [2]. Let us assume an interaction Lagrangian density of the current-current type; a lepton current

$$j_\lambda = i\bar{u}_\mu \gamma_\lambda (1 + \gamma_5) v_\nu\,, \tag{II.2}$$

and the isobar described by four spinor quantities $\overline{W}_\alpha$ satisfying the subsidiary conditions

$$\gamma_\lambda \overline{W}_\lambda(k) = 0\,, \tag{II.3}$$

$$k_\lambda \overline{W}_\lambda(k) = 0\,, \tag{II.4}$$

to exclude transitions with spin-$\frac{1}{2}$ states. Then the most general expression of the hadronic current to be coupled to (II.2) is

$$J_\lambda(Q, K) = J^V_\lambda + J^A_\lambda\,,$$

where (see also Fig. 3)

$$J^V_\lambda = \overline{W}_\beta(k)[a_1 \delta_{\lambda\beta} + Q_\beta(ia_2\gamma_\lambda + a_3(\gamma_\lambda\gamma_\alpha - \gamma_\alpha\gamma_\lambda)Q_\alpha + a_4 Q_\lambda)]\gamma_5 V(P_1)\,, \tag{II.5}$$

$$J^A_\lambda = \overline{W}_\beta(k)[b_1 \delta_{\lambda\beta} + Q_\beta(ib_2\gamma_\lambda + b_3(P_1 + k)_\lambda + b_4 Q_\lambda)]\, V(P_1)\,. \tag{II.6}$$

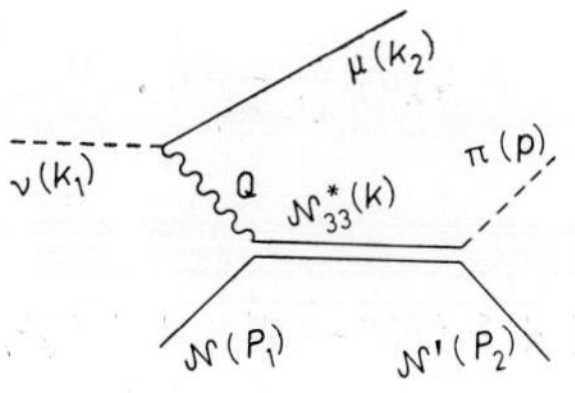

Fig. 3.

These are the only terms which can be constructed using the independent kinematical parameters and the gamma matrices, the Kronecker symbol $\delta_{\lambda\mu}$ and totally antisymmetric tensor $\varepsilon_{\lambda\mu\delta\varrho}$. Any other term which one may construct can be transformed into a combination of those in (II.5) and (II.6) using the Dirac equation and the identity

$$0 = \delta_{\mu\nu}\gamma_\lambda + \delta_{\nu\lambda}\gamma_\mu - \delta_{\mu\lambda}\gamma_\nu + \varepsilon_{\mu\nu\lambda\alpha}\gamma_5\gamma_\alpha - \gamma_\mu\gamma_\nu\gamma_\lambda\,. \tag{II.7}$$

The quantities a_i's and b_i's are Lorentz-invariant quantities depending on the scalars which one can build from the available kinematical parameters.

The choice of the values of the parameters a's and b's determines the physical behaviour of the above formulae. BERMAN e VELTMAN [2] take the vector coupling from photoproduction, assuming (CVC), which is fitted on experimental data, *i.e.*

$$a_3 = a_4 = 0$$

and

$$a_1/a_2 = -(m + M)\,.$$

Thus the vector part of the weak hadronic current is

$$J_\lambda^V = 4.5\,\frac{G}{\sqrt{2}}\sqrt{\frac{3}{2}}\left[\overline{W}_\beta\left(\delta_{\lambda\beta}\gamma_5 - \frac{i}{m+M}\,Q_\beta\gamma_\lambda\right)\gamma_5\,V\right]. \tag{II.8}$$

The axial-vector part is defined by the requirements of the partially conserved axial-vector current hypothesis. Taking the divergence of (II.6) one has

$$Q_\lambda J_\lambda^A = \overline{W}_\beta Q_\beta[b_1 + b_2(M^* - M) + b_3(M^{*2} - M^2) + b_4 Q^2]\,V(P_1)\,. \tag{II. 9}$$

It is assumed [2] that the b_4-term in (II.6) is dominated by a 1π-pole diagram. In this case it should be represented by the expression

$$b_4 Q_\beta Q_\lambda \to b_0(Q^2)\,\frac{g_{\pi\mu\nu}\,g_{\pi\mathcal{N}^*\mathcal{N}}}{m_\pi^2}\,\frac{1}{Q^2 + m_\pi^2}\,Q_\beta Q_\lambda\,.$$

For large Q^2 ($\gg m_\pi^2$) PCAC requires (II.9) going to zero

$$Q_\lambda J_\lambda^A \xrightarrow[Q^2 \gg m_\pi^2]{} 0\,.$$

Thus we can write

$$b_1 + b_2(M^* - M) + b_3(M^* - M^2) + b_0(Q^2)\,\frac{g_{\pi\mu\nu}\,g_{\pi\mathcal{N}^*\mathcal{N}}}{m_\pi^2} = 0\,.$$

Unless b_2 and b_3 are assumed to be singular for $M^* = M$, the second and third term are small compared with the others. Neglecting them both

$$b_1 = -b_0 \frac{g_{\pi\mu\nu} g_{\pi\mathcal{N}^*\mathcal{N}}}{m_\pi^2} = -b_0 \frac{2.2}{\sqrt{2}} \cdot \frac{10^{-5}}{M^2} = -b_0 2.2 \frac{G}{\sqrt{2}} .$$

In conclusion, one obtains

$$\text{(II.10)} \qquad J_\lambda^{\mathrm{V}} = c_I G_{\mathrm{V}} \left(\overline{W}_\beta \left(\delta_{\beta\lambda} - \frac{i}{M + M^*} Q_\beta \gamma_\lambda \right) \gamma_5 V \right), \qquad G_{\mathrm{V}} = 4.5 \frac{G}{\sqrt{2}} F_{\mathrm{V}}(Q^2) ,$$

$$\text{(II.11)} \qquad J_\lambda^{\mathrm{A}} = c_I G_{\mathrm{A}} \left(\overline{W}_\beta \left(\delta_{\beta\lambda} - \frac{1}{Q^2 + m_\pi^2} Q_\beta Q_\lambda \right) V \right), \qquad G_{\mathrm{A}} = \frac{2.2}{\sqrt{3}} \frac{G}{\sqrt{2}} F_{\mathrm{A}} ,$$

c_I being an isospin coefficient and $F_{\mathrm{V}}(0) = F_{\mathrm{A}}(0) = 1$. The relative sign between the axial-vector and vector parts is determined by comparison with the static model and is negative.

The total cross-section thus obtained, using Hofstadter's dipole form factors, and averaging over proton and neutron target, predicts a cross-section $\sim (4 \div 6) \cdot 10^{-39}\ \mathrm{cm}^2$, *i.e.* of about the same order of magnitude of the elastic one—a result which we shall discuss again later.

Similar methods can be used to calculate higher resonances [3]. The 1512-baryon resonance is a $J = \frac{3}{2}$, $J = \frac{1}{2}$ system. In this case the couplings can be chosen as

$$\text{(II.12)} \qquad J_\lambda = J_\lambda^{\mathrm{V}} + J_\lambda^{\mathrm{A}} , \quad J_\lambda^{\mathrm{V}} = C_{\mathrm{V}} \frac{m_\rho^2}{Q^2 + m_\rho^2} Q_\mu (T_{\mu\lambda} - T_{\lambda\mu}) ,$$

where

$$T_{\lambda\mu} = \overline{W}_\lambda \left(P_{1\mu} + k_\mu - \frac{D m_\pi}{D'} \gamma_\mu \right) U .$$

The constants are taken directly from photoproduction (as well as the idea that the coupling is mediated by the ρ-meson pole!). One gets

$$C_{\mathrm{V}} = \frac{\sqrt{3}}{2} \frac{1}{m_\pi^2} \frac{2}{\sqrt{3}} D' ,$$

$$D = 0.033 , \qquad D' = 0.0117 .$$

The axial-vector part, assumed to be mediated by a π and an A_1' is taken to be

$$\text{(II.13)} \qquad J_\lambda^{\mathrm{A}} = -c_{\mathrm{A}} \left[\frac{Q_\lambda Q_\mu}{Q^2 + m_\pi^2} + a \frac{M_{\mathrm{A}_1}^2}{Q^2 + M_{\mathrm{A}_1}^2} \left(-\delta_{\mu\nu} + \frac{Q_\lambda Q_\mu}{M_{\mathrm{A}}^2} \right) \right] \overline{W}_\lambda \gamma_5 U ,$$

with

$$c_{\mathrm{A}} = g_{\pi\mu\nu} g_{\pi\mathcal{N}\mathcal{N}^*} = 1.98\ \mathrm{G} .$$

In the spirit of PCAC $\left(i.e.\ \lim_{Q^2 \gg m_\pi^2} Q_\lambda J_\lambda^{\mathrm{A}} = 0 \right)$, $a = 1$.

Using (II.12) and (II.13) also this contribution can be calculated, thus completing the picture up to $M^* \sim 1.5\ (\text{GeV}/c)^2$. Due to the relatively smaller coupling as compared to $\mathcal{N}^*_{33}$ and to the drop in the neutrino flux at an energy at which the $\mathcal{N}_{31}(1512)$ becomes substantial, an experimental test of this model in the 1π-production channel has not been possible so far. However, one could test it in the 2-π channel, as $\mathcal{N}^*(1512)$ is known also to decay in to the mode

$$\mathcal{N}^*(1512) \to \pi^- + \mathcal{N}^{*++}(1240)$$
$$\hookrightarrow p + \pi^+ .$$

This has been done [3, 7] and the results are consistent with the theory. Due to the poor statistics available, this is a test of the theory to within $\pm 30\,\%$.

3. – The « isobaric model » for 1π production.

A somewhat more general treatment of the same model has been given by several other authors [4, 5]. The difference consists essentially in including more diagrams and in the use of dispersion relations.

We shall refer here to the work of SALIN and that of ADLER which have been more extensively used for the discussion of the experimental data.

The most general form of the hadronic current which one can write using the available vectors (see Fig. 3)—*i.e.*

$$Q,\ P_1,\ P_2,\ p = Q + P_1 - P_2\,,$$

and the γ's, γ_5—is formed by the following eight pseudovectors:

$$\text{(II.14)} \qquad V_i \equiv \begin{cases} i\gamma_5 P_{1\lambda}\,, & i\gamma_5 P_{2\lambda}\,, \quad i\gamma_5 Q_\lambda\,, \quad \gamma_5\gamma_\lambda\,, \\ \gamma_5 P_{1\lambda}(\gamma\cdot Q)\,, & \gamma_5 P_{2\lambda}(\gamma\cdot Q)\,, \quad \gamma_5 Q_\lambda(\gamma\cdot Q)\,, \quad i\gamma_5\gamma_\lambda(\gamma\cdot Q)\,, \end{cases}$$

and by the vectors which are derived from (4.14) omitting the γ_5's.

Thus the hadronic current will be of the type

$$\text{(II.15)} \qquad J = \sum^{1\ldots 8} V_i \mathscr{F}_i(\nu, \nu_\beta, Q^2) + \sum^{1\ldots 8} A_i \mathscr{G}(\nu, \nu_\beta, Q^2)\,,$$

where the $\mathscr{F}$'s and the $\mathscr{G}$'s are functions of three independent scalars which we can build out of the hadron vertex

$$\nu = -\frac{PQ}{M}\,, \qquad \nu_\beta = \frac{Q\cdot k}{2M}\,, \qquad Q^2\,,$$

being

$$P = \tfrac{1}{2}(P_1 + P_2)\,.$$

An alternative form would be

$$(\text{II.16})\qquad \begin{cases} V_1 = i\gamma_5\gamma_\lambda(\gamma Q)\,, & V_5 = \gamma_5\gamma_\lambda\,, \\ V_2 = 2i\gamma_5 P_\lambda\,, & V_6 = \gamma_5(\gamma\cdot Q)P_\lambda\,, \\ V_3 = 2i\gamma_5 Q_\lambda\,, & V_7 = \gamma_5(\gamma\cdot Q)Q_\lambda\,, \\ V_4 = 2i\gamma_5 k_\lambda\,, & V_8 = \gamma_5(\gamma\cdot Q)k_\lambda\,. \end{cases}$$

If we require conservation of the vector current, two of the $\mathscr{F}$'s can be expressed as functions of other $\mathscr{F}$'s. In fact, the equation

$$J^{\mathrm{V}}_\lambda = Q_\lambda = 0\,,$$

in this

$$(\text{II.17})\qquad i\gamma_5(Q^2\mathscr{F}_1 + 2(PQ)\mathscr{F}_2 + 2Q^2\mathscr{F}_3 + 2(QK)\mathscr{F}_4) + \\ + \gamma_5(\gamma\cdot Q)(\mathscr{F}_5 + \mathscr{F}_6(PQ) + \mathscr{F}_7 Q^2 + \mathscr{F}_8 kQ) = 0\,.$$

Both terms must be separately equal to zero because the γ's operate on the spin state of the initial spinors. Thus (II.17) splits into two equations.

In calculating the $\mathscr{F}$'s one can use the isotriplet hypothesis and thus determine the vector part completely.

SALIN considers the contribution of the four diagrams

1) Nucleon exchange

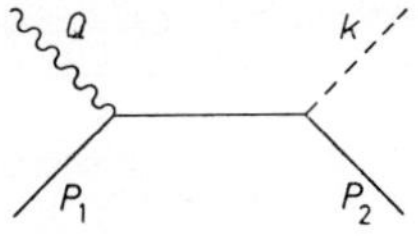

2) Crossed nucleon exchange

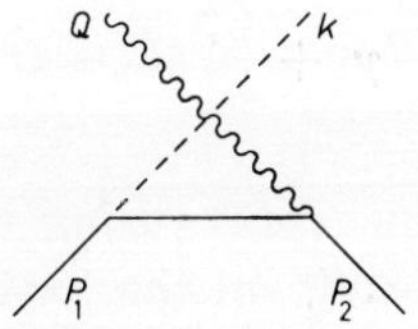

3) Pion exchange

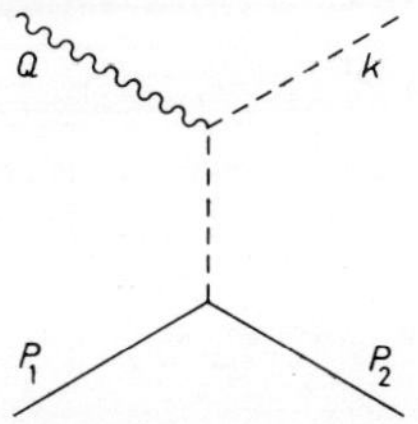

4) Resonance exchange

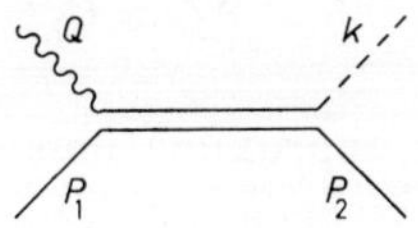

The first three diagrams define the contribution of the Born terms. For the vector part of the «isobaric term» SALIN takes the coupling already determined in photoproduction and electron production. Using the notations of ref. [6], the already quoted coupling for the $\mathcal{N}\mathcal{N}^*$ vertex, is

$$\overline{W}_\beta\left\{\left[\frac{C_3^V}{m_\pi}\gamma_\lambda+\frac{C_4^V}{m_\pi^2}(P_2+k)_\lambda+\frac{C_5^V}{m_\pi^2}P_{1\lambda}\right]\gamma_5\,(Q_\lambda j_\beta-j_\lambda Q_\beta)\right\}. \tag{II.18}$$

The C_i's are functions of Q^2. From photoproduction one has

$$C_4^V(0)+C_5^V(0)=0\,,\qquad C_3^V(0)/m_\pi=80\ (\text{GeV})^{-2}\,,$$

thus one can retain C_3^V only.

To find the Q^2-dependence of C_3^V, SALIN assumes that each amplitude of the multipole expansion of the vector amplitude satisfies an unsubtracted dispersion relation and that the imaginary part of the dispersive integral is dominated by the 33 resonance:

$$M^i(s)=M^i_{\text{Born}}+\frac{1}{\pi}\int\frac{\text{Im}\,M_{33}}{s'-s}\,\text{d}s'\,,$$

where

$$\text{Im}\,M^i_{33}=M_i\exp[-i\delta_{33}]\sin\delta_{33}\,.$$

An approximate solution of this equation has been obtained by SALIN and fitted on the experimental data.

For the axial part he uses the expression

$$\overline{W}_\beta\left\{\left[\frac{C_3^A}{m_\pi}\gamma_\lambda+\frac{C_4^A}{m_\pi^2}(P_2+Q)_\lambda\right](Q_\lambda j_\beta-j_\lambda Q_\beta)\right\}u(p_1)\,, \tag{II.19}$$

C_3^A and C_4^A are determined only using dispersion relations. He neglects the magnetic axial multipole $\mathcal{M}_1^+$ on the basis that $\mathcal{M}^+_{1\text{Born}}\ll\mathcal{E}^+_{1\text{Born}}$. He finds that the contribution of the first term is small and neglects it. Thus the total amplitude reduces to

$$\overline{W}_\beta\left\{\left[\frac{C_3^V}{m_\pi}\gamma_\lambda+\frac{C_4^A}{m_\pi^2}(P_2+Q)_\lambda\right]F_{\lambda\beta}\right\}U(P_1)\,, \tag{II.20}$$

where

$$F_{\lambda\beta}=Q_\lambda j_\beta-j_\lambda Q_\beta\,. \tag{II.21}$$

Apart from a number of approximations, this is the basis of his computations. It is worth noting that it does not satisfy PCAC.

ADLER assumes that the invariant amplitudes satisfy one-variable dispersion relations at fixed momentum transfer from which he obtains equations for the multipoles analogous to those above.

The same equation is satisfied by the elastic 33–partial-wave pion-nucleon scattering. Thus, making on both sides some approximations (and hoping for future cancellations!) ADLER finds for the relevant multipole

$$M_1^+ \simeq M_1^{+B}\left(1+\frac{a^2(Q^2)}{\omega\omega_{33}}\right)\frac{(f_{\pi 1}^+)_{33}}{(f_{\pi 1^+})_{33}^{\text{Born}}}\,, \tag{II.22}$$

(where $\omega = M^* - M$ and $f(\omega)$ represents a multipole amplitude); and similar expressions for $\mathscr{E}_1^+$ and $\mathscr{L}_1^+$ (a being a correction factor adjusted empirically).

The two calculations are not very different but give different results. Tables I and II are clearly indicative in this respect.

TABLE I. – *Vector and axial couplings in Salin's and Adler's models.* (BIJTEBIER, ref. [6]).

	SALIN		ADLER	
	Value for $Q^2 = 0$, $\omega = \omega_{33}$	Q^2-variation (*) for $\omega = \omega_{33}$	Value for $Q^2 = 0$, $\omega = \omega_{33}$	Q^2-variation (*) for $\omega = \omega_{33}$
C_3^V/m_π	80	$P_1(Q^2)$ (**)	80/0.8744	$P_1(Q^2)(1 + a^2/\omega_{33}^2)$ (**)
C_4^V/m_π^2	0	—	$-(C_3^V(0)/m_\pi)(1/\omega_{33})$	$P_1(Q^2)(1 + a^2/\omega_{33}^2)$ (**)
C_4^A/m_π^2	−115	1	15	$P_1(Q^2)(1 + a^2/\omega_{33}^2)$ (**)
C_5^A	0	—	46	$P_1(Q^2)(1 + a^2/\omega_{33}^2)$ (**)

(*) Energy unit: 1 GeV. The Q^2-variation does not include the variation of F_m and G_A. We took $G_A(0) = 1.25$.
(**) $P_1(Q^2) = 1 - 1.215Q^2/(100 m_\pi^2 + Q^2)$.

TABLE II. – *Multipoles contributions to the total cross-section for* $k_{10} = 5$ GeV (*in* 10^{-39} cm^2).

	M	L	$\mathscr{E}$	$\mathscr{L}$	Interference M	Total
SALIN	1.401	0.413	0.676	7.305	0.089	9.902
ADLER	1.633	0.019	1.184	0.776	0.476	4.095

They agree for the vector part.

For the axial part Salin's calculation does not satisfy PCAC and leads to a much bigger contribution from the longitudinal multipole $\mathscr{L}_1^\dagger$ for $Q^2 > 0$. Apart from that they agree.

The addition of a C_5^A coupling would bring the two computations to close analogy (BIJTEBIER [6]).

BERMAN and VELTMAN (when corrected for a numerical error) give analogous results to that of ADLER. Thus the only calculation which is close to the experimental data is that of SALIN in the version which is probably not correct.

BIJTEBIER and SALIN have tried to correct the latter calculation introducing a « C_5^A coupling » but varying C_4^A so as to give the same final result. The price to pay (BIJTEBIER [6]) « is high because it untroduces an innatural procedure in the handling of the dispersion relations ».

It is worth noting that the charge ratio π^+/π^0 is widely different in the case of 1-channel resonance production ($\frac{3}{2}\frac{3}{2}$ only) and in the isobaric model where the Born term contributes with a different ratio.

Experiments in this respect are not very precise due to the difficulty in detecting π^0's in small bubble chambers.

4. – The failure of the « peripheral models ».

Peripheral models have been used assuming that the relevant diagrams for inelastic processes are of the type.

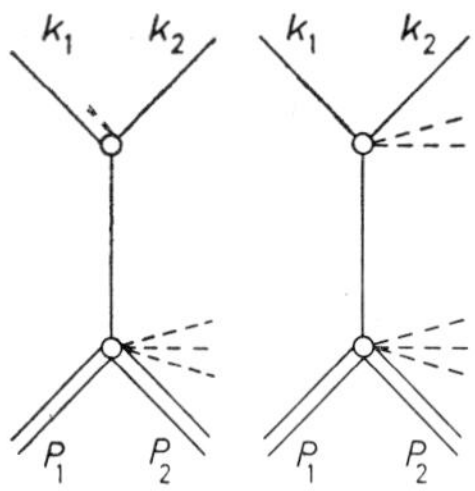

Fig. 4.

The propagator is assumed to be a pion which in most cases has a small Q^2. The upper vertices are computed from known or assumed couplings and the lower ones deduced from experimental « partial cross-sections » measured experimentally using real pions.

Both in single-pion production channels as in others, these models have proved utterly inadequate for the following reasons:

1) their predictions of the total cross-section are in general at least one order of magnitude smaller than the observed ones;

2) they predict a Q^2-distribution which is much too soft;

3) they predict a wrong charge ratio.

A detailed discussion of this effect was given by BOTTINO, CIOCCHETTI and he author [7].

5. – The « meson dominance » model.

The success of the « vector-meson dominance » model in describing high energy photoproduction processes and—to some extent—electroproduction ($Q^2 \sim 0$) suggests the use of a similar theory to describe high-energy neutrino reactions.

According to this model, the production of high-energy hadronic systems by neutrinos should have the general characters of high-energy meson-nucleon collisions.

Weak vector currents should be dominated by ρ-meson exchange. Then the « vector » cross-section for the production of a given final state F by a ρ-meson of given Q^2 is [8]

$$\frac{d\sigma^V}{dQ^2 dQ_0} = \frac{G^2}{4\pi^2} g_{l\rho}^2 \frac{1}{E_\nu^2} \frac{Q^2}{(Q^2+m_\rho^2)^2} |\boldsymbol{Q}| \left\{ \sigma_T^{\rho \to F} + \frac{(\sigma_T + \sigma_L)^{\rho \to F}}{2|\boldsymbol{Q}|^2} (4E_\nu E_\mu - Q^2) \right\},$$

where $g_{l\rho}$ is the coupling constant of the ρ-meson to the lepton current.

In the high-energy region, where ρ production should be mainly diffractive, the weak vector cross-section for $\rho^\pm$ production should be directly related to the ρ^0 electroproduction cross-section by the relation

$$\frac{d^2\sigma^{(\nu/\bar{\nu})}}{dQ^2 dQ_0} = \frac{4q^4 G^2}{e^4} \frac{d^2\sigma^{(e)}}{dQ^2 dQ_0}.$$

Analogously, the axial-vector part of the weak currents should be dominated by $A_1^\pm$ and exchange. The partially-conserved axial current hypothesis relates, at $Q^2 = 0$, the longitudinal A_1 and the π scattering amplitudes.

An experimental test of the predictions of this model poses some problems. The computation of the « weak cross-sections », predicted by it, requires the extrapolation of the cross-sections $\sigma^{\rho \to F}$, $\sigma^{\pi \to F}$, $\sigma^{A_1 \to F}$ from the physical to the unphysical states.

For the transverse components, photoproduction experiments seem to indicate that, at least at low Q^2, no strong Q^2-dependence of the related cross-section is expected.

But it is not so for longitudinal cross-sections. For them, at low Q^2, a strong Q^2-dependence *is* expected. Moreover (see ref. [8]), for physical longitudinal $A_1 (Q^2 = -m_{A_1}^2)$, the constraints imposed by PCAC are probably not valid. Thus, away from $Q^2 = 0$ the model is not of great use.

Only new experimental data might help to solve these problems.

6. – General considerations on total cross-sections.

Let us consider now an «inclusive» expression of the cross-section for neutrino interaction

$$\nu + \mathcal{N} \to \ell^- + \text{anything}\,, \tag{II.23}$$

where the «anything» is integrated over all its internal degrees of freedom. The differential cross-section $d\sigma/dQ^2$ of this process has the form

$$\frac{d\sigma}{dQ^2} = \frac{G^2}{8\pi}\frac{1}{\varepsilon^2}\mathcal{W}_{\lambda\mu}\,t_{\lambda\mu}\,, \tag{II.24}$$

where ε is the primary neutrino energy, $t_{\lambda\mu}$ is the lepton trace tensor and $\mathcal{W}_{\lambda\mu}$ is a tensor describing the hadron part. For elastic or quasi-elastic reactions, *i.e.*

$$\text{«anything»} \equiv \text{one definite particle}\,,$$

no integration is needed (except summation on spin states) as all the kinematical parameters are uniquely defined.

In both cases $\mathcal{W}_{\lambda\mu}$ can depend only on Q_λ and $P_{1\lambda}$. Thus the most general form which it can have is

$$\begin{aligned}\mathcal{W}_{\lambda\mu} = W_1\delta_{\lambda\mu} + W_2 P_\lambda P_\mu + W_3\varepsilon_{\lambda\mu\alpha\beta}Q_\alpha P_\beta + \\ + W_4 Q_\lambda Q_\mu + W_5(P_\lambda Q_\mu + Q_\lambda P_\mu) + iW_6(P_\lambda Q_\mu - Q_\lambda P_\mu)\,,\end{aligned} \tag{II.25}$$

where

$$P = P_1 + P_2\,.$$

It is usually stated that the locality of the lepton current requires $\mathcal{W}_{\lambda\mu}t_{\lambda\mu}$ to be a quadratic polynomial in ε [9]. This statement is rather vague and must be qualified. The quadratic dependence stems out of the fact that the interaction is of current-current type [10] and that the lepton current is of the form (2.2), *i.e.*

$$j_\lambda(x) = i\bar{\psi}_e(x)\gamma_\lambda(1+\gamma_5)\psi_\nu(x) \tag{II.26}$$

—which *is* a local current. Thus the trace is of the form

$$t_{\lambda\mu} \approx n_\lambda n_\mu - Q_\lambda Q_\mu + \delta_{\lambda\mu}(Q^2 + m_\mu^2) \mp \varepsilon_{\lambda\mu\sigma\varrho}n_\sigma Q_\varrho\,, \tag{II.27}$$

where $Q_\lambda = (k_1 - k_2)_\lambda$ and $n_\lambda = (k_1 + k_2)_\lambda$.

Thus [9], because of the form of $\mathscr{W}_{\lambda\mu}$, making all possible products in the contraction of the two tensors, all terms result proportional either to ε^2 or ε or independent of t. This is a consequence of at least two hypotheses, which in fact amount to saying [10] that a particle of spin 1 is exchanged in the interaction.

Neglecting the lepton mass and using (II.25), (II.24) becomes

$$\text{(II.28}a\text{)}\qquad \frac{\mathrm{d}\sigma}{\mathrm{d}Q^2} = \frac{G^2}{32\pi M^2}\frac{1}{\varepsilon^2}[8M^2Q^2W_1 + \\ + \{(P\cdot n)^2 - Q^2(Q^2+4M^2)\}W_2 \pm 2(P\cdot n)Q^2W_3],$$

or, expliciting the angle

$$\text{(II.28}b\text{)}\qquad \frac{\mathrm{d}\sigma}{\mathrm{d}Q^2} = \frac{G^2}{2\pi}\frac{\varepsilon'_\mu}{\varepsilon}\left(W_2\cos^2\frac{\theta}{2} + 2W_1\sin^2\frac{\theta}{2} \mp \frac{\varepsilon+\varepsilon_\mu}{M}W_3\sin\frac{\theta}{2}\right).$$

One can derive immediately some interesting relations.

In a purely «elastic» case, for example, one has

$$\lim_{\varepsilon\to\infty}\frac{\mathrm{d}\sigma}{\mathrm{d}Q^2} = \frac{G^2}{2\pi}W_2$$

and

$$\frac{\mathrm{d}\sigma(\nu)}{\mathrm{d}Q^2} - \frac{\mathrm{d}\sigma(\bar\nu)}{\mathrm{d}Q^2} = -\frac{G^2}{4\pi M}\frac{Q^2}{\varepsilon^2}(\varepsilon+\varepsilon')W_3.$$

Analogous, although more complicated, rules can be obtained for the inelastic cases.

For each channel one can calculate the cross-sections related to a definite polarization state of the virtual mediator. For example, for the transverse right-handed cross-section for a conserved vector current

$$\sigma_R \approx \varepsilon^*_\lambda\varepsilon_\mu\mathscr{W}_{\lambda\mu} \to \simeq \frac{1}{2}(\mathscr{W}_{11}+\mathscr{W}_{22}),$$

(ε_λ being a polarization vector).
Remembering $\varepsilon_\mu Q_\mu$, one gets from (II.25)

$$\sigma_R \approx \frac{1}{2}\left(W_1 + \frac{1}{M^2}p_1^2W_2 + W_1 + \frac{1}{\pi^2}p_1^2W_2\right) \simeq W_1 + \frac{P_1^2+P_2^2}{\pi^2}W_2,$$

taking $P_1 \equiv (\bar 0, M_1)$ $p_1^2 + p_2^2 = 0$, one has $\sigma_R \approx W_1$.

Analogously, one can find σ_L and σ_S. One obtains in fact

$$\frac{W_1}{W_2} = \left(1+\frac{Q_0^2}{Q^2}\right)\frac{\sigma_T}{\sigma_S+\sigma_T}\,, \qquad \sigma_T = \sigma_L + \sigma_R\,,$$

$$\frac{W_3}{W_2} = \frac{M}{Q^2}\sqrt{Q^2+Q_0^2}\,\frac{\sigma_L-\sigma_R}{\sigma_S+\sigma_T} \qquad \left(\text{for antineutrins}\begin{Bmatrix}\sigma_\nu^\nu \to \sigma_R^{\bar\nu}\\ \sigma_R^\nu \to \sigma_\nu^{\bar\nu}\end{Bmatrix}\right)$$

Since, by definition all the σ's are positive, it follows

$$W_1 \leqslant \left(1+\frac{Q_0^2}{Q^2}\right)W_2\,,$$

$$|W_3| \leqslant W_2\frac{M}{Q^2}\sqrt{Q^2+Q_2^2}\,.$$

Consider now the reaction

(II.29) $$\nu + p \to \ell^- + p + X^+$$

where X is anything with baryon number $B=0$ (and for example $S=0$). Integrating over all the internal degrees of freedom of X the structure functions which define the above transitions are functions of three independent scalars as if X was a bound system of definite mass $\sqrt{-\boldsymbol{P}^2}$.

The square, summed, averaged matrix element then is a function of some structure functions [9] W_i

$$\mathscr{W} = \mathscr{W}_{\lambda\mu}\, t_{\lambda\mu}$$

with

$$\mathscr{W}_{\lambda\mu} = \sum_i^{1\dots 9} W_i\, \omega_{\lambda\mu}^{(i)}\,,$$

where the $\omega_{\lambda\mu}^{(i)}$ are functions of the momenta associated with the hadronic vertex (see ref. [9] for details).

As Pais has pointed out, a number of hypotheses related to the mechanism of the processes of the type (II.29) can be tested by studyng experimentally the dependence of $\mathscr{W}$ on one or more dynamical variables. In particular one can investigate the question of Regge-pole dominance. In fact, if the process is dominated by a simple Regge trajectory—in particular by a pomeron trajectory—the cross-section might exhibit a power-law behaviour as $\nu \to \infty$. This and other properties are part of the programme of the forthcoming neutrino physics.

REFERENCES

[1] I. BUDAGOV, D. C. CUNDY, C. FRANZINETTI, W. B. FRETTER, H. W. K. HOPKINS, C. MANFREDOTTI, G. MYATT, F. A. NEZRICK, M. NIKOLIĆ, T. B. NOVEY, R. B. PALMER, J. B. M. PATTISON, D. H. PERKINS, C. A. RAMM, B. ROE, R. STUMP, W. VENUS, H. W. WACHSMUTH and H. YOSHIKI: *Phys. Lett.*, **29** B, 524 (1969). See also: D. H. PERKINS: *Topical Conference on Weak Interactions* (Geneva, 1969); C. FRANZINETTI: *Topical Conference on Weak Interactions* (Geneva, 1969).

[2] We follow the treatment given by S. BERMAN and M. VELTMAN: *Nuovo Cimento*, **38**, 993 (1965).

[3] A. BOTTINO and G. CIOCCHETTI: *Lett. Nuovo Cimento*, **2**, 780 (1969).

[4] A detailed list of references on this subject (up to Jan. 1966) is contained in the CERN report CERN 66-13.

[5] After the preparation of these notes the author has been informed of a recent work by P. A. ZUCKER recently submitted for publication.

[6] J. BIJTEBIER: Reprint of Brussel University, Institut Interuniversitaire des Sciences (Brussel, 1969).

[7] A. BOTTINO, G. CIOCCHETTI and C. FRANZINETTI: *Rendiconti del Seminario Matematico e Fisico di Milano*, Vol. **39** (Milano, 1969).

[8] C. A. PIKETTY and L. STODOLSKY: *Topical Conference on Weak Interactions* (Geneva, 1969).

[9] A. PAIS: *Madison Conference on Expectations for Particle Reactions at the New Accelerators* (Madison, 1970).

[10] This remark is due to J. S. BELL.

Lectures on Chiral Symmetry Breaking.

R. DASHEN (*)

Institute for Advanced Study - Princeton, N. J.

1. – Introduction.

All the so-called internal symmetries of hadrons are, in reality, broken symmetries. The general belief is that isospin is broken by the more-or-less understood electromagnetic interaction. I say « belief » because in the absence of being able to compute anything from first principles, *e.g.*, the proton-neutron mass difference, one cannot rule out the possibility that there is some small purely-hadronic interaction that also breaks isospin. With our present calculational ability, the only general thing we can say about electromagnetic breaking of isospin is that (to order α) the effective symmetry-breaking Hamiltonian will contain pieces with isospin equal to zero, one and two, but no others. Unfortunately this does not do much good. For example, this places no restriction on mass differences within a multiplet of isospin less than $\frac{3}{2}$.

The situation with regard to SU_3 is quite different. In this case no one has the slightest idea about what, on a fundamental level, is the source of the symmetry breaking. On the other hand, we do have a very good phenomenology for SU_3 breaking. There is good evidence that the piece of the strong Hamiltonian which breaks SU_3 is the eighth component of some eight-dimensional (octet) representation of SU_3. If one then assumes that a first-order (in symmetry breaking) calculation is valid, the result is the marvelous Gell-Mann–Okubo mass formula [1].

I am going to be talking about $SU_3 \otimes SU_3$ breaking. Here we have the worst of both worlds. Neither do we have any fundamental ideas as to why or how $SU_3 \otimes SU_3$ is broken, nor have we yet developed a useful phenomenology. There has been however one particularly interesting theoretical input. This is the $(3, \bar{3}) \oplus (\bar{3}, 3)$ model of GELL-MANN, OAKES and RENNER [2] and GLASHOW and WEINBERG [3]. These authors suggest that the piece of the strong interaction which breaks $SU_3 \otimes SU_3$ can be written as $u_0 + cu_8$ where u_0 and u_8 are an SU_3 singlet and eight component of an SU_3 octet both belonging

(*) Alfred P. Sloan Foundation Fellow.

to the $(3, \bar{3}) \oplus (\bar{3}, 3)$ representation fo $SU_3 \otimes SU_3$. The object c is a number whose determination will be discussed later. The originators of this model make a very strong argument that it is by far the most theoretically pleasing of any as yet conceived of model. So, we have a possible group-theoretic statement about $SU_3 \otimes SU_3$ breaking which is quite analogous to Gell-Mann's original suggestion that SU_3 breaking transforms like the eighth component fo an octet. Why, then, can we not develop the same kind of phenomenology that we have in SU_3 breaking? Hopefully we can but it is hard and has not yet been done. The reasons why have to do with the nature of $SU_3 \otimes SU_3$ symmetry [4].

Exact SU_3 symmetry means degenerate multiplets of particles, the baryon octet for example. Broken SU_3 means mass splittings and octet symmetry breaking means a certain pattern of mass differences. Masses are generally well known experimentally and it is easy to spot the octet pattern. On the other hand, exact $SU_3 \otimes SU_3$ symmetry does *not* mean $SU_3 \otimes SU_3$ multiplets of particles. Rather, in the exact $SU_3 \otimes SU_3$ limit we have eight massless Goldstone particles, π, K and η, which allow the symmetry to be realized without multiplets of particles. The consequences of this symmetry are, of course, more than just the presence of massless (low mass in the real, broken-symmetry world) pseudoscalar mesons. The symmetry also tells us that these mesons satisfy low-energy theorems which lead to things like the Goldberger-Treiman [5] relation and the Adler-Weisberger [6] sum rule. In the limit of $SU_3 \otimes SU_3$ symmetry these low-energy theorems would be exact statements about on-mass-shell processes. For the real world where symmetry breaking occurs, these theorems are only approximate when regarded as statements about on-shell hadronic processes. Evidently, there are two sources of information about $SU_3 \otimes SU_3$ breaking. First, the entire mass of a pseudoscalar meson comes from the symmetry-breaking interaction. What can we learn from the known masses of π, K and η? The answer is, not much. In the $(3, \bar{3}) \oplus (\bar{3}, 3)$ model mentioned above, all one can learn from studying these masses is the value of the parameter c. Secondly, one could study corrections to low-energy theorems. This can be indeed be done and will be a major part of what I will be talking about in the following lectures. Some progress has been made and I think that it is fair to say that the subject is in much better shape now than it was, say, three years ago. However, it is also fair to say that no clear pattern or really useful phenomenology has yet emerged. The reasons for this are twofold. First, the theoretical problem of finding out how to compute corrections to low-energy theorems is a difficult and often tedious one. Secondly, one often finds that the relevant experimental numbers have not been determined with sufficient accuracy. Most of you know, I am sure, of the confused state of our knowledge of the parameters in K_{l3} decay, but how many of you know that the error in the Goldberger-Treiman relation could be as small as 1% or as large as (15÷20)%?

Let me be more specific about what I mean by « computing » corrections to low-energy theorems. Neither myself nor anyone else I know is about to go out and compute a reliable number for the error in the Goldberger-Treiman relation. What one can hope to do is to relate various $SU_3 \otimes SU_3$-symmetry breaking effects in the same way that the Gell-Mann–Okubo mass formula relates mass splittings within an SU_3 multiplet. This can be done if: i) we know something about the group-theoretic properties of the symmetry-breaking interaction and ii) we are willing to work to lowest order in symmetry breaking. As far as i) is concerned we have the $(3, \overline{3}) \oplus (\overline{3}, 3)$ model which is at least a good place to start. As for ii), who knows? I pointed out above that a first-order symmetry-breaking calculation in SU_3 yields the remarkably successful Gell-Mann–Okubo mass formula. This does not, of course, have to mean that in a real calculation of, say, M_Λ-M_N it would be sufficient to work only to lowest order in SU_3 breaking. It may very well be that higher-order terms are large but just happen to display the same characteristic octet pattern as does the lowest-order term. In the absence of the ability to compute the magnitude, as opposed to pattern of mass splittings it is not possible to rigorously say whether or not the observed deviations from SU_3 are predominantly a first-order effect. On the basis of the mass formula it does seem reasonable to suppose, however, that first-order calculations are adequate for SU_3 breaking. What about $SU_3 \otimes SU_3$? One expects the breaking of SU_3 and $SU_3 \otimes SU_3$ to be comparable in magnitude, so that what works for one ought to work for the other. On the other hand, there have been some hints that the situation may not be as simple as this [6, 7]. When I talk about specific problems, I will try to give an indication of what neglecting second-order terms means in practice.

Finally, let me end this introduction with a little pep talk. I have said that worrying about $SU_3 \otimes SU_3$ breaking is difficult and that, at present, the situation is both theoretically and experimentally confusing. Why, then, worry about it at all? My answer to this is somewhat round-about. I tend to think that neither the weak nor strong-interaction problem will be solved alone. Rather, I suspect that a real understanding of either will come only when we understand how they fit together. The reason for thinking this is that the one-to-one correspondence between Gell-Mann's algebra of weak and electromagnetic currents, on one hand, and the strong-interaction symmetries, on the other hand, is far too remarkable to be of anything but the deepest significance. In a manner of speaking, the weak and strong interactions seem to « know about each other » in considerable detail. It is hard to see how this could be unless they are very closely intertwined if not, in fact, the same thing. Of course, these are just words but I think that it is sometimes useful to say them. What this has to do with symmetry breaking is that one of the key submysteries to the general mystery of weak and strong interactions, is why and how are the strong interaction symmetries broken? At our present level of knowledge one probably cannot expect to learn much more than just certain group-theoretical properties of the symmetry breaking. This could, however, be an important piece to the puzzle. Later you will see that we are also confronted with a very interesting submystery, in connection with the role of electromagnetism in $SU_3 \otimes SU_3$.

2. – Models of chiral symmetry.

It would be well to begin our discussion of symmetry breaking with a review of how chiral symmetry works. The best way to do this is to look at some models, notably the σ-model [5]. Before plunging ahead, however, let us orient ourselves. Lagrangian models of chiral symmetry are usually treated in tree approximation where, to obtain any quantity of interest, one calculates all relevant graphs with no closed loops. The beauty of the tree approximation is that it is a simple scheme which is guaranteed to preserve all symmetry properties of the Lagrangian. It also turns out that tree approximation is a type of semi-classical approximation, *i.e.* one where we let $\hbar$ go to zero. Thus, in talking about models we can switch back and forth between classical and « tree » language without running into trouble. Lagrangian models studied with this classical-tree method are a marvelously rich laboratory for obtaining insight into $SU_3 \otimes SU_3$ symmetry and its breaking. Unfortunately, the models sometimes tend to pick up a life of their own. That is, we sometimes forget that they are just laboratories and begin to take their detailed properties seriously. This is dangerous because such details might be an artifact of the approximation or a peculiarity of some Lagrangian model which we would never dream of suggesting as a fundamental theory of strong interactions. Having said this, let us go onto the σ-model. The next few paragraphs are pirated, essentially verbatim from ref. [6].

The Lagrangian

$$\mathscr{L} = \bar{N} i\gamma\cdot\partial N + g\bar{N}(\sigma' + i\tau\cdot\pi\gamma_5)N - \tfrac{1}{2}(\partial_\mu\sigma'^2 + \partial_\mu\pi^2) - B^2(\sigma'^2 + \pi^2 - A)^2\,, \tag{2.1}$$

where N and π are fields with the quantum numbers of the nucleon and pion, and σ' is an isoscalar 0^+-meson field, is well known to be invariant under the group $SU_2 \otimes SU_2$ with corresponding conserved vector and axial-vector currents $\mathscr{V}_i^\mu = \bar{N}\gamma^\mu \tfrac{1}{2}\tau_i N + (\pi\times\partial_\mu\pi)_i$ and $\mathscr{A}_i^\mu = \bar{N}\gamma^\mu\gamma_5\tfrac{1}{2}\tau_i N + \sigma'\partial_\mu\pi_i - \partial_\mu\sigma'\pi_i$.

Let us ignore, for the moment, the first two terms in $\mathscr{L}$ and think of the remaining mesonic terms as a classical theory. In the ground state of this classical theory the fields σ' and π will be constants (independent of x and t), such that the potential energy $B^2(\sigma'^2 + \pi^2 - A)^2$ is a minimum. For positive A we have $\sigma'^2 + \pi^2 = A$ at the minimum. Evidently, a straightforward perturbation expansion can make sense only for $A < 0$. Classically, perturbation theory corresponds to expanding around $\sigma'=0$, $\pi=0$, which is a stable minimum of potential energy for $A<0$, but is a local maximum for $A>0$. To do perturbation theory for $A>0$ we need to define new fields which vanish at the minimum of energy. To this end we choose as the «physical» ground state $\pi=0$, $\sigma'=-\sqrt{A}$, and introduce $\sigma=\sigma'+\sqrt{A}$. In terms of σ the Lagrangian is

$$(2.2)\qquad \mathscr{L}=\bar{N}i\gamma\cdot\partial N-(g\sqrt{A})\bar{N}N+g\bar{N}(\sigma+i\tau\cdot\pi\gamma_5)N-\tfrac{1}{2}[(\partial_\mu\sigma)^2+(\partial_\mu\pi)^2]-\\ -4B^2A\sigma^2+(2\sqrt{A})B^2\sigma(\sigma^2+\pi^2)-B^2(\sigma^2+\pi^2)^2\,.$$

We may now forget about our classical arguments and think of eqs. (2.1) and (2.2) as defining a quantum theory which can be studied in perturbation theory. The rule is, of course, to use (2.1) for $A>0$. Let us compare the two cases $A<0$ and $A>0$. If we neglect the interaction terms, Lagrangian (2.1) describes massless nucleons and π and σ mesons with a common mass of $(-4AB^2)^{\frac{1}{2}}$, while Lagrangian (2.2) describes nucleons with mass $g\sqrt{A}$, a σ-meson with mass $(8AB^2)^{\frac{1}{2}}$ and massless pions. These qualitative features are not changed by the interaction terms. To all orders, Lagrangian (2.1) yieds massless nucleons and a degenerate multiplet of π and σ, while Lagrangian (2.2) gives massive nucleons, a massive σ and massless pions.

Evidently, the $SU_2\otimes SU_2$ symmetry of the model manifests itself in very different ways in the two cases $A<0$ and $A>0$. In the former case we have a «normal» symmetrical theory with π and σ forming a multiplet, while in the latter case there is no $SU_2\otimes SU_2$ multiplet structure, but we have instead a massless pion, *i.e.* a Goldstone boson. The transition between the two cases is interesting. If we approach $A=0$ from the negative side, the degenerate π and σ masses both approach zero. Then as A increases through positive values, σ picks up a mass, but the pion mass sticks at zero. Although this transition through $A=0$ is smooth and physically sensible, it is certainly not analytic. The origin of this nonanalyticity at $A=0$ are, of course, the factors of $\sqrt{A}$ in eq. (2.2).

According to the Goldstone theorem, a symmetry realized through massless bosons implies a degenerate vacuum (defined as the state of lowest energy). That this should be the case is easily seen in the classical theory where there are a threefold infinity of solutions to $\sigma'^2+\pi^2=A$, and hence a three-fold infinity of lowest-energy states. Physically, these extra vacua are obtained by adding zero-energy pions to the particular vacuum which we have chosen to be the physical one.

Let us now see what happens if we break the $SU_2\otimes SU_2$ symmetry of the model by adding a term such as $\varepsilon\sigma'$ or $\varepsilon\bar{N}N$ to the Lagrangian. For small ε it is easy to see what such a term will do. If A is negative, the π-σ multiplet will split slightly and the nucleon will pick up a small mass. On the other hand, if A is positive, the main quantitative effect of the perturbation will be to give the pion a small mass. A perturbation will also remove the degeneracy of the vacuum. This is particularly easy to see when the breaking term is $\varepsilon\sigma'$. The classical ground state is then unique and corresponds to $\pi = 0$, $\sigma' = -\sqrt{A} + \varepsilon/2B^2A + O(\varepsilon^2)$.

It is interesting to look at the Goldberger-Treiman relation in the σ-model. To do this we need to be more specific about the symmetry breaking. The operators σ' and $\bar{N}N$ have the same abstract group-theoretical properties. Specifically, they both belong to the $(\frac{1}{2}, \frac{1}{2})$ representation of $SU_2 \otimes SU_2$. In the model they look different because they are made out of different « fundamental fields ». However, according to our philosophy regarding models we should not take details like this seriously. Let us then, consider the general symmetry-breaking term $\varepsilon_1\sigma' + \varepsilon_2\bar{N}N$. A simple computation then gives

$$\partial_\mu \mathscr{A}^\mu = \varepsilon_1 \boldsymbol{\pi} - i\varepsilon_2 \bar{N}\gamma_5\tau_i N \tag{2.3}$$

for the divergence of the axial current. To calculate the matrix element of $\partial_\mu\mathscr{A}^\mu$ between nucleons in tree approximation we need only consider two graphs, a pion-pole graph from the first term in $\partial_\mu\mathscr{A}^\mu$ and the contact term of from the second part of $\partial_\mu\mathscr{A}^\mu$. If only the pion-pole term were present, it is obvious that the Goldberger-Treiman relation would be satisfied exactly. However, the contact term results in a failure of pole dominance and gives a correction to the Goldberger-Treiman relation. Of course, this correction, which comes from the $\varepsilon_2\bar{N}\gamma_5\tau_i N$ term in $\partial_\mu\mathscr{A}^\mu_i$, will be small if ε_2 is and in fact vanishes in the exact $SU_2 \otimes SU_2$ limit. Evidently, in this model the Goldberger-Treiman relation is a consequence of $SU_2 \otimes SU_2$ symmetry and fails to the extent that $SU_2 \otimes SU_2$ is broken.

The above discussion of the Goldberger-Treiman relation would seem to be independent of whether the symmetry is realized by multiplets ($A < 0$) or massless points ($A > 0$). This is, in fact, true but the result is nontrivial only if A is greater than zero and we are supposed to use the Lagrangian (2.2) which has degenerate vacuum and massless pion. To see this we recall that the Goldberger-Treiman relation relates the nucleon mass $M_{\mathscr{N}}$, the π-$\mathscr{N}$ coupling constant G, the axial-vector coupling g_A and the pion-decay constant $(2f_\pi)^{-1}$ by

$$2Mg_A \approx \frac{G}{f_\pi}\,. \tag{2.4}$$

Clearly, to get a nontrivial result we must have both M and f_π^{-1} nonzero in the symmetry limit. In our model M is nonvanishing in the symmetry limit only if $A > 0$. If $A < 0$, M is of order ε_1 and ε_2. What about f_π^{-1}? It is easy to see that for $A < 0$ the current $\sigma'\partial_\mu\boldsymbol{\pi} - \boldsymbol{\pi}\partial_\mu\sigma'$ will give $f_\pi^{-1} = O(\varepsilon_1)$ in tree approximation. On the other hand, if A is positive, we must convert from σ' to σ which gives $f_\pi^{-1} = \sqrt{A} + O(\varepsilon_1)$. Therefore, if $A > 0$, both sides of eq. (2.4) are of order unity as ε_1 and ε_2 go to zero and, as pointed out above, the error is $O(\varepsilon_2)$. Hence we have a nontrivial result. However, if $A < 0$ then both sides of eq. (2.4) are $O(\varepsilon_1)$ or $O(\varepsilon_2)$ and the error is $O(\varepsilon_2)$, which makes the result trivial.

To summarize, we have learned two things. First, in this model the Goldberger-Trieman relation can be thought of as the consequence of an $SU_2 \otimes SU_2$ symmetry which is realized by way of a degenerate vacuum. Secondly, the symmetry breaking will cause the Goldberger-Treiman relation to become an approximate statement. In our model it appears that by setting $\varepsilon_2 = 0$ but keeping ε_1 finite, one could have an exact Goldberger-Treiman relation without exact $SU_2 \otimes SU_2$ symmetry. This is however an artifact of tree approximation. The term $\varepsilon_1 \sigma'$ itself will keep eq. (2.4) from being exact as soon as closed loop diagrams are included.

There are many other interesting models of chiral symmetry. The most elegant and probably most useful of these are the nonlinear Lagrangians invented by WEINBERG [8] and others. Unfortunately, I do not have time to discuss them but they have been amply reviewed in the literature. I should also mention an old model due to Nambu where the pion is not « elementary » but appears as an $\mathcal{N}$-$\bar{\mathcal{N}}$ bound state. The symmetry of Nambu's model is the same as that of the more « conventional » σ-model. Finally, I would like to suggest that someone try to invent yet another model. This would be a model in which the pion is both a Goldstone boson and lies on a Regge trajectory. My prejudice is that the real pion does both these things, but I have never seen a model where it happens explicitly. Perhaps some sort of Van Hove-type scheme where a trajectory is built out of a tower of elementary particles could be made to work.

3. – Chiral symmetry in the strong interactions.

Chiral symmetry, by which I mean $SU_3 \otimes SU_3$ or its subgroup $SU_2 \otimes SU_2$, grew naturally out of the large body of work on current algebra and PCAC which was done four to five years ago [5]. I will assume that everyone is familiar with the sorts of manipulations one does in current algebra, *i.e.* moving derivatives in and out of time-ordered products and pole dominance methods. We will concentrate here on the idea of the symmetry which is believed (at least by some fraction of us) to underly the very successful results of current algebra and PCAC.

In current algebra one assumes eight vector currents V^μ_a $a = 1, \dots, 8$ and eight axial-vector currents A^μ_a $a = 1, \dots, 8$ which satisfy Gell-Mann's famous equal-time commutation relations:

$$(3.1)\quad \begin{cases} [V^0_a(\boldsymbol{x}, t),\ V^0_b(\boldsymbol{y}, t)] = i\delta^3(\boldsymbol{x}-\boldsymbol{y}) f_{abc} V^0_c(\boldsymbol{x}, t)\,, \\ [V^0_a(\boldsymbol{x}, t),\ A^0_b(\boldsymbol{y}, t)] = i\delta^3(\boldsymbol{x}-\boldsymbol{y}) f_{abc} A^0_c(\boldsymbol{x}, t)\,, \\ [A^0_a(\boldsymbol{x}, t),\ A^0_c(\boldsymbol{y}, t)] = i\delta^3(\boldsymbol{x}-\boldsymbol{y}) f_{abc} V^0_c(\boldsymbol{x}, t)\,, \end{cases}$$

where f_{abc} are the structure constants of SU_3. The charges associated with these currents

$$\left\{\begin{aligned} F_a(t) &= \int d^3\boldsymbol{x}\, V^0_a(\boldsymbol{x}, t)\,, \\ F^5_a(t) &= \int d^3\boldsymbol{x}\, A^0_a(\boldsymbol{x}, t)\,, \end{aligned}\right. \tag{3.2}$$

generate the Lie algebra of $SU_3 \otimes SU_3$, *i.e.*

$$\left\{\begin{aligned} [F_a(t), F_b(t)] &= if_{abc}F_c(t)\,, \\ [F_a(t), F^5_b(t)] &= if_{abc}F^5_c(t)\,, \\ [F^5_a(t), F^5_b(t)] &= if_{abc}F_c(t)\,. \end{aligned}\right. \tag{3.3}$$

The weak and electromagnetic currents of hadrons are supposed to be built out of the sixteen currents V^μ_a and A^μ_a. We will not be particularly interested in this aspect of the subject, but just let me remind you of what I said in my introductory remarks about the fascinating interrelationships between strong-interaction symmetries and weak interactions.

The nonstrange vector currents V^μ_a, $a = 1, 2, 3$ are the currents of isospin and (in the absence of electromagnetism) are conserved. This means that the generators F_a for $a = 1, 2$ and 3 are time-independent and lead to a symmetry of strong interactions which, as we all know, is realized by isospin multiplets. In SU_3 symmetry we assume that the remaining vector currents are almost conserved and consequently the remaining generators F_a have a weak time dependence. This gives an approximate symmetry which is again realized by multiplets of particles.

The original statement of PCAC was that the divergences of the nonstrange axial currents $\partial_\mu A^\mu_a$, $a = 1, 2, 3$ are dominated by their pion pole. Sometimes this was extended to dominance of the remaining axial-vector divergences by kaon and eta poles. In chiral symmetry we replace the pole-dominance hypothesis with the following two statements:

i) We assume that the strong-interaction Hamiltonian H can be written as

$$H = H_0 + \varepsilon H'\,, \tag{3.4}$$

where H_0 is $SU_3 \otimes SU_3$-symmetric and the symmetry-breaking term $\varepsilon H'$ is in some sense small. (We introduce the «small» scale parameter ε as a formal device for keeping track of powers of symmetry breaking.)

ii) In the limit of $SU_3 \otimes SU_3$ symmetry, *i.e.* when $\varepsilon \to 0$, we assume that the vacuum is SU_3-invariant so that particles fall into SU_3 multiplets but that it is not $SU_3 \otimes SU_3$-invariant. The full $SU_3 \otimes SU_3$ symmetry must then be realized by an octet of massless pseudoscalar mesons which we identify with π, K and η.

Two obvious questions now arise. First, are these statements consistent in a general physical or mathematical sense? Secondly, what is the matter with just pole dominance, why do we need the symmetry? Let me answer the first question first. We have just seen how the σ-model can describe a world which essentially satisfies i) and ii) above. The only real difference is that there we are talking about $SU_3 \otimes SU_3$, while in the σ-model we had only $SU_2 \otimes SU_2$ symmetry. (Recall that there we had an isospin-invariant vacuum, and hence SU_2 multiplets of particles but the full $SU_2 \otimes SU_2$ symmetry was realized by massless pions.) The σ-model is easily generalized to $SU_3 \otimes SU_3$ and, not surprisingly one obtains a model with SU_3 multiplets of states and an octet of massless pseudoscalar mesons. Of course, there are many other models, *e.g.*, nonlinear Lagrangians, which satisfy i) and ii).

Turning now to the question of why replace pole dominance by symmetry, one answer is simply aesthetics. The idea of chiral symmetry is to my mind at least, much prettier than pole dominance. Secondly, one feels that an understanding of how this symmetry works might contain a real clue to the basic nature of the strong interactions. Especially, in view of this thing which I keep coming back to about the interlocking of symmetry and weak interactions. On a more practical level, one can hope to use the symmetry idea to do better than just pole dominance. The idea here is that we are relating PCAC to SU_3 and hence relating deviations from PCAC, *i.e.* $SU_3 \otimes SU_3$ breaking effects, to SU_3 breaking. One hopes in this way to make some progress in understanding corrections to PCAC.

If we assume that the symmetry-breaking term $\varepsilon H'$ is not too singular, then the equations

$$\partial_\mu V^\mu_a(\boldsymbol{x}, t) = i[\varepsilon\mathscr{H}'(\boldsymbol{x}, t), F_a(t)]\,, \qquad \partial_\mu A^\mu_a(\boldsymbol{x}, t) = i[\varepsilon\mathscr{H}'(\boldsymbol{x}, t), F^5_a(t)]\,, \tag{3.5}$$

hold where $\mathscr{H}'$ is the density of H', *i.e.* $H' = \int d^3x\mathscr{H}'$. We will find these formulae useful later.

Establishing some more notation we define

$$\langle p_a(q)|A^\mu_a(0)|0\rangle = -iq^\mu/2f_a\,, \tag{3.6}$$

where $\langle p_a(q)|$ is a pseudoscalar-meson state (covariantly normalized) with four-momentum q and the f_a are constants. The numbers f_π and f_K are measured $\pi \to \mu + \nu$ and $K \to \mu + \nu$. Their approximate values are

$$\begin{cases} \dfrac{1}{2f_\pi} \approx 0.67\, m_\pi\,, \\[2mm] \dfrac{1}{2f_K} \approx 1.25\, \dfrac{1}{2f_\pi} \approx 0.83\, m_\pi\,, \end{cases} \tag{3.7}$$

where there is some uncertainty in f_K due to the fact that the Cabibbo angle is not known all that well. Taking the divergence of eq. (3.6) gives the identity

$$\langle p_a(q)|\partial_\mu A^\mu_a(0)|0\rangle = m_a^2/2f_a \,. \tag{3.8}$$

We can use this equation to make statement ii) above more precise. As we let ε go to zero, it is clear from eq. (3.5) that $\partial_\mu A^\mu_a$ must vanish. Hence, according to (3.8) either m_a^2 or $(2f_a)^{-1}$ must be zero in the symmetry limit. We assume that m_a vanishes and that $(2f_a)$ remains finite. Remember how this worked in the σ-model? Since we have exact SU_3 when $\varepsilon = 0$, all the f_a's are the same in this limit. Thus we expect f_π and f_K to differ by not more than the $(20\div30)\%$ which is typical of SU_3 breaking. Evidently, this is born out by the data.

Often we will be talking about $SU_2 \otimes SU_2$ symmetry rather than the full $SU_3 \otimes SU_3$. With the obvious modifications all of the above remarks apply to this subsymmetry. In fact, one hears a lot more about $SU_2 \otimes SU_2$ than about $SU_3 \otimes SU_3$. There are good reasons for this. First, it is clear that the masses of our approximate Goldstone bosons are some measure of the size of symmetry breaking. Since the π mass is so much smaller than that of the kaon or eta, one therefore suspects that $SU_2 \otimes SU_2$ might be a much better symmetry than $SU_3 \otimes SU_3$. Furthermore, the experimental data are much better for pion interactions than for kaon or eta interactions.

4. – The $(\bar{3}, 3)\oplus(3, \bar{3})$ and $(8, 8)$ models of symmetry breaking.

The $(\bar{3}, 3)\oplus(3, \bar{3})$ representation of SU_3 contains 18 objects, an even-parity SU_3 singlet u_0, an odd-parity singlet v_0, an even-parity SU_3 octet u_a, $a = 1, 2, \ldots, 8$, and an odd-parity octet v_a, $a = 1, 2, \ldots, 8$. The commutators with the F's are

$$\begin{cases} \left[F_a, \begin{matrix} u_0 \\ v_0 \end{matrix}\right] = 0 \,, \\ \left[F_a, \begin{matrix} u_b \\ v_b \end{matrix}\right] = if_{abc}\begin{pmatrix} u_c \\ v_c \end{pmatrix}, \end{cases} \tag{4.1}$$

in an obvious notation. The commutation relations with F^5's are more complicated:

$$\begin{cases} \left[F^5_a, \begin{matrix} u_b \\ v_b \end{matrix}\right] = id_{abc}\begin{pmatrix} v_c \\ u_c \end{pmatrix} - i\sqrt{\dfrac{2}{3}}\,\delta_{ab}\begin{pmatrix} v_0 \\ u_0 \end{pmatrix}, \\ \left[F^5_a, \begin{matrix} u_0 \\ v_0 \end{matrix}\right] = -i\sqrt{\dfrac{2}{3}}\begin{pmatrix} v_a \\ u_a \end{pmatrix}. \end{cases} \tag{4.2}$$

As I mentioned before, GELL-MANN, OAKES, and RENNER and GLASHOW and WEINBERG have suggested that the symmetry-breaking Hamiltonian belongs to a $(3, \bar{3}) \oplus (\bar{3}, 3)$ representation and has the specific form

$$\varepsilon\mathscr{H}' = u_0 + c u_8 \,, \tag{4.3}$$

where c is a constant to be determined.

In this model the axial-vector divergences are

$$\partial_\mu A^\mu_a = -\frac{1}{\sqrt{3}}(\sqrt{2}+c)\, v_a \qquad \text{for } a = 1, 2 \text{ and } 3, \tag{4.4a}$$

$$\partial_\mu A^\mu_a = -\frac{1}{\sqrt{3}}\left(\sqrt{2}-\frac{c}{2}\right) v_a \qquad \text{for } a = 4, 5, 6 \text{ and } 7 \text{ and,} \tag{4.4b}$$

$$\partial_\mu A^\mu_8 = -\frac{1}{\sqrt{3}}(\sqrt{2}-c)\, v_8 - \sqrt{\frac{2}{3}}\, v_0 \,. \tag{4.4c}$$

With these formulae in hand we can use eq. (3.8) to estimate the parameter c. Combining eqs. (4.4) and (3.8) yields

$$\begin{cases} \dfrac{m_\pi^2}{2f_\pi} = -\dfrac{1}{\sqrt{3}}(\sqrt{2}+c)\,\langle\pi|v_\pi|0\rangle \,, \\[2ex] \dfrac{m_K^2}{2f_K} = -\dfrac{1}{\sqrt{3}}\left(\sqrt{2}-\dfrac{c}{2}\right)\langle K|v_K|0\rangle \,, \\[2ex] \dfrac{m_\eta^2}{2f_\eta} = -\dfrac{1}{\sqrt{3}}(\sqrt{2}-c)\,\langle\eta|v_\eta|0\rangle - \sqrt{\dfrac{2}{3}}\langle\eta|v_0|0\rangle \,, \end{cases} \tag{4.5}$$

where the meaning of the particle subscripts on the octet operators v_α should be obvious. Now let us use SU_3 to evaluate the unknown matrix elements in eq. (4.5). In the SU_3 limit $\langle\eta|v_0|0\rangle = 0$ and $\langle a|v_a|0\rangle = R =$ (constant independent of a). Furthermore, in the same approximation we can neglect the difference between f_π, f_K and f_η, setting them all equal to a common value f. We then find

$$\begin{cases} m_\pi^2 \approx (\sqrt{2}+c)\left(-\dfrac{2fR}{\sqrt{3}}\right), \\[2ex] m_K^2 \approx \left(\sqrt{2}+\dfrac{c}{2}\right)\left(-\dfrac{2fR}{\sqrt{3}}\right), \\[2ex] m_\eta^2 \approx (\sqrt{2}-c)\left(-\dfrac{2fR}{\sqrt{2}}\right). \end{cases} \tag{4.6}$$

Since these equations contain only two unknown parameters c and R there is obviously one relation among the three masses, but it is just the Gell-Mann–Okubo formula

$$\frac{m_\pi^2}{3} + m_\eta^2 - \frac{4}{3} m_K^2 = 0\,, \tag{4.7}$$

which is no surprise since we built octet breaking of SU_3 into our choice of $\mathscr{H}'$. The only really interesting thing about eq. (4.6) is that it determines c from

$$\frac{m_\pi^2}{m_K^2} \approx (\sqrt{2}+c)\left(\sqrt{2}-\frac{c}{2}\right)^{-1}, \tag{4.8}$$

which gives $c \approx -1.25$. It is clear that with this value of c, ∂A^μ_a is very small for $a = 1, 2$ and 3, implying that $SU_2 \otimes SU_2$ is a particularly good symmetry.

I would like now to point out some features of another model where the symmetry-breaking term belongs to the (8, 8) representation of $SU_3 \otimes SU_3$. The (8, 8) representation contains an even-parity SU_3 singlet S_0, an even-parity SU_3 octet S_α, $\alpha = 1, \dots, 8$ and an odd-parity SU_3 octet P_α, $\alpha = 1, \dots, 8$. It also contains a 27-plet of even parity operators and twenty more odd parity operators belonging to the 10 and $\overline{10}$ representations of SU_3. We will need the following commutation relations:

$$\begin{cases} [F^5_a, S^0] = i\sqrt{\dfrac{3}{8}}\,P_a\,, \\[2ex] [F^5_a, S_b] = i\,\dfrac{3}{\sqrt{5}}\,d_{abc}P_c + (\text{operators belonging to 10 and }\overline{10})\,, \end{cases} \tag{4.9}$$

where as noted the commutator of F^5_a with S_b contains **10** and $\overline{\mathbf{10}}$ operators as well as the explicitly shown octet terms. Let us now suppose that the symmetry-breaking Hamiltonian is of the form

$$\varepsilon\mathscr{H}' = S_0 + CS_8\,, \tag{4.10}$$

where the new constant C is the analog of c in the $(\overline{3}, 3)$ model. In this case the divergences of the axial currents will be

$$\partial_\mu A^\mu_a = -\left(\sqrt{\frac{3}{8}} + \frac{3}{\sqrt{5}}\,Cd_{8aa}\,P_a\right) + (\text{operators belonging to 10 and }\overline{10})\,, \tag{4.11}$$

where we have used the fact that the matrix d_{8ab} is diagonal. If we sandwich both sides of eq. (4.11) between pseudoscalar-meson states and the vacuum and make the same approximations as before, the 10 and $\overline{10}$ terms can be dropped. (Because the matrix element of a 10 or $\overline{10}$ operator between an octet state and the vacuum vanishes in the SU_3 limit.) Therefore we can write

$$(4.12)\qquad \begin{cases} m_\pi^2 \approx \left(\sqrt{\frac{3}{8}} + C\sqrt{\frac{3}{5}}\right)(-2fR')\,, \\ m_K^2 \approx \left(\sqrt{\frac{3}{8}} - \frac{C}{2}\sqrt{\frac{3}{5}}\right)(-2fR')\,, \\ m_\eta^2 \approx \left(\sqrt{\frac{3}{8}} - C\sqrt{\frac{3}{5}}\right)(-2fR')\,, \end{cases}$$

in the (8, 8) model where R', the analog of R, is an SU_3-symmetric approximation to the matrix elements of P_a between meson and vacuum. It is easy to verify that the masses in eq. (4.12) satisfy the Gell-Mann–Okubo mass formula, as they should. Also, C can be determined from

$$(4.13)\qquad \frac{m_\pi^2}{m_K^2} \approx \frac{\sqrt{\frac{3}{8}} + C\sqrt{\frac{3}{5}}}{\sqrt{\frac{3}{8}} - (C/2)\sqrt{\frac{3}{5}}}\,.$$

Since m_π^2/m_K^2 is a small number, it is clear that C will turn out to be near $-\sqrt{\frac{5}{8}}$. Consequently, the coefficient of the P_a term in eq. (4.11) for $\partial_\mu A_a^\mu$ will be small when a is equal to 1, 2 or 3. This will not, however, mean that $SU_2 \otimes SU_2$ is a better symmetry than $SU_3 \otimes SU_3$ or SU_3 itself. There are still the 10 and $\overline{10}$ terms which contribute to this divergence. An explicit but tedious calculation shows that the coefficients of these terms are nonvanishing when $C = -\sqrt{\frac{5}{8}}$. Thus in an (8, 8) model we can easily get the observed small pion mass but $SU_2 \otimes SU_2$ will not be a particularly good symmetry.

It may be worth asking yourself what evidence, apart from the smallness of the pion mass, there is that $SU_2 \otimes SU_2$ is a better symmetry than SU_3. We will come back to this point later.

5. – The Goldberger-Treiman relations.

The general form of the matrix element of the nonstrange axial currents between nucleons is

$$(5.1)\qquad \langle N(p')|A^\mu(0)|N(p)\rangle = \bar{u}(p')(\tfrac{1}{2}\boldsymbol{\tau} g(q^2)\gamma^\mu\gamma_5 + \tfrac{1}{2}\boldsymbol{\tau} q^\mu h(q^2)\gamma_5 u(p)\,,$$

where $q = p' - p$ and g and h are form factors. At zero momentum transfer, $g(0) = g_A \approx 1.2$. Taking the divergence of both sides of eq. (5.1) gives

$$(5.2)\qquad \langle N(p')|\partial_\mu \boldsymbol{A}^\mu(0)|N(p)\rangle = iD(q^2)\bar{u}(p')\,\tfrac{1}{2}\boldsymbol{\tau}\gamma u(p)\,,$$

where

$$(5.3)\qquad D(q^2) = 2M_N g(q^2) + q^2 h(q^2)\,.$$

Chiral symmetry says that the divergence $\partial_\mu A^\mu(0)$ is in some sense small. To see precisely what this means, let us write the usual dispersion relation for $D(q^2)$, which for convenience we assume needs no subtraction. Using eq. (3.8) to evaluate the pion-pole term, we have

$$D(q^2) = \frac{m_\pi^2 G}{2f_\pi(m_\pi^2 - q^2)} + \int_{Im_\pi^2}^{\infty} \frac{\varrho(m^2)}{m^2 - q^2}, \tag{5.4}$$

where G is the π-$\mathcal{N}$ coupling constant ($G^2/4\pi \approx 14.7$). Notice that the residue of the pion-pole term is $(m_\pi^2/2f_\pi)G$ and is indeed small because of the factor of m_π^2. In fact, in the limit of chiral symmetry this residue vanishes as it should since $\partial_\mu A^\mu$ vanishes identically in this limit. It should be obvious now that $\varrho(m^2)$ is also a small quantity which vanishes when $SU_2 \otimes SU_2$ is exact.

Let us now evaluate eq. (5.4) at $q^2 = 0$, using eq. (5.3) to obtain

$$D(0) = 2M_{\mathcal{N}} g_A = \frac{G}{2f_\pi} + \int_{9m_\pi^2}^{\infty} \frac{\varrho(m^2)}{m^2}\, dm^2 . \tag{5.5}$$

We see that the factor of m_π^2 in the residue of the pion-pole has been cancelled by an m_π^2 in the denominator, so that the contribution of the pion-pole to $D(0)$ is actually of order unity as $SU_2 \otimes SU_2$ symmetry breaking is turned off. The spectral integral, on the other hand, does not have this compensating factor of m_π^{-2} and therefore vanishes in the symmetry limit. For small symmetry breaking we therefore have

$$2M_{\mathcal{N}} g_A \approx \frac{G}{2f_\pi} . \tag{5.6}$$

One can also write Goldberger-Treiman relations for the matrix elements of the strange axial currents between Σ and $\mathcal{N}$ and Λ and $\mathcal{N}$. This is of more than academic interest since these matrix elements are measured in decays $\Sigma \to \mathcal{N} + e + \nu$ and $\Lambda \to \mathcal{N} + e + \nu$. Denoting the analogs of g_A for these decays by $g_{A,\Sigma}$ and $g_{A,\Lambda}$ and the K-Σ-$\mathcal{N}$ and K-Λ-$\mathcal{N}$ coupling constants by G_Σ and G_Λ, the Goldberger-Treiman relations are

$$\begin{cases} (M_{\mathcal{N}} + M_\Sigma)\, g_{A\,\Sigma} \approx \dfrac{G_\Sigma}{2f_K}, \\[2ex] (M_{\mathcal{N}} + M_\Lambda)\, g_{A,\Lambda} \approx \dfrac{G_\Lambda}{2f_K} . \end{cases} \tag{5.7}$$

In the $(\overline{3}, 3) \oplus (3, \overline{3})$ model we can relate the errors in these generalized Goldberger-Treiman relations to the errors in the ordinary relation [9] of eq. (5.6). This is done as follows. Noting that in this model $\partial_\mu A^\mu_a$ is equal to $(-1/\sqrt{3})(\sqrt{2}+c)v_a$ for $a=1,2,3$, we can write

$$2M_{\mathcal{N}}g_A-\frac{G}{2f_\pi}=-\frac{1}{\sqrt{3}}(\sqrt{2}+c)\int\frac{\mu_{\pi\mathcal{N}\mathcal{N}}m^2}{m^2}\,\mathrm{d}m^2\,, \tag{5.8}$$

where $\mu_{\pi\mathcal{N}\mathcal{N}}$ is the weight function in a dispersion relation for the matrix element $\langle N|v_\pi(0)|N\rangle$. Similarly, the exact form of eq. (5.7) is

$$\begin{cases}(M_{\mathcal{N}}+M_\Sigma)g_{A,\Sigma}-\dfrac{G_\Sigma}{2f_K}=-\dfrac{1}{\sqrt{3}}\left(\sqrt{2}-\dfrac{c}{2}\right)\displaystyle\int\frac{\mu_{K\Sigma\mathcal{N}}(m^2)}{m^2}\,\mathrm{d}m^2\,,\\[2ex] (M_{\mathcal{N}}+M_\Lambda)g_{A,\Lambda}-\dfrac{G_\Lambda}{2f_K}=-\dfrac{1}{\sqrt{3}}\left(\sqrt{2}-\dfrac{c}{2}\right)\displaystyle\int\frac{\mu_{K\Lambda\mathcal{N}}(m^2)}{m^2}\,\mathrm{d}m^2\,,\end{cases} \tag{5.9}$$

where $\mu_{K\Sigma\mathcal{N}}$ and $\mu_{K\Lambda\mathcal{N}}$ are weight functions for the matrix elements $\langle N|v_K(0)|\Sigma\rangle$ and $\langle N|v_K(0)|\Lambda\rangle$. The V's are an octet of operators, so one can use SU_3 to relate the right-hand sides of eqs. (5.8) and (5.10). Since the right-hand sides were already supposed to be small, the error incurred by this approximation should be of second order in smallness. In this way one finds

$$\mu_{K\Sigma\mathcal{N}}\approx(1-2\lambda)\mu_{\pi\mathcal{N}\mathcal{N}}\,,\qquad \mu_{K\Lambda\mathcal{N}}\approx\sqrt{\tfrac{3}{2}}(1-2\lambda/3)\mu_{\pi\mathcal{N}\mathcal{N}}\,, \tag{5.10}$$

where $\lambda/(\lambda-1)$ is the usual (*a priori* unknown) F/D ratio which appears in the SU_3 symmetric coupling of three octets. Taking the linear combination of eqs. (5.8) and (5.9) which eliminates λ and using eq. (4.6) to evaluate c then yields

$$\sqrt{6}[(M_{\mathcal{N}}+M_\Lambda)g_{A,\Lambda}-G_\Lambda/2f_K]-[(M_\Sigma+M_{\mathcal{N}})g_{A,\Sigma}-G_\Sigma/2f_K]\approx \approx\frac{2m_K^2}{m_\pi^2}[2M_{\mathcal{N}}g_A-G_\pi/2f]\,. \tag{5.11}$$

Thus we have obtained a relation between different deviations from perfect PCAC. Note that eq. (5.11) is a consequence simply of the assumed group-theoretic properties of symmetry breaking. It is, therefore, quite analogous to the Gell-Mann–Okubo sum rule.

The factor $m_K^2/m_\pi^2\sim12$ on the right-hand side of (5.11) is large. Therefore, at least one of the errors in the generalized Goldberger-Treiman relations on the left-hand side of (5.11) must be large compared to $(2M_{\mathcal{N}}g_A-G/2f_\pi)$ if the $(\overline{3},3)\oplus(3,\overline{3})$ model which leads to (5.11) is to be valid. Unfortunately, G and g_A are not known accurately enough to say much more than that $2M_{\mathcal{N}}g_A\sim G/2f_\pi$ with an error of not more than about 10%. The error could be as small as 1% which would imply an error of (10÷20)% in the generalized

Goldberger-Treiman relations. This would certainly be an acceptable state of affairs. However, if $2M_{\mathcal{N}}g_A$ and $G/2f_\pi$ differ by 10%, then eq. (5.11) predicts a 100% error in the generalized Goldberger-Treiman relations. I doubt that this is possible. The quantities $g_{A,\Lambda}$ and $g_{A,\Sigma}$ are known to roughly satisfy the SU_3 prediction:

$$g_{A,\Sigma} \sim (1-2\alpha)\,g_A\,, \qquad g_{A,\Lambda} \sim \sqrt{\tfrac{3}{2}}\,(1-2\alpha/3)\,g_A\,, \tag{5.12}$$

where $\alpha/(\alpha-1)$ is an F/D ratio. Using this and neglecting the differences between $M_{\mathcal{N}}+M_\Lambda$, $M_{\mathcal{N}}+M_\Sigma$ and $2M_{\mathcal{N}}$ and between f_K and f_π leads to the *very rough* formula

$$\frac{\sqrt{6}\,G_\Lambda - G_\Sigma + 2G}{2G} \sim \left(\frac{m_K^2}{m_\pi^2}+1\right)\left(\frac{4M_{\mathcal{N}}g_A f_\pi}{G}-1\right). \tag{5.13}$$

Now, since the SU_3 predictions for G_Λ and G_Σ are

$$G_\Sigma \sim (1-2\alpha')\,G\,,$$
$$G_\Lambda \sim \sqrt{\tfrac{3}{2}}\,(1-2\alpha'/3)\,G\,,$$

where $\alpha'/(\alpha'-1)$ is again an F/D ratio, it is easy to see that the left-hand side of (5.13) vanishes in the SU_3 limit and is thus a measure of SU_3 breaking for the meson-baryon coupling constants. Therefore, if $4M_{\mathcal{N}}g_A g_\pi/G$ differs from unity by 10%, SU_3 breaking in the coupling constants G, G_Λ and G_Σ will have to be on the order of 100%. The constants G_Σ and G_Λ have not yet been determined with sufficient accuracy to rule out this possibility. One would certainly hope however that SU_3 breaking effects are not this large.

The above discussion can be summarized by saying that *in the* $(3,\overline{3})\oplus(\overline{3},3)$ *model, either the Goldberger-Treiman relation* $2Mg_A \approx G/2f_\pi$ *must be accurate to better than about 3% or there must be very large* (~100%) SU_3 *symmetry breaking in the meson-baryon coupling constants.*

What about the (8, 8) model? In this model the divergences of the axial currents contain $\overline{\mathbf{10}}$ and **10** pieces as well as an octet piece and eq. (5.14) does not hold. Thus for (8, 8) breaking, there is no conflict between a reasonably large (~10%) error in the Goldberger-Treiman relation and reasonably accurate (~30%) SU_3 for meson baryon couplings. The basic reason for this is that with (8, 8) symmetry breaking, $SU_2\otimes SU_2$ is broken by an amount comparable to SU_3 even though m_π^2 is small. In $(3,\overline{3})\oplus(\overline{3},3)$, on the other hand, the breaking of $SU_2\otimes SU_2$ is on the order of m_π^2/m_K^2 times the breaking of SU_3. If this is the case, a quantity like $(4Mg_A f_\pi/G-1)$, which is a measure of $SU_2\otimes SU_2$ breaking, must be very small if SU_3 breaking is to be of a reasonable size.

6. – More on Goldberger-Treiman relations.

The weight functions $\mu_{\pi\mathcal{N}\mathcal{N}}$, $\mu_{K\Sigma\mathcal{N}}$ and $\mu_{K\Lambda\mathcal{N}}$ are all of order ε as $\varepsilon \to 0$. (Recall that ε is a formal parameter which we use to keep track of orders of symmetry breaking.) Therefore if the integrals on the right-hand sides of eqs. (5.10) and (5.11) are convergent as $\varepsilon \to 0$ the errors in the Goldberger-Treiman relations will be strictly $O(\varepsilon)$. It is easy to see, however, that the possibility of divergent integrals exists. As ε goes to zero, the three-pion threshold moves down to $m^2 = 0$ which could cause the integral to diverge at the lower end. Li and Pagels [10] recently pointed out that this does, in fact, happen and that

$$\int_{(3\text{-pion state})} \frac{\mu_{\pi\mathcal{N}\mathcal{N}}}{m^2}\, dm^2 = \varepsilon \log \varepsilon \tag{6.1}$$

as $\varepsilon \to 0$. These $\varepsilon \log \varepsilon$ terms do not have to satisfy eq. (5.11). However, the magnitude of these terms can be estimated (since they come only from the three-meson intermediate state near threshold) and they are known to be small. Therefore, for purposes of phenomenology, the bothersome $\varepsilon \log \varepsilon$ terms can be ignored. Their presence is of considerable theoretical interest however.

Next let us ask how accurate eq. (5.11) should be, assuming that we have already agreed to ignore the $\varepsilon \log \varepsilon$ terms. Since the μ's are separately of order ε, it is clear that the error in eq. (5.11) is of order ε^2. Therefore, we can say that the error in eq. (5.11) is of order ε^2, that is of second order in symmetry breaking. Thus eq. (5.11) is a sum rule which relates quantities of first order in symmetry breaking with an error which is of second order.

In actuallity, the error is probably not ε^2 but $\varepsilon^2 \log \varepsilon$ or $\varepsilon^2 \log^2 \varepsilon$ since we may anticipate that there will be further divergences at threshold. Again these logarithmic terms will come only from states containing a few mesons and are expected to be small.

7. – More on meson masses.

It is interesting to study the vacuum expectation value

$$\langle 0 | i \int d^4x\, T(\partial_\mu A^\mu_a(x)\, \partial_\nu A^\nu_a(0) | 0 \rangle = \frac{m_a^2}{(2f_a)^2} + \int \frac{\varrho_a(\mu^2)}{\mu^2}\, d\mu^2 , \tag{7.1}$$

where the first term on the right is the contribution of the single pseudoscalar-meson state and ϱ_a is the spectral weight function of the higher states, *e.g.* 3-meson, baryon-antibaryon, etc. Since ϱ_a is the weight function for two divergences, it is clear that it is of order ε^2. Note that the first term on the right-hand side of eq. (7.1) is proportional to m_a^2 and is of order ε rather than ε^2. This comes about because of a factor $m_a^{-2} \sim \varepsilon^{-1}$ from the denominator of the meson pole. Remember how essentially the same thing happened in our discussion of the Goldberger-Treiman relation.

A straightforward current algebra calculation gives

$$-\langle 0|[F^5_a, [F^5_a, \varepsilon\mathscr{H}'(0)]]|0\rangle = \frac{m_a^2}{(2f_a)^2} + \int \frac{\varrho_a(\mu)^2}{\mu^2}\, d\mu^2 , \tag{7.2}$$

where the so-called «sigma-commutator» $[F^5_a, [F^5_a, \varepsilon\mathscr{H}'(0)]]$ has shown up in the process of moving derivations outside the time-ordering symbol in eq. (7.1). Equation (7.2) is still exact. To get something useful we drop the spectral integral which is second order in ε to obtain the simple result

$$m_a^2 = -(2f_a)^2\langle 0|[F^5_a, [F^5_a, \varepsilon\mathscr{H}'(0)]]|0\rangle + O(\varepsilon^2) , \tag{7.3}$$

which relates the meson masses to vacuum expectation values of a double commutator of F's with $\varepsilon\mathscr{H}'$. Actually, the error in eq. (7.3) is probably not strictly $O(\varepsilon^2)$ but rather of order $\varepsilon^2\log^2\varepsilon$ due to infra-red divergences of the type described above. Possible $\varepsilon^2\log^2\varepsilon$ terms are not, however, believed to be of any particular consequence.

In the $(3, \bar{3}) \oplus (\bar{3}, 3)$ model the double commutator in eq. (7.3) can be evaluated explicitly. It is almost trivial to see that $[F^5_a, [F^5_a, \varepsilon\mathscr{H}']]$ will be a sum of u_0 and various u_a's. The vacuum expectation value of one of the octet of operator u_a will be of order ε relative to the expectation value of u_0. This is simply a consequence of the fact that the vacuum becomes SU_3-symmetric in the limit $\varepsilon \to 0$. Therefore, we need only the coefficient of u_0 in $[F^5_a, [F^5_a, \varepsilon\mathscr{H}']]$. A simple calculation gives

$$\begin{aligned}[F^5_a, [F^5_a, \varepsilon\mathscr{H}']] &= \\ &= -\left(\tfrac{2}{3} + \sqrt{\tfrac{2}{3}}\, c d_{aa8}\right) u_0 + [\text{other members of } (3, \bar{3}) \oplus (\bar{3}, 3)] ,\end{aligned} \tag{7.4}$$

and we have

$$m_a^2 = 4f_a^2\left(\tfrac{2}{3} + \sqrt{\tfrac{2}{3}}\, c d_{aa8}\right)\langle 0|u_0|0\rangle + O(\varepsilon^2) . \tag{7.5}$$

Note that it would not have been consistent to keep $\langle 0|u_8|0\rangle$ which is of order ε^2 without keeping the spectral integral in (7.2). Therefore, the meson masses only give us information about $\langle 0|u_0|0\rangle$. Explicitly, one finds

$$\left\{\begin{aligned} m_\pi^2 &= 4f_\pi^2\frac{\sqrt{2}}{3}(\sqrt{2}+c)\langle 0|u_0|0\rangle , \\ m_K^2 &= 4f_K^2\left(\sqrt{2}-\frac{c}{2}\right)\langle 0|u_0|0\rangle , \\ m_\eta^2 &= 4f_\eta^2(\sqrt{2}-c)\langle 0|u_0|0\rangle . \end{aligned}\right. \tag{7.6}$$

If we equate f_π, f_K and f_η, which can be done at the present level of accuracy, eqs. (7,6) become the same as eqs. (4.5) with the unknown R replaced by the new unknown $\langle 0|u_0|0\rangle$. Thus we get the same value for c, which is to be expected if one is working in a consistent approximation.

In the $(3, \overline{3}) \oplus (\overline{3}, 3)$ model, then, the net information content of the meson masses is the value of the parameter c and the value of $\langle 0|u_0|0\rangle$. This circumstance is, in fact, quite general. Suppose that $\varepsilon\mathscr{H}'$ belongs to a single $SU_3 \otimes SU_3$ representation of the form $(X, \overline{X}) \oplus (\overline{X}, X)$ such as $(3, \overline{3}) \oplus (\overline{3}, 3)$, $(6, \overline{6}) \oplus (\overline{6}, 6)$ or $(8, 8)$. Any such representation contains a unique even parity SU_3 singlet, call it O_0. It will also contain at least one even-parity eighth-component of octet O_8. We can then try $\varepsilon\mathscr{H}' = O_0 + CO_8$ where C is a parameter. Then to get the meson masses we need the coefficient of O_0 in the double commutator $[F_a, F_a, \varepsilon\mathscr{H}']$. It will have the general form

$$(7.7) \qquad -[F_a, [F_b, O_0 + CO_8]] = \\ = A_x \delta_{ab} O_0 + CB_x d_{aa8} O_0 + (\text{other members of } (X, \overline{X}) \oplus (\overline{X}, X)) ,$$

where A_x and B_x are Clebsch-Gordan coefficients which depend on the representation. The meson masses will then be

$$(7.8) \qquad m_a^2 = (2f_a)^2 (A_x + CB_x d_{aa8}) \langle 0|O_0|0\rangle .$$

Now comparing with eq. (7.5) it is clear that one can always choose C and $\langle 0|O_0|0\rangle$ in such a way as to reproduce the observed masses. Conversely, a knowledge of the masses leads to nothing more than a determination of these parameters.

Evidently, we have learned that the meson masses can be fitted with $\varepsilon\mathscr{H}'$ belonging to almost any representation of $SU_3 \otimes SU_3$. In order to check the $(3, \overline{3}) \oplus (3, \overline{3})$ model, we have to look at something else.

8. – π-π scattering.

In Weinberg's [5, 11] analysis of π-π scattering he showed that a particular term in the π-π amplitude is proportional to $\langle \pi_a|[F_b^5, [F_c^5, \varepsilon\mathscr{H}']]|\pi_d\rangle$. A further reduction of pions gives

$$(8.1) \qquad \langle \pi_a|[F_b^5, [F_c^5, \varepsilon\mathscr{H}']]|\pi_d\rangle = \langle 0|\Big[F_a^5, \Big[F_d^5, [F_b^5, [F_c^5, \varepsilon\mathscr{H}']]\Big]\Big]|0\rangle ,$$

so that π-π scattering measures the vacuum expectation value of a sort of « generalized σ-term » involving four commutators with F^5's. WEINBERG also pointed out that if $[F_b^5, [F_c^5, \varepsilon\mathscr{H}']]$ contains no $I = 2$ piece, then the π-π matrix element of this operator is determined. This comes about for the following reason. One can show that if $[F^5_b, [F^5_c, \varepsilon\mathscr{H}']]$ is pure $I=0$, then $\varepsilon\mathscr{H}'$ must belong to the $(\frac{1}{2}, \frac{1}{2})$ representation of $SU_2 \otimes SU_2$. In the $(\frac{1}{2}, \frac{1}{2})$ representation there is a unique even parity operator which we can call σ_0. By parity, then $\varepsilon\mathscr{H}'$ is proportional to σ_0 and so is any operator formed by commuting $\varepsilon\mathscr{H}'$ with an *even* number of F^5's. Therefore one would have

$$F^5_a, F^5_d, [F^5_b, [F^5_c, \varepsilon\mathscr{H}']] [F^5_a, [F^5_b, \varepsilon\mathscr{H}']] \sigma_0 , \tag{8.2}$$

so that the right-hand side of eq. (8.1) can be evaluated in terms of the pion mass by using eq. (7.5).

In the $(3, \overline{3}) \oplus (\overline{3}, 3)$ model, Weinberg's hypothesis of pure $I=0$ sigma-terms is valid. For (8, 8), on the other hand, it is not. Therefore, analysis of π-π scattering could in principle, distinguish between $(3, \overline{3}) \oplus (\overline{3}, 3)$ and (8, 8) symmetry breaking.

If $\varepsilon\mathscr{H}'$ belongs to (8, 8) or to the more general representation $(X, \overline{X}) \oplus \oplus (\overline{X}, X)$, the right-hand side of (8.1) can still be calculated. As long a commuting F^5's with $\varepsilon\mathscr{H}'$ cannot produce more than one independent even-parity SU_3 singlet operator O_0, vacuum expectation values like that on the right-hand side of eq. (8.1) can always be expressed as Clebsch-Gordan coefficients times $\langle 0|O_0|0\rangle$. The unknown $\langle 0|O_0|0\rangle$ is then eliminated by using the mass formula. The approximations and methods used to do this are basically the same as were used to obtain eq. (7.5).

Recent analyses [12] of π-π scattering tend to strongly favor $(3, \overline{3}) \oplus (\overline{3}, 3)$ over (8, 8) or more complicated representations. However, it must be born in mind that the $\pi\pi$ amplitude is related to the sigma-term only at very low energies where direct experimental data are essentially nonexistent. Theoretical models of π-π scattering which are constrained to fit the experimental inferences about π-π interactions at higher energies may contain serious biases.

9. – K_{l3} decay.

The matrix element of the strangeness-changing vector current between K and π is conventionally written as

$$\langle K(q)|V^\mu(0)|\pi(p)\rangle = (p^\mu + q^\mu)F_+(t) + (p^\mu - q^\mu)F_-(t) ,$$

where $t = (p-q)^2$. Our normalization is such that $F_+(0)$ would be equal to unity if SU_3 were exact. Also exact SU_3 would imply $F_- = 0$.

The original « soft-pion » theorem [5] of Callan and Treiman and Mathur, Okubo and Pandit was

$$F_+(m^2_K) + F_-(m^2_K) \approx f_\pi/f_K . \tag{9.1}$$

This is an interesting relation because it uses PCAC to relate SU_3 breaking effects. To make this explicit we rewrite (9.1) as

$$[F_+(0)-1]+[F_+(m_K^2)-F_+(0)]+F_-(m_K^2)+[1-f_\pi/f_K]=0\,. \tag{9.2}$$

Now a soft-pion limit is equivalent to setting $m_\pi^2 \approx 0$ which would then give $m_K^2 \approx 0$ in an SU_3 limit. With this is mind, one easily sees that each term in (9.2) is separately of order SU_3 breaking.

WEINSTEIN and myself [13] showed this more explicitly when we obtained the only slightly different result:

$$\left(\frac{\lambda_+}{m_\pi^2}\right)(m_K^2-m_\pi^2)+\xi(0)=\frac{1}{2}\left(\frac{f_\pi}{f_K}-\frac{f_K}{f_\pi}\right)+O(\varepsilon^2)\,, \tag{9.3}$$

where $\lambda_+/m_\pi^2=(\partial/\partial t)\log F_+|_{t=0}$ and $\xi(0)=F_-(0)/F_+(0)$.

Each term in eq. (9.3) is explicitly an SU_3 breaking effect and the error is explicitly of order ε^2. The discovery of $\log\varepsilon$ terms in symmetry breaking by LI and PAGELS suggested that eq. (9.3) might, however, be incorrect. LI, PAGELS, WEINSTEIN and I have looked into this and have found that the correct version of eq. (9.3) is

$$\frac{\lambda_+}{m_\pi^2}(m_K^2-m_\pi^2)+\xi(0)=\frac{1}{2}\left(\frac{f_\pi}{f_K}-\frac{f_K}{f_\pi}\right)+\Delta+O(\varepsilon^2\log\varepsilon)\,. \tag{9.4}$$

Where Δ is a quantity which can be calculated but which is model-dependent. In the $(3,\bar{3})\oplus(\bar{3},3)$ model, Δ turns out to be small compared to $\frac{1}{2}(f_\pi/f_K-f_K/f_\pi)$ and therefore has little practical effect. If $\varepsilon\mathscr{H}'$ belongs to a different representation, (8, 8) say, Δ is expected to be of a similar size.

The experimental situation in K_{l3} decay is still too confused to tell whether or not eq. (9.4) holds. Hopefully we will know within a year. The importance of this test is not that it differentiates between models of symmetry breaking, but that it is an (essentially) model-independent test of the idea of expanding in powers of symmetry breaking.

10. – π-$\mathcal{N}$ scattering.

Pion-nucleon scattering played an important role in the early days of current algebra and PCAC [5]. In fact it was Adler's « consistency condition » and the Adler-Weisberger sum rule which really got the subject going in the first place. Our particular interest in π-$\mathcal{N}$ scattering has to do with what it can tell us about symmetry breaking.

To this end, let us define the usual off-shell amplitude by using $\varphi_\pi = (2f_\pi/m_\pi^2)\,\partial_\mu A^\mu$ for the pion field. The off-shell amplitude for $\pi_a(q)+N(p)\rightarrow$

$\to \pi_b(q') + N(p')$ can then be written as

$$(10.1)\qquad \bar{u}(p')\left[\left(A^+ + \frac{q+q'}{2}B^+\right)\frac{\{\tau_a, \tau_b\}}{2} + \left(A^- + \frac{q+q'}{2}B^-\right)\frac{[\tau_a, \tau_b]}{2}\right]u(p)\,,$$

where the four function $A^\pm$ and $B^\pm$ depend on the variables q^2, q'^2 and

$$(10.2)\qquad \nu = (p+p')\cdot(q+q')/4M_{\mathcal{N}}\,, \qquad \nu_B = -q\cdot q'/2M_{\mathcal{N}}\,.$$

We will concentrate on the particular amplitude

$$(10.3)\qquad T(q^2, q'^2, \nu, \nu_B) = A^+(q^2, q'^2, \nu, \nu_B) + \nu B^+(q^2, q'^2, \nu, \nu_B)\,.$$

The nucleon-pole term in this amplitude can be written as

$$(10.4)\qquad T_{\text{pole}} = K_{\pi\mathcal{N}\mathcal{N}}(q^2)\,K_{\pi\mathcal{N}\mathcal{N}}(q'^2)\,\frac{G^2}{M}\,\frac{\nu_B}{\nu_B^2 - \nu^2}\,,$$

where $K_{\pi\mathcal{N}\mathcal{N}}$ is the pionic form factor of the nucleon which is, of course, determined by our choice of pion field. Note that at $\nu_B = 0$, the pole term vanishes.

The usual current algebra calculation gives the Adler conditions

$$(10.5)\qquad T(m_\pi^2, 0, 0, 0) = 0\,, \qquad T(0, m_\pi^2, 0, 0) = 0\,,$$

where we always define the point $\nu = 0$, $\nu_B = 0$ by taking the limit $\nu_B \to 0$ followed by $\nu \to 0$; this eliminates the pole term. Another standard calculation gives

$$(10.6)\qquad T(0, 0, 0, 0) = -4f_\pi^2 \sigma_{\mathcal{N}\mathcal{N}}\,,$$

where

$$(10.7)\qquad \sigma_{\mathcal{N}\mathcal{N}} = \tfrac{1}{3}\sum_{i=1}^{3}\langle N|[F_i^5, [F_i^5, \varepsilon\mathcal{H}']]|N\rangle\,.$$

It would clearly be a good thing to know $\sigma_{\mathcal{N}\mathcal{N}}$. The value of this matrix element might tell us something interesting about symmetry breaking. Unfortunately, eq. (10.6) is not very useful since it is an off-shell amplitude. You might ask why not just approximate the off-shell amplitude by its on-shell value? To see why let us write

$$(10.8)\qquad T(m_\pi^2, m_\pi^2, 00) = T(0, 0, 0, 0) + \frac{\partial}{\partial q^2}T(0, 0, 0, 0)\,m_\pi^2 + {} + \frac{\partial}{\partial q'^2}T(0, 0, 0, 0)\,m_\pi^2 + O(\varepsilon^2)\,,$$

where we have dropped terms of order m_π^4 since they are of order ε^2. Now from eq. (10.6) it is clear that $T(0, 0, 0, 0)$ is of order ε but so is a term like $(\partial/\partial q^2)T(0, 0, 0, 0)m_\pi^2$. Therefore, $T(m_\pi^2, m_\pi^2, 0, 0)$ is order ε but has no particular reason to be equal to $T(0, 0, 0, 0)$, except that both must be small. In fact, we will see that

$$T(m_\pi^2, m_\pi^2, 0, 0) \approx -T(0, 0, 0, 0)\,, \tag{10.9}$$

where the minus sign is to be noted. To see this, we expand eq. (10.5) in powers of ε to obtain

$$\begin{cases} T(m_\pi^2, 0, 0, 0) = T(0, 0, 0, 0) + \dfrac{\partial}{\partial q^2} T(0, 0, 0, 0)m_\pi^2 + O(\varepsilon^2) = 0\,, \\[2ex] T(0, m_\pi^2, 0, 0) = T(0, 0, 0, 0) + \dfrac{\partial}{\partial q'^2} T(0, 0, 0, 0)m_\pi^2 + O(\varepsilon^2) = 0\,. \end{cases} \tag{10.10}$$

Thus, up to $O(\varepsilon^2)$ each of the derivative terms in eq. (10.8) is equal to $-T(0, 0, 0, 0)$ which verifies eq. (10.9) and gives

$$T(m_\pi^2, m_\pi^2, 0, 0) = 4f_\pi^2\sigma_{\mathcal{N}\mathcal{N}} + O(\varepsilon^2)\,. \tag{10.11}$$

What we have achieved is a way to determine the matrix element $\sigma_{\mathcal{N}\mathcal{N}}$ from *on-shell* π-$\mathcal{N}$ scattering data. If we agree to neglect the $O(\varepsilon^2)$ term in eq. (10.11), the determination of $\sigma_{\mathcal{N}\mathcal{N}}$ from experiment is now completely unambiguous.

Before proceeding to the experimental picture, let us estimate the size of $\sigma_{\mathcal{N}\mathcal{N}}$ in the $(3, \overline{3}) \oplus (\overline{3}, 3)$ model. In this model we have

$$\sigma_{\mathcal{N}\mathcal{N}} = \tfrac{1}{3}(\sqrt{2} + c)\langle N|\sqrt{2}\,u_0 + u_8|N\rangle\,. \tag{10.12}$$

Taking the value $c \approx -1.25$ which gives the proper π to K mass ratio, the numerical factor $\frac{1}{3}(\sqrt{2} + c)$ is about equal to 0.05. Also from the mass differences in the baryon octet, we know that $\langle N|u_8|N\rangle$ is approximately equal to -200 MeV. Therefore, in the $(3, \overline{3}) \oplus (\overline{3}, 3)$ model, one has

$$\sigma_{\mathcal{N}\mathcal{N}} \approx -10\ \text{MeV} + 0.07\langle N|u_0|N\rangle\,. \tag{10.13}$$

The matrix element $\langle N|u_0|N\rangle$ is not known, but the naive guess would be that its magnitude would be similar to that of $\langle N|u_8|N\rangle$. The reason for this is that the whole chiral symmetry game is based on the assumption that $SU_3 \otimes SU_3$ is about as good a symmetry as SU_3. Since u_0 breaks $SU_3 \otimes SU_3$ and u_8 breaks SU_3 (as well as $SU_3 \otimes SU_3$), we cannot allow $\langle N|u_0|N\rangle$ to be an order of magnitude bigger than $\langle N|u_8|N\rangle$ and still have the two symmetries broken by a comparable amount. Therefore in the $(\bar{3}, 3) \oplus (3, \bar{3})$ model one estimates that

$$|\sigma_{\mathcal{N}\mathcal{N}}| \sim 10 \text{ to } 20 \text{ MeV}. \tag{10.14}$$

Another way to arrive at eq. (10.14) is to note that in the $(3, \bar{3}) \oplus (\bar{3}, 3)$ model $SU_2 \otimes SU_2$ is a much better symmetry than SU_3. Therefore $\sigma_{\mathcal{N}\mathcal{N}}$ which is a measure of $SU_2 \otimes SU_2$ breaking should be small compared to $\langle N|u_8|N\rangle$ which is the analogous measure of SU_3 breaking.

What do the experiments say? To reach the unphysical (but on-shell) point $\nu = \nu_B = 0$ one has to use a dispersion relation. Fortunately, there are π-π-$\mathcal{N}$ phase-shift analyses which make this possible. CHENG and I [14] carried out a thorough analysis of the dispersion relation and the best available data. The number we found was

$$\sigma_{\mathcal{N}\mathcal{N}} \approx 110 \text{ MeV}, \tag{10.15}$$

which is clearly inconsistent with the above interpretation of the $(3, \bar{3}) \oplus (\bar{3}, 3)$ model. It appears to imply that the symmetry breaking belongs to some other representation of $SU_3 \otimes SU_3$. Note that this value of $\sigma_{\mathcal{N}\mathcal{N}}$ is not large on an absolute scale, after all SU_3 breaking is typically 100 to 200 MeV. What it means is that $SU_2 \otimes SU_2$ breaking is comparable to SU_3 breaking. This, as noted before, happens in models like (8, 8).

Since eq. (10.15) appears to contradict everyone's favorite model of symmetry breaking, it is important to see if there is a way out. I will list the various possibilities and comment on them.

i) The phase-shift analyses that were used to obtain the number $\sigma_{\mathcal{N}\mathcal{N}} \approx 110$ MeV may have serious systematic errors at low energies. It is marginally possible that when new, accurate low-energy data are available, the experimental value of $\sigma_{\mathcal{N}\mathcal{N}}$ will become compatible with $(3, \bar{3}) \oplus (\bar{3}, 3)$. Some calculations by the Karlsruhe group [15] suggest this possibility. Using a different method which places more emphasis on the (also uncertain) medium energy data they found something like 60 to 70 MeV for $\sigma_{\mathcal{N}\mathcal{N}}$. At this point one can only wait for a new phase-shift analysis.

ii) The neglected $O(\varepsilon^2)$ terms may not be small. In the $(3, \bar{3}) \oplus (\bar{3}, 3)$ model where $SU_2 \otimes SU_2$ is supposed to be a very good symmetry it is hard to see how this could happen: BROWN, PARDEE and PECCEI [16] have looked at this question in some detail and cannot find any obvious mechanism for producing a large $O(\varepsilon^2)$ term. The infra-red log terms will only add some small $O(\varepsilon^2 \log^2 \varepsilon)$ corrections.

iii) It could be that the $(3, \bar{3}) \oplus (\bar{3}, 3)$ model is correct but the matrix elements of u_0 are large. (One should have $\langle N|u_0|N\rangle \sim 1500$ MeV to get a number like that in eq. (10.15).) If so we will have to change our interpretation of $SU_3 \otimes SU_3$. An attractive possibility exists. In $SU_3 \otimes SU_3$ symmetry the divergences $\partial_\mu A^\mu_a$ of the axial currents are in some sense small operators. On the other hand, they can have large matrix elements at small momentum transfer. This is because of what low-mass poles do to the pseudoscalar mesons which are the Goldstone bosons of the theory. Evidently, we could make u_0 a small operator with large matrix elements if we could find an (approximate) Goldstone boson to go with it. Such a particle would be a 0^+, SU_3 singlet. Several people have suggested that scale invariance is realized by a Goldstone boson. It turns out that this would nicely explain [17, 18] our large value of $\sigma_{\mathcal{N}\mathcal{N}}$. However, there is no further hard experimental evidence for a so-called « dilation ». It remains a challenge to theorists to make some *solid* predictions which can be tested experimentally.

Previous estimates of $\sigma_{\mathcal{N}\mathcal{N}}$ have either used off-shell extrapolations or have attempted to relate $\sigma_{\mathcal{N}\mathcal{N}}$ to the on-shell amplitude at threshold rather than at our point $\nu = \nu_B = 0$. Off-shell extrapolations are necessarily based on the faith that some dispersion relation is dominated by a few low-mass states. Therefore, one simply cannot make any *a priori* statement about their expected accuracy. It is important to note that extrapolating off-shell to find $\sigma_{\mathcal{N}\mathcal{N}}$ is a different game than using PCAC to make « smooth » off-shell extrapolations of pion amplitudes. Calculating $\sigma_{\mathcal{N}\mathcal{N}}$ is calculating a correction to $SU_2 \otimes SU_2$ symmetry and hence a correction to PCAC. What I want to discuss now is the relation of the σ-term to the on-shell amplitude at threshold.

The amplitude T is even in ν and hence a function only of ν^2. In the region between ν and $\nu_B = 0$ and physical threshold, both ν^2 and ν_B are of order ε, that is they are proportional to m_π^2. Therefore we can approximate the very low-energy π-$\mathcal{N}$ amplitude by expanding around $\nu = 0$ and $\nu_B = 0$ and neglecting terms higher than ν^2 and ν_B. The error will evidently be $O(\varepsilon^2)$ with the result that

$$T(m_\pi^2, m_\pi^2, \nu, \nu_B) = \frac{G^2}{m}\frac{\nu_B^2}{\nu_B^2 - \nu^2} + 4f_\pi^2\sigma_{\mathcal{N}\mathcal{N}} + a\nu^2 + b\nu_B + O(\varepsilon^2)\,, \tag{10.16}$$

where the pole term has been explicitly included. The coefficients a and b of ν^2 and ν_B can be computed by feeding phase shifts into a dispersion integral. CHENG and I did this and found that eq. (10.16) gave a very good representation of T in the region between $\nu = \nu_B = 0$ and threshold. One would expect this to be the case since eq. (10.16) is a consistent approximation. The pole terms takes care of the singularity and all terms of $O(\varepsilon)$ have been included. At physical threshold a very peculiar thing happens. The pole term is small there and the two terms $a\nu^2$ and $b\nu_B$ combine in such a way as to almost exactly cancel $4f_\pi^2\sigma_{\mathcal{N}\mathcal{N}}$. The smallness of the isospin-even scattering length is, of course, well known. What is surprising is that this extremely small scattering length is not the result of a small σ-term.

Let me conclude this Section on π-$\mathcal{N}$ scattering by saying that I do not know of any other simple information about symmetry breaking that can be obtained from π-$\mathcal{N}$ scattering. This is not to say that there is not any.

11. – Nonleptonic weak interactions.

Analyses of the decays $Y \to \pi + \mathcal{N}$, $K \to 3\pi$ and $K \to 2\pi$ were one of the major triumphs of PCAC and current algebra. If we assume the usual current-current theory of nonleptonic weak interactions, a study of these decays could produce information about strong chiral-symmetry breaking.

It has been known for sometime that neglecting *CP* violation, the decay amplitude for $K \to 2\pi$ is of order $G\lambda$ where G is the (weak interaction) Fermi constant and λ is a parameter which measures SU_3 breaking in the strong interactions. This is a mild embarrassment since it says that in the rate for $K \to 2\pi$ should be suppressed a factor of λ^2 relative to an « allowed » decay like $\Lambda \to \pi + \mathcal{N}$. Experimentally this suppression does not appear to be there, but it is hard to know how to compare the decay rate of a light boson to that of a heavy fermion.

It can be shown [19] that in the $(3, \overline{3}) \oplus (\overline{3}, 3)$ model for $SU_3 \otimes SU_3$ breaking, the amplitude for $K \to 2\pi$ is of order $\varepsilon^2 G$; that is, the decay is doubly forbidden. In other models like $(8, 8)$, the amplitude will be of order εG which is just the old SU_3 result (ε and λ are essentially the same). Is the ε^4 suppression in the rate for $K \to 2\pi$ a real embarrassment for the $(3, \overline{3}) \oplus (\overline{3}, 3)$ model? I do not really know, again because it is hard to compare different decays.

To my knowledge, the role of chiral symmetry breaking in $K \to 3\pi$ has not been studied thoroughly. Perhaps something can be learned here. Also, it might be worth looking at the K_S-K_L mass difference.

12. – Electromagnetic mysteries.

Assuming that the electromagnetic current is a U-spin singlet, as is required by Gell-Mann's current algebra, COLEMAN and GLASHOW [1] derived the sum rule

$$(p - n) - (\Xi^- - \Xi^0) + \Sigma^- - \Sigma^+) = 0 + O(\varepsilon\alpha)\,, \tag{12.1}$$

where the particle symbols stand for their mass. I have indicated that the error is of order $\varepsilon\alpha$, that is of order of SU_3 breaking times electromagnetism. Since the individual mass differences, *e.g.*, $(\Sigma^- - \Sigma^+)$, are of order α, one expects the sum rule to hold to roughly $20\,\%$. The experimental masses do, in fact, satisfy the sum rule very well.

SU_3 by itself does not give any information about the electromagnetic mass differences of pseudoscalar mesons. However, $SU_3 \otimes SU_3$ yields a sum rule which is essentially the exact analog of the Coleman-Glashow sum rule. It is [6]

$$m^2_{\pi^+} - m^2_{\pi^0} = m^2_{K^+} - m^2_{K^0} + O(\varepsilon\alpha)\,, \tag{12.2}$$

which is a disaster since the right-hand side is a factor of five larger than the left-hand side and of opposite sign.

There is another electromagnetic difficulty in chiral symmetry. As you might have guessed this is the decay $\eta \to 3\pi$ which has been causing trouble since the beginnings of current algebra [5].

The difficulties with both the electromagnetic mass differences and η decay can be traced to the fact that in Gell-Mann's current algebra scheme the electromagnetic current is not only a U-spin singlet in SU_3 but is also a U-spin $\otimes$ $\otimes$ U-spin singlet in $SU_3 \otimes SU_3$. Among the consequences of this is the fact that the electromagnetic current is supposed to commute with F_3^5, the axial charge with the quantum numbers of the π^0. Those of you who are familiar with the problem will remember that this is the root of the trouble in η decay. WEINSTEIN [20] has particularly stressed the fundamental nature of the troubles encountered here and the close relation between the mass formula in eq. (12.2) and the failure of PCAC in η decay.

An obvious way out is to abandon the hypothesis that the electromagnetic current is a U-spin $\otimes$ U-spin singlet. This would spoil the beautiful relationship between strong-interaction symmetries and weak and electromagnetic currents, but it may be necessary. Alternatively, one can imagine that the observed isospin violations in nuclear and particle physics are not entirely due to electromagnetism. In either case, if one still assumes U-spin invariance of the effective isospin violation, as is suggested by the Coleman-Glashow sum rule, then a reasonable estimate for the $\eta \to 3\pi$ can be made. The input is the observed value of $(m^2_{K^+} - m^2_{K^0}) - (m^2_{\pi^+} - m^2_{\pi^0})$ and the agreement with experiment is good [20].

There is a fairly direct way to test the hypothesis that the actual electromagnetic current is not a U-spin$\otimes$$U$-spin singlet. If this is the case, the old PCAC prediction for low-energy pion photoproduction [5] will be wrong. CHENG and I [21] made a dispersion relation analysis of $\gamma + p \to p + \pi^0$ at low energies, both inside and outside the physical region. The presently available multipole analyses give an amplitude which is completely consistent with the PCAC prediction. The experimental errors, however, are rather large in π^0 photoproduction and the question is by no means closed.

My pet theory of what goes wrong with electromagnetism runs as follows. If we were to mathematically separate the strong interactions from weak, electromagnetic and gravitational interactions and then calculate electromagnetic effects to order α, they would turn out to be logarithmically divergent.

With a logarithmic divergence the cut-off mass can be very high, 10^{20} GeV, for example. We can suppose, then, that something like gravity, weak interactions, a true nonlocality in the world, or what-have-you, acts as a cut-off. This cut-off has no reason to be $SU_3 \otimes SU_3$ symmetric. Hence, the effective second-order electromagnetic interaction will not be a U-spin $\otimes$ U-spin singlet. To save the Coleman-Glashow sum rule one can suppose either that the cut-off respects SU_3 or that there is some «octet enhancement» mechanism working, which, among other things, would give the sum rule.

REFERENCES

[1] See, for example, M. GELL-MANN and Y. NE'EMAN: *The Eightfold Way* (New York, 1964).

[2] M. GELL-MANN, R. OAKES and B. RENNER: *Phys. Rev.*, **175**, 2195 (1968).

[3] S. GLASHOW and S. WEINBERG: *Phys. Rev. Lett.*, **20**, 224 (1968).

[4] The history of «spontaneously» broken symmetries from superconductivity, through Goldstone's work to Nambu's suggestion that the pion is a Goldstone particle would be a lecture in itself. Let me simply say that many people contributed to the basic idea of $SU_3 \times SU_3$ symmetry in strong interactions.

[5] See, for example, S. ADLER and R. DASHEN: *Current Algebra and Applications to Particle Physics* (New York, 1968).

[6] R. DASHEN: *Phys. Rev.*, **183**, 1245 (1968).

[7] P. CARRUTHERS and R. HAYMAKER: Cornell preprint.

[8] S. WEINBERG: *Phys. Rev. Lett.*, **18**, 188 (1967).

[9] R. DASHEN and M. WEINSTEIN: *Phys. Rev.*, **188**, 2330 (1969).

[10] L.-F. LI and H. PAGELS: *Phys. Rev. Lett.*, **26**, 1204 (1971).

[11] S. WEINBERG: *Phys. Rev. Lett.*, **17**, 616 (1966).

[12] J. BREHM: University of Massachusetts (Amherst) preprint.

[13] R. DASHEN and M. WEINSTEIN: *Phys. Rev. Lett.*, **22**, 1337 (1969).

[14] T. CHENG and R. DASHEN: *Phys. Rev. Lett.*, **26**, 594 (1971).

[15] G. HÖHLER, H. JAKOB and R. STRAUSE: *Phys. Lett.*, **35** B, 445 (1971).

[16] L. BROWN, W. PARDEE and R. PECCEI: University of Washington preprint.

[17] R. CREWTHER: *Phys. Rev. D*, **3**, 3152 (1971).

[18] G. ALTARELLI, N. CABIBBO and L. MAIANI: *Phys. Lett.*, **35** B, 415 (1971).

[19] R. DASHEN: *Phys. Rev. D*, **3**, 1879 (1971).

[20] M. WEINSTEIN: New York University preprint.

[21] T. CHENG and R. DASHEN: IAS preprint.

Some Features of $(3, \bar{3}) \otimes (\bar{3}, 3)$ Breaking of Chiral Symmetry.

L. MAIANI

Istituto Superiore di Sanità - Roma

In this lecture I will illustrate some general features of Lagrangian theories of strong interactions where the Lagrangian can be meaningfully separated into two terms: a term $\mathscr{L}_0$ symmetrical under the chiral $SU_3 \otimes SU_3$ group, and a breaking term $\mathscr{L}_B$ which is assumed to transform, under the same group, according to the representation $(3, \bar{3}) \oplus (\bar{3}, 3)$. This is the kind of theory one abstracts from simple quark models (in these models $\mathscr{L}_B$ representing just a quark-mass term) or from more elaborate σ-models (where $\mathscr{L}_B$ is related to fundamental scalar fields). As we shall see, the theory displays a remarkable degree of symmetry and allows some general features of hadron spectrum to be interpreted in a very simple way.

In the following I will restrict myself to rather general considerations, without going much into a detailed connection of the parameters appearing in the theory with physically measurable quantities (such as masses, decay amplitudes, etc.). These aspects are fully covered in the lectures by DASHEN at this School, where also a comparison with other schemes, different from the one discussed here, is given.

Let me conclude this introduction apologizing for not presenting any particularly new material, except for some speculation on scale invariance discussed in the end.

This lecture has a rather pedagogical character, and I hope can provide some useful introduction to the wide literature appeared in the last few years on the subject.

The relevance of chiral $SU_3 \otimes SU_3$ was pointed out by GELL-MANN since 1962 [1], in the frame of what has afterwards been called the free-quark model. So, let me start by considering a free, massive-quark Lagrangian:

$$\mathscr{L} = i\bar{\psi}_\alpha \not{\partial} \psi_\alpha + \bar{\psi}_\alpha \varepsilon_{\alpha\beta} \psi_\beta \equiv i\bar{\psi}\not{\partial}\psi + \bar{\psi}\varepsilon\psi\,, \tag{1}$$

ε is a 3×3 real diagonal matrix, representing quark masses, and α and β run from 1 to 3.

It is useful to introduce the left- and right-handed quark fields

$$\psi_{L\alpha} = a_+ \psi_\alpha , \qquad \psi_{R\alpha} = a_- \psi_\alpha , \qquad a_\pm = \frac{1 + \gamma_5}{2} .$$

If the quarks were massless, so that $\mathscr{L}$ would reduce to the kinetic energy term, $\mathscr{L}$ would be fully symmetrical under the set of transformations

$$\text{(2)} \qquad \psi_{L\alpha} \to U_{\alpha\beta} \psi_{L\beta} , \qquad \psi_{R\alpha} = V_{\alpha\beta} \psi_{R\beta} ,$$

U and V being 3×3 unitary, unimodular matrices.

This set of transformations, each of which is evidently characterized by the pair of matrices (U, V), is the chiral $SU_3 \otimes SU_3$ group.

Actually, the massless version of eq. (1) is symmetrical under the transformations eq. (2) even when U and V are unitary but not unimodular, this corresponding to symmetry under $U_3 \otimes U_3 = U_1 \otimes U_1 \otimes SU_3 \otimes SU_3$.

One of the two U_1 groups can be identified with quark-number conservation (corresponding in the real world to baryon-number conservation) and the other one with the quark-helicity conservation.

The relevance to the real world of this latter symmetry is, at the moment, more dubious than anything I will say in this lecture, so that I will forget about it in the following [2]. Also, I will not mention quark-number conservation anymore, it being implicit in all I will say. $SU_3 \otimes SU_3$ symmetry brings with it sixteen conserved currents, which can be written as

$$\text{(3)} \qquad \begin{cases} L^\mu_{\alpha\beta}(x) = \bar{\psi}_{L\beta}(x) \gamma_\mu \psi_{L\alpha}(x) , \\ R^\mu_{\alpha\beta}(x) = \bar{\psi}_{R\beta}(x) \gamma_\mu \psi_{R\alpha}(x) , \end{cases}$$

or, in terms of the more familiar vector and axial-vector currents

$$\text{(4)} \qquad \begin{cases} V^i_\mu = \mathrm{Tr}\,(L^\mu + R^\mu) \dfrac{\lambda_i}{2} = \bar{\psi} \gamma_\mu \dfrac{\lambda_i}{2} \psi , \\ A^i_\mu = \mathrm{Tr}\,(L^\mu - R^\mu) \dfrac{\lambda_i}{2} = \bar{\psi} \gamma_\mu \gamma_5 \dfrac{\lambda_i}{2} \psi , \end{cases}$$

λ_i being the eight Gell-Mann matrices.

In terms of quark fields we also construct the operators

$$\text{(5)} \qquad \begin{cases} M_{\alpha\beta}(x) = \bar{\psi}_{R\beta}(x) \psi_{L\alpha}(x) , \\ (M^+)_{\alpha\beta}(x) = \bar{\psi}_{L\beta}(x) \psi_{R\alpha}(x) , \end{cases}$$

which again are connected to the familiar scalar and pseudoscalar densities [1] according to

$$(6)\qquad \begin{cases} u_i = \mathrm{Tr}\,(M + M^+)\lambda_i = \bar{\psi}\lambda_i\psi\,, \\[2ex] v_i = \mathrm{Tr}\,\dfrac{M - M^+}{i}\lambda_i = i\bar{\psi}\lambda_i\gamma_5\psi\,, \end{cases}$$

here i goes from 0 to 8.

The matrices M and M^+ have simple transformation properties under the transformations of the quark fields given by eq. (2)

$$(7)\qquad \begin{cases} M \xrightarrow[(U,V)]{} UMV^+\,, \\[2ex] M^+ \xrightarrow[(U,V)]{} VM^+U^+\,, \end{cases}$$

which reveal that M transforms according to the $(3, \bar{3})$ representation, and M^+ according to the $(\bar{3}, 3)$. We will also consider infinitesimal transformations, both left- and right-handed:

$$(8)\qquad (U, V) = (1 + i\alpha^i\lambda_i, 1)\,, \qquad \delta_L M = i\alpha^i\lambda_i M\,, \qquad \delta_L M^+ = -i\alpha^i M^+\lambda_i\,;$$

$$(9)\qquad (U, V) = (1, 1 + i\beta^i\lambda_i)\,, \qquad \delta_R M = -i\beta^i M\lambda_i\,, \qquad \delta_R M^+ = i\beta^i\lambda_i M^+\,.$$

We conclude these preliminaries by giving the transformation properties of the currents, which can be derived from eq. (3):

$$(10)\qquad L_\mu \xrightarrow[(U,V)]{} UL_\mu U^+\,,$$

$$(11)\qquad R_\mu \xrightarrow[(U,V)]{} VR_\mu V^+\,.$$

Equations (10) and (11) indicate that L^μ transforms as the (8.1) and R^μ as the (1.8) representations.

We now introduce e.m. and weak interactions in the Lagrangian eq. (1), writing

$$\begin{aligned} \mathscr{L} &= i\bar{\psi}\not{\partial}\psi + \bar{\psi}\varepsilon\psi + g[W^\mu\bar{\psi}\gamma_\mu(1+\gamma_5)\lambda_w\psi + \text{h.c.}] + eA^\mu\bar{\psi}\gamma_\mu\lambda_Q\psi = \\ &= i(\bar{\psi}_L\not{\partial}\psi_L + \bar{\psi}_R\not{\partial}\psi_R) + (\bar{\psi}_R\varepsilon\psi_L + \bar{\psi}_L\varepsilon\psi_R) + 2g[W^\mu\bar{\psi}_L\gamma_\mu\lambda_w\psi_L + \text{h.c.}] + \\ &\quad + eA^\mu(\bar{\psi}_L\gamma_\mu\lambda_Q\psi_L + \bar{\psi}_R\gamma_\mu\lambda_Q\psi_R) = \\ &= \mathscr{L}_0 + \mathrm{Tr}\,(M\varepsilon + M^+\varepsilon) + 2g[W^\mu\,\mathrm{Tr}\,(L_\mu\lambda_w) + \text{h.c.}] + eA^\mu\,\mathrm{Tr}\,(L_\mu + R_\mu)\lambda_Q\,. \end{aligned}$$

We fix the charge spectrum by requiring

$$\lambda_Q = \begin{pmatrix} \frac{2}{3} & 0 & 0 \\ 0 & -\frac{1}{3} & 0 \\ 0 & 0 & -\frac{1}{3} \end{pmatrix}, \tag{13}$$

λ_w is then determined (by requirements of having charge $+1$ and of generating with λ_w^+ an SU_2 group [3]) to have the form

$$\lambda_w = \begin{pmatrix} 0 & \cos\theta & \sin\theta \\ 0 & 0 & 0 \\ 0 & 0 & 0 \end{pmatrix}. \tag{14}$$

We see that the Lagrangian eq. (12) can be thus split into

i) a symmetric term $\mathscr{L}_0$;

ii) an $SU_3 \otimes SU_3$ breaking term $\mathscr{L}_B$, transforming as $(3, \bar{3}) \oplus (\bar{3}, 3)$;

iii) e.m. and weak terms, transforming according to $(8, 1) \oplus (1, 8)$ and given in terms of the $SU_3 \otimes SU_3$ currents via the matrices λ_Q and λ_w.

θ is (with some qualification to be given later) the Cabibbo angle [4].

Properties i), ii) and iii) are those which we want to abstract from the quark model, and constitute the basis of our considerations.

However, the breaking term $\mathscr{L}_B$ as given by eq. (12) is not the most general $(3, \bar{3}) \oplus (\bar{3}, 3)$ element, compatible with hermicitity of $\mathscr{L}_B$. This is rather given by

$$\mathscr{L}_B = \mathrm{Tr}\,(M\varepsilon^+ + M^+\varepsilon)\,, \tag{15}$$

ε being any 3×3 matrix. If ε is real and diagonal we get back eq. (12).

In conclusion, let us write our Lagrangian as

$$\mathscr{L} = \mathscr{L}_0 + \mathscr{L}_B(\varepsilon) + \mathscr{L}_{\text{e.m.}}(\lambda_Q) + \mathscr{L}_w(\lambda_w)\,, \tag{16}$$

λ_Q and λ_w are given by eqs. (13) and (14), $\mathscr{L}_B$ is given by eq. (15), ε being any matrix consistent with charge conservation:

$$\varepsilon = \begin{pmatrix} \varepsilon_1 & 0 & 0 \\ 0 & a & b \\ 0 & c & d \end{pmatrix}, \tag{17}$$

ε_1, a, b, c, d = arbitrary complex numbers.

In the free-quark model $\mathscr{L}_0$ and $\mathscr{L}_B$ are trivial, but it is easy to construct more complicated models in which $\mathscr{L}_0$ and $\mathscr{L}_B$ both contain nontrivial interactions, like the σ-model or the gluon model.

Our first problem will be that of determining what symmetries can $\mathscr{L}_B$ retain or violate [5].

The first observation is that, if ε has the general form eq. (17) $\mathscr{L}_B$ seems to violate strangeness and parity. Why? Let us rewrite eq. (15) as

$$\mathscr{L}_B = \mathrm{Tr}\left(\frac{M+M^+}{2}\right)(\varepsilon+\varepsilon^+) + \mathrm{Tr}\left(\frac{M-M^+}{2i}\right)\left(\frac{\varepsilon-\varepsilon^+}{i}\right).$$

Comparing with eq. (6), we see that $\mathscr{L}_B$ contains terms of the form $(\bar{n}\lambda)$ and $i(\bar{n}\gamma_5\lambda)$, *i.e.* both parity ($P$) and strangeness ($S$) violating. Actually, there have been attempts to connect at least parts of the S-violating nonleptonic decay amplitudes to an «effective» Lagrangian transforming as a piece of a $(3,\bar{3})\oplus(\bar{3},3)$. What we will show now is that, in the scheme represented by eq. (16), these pieces are illusory because

a) $\mathscr{L}_B$ can violate P only if it violates also CP;

b) $\mathscr{L}_B$ is always strangeness conserving.

If $\mathscr{L}_B$ contains other pieces besides the $(3,\bar{3})\oplus(\bar{3},3)$ term, then *b*) does not hold any more, whereas *a*) is still true. Since the main part of the observed nonleptonic decays is CP-conserving, then a $(3,\bar{3})$ can represent at most P-conserving, $\Delta S=1$ nonleptonic amplitudes, *provided* $\mathscr{L}_B$ contains more terms than in our scheme (*e.g.*, an $(8.1)\oplus(1.8)$ piece). Elementary considerations, which I will not report here [6] show that any 3×3 matrix ε can always be written as

$$\varepsilon = U\varepsilon_D V^+ \exp[i(\varphi/3)], \tag{18}$$

U and V being suitable unitary and unimodular matrices, ε_D a real diagonal matrix and φ some real phase (eq. (18) holds in general for $n\times n$ matrices, with the substitution $\frac{1}{3}\varphi\to(1/n)\varphi$).

Suppose now to apply to the basic fields in $\mathscr{L}$ (whatever they are) the

$SU_3 \otimes SU_3$ transformation (U, V). Then

$$\mathscr{L}_0 \to \mathscr{L}_0\,,$$

$$\mathscr{L}_B \to \mathrm{Tr}\,(MV^+\varepsilon^+U + M^+U^+\varepsilon V) = \mathrm{Tr}\,(M\varepsilon_D \exp[-i(\varphi/3)] + M^+\varepsilon\exp[+i(\varphi/3)])\,,$$

$$\mathscr{L}_{\text{e.m.}} \to eA^\mu\,\mathrm{Tr}\,(L_\mu U^+\lambda_Q U + R_\mu V^+\lambda_Q V)\,,$$

$$\mathscr{L}_w \to 2g[W^\mu\,\mathrm{Tr}\,(L_\mu U^+\lambda_w U) + \text{h.c.}]\,.$$

Since $[\varepsilon, \lambda_Q] = 0$, we can furthermore choose U and V such that they commute with λ_Q, so that $\mathscr{L}_{\text{e.m.}}$ remains unchanged.

Then we see that the Lagrangian eq. (16) is equivalent to

$$\mathscr{L} = \mathscr{L}_0 + \mathscr{L}_B\,(\varepsilon_D \exp[i\,(\varphi/3)]) + \mathscr{L}_{\text{e.m.}}(\lambda_Q) + \mathscr{L}_w(U^+\lambda_w U)\,,$$

where

a) ε is replaced by a diagonal matrix, so that it has no more S-violating terms;

b) $\lambda_w \to U^+\lambda_w U$, this corresponding just to a redefinition of the Cabibbo angle.

In fact, since $[U, \lambda_Q] = 0$

$$U = \begin{pmatrix} 1 & 0 & 0 \\ 0 & \cos\varphi & \sin\varphi \\ 0 & -\sin\varphi & \cos\varphi \end{pmatrix},$$

$$U^+\lambda_w U = \begin{pmatrix} 0 & \cos(\theta+\varphi) & \sin(\theta+\varphi) \\ 0 & 0 & 0 \\ 0 & 0 & 0 \end{pmatrix}.$$

This disposes-of S-violation.

As for parity, we observe that the new $\mathscr{L}_B$, if $\varphi \neq 0$, contains both terms like $\bar{p}p$ and terms like $i\bar{p}\gamma_5 p$ which have not only opposite P, but also opposite CP (having the same C).

In conclusion, the best eq. (16) can do is to provide us with CP- and P-violation in $\Delta S = 0$ channels (*i.e.*, for example, in nuclear levels). This restricts φ to be extremely small, and from now on we will assume that ε is such that $\varphi = 0$. Actually, it is very simple to see if a given ε will give a *CP*-conserving theory. From eq. (18)

$$\det \varepsilon = \exp[i\varphi] \det \varepsilon_D ,$$

i.e. if $\det \varepsilon$ is real $\varphi = 0$ and *vice versa*. Two comments are in order.

First observe that if $\mathscr{L}_0$ were invariant under $U_3 \otimes U_3$ we could put the factor $\exp[i(\varphi/3)]$ into U or V^+, and in this case we could always eliminate *CP*-violation just as we have done for *S*-violation.

This means that in free-quark models (or even in the gluon model) one can never get a *CP*-violation out of eq. (16). The second comment is this. If ε_D is proportional to the unit matrix, we can make an additional transformation (U_θ, U_θ) of the basic fields such that

$$U_\theta^+ \lambda_w U_\theta = \begin{pmatrix} 0 & 1 & 0 \\ 0 & 0 & 0 \\ 0 & 0 & 0 \end{pmatrix},$$

i.e. in absence of SU_3 breaking, we can always rotate away the Cabibbo angle eliminating *S*-violation also from weak interactions. This points to a deep connection between the actual value of θ ($\theta \sim 0.22$) and the breaking of SU_3 and possibly of $SU_3 \otimes SU_3$, and to an interplay of weak and strong interactions, which must co-operate somehow to produce the bizarre angle observed in nature.

This idea has been pursued by various authors [6-8] even with encouraging results, but a real breakthrough has not yet been achieved.

The value of θ remains still as one of the most challenging problems for theorists.

From now on we will understand ε as to be of the form

$$\varepsilon = \begin{pmatrix} \varepsilon_1 & 0 & 0 \\ 0 & \varepsilon_2 & 0 \\ 0 & 0 & \varepsilon_3 \end{pmatrix}. \tag{19}$$

Using eq. (6), $\mathscr{L}_B$ can be written alternatively, as (see, *e.g.*, ref. (9))

$$\mathscr{L}_B = \alpha_0 u_0 + \alpha_8 u_8 + \alpha_3 u_3 ,$$

$\alpha_{0,8,3}$ being related to the ε_i's according to

$$\left\{\begin{aligned} \alpha_0 &= \frac{1}{\sqrt{6}}(\varepsilon_1 + \varepsilon_2 + \varepsilon_3)\,, \\ \alpha_8 &= \frac{1}{2\sqrt{3}}(\varepsilon_1 + \varepsilon_2 - 2\varepsilon_3)\,, \\ \alpha_3 &= \frac{1}{2}(\varepsilon_1 - \varepsilon_2)\,. \end{aligned}\right. \tag{20}$$

Observe that we have left open the possibility that $\mathscr{L}_B$ contains some I-spin violating term.

Before connecting $\mathscr{L}_B$ to experiments (*e.g.*, ratios of ε_i or α_i to meson spectrum) one has to be sure that the parametrization eq. (19) is unique, and that there are no ambiguities.

Actually this is not so [10]. The requirement that ε be diagonal does not fix uniquely the ε_i's. We can still perform $SU_3 \otimes SU_3$ transformations which 1) exchange two ε_i's (actually only ε_2 and ε_3, if we want to keep the spectrum of λ_Q fixed); 2) flip the sign of any two of them. Any of these transformations do not change $\mathscr{L}_{\text{e.m.}}$ or $\mathscr{L}_w$ (apart from trivial redefinitions) but drastically change the pattern of the parameters ε_i or α_i so that it does not make sense at the present level to attach, say, to α_3 the meaning of an I-spin violation. In quark language, these transformations correspond to exchange n and λ quarks and/or change the intrinsic parities of any two quarks.

We will fix this ambiguity in the following [11] and will see that its elimination is obtained only when one takes into account the fact that ε_D by itself *does not* give a complete description of the symmetry breaking.

From now on let us neglect $\mathscr{L}_{\text{e.m.}}$ and $\mathscr{L}_w$, and restrict to $\mathscr{L} = \mathscr{L}_0 + \mathscr{L}_D$.

We introduce a very important quantity, which is the vacuum expectation value (VEV) of the fields contained in $M(x)$ [12, 5]:

$$\langle 0|M(x)|0\rangle = \langle 0|M(0)|0\rangle = \eta\,,$$

η is, analogously to ε, a 3×3 matrix.

If our theory were exactly invariant under $SU_3 \otimes SU_3$, η would vanish. To see this, recall that it can be shown that all symmetries of the vacuum are symmetries of the world [13], so that if $\mathscr{U}(U, V)$ is the Hilbert-space operator corresponding to the element (U, V) of $SU_3 \otimes SU_3$, then $\mathscr{U}(U, V)|0\rangle = |0\rangle$ implies

$$\eta = \langle 0|M|0\rangle = \langle 0|\mathscr{U}(U, V)M\mathscr{U}^+(U, V)|0\rangle = U\langle 0|M|0\rangle V^+ = U\eta V^+ \tag{21}$$

for any (U, V). This can be satisfied only if $\eta = 0$.

When $\mathscr{L}_B \neq 0$ we have then $\eta \neq 0$. However, in certain models it happens that η does not vanish even in the limit $\mathscr{L}_B = 0$. This situation is usually referred to as « spontaneous breaking » of $SU_3 \otimes SU_3$ and is the one I will discuss here. It corresponds to the presence of stable solutions for the vacuum which display a lower degree of symmetry than the Lagrangian. In fact, in the limit $\mathscr{L} = \mathscr{L}_0$ the Lagrangian is symmetric under the full $SU_3 \otimes SU_3$ whereas the vacuum is left invariant only by those transformations such that eq. (21) holds. This phenomenon in turn is connected with the appearance of massless spin-zero bosons (Goldstone bosons), one for each generator of $SU_3 \otimes \otimes SU_3$ which does not leave invariant η [14, 9].

Before touching upon the argument of Goldstone bosons, however, let me consider in some detail the symmetry structure of η.

To visualize the situation let us consider first a classical example, that of a ferromagnet.

Consider a system of spins in an infinite volume. In absence of external fields, the system is described by a rotationally invariant Hamiltonian $\mathscr{H}_0$. Call $|0\rangle$ the ground state of $\mathscr{H}_0$.

Usually, the magnetization of the ground state

$$\langle 0|\mathscr{M}|0\rangle ,$$

vanishes. This is certainly so if $|0\rangle$ itself is rotationally invariant.

However, in the case of a ferromagnet $\mathscr{H}_0$ is such that the stable ground state has a nonvanishing magnetization $\boldsymbol{m} = \langle 0|\mathscr{M}|0\rangle$, so that $|0\rangle$ is not symmetrical under the full rotation group, but only under rotations around $\boldsymbol{m}$. Of course, we can orientate $\boldsymbol{m}$ in whatever direction we want, and ground states with different orientation of $\boldsymbol{m}$ are degenerate with respect to energy.

These states do not lie in the same Hilbert space, but rather each of them corresponds to mathematically inequivalent though physically identical theories.

Suppose we introduce now a weak external magnetic field $\boldsymbol{H}$. The Hamiltonian is changed into

$$\mathscr{H} = \mathscr{H}_0 + \mathscr{M} \cdot \boldsymbol{H} ,$$

and out of all the infinite number of degenerate ground states one is selected as the lowest-energy state, this being that one in which $\boldsymbol{m}$ is parallel to $\boldsymbol{H}$. How are $\boldsymbol{m}$ and $\boldsymbol{H}$ connected? For weak fields $|\boldsymbol{m}|$ depends little upon $\boldsymbol{H}$, and is mainly determined by $\mathscr{H}_0$. However, no matter how weak is the external « breaking » $\boldsymbol{H}$, there is always a strong correlation in that the stable ground state has to have $\boldsymbol{m}$ parallel to $\boldsymbol{H}$.

Let us now go back to our case. Here η is the analog of $\boldsymbol{m}$ and in the limit $\varepsilon = 0$ identical theories correspond to matrices related by $SU_3 \otimes SU_3$ transformations

$$\eta \to U\eta V^+ .$$

Using eq. (18) then we can always choose a frame where η is diagonal so that we have always S-conservation. However, when $\varepsilon \neq 0$ the question arises whether η is diagonal in the same frame where ε is such, *i.e.* whether ε and η are in some way constrained to be « parallel » as in the ferromagnet analogy. This is actually the case. We insert eqs. (8), (9) and (15) in the divergence formulae

$$\partial^\mu L^i_\mu = \frac{\delta \mathscr{L}}{\delta \alpha_i} = \frac{\delta \mathscr{L}_B}{\delta \alpha_i},$$

$$\partial^\mu R^i_\mu = \frac{\delta \mathscr{L}}{\delta \beta_i} = \frac{\delta \mathscr{L}_B}{\delta \beta_i},$$

and we obtain (ε is real and diagonal)

$$\begin{cases} \partial^\mu L^i_\mu = i \operatorname{Tr}(\lambda^i M - M^+ \lambda^i)\varepsilon, \\ \partial^\mu R^i_\mu = -\operatorname{Tr}(M\lambda^i - \lambda^i M^+)\varepsilon. \end{cases} \tag{22}$$

The VEV of eq. (22) must vanish by translation invariance and this gives us a set of relations between ε and η. It is then a simple matter of algebra to show that these equations imply precisely that ε and η can be simultaneously diagonalized by an $SU_3 \otimes SU_3$ rotation.

It remains the possibility that ε and η may not be relatively real. This would give rise to a kind of « spontaneous breakdown of P and CP [15] » which has the same features and disadvantages as the one previously discussed. I will not insist therefore on this, and assume that η and ε can be both brought into a real diagonal form. This form does in general suffer from the ambiguities we mentioned above, but now we are in the position to fix them.

We have already mentioned that η must possess all the symmetries of the vacuum, *i.e.* of the real world. Now the particle spectrum clearly displays a very good I-spin symmetry and a more approximate SU_3 symmetry. This implies that it must be possible to choose an $SU_3 \otimes SU_3$ frame such that η takes the form

$$\eta = \begin{pmatrix} \eta_1 & 0 & 0 \\ 0 & \eta_2 & 0 \\ 0 & 0 & \eta_3 \end{pmatrix},$$

with

$$\eta_1 \sim \eta_2 \qquad \text{up to } I\text{-spin corrections},$$

$$\eta_{1,2} \sim \eta_3 \qquad \text{up to } SU_3 \text{ corrections}.$$

Note that this frame is now uniquely defined. We are not allowed any more to exchange η_2 with η_3 or to change sign to any two η_i's, as this would spoil our approximate equalities [11]. In this frame then the diagonal elements of both η and ε have a physical meaning and can be compared to physical quantities. Also, this frame defines the physical Cabibbo angle.

Moreover, if the symmetries of η are also symmetries of the vacuum (*i.e.* if η gives a complete description of symmetry breaking in the vacuum) then if we want chiral symmetry to be much more badly broken in particle spectrum than SU_2 or SU_3 (as is indicated by the absence of parity doublets) we must also require

$$|\eta_i| \gg |\eta_i - \eta_j| \qquad \text{for any } i \text{ and } j\,.$$

This statement, however, is more model-dependent than the others, and we are unable to prove it in general. It is in fact verified in the σ-model and we shall assume its validity.

The ferromagnet analogy also clearly indicates what is the connection between η and ε. In the limit $\varepsilon \to 0$ we expect η to tend to a finite value which will display all those symmetries which are not spontaneously broken, *i.e.* realized with particle multiplets. SU_2 and SU_3 symmetries appear indeed to be of such a type, so that we expect

$$\lim_{\varepsilon \to 0} \eta = \eta_0 \propto \mathbf{1}\,, \tag{23}$$

η_0 is determined by $\mathscr{L}_0$ only, and we have an octet of pseudoscalar massless bosons. When $\varepsilon \neq 0$, we expect

$$\eta = \eta_0 + O(\varepsilon)\,. \tag{24}$$

If $\mathscr{L}_B$ can be considered in some sense a small perturbation, η will depend very little upon ε and be mainly determined by $\mathscr{L}_0$. In this scheme it may be meaningful to apply perturbation theory in $\mathscr{L}_B$, starting from the spontaneously broken solution η_0, as discussed by DASHEN [9].

Let me now very briefly discuss pseudoscalar meson-mass formulae [12, 16] as an illustration of the arguments presented above. I will restrict to π and K masses, and, for the sake of brevity, will not give any derivation, but simply quote from ref. [2]. Then (neglecting I-spin violations, *i.e.* putting $\varepsilon_1 = \varepsilon_2$, $\eta_1 = \eta_2$) we have

$$m_\pi^2 = -Z_\pi^2 \frac{\varepsilon_1}{\eta_1}\,, \tag{25}$$

$$m_K^2 = -Z_K^2 \frac{\varepsilon_1 + \varepsilon_3}{\eta_1 + \eta_3}\,, \tag{26}$$

$$m_\pi^2 F_\pi = -2\varepsilon_1 Z_\pi\,, \tag{27}$$

$$m_\pi^2 F_K = -(\varepsilon_1 + \varepsilon_3) Z_K\,, \tag{28}$$

where we have defined

$$Z_\pi = \langle \pi | v^\pi | 0 \rangle , \qquad Z_K = \langle K | v^K | 0 \rangle ,$$

$$\langle \pi | \partial^\mu A^\pi_\mu | 0 \rangle = m^2_\pi F_\pi , \qquad \langle K | \partial^\mu A^K_\mu | 0 \rangle = m^2_K F_K .$$

Equations (27) and (28) can be rewritten as

$$Z_\pi F_\pi = 2\eta_1 , \qquad Z_K F_K = \eta_1 + \eta_3 . \tag{29}$$

These equations clearly display the Goldstone phenomenon: if η remains finite when $\varepsilon \to 0$, then by eqs. (25), (26) and (29) both m_π and m_K vanish in this limit. Also, if we assume eq. (24) to hold, then

$$Z_\pi = Z_K + O(\varepsilon) , \qquad F_\pi = F_K + O(\varepsilon) ,$$

so that to lowest order in ε

$$\frac{m^2_\pi}{m^2_K} = \frac{2\varepsilon_1}{\varepsilon_1 + \varepsilon_3} ,$$

which indicates that $\varepsilon_1 \ll \varepsilon_3$ (this corresponds in Dashen's notations to $c \sim -\sqrt{2}$).

Finally, these equations show that the vacuum breaking η is connected to F_π and F_K (which in fact display an approximate SU_3 symmetry) whereas ε is related to the meson masses, the smallness of pion mass indicating that $\mathscr{L}_B$ is, to a good approximation, $SU_2 \otimes SU_2$-invariant.

A complete analysis of the relations linking ε and η to the observed p.s. meson masses, and to the decay coupling constants F_π and F_K is beyond the scope of this lecture and I will not elaborate on this any further (see, *e.g.*, ref. [2, 9, 16] and also ref. [17]).

Rather I will conclude by presenting some further speculations on the symmetry structure of $\mathscr{L}$, which have recently received some attention.

If we go back to eq. (1), we see that the quark kinetic-energy terms is not only invariant under $SU_3 \otimes SU_3$, but also under *scale transformations*, as discussed in Callan's lectures [18]. It is then interesting to see what happens if we conjecture the same to be true for the symmetric lagrangian $\mathscr{L}_0$ appearing in eq. (12), *i.e.* assume that scale invariance is broken only by the same term which breaks $SU_3 \otimes SU_3$ (apart from c-numbers) [19, 20].

It is very easy to see qualitatively what is going on in this case. What happens is that, if in the limit $\varepsilon \to 0$, η stays finite, we will have, in addition to $SU_3 \otimes SU_3$ breaking, a spontaneous breaking of scale invariance. Correspondingly, a scalar Goldstone boson (called « dilaton ») appears. If in the same limit eq. (23) holds, the dilaton is coupled mainly to the SU_3, singlet scalar density u_0. When $\varepsilon \neq 0$, the $(\text{mass})^2$ of the dilaton is of order $\varepsilon (i.e. \sim \sim m_K^2)$ and, as a consequence, we will have a systematic enhancement [20, 21] of the matrix elements of u_0 with respect to the matrix elements of u_8.

Another consequence of this assumption is that, since [18]

$$\theta^\mu_\mu = -(4-d)\mathscr{L}_B ,$$

θ^μ_μ being the «improved» energy-momentum tensor [22] and d the dimension of $\mathscr{L}_B$, the mass of any hadron A is proportional to the matrix element of $\mathscr{L}_B$:

$$M_A = -(4-d)\langle A|\mathscr{L}_B|A\rangle . \tag{30}$$

Equation (30) may look rather strange since we expect in general M_A to be approximately SU_3-invariant, whereas we have seen that $\mathscr{L}_B$ is far from being so. However, the enhancement of $\langle A|u_0|A\rangle$ with respect to $\langle A|u_8|A\rangle$ that we have just mentioned compensates for the unsymmetrical nature of $\mathscr{L}_B$, and makes M_A to have a large SU_3 singlet part.

It is very unclear at present if these considerations are of any value especially in view of the anomalies found to appear (in perturbation theory) in the Ward identities derived from approximate scale invariance [23]. It is however interesting that they are qualitatively supported by recent calculations of the σ-term in π-$\mathcal{N}$ scattering, by CHENG and DASHEN [24] and by others [25]. The large value for σ found there is qualitatively in agreement with the nucleon mass [20, 25], as given by eq. (30) with $d = 3$ (as in the quark model). It also gives evidence for the presence of an enhancement of $\langle N|u_0|N\rangle$ with respect to $\langle N|u_8|N\rangle$ of the right order of magnitude.

REFERENCES

[1] M. GELL-MANN: *Phys. Rev.*, **125**, 1067 (1962).

[2] See however, *e.g.*, S. GLASHOW: *International School of Physics «E. Majorana»* (Erice, 1967).

[3] N. CABIBBO: *International School of Physics «E. Maiorana»* (Erice, 1963); M. GELL-MANN: *Physics*, **1**, 1 (1965).

[4] N. CABIBBO: *Phys. Rev. Lett.*, **10**, 531 (1963).

[5] N. CABIBBO and L. MAIANI: *Phys. Rev. D*, **1**, 707 (1970); *Evolution in Particle Physics*, edited by M. CONVERSI (New York, 1970).

[6] See, *e.g.*, N. CABIBBO, R. GATTO and C. ZEMACH: *Nuovo Cimento*, **16**, 168 (1960).

[7] N. CABIBBO: *International School of Physics «E. Majorana»* (Erice, 1968); N. CABIBBO and G. DE FRANCESCHI: *Proceedings of the VIII Nobel Symposium on Elementary Particle Theory* (Aspenåsgården, 1968).

[8] R. GATTO, G. SARTORI and M. TONIN: *Phys. Lett.*, **28** B, 128 (1968).

[9] R. DASHEN: this Volume, p. 204.
[10] T. K. KUO: *Phys. Rev. D*, **2**, 394 (1970).
[11] L. MAIANI and K. T. MAHANTHAPPA: *Phys. Lett.*, **33** B, 499 (1970).
[12] S. L. GLASHOW and S. WEINBERG: *Phys. Rev. Lett.*, **20**, 224 (1968). See also ref. [2].
[13] S. COLEMAN: *International School of Physics « E. Majorana »* (Erice, 1966).
[14] J. GOLDSTONE, A. SALAM and S. WEINBERG: *Phys. Rev.*, **127**, 965 (1962).
[15] R. DASHEN: *Phys. Rev. D*, **3**, 1879 (1971).
[16] M. GELL-MANN, R. J. OAKES and B. RENNER: *Phys. Rev.*, **175**, 2195 (1968).
[17] G. PARISI and M. TESTA: *Nuovo Cimento*, **67** A, 13 (1970).
[18] C. CALLAN: this Volume, p. 264.
[19] J. ELLIS, P. H. WEISZ and B. ZUMINO: *Phys. Lett.*, **34** B, 91 (1971).
[20] G. ALTARELLI, N. CABIBBO and L. MAIANI: *Phys. Lett.*, **35** B, 415 (1971).
[21] J. ELLIS: *Phys. Lett.*, **33** B, 591 (1970).
[22] C. CALLAN, S. COLEMAN and R. JACKIW: *Ann. of Phys.*, **59**, 42 (1970).
[23] See, *e.g.*, S. COLEMAN: this Volume, p. 280.
[24] T. P. CHENG and R. DASHEN: *Phys. Rev. Lett.*, **26**, 549 (1971).
[25] This point is discussed in detail by: G. ALTARELLI: this Volume, p. 253, where all the relevant references are quoted.
[26] R. CREWTHER: *Phys. Rev. D*, **3**, 3152 (1971).

An Algebraic Approach to the Saturation of Chiral Algebra.

C. A. SAVOY
CERN - Geneva

In this seminar I will discuss some recent work on the saturation of chiral algebra [1].

An algebraic saturation scheme consists of two steps: first we look for a reliable set of commutation relations and the quantum numbers of the operators that appear there, then we postulate a complete set of states for these operators. We will try to use empirical facts as a guide to our choices.

Our first approximation is a natural but strong hypothesis: we consider only one-particle hadronic states, *i.e.* meson and baryon resonances (and bound states), in the $p \to \infty$ frame. The usefulness of this approximation in the saturation of current algebra has been emphasized [2]. These states are labelled by their quantum numbers (SU_3 index, spin, helicity, parity, etc.).

We have been looking for the matrix elements of the axial charge Q_5^α ($\alpha = 1, \ldots, 8$). Together with the vector charges, Q^α, they are assumed to generate the $SU_3 \times SU_3$ chiral algebra

(1) $$[Q^\alpha, Q^\beta] = if_{\alpha\beta\gamma} Q^\gamma ,$$

(2) $$[Q^\alpha, Q_5^\beta] = if_{\alpha\beta\gamma} Q_5^\gamma ,$$

(3) $$[Q_5^\alpha, Q_5^\beta] = if_{\alpha\beta\gamma} Q^\gamma .$$

Equation (1) corresponds to the fact that the vector charges generate the SU_3 group. The axial charges are measurable quantities and we believe them to form an SU_3 octet: this is the content of eq. (2). The last commutation relation, eq. (3), normalizes the Q_5^α matrix elements; it gives rise to the Adler-Weisberger relation, a successful test of current algebra.

By the use of PCAC the Q_5 matrix elements are related to the pseudoscalar meson strong coupling constants. For example, from eq. (3) we get the Adler-Weisberger relation, a sum rule for the π-hadron amplitudes with isospin one in the t-channel ($I_t = 1$) [2].

But we can go the other way and this is how WEINBERG wrote a new important algebraic condition [3]. He considered $I_t = 2$ processes $\pi^+\alpha \to \pi^-\beta$

and assumed that only low-lying objects (Regge poles and cuts) could be exchanged in the t-channel, which is exotic. Therefore we are allowed to write unsubtracted dispersion relations for those amplitudes, and we saturate them in the narrow width resonance appoximation. By the use of PCAC and the one-particle state completeness, WEINBERG deduced the algebraic relation [3]

$$[Q_5^+, [Q_5^+, m^2]] = 0 , \tag{4}$$

where m^2 is the hadronic (mass)^2 operator, which is obviously diagonal in all quantum numbers of the hadronic states and independent of helicity and third component of isospin. For simplicity we will not consider SU_3 breaking, so that m^2 is an SU_3 singlet.

If we assume that eqs. (3) and (4) are saturated in the $p_z \to \infty$ one-particle state space, the result is almost trivial. Equation (3) tells us that hadronic states build up a *reducible* representation of the $SU_3 \times SU_3$ chiral algebra. This means that we can get axial charge matrix elements (and coupling constants) once we know the chiral content of each state, *i.e.* the way it develops into chiral irreducible representations. The information one gets from eq.(4) is the chiral behaviour of the m^2-operator: it has to behave like an $(1, 1) + [(3, \bar{3}) + (\bar{3}, 3)]$ representation under $SU_3 \times SU_3$; these are the only ones with an SU_3 singlet and no exotics.

We get more restrictions from the transformation properties of Q_5^α under angular momentum and parity. It is a negative normality zero-helicity operator and it behaves like a $J^P = 1^+$ object between equal mass states [2].

Therefore, what we are looking for is a model for the one-particle hadronic states where chiral contents are known. In such a model the π couplings are determined by PCAC and one could also get some information on the (mass)^2 operator.

It has been empirically observed that the experimental resonances (both baryons and mesons) are very confortably classified into supermultiplets of the $SU_{6S} \times O_{3L}$ group. The generators of SU_6 are: $Q = A(\lambda_\alpha)$ (unitary-spin operator), $S_i = A(\sigma_i)$ (internal spin operator) and $A(\lambda_\alpha \sigma_i)$. The angular momentum of the states is defined as $\boldsymbol{J} = \boldsymbol{L} + \boldsymbol{S}$, the parity is $(-)^{L+1}$ for mesons and $(-)^L$ for baryons and so on. Mesons are assigned to $\underline{35}$-plets of SU_6 excited to any L, while for baryons the best choice is the $\underline{56}$ with even L-excitation and the $\underline{70}$ for odd L [4]. For example, if we restrict ourselves to nonstrange mesons, we have (π, ϱ, ω) for $L = 0$ and, correspondently $(B; A_2, A_1, \delta; f, D, \sigma)$ when they are excited to $L = 1$.

In oder to get the chiral content of the states we notice that the operators $A(\lambda_\alpha) = Q^\alpha$ and $A(\lambda_\alpha \sigma_z)$ generate an $SU_3 \times SU_3$ algebra, and furthermore, $A(\lambda_\alpha \sigma_z)$ has $J^P = 1^+$, so that it is a good candidate for the physical Q_5^α. Each one of the $SU_6 \times O_3$ supermultiplets is a reducible representation of the $A(\lambda_\alpha)$, $A(\lambda_\alpha \sigma_z)$ chiral algebra. Therefore, pionic couplings would only occur inside the supermultiplets, and should be pure p-wave. For example, the ρ would only be coupled to π and ω, while the nucleon would couple to itself and to the Δ (1236). These couplings come out with larger values than the experimental ones. Then, a m^2-operator which satisfies eq. (4), must be of the form [5], $A(L)+B(L)(\boldsymbol{L}\cdot\boldsymbol{S})$, because $(\boldsymbol{L}\cdot\boldsymbol{S})$ is the only $(1,1)+[(3,\overline{3})+(\overline{3},3)]$ object we can construct which is diagonal in J. It gives no splitting between π and ρ, or between $\mathcal{N}$ (940) and Δ (1236), but anyway the (mass)^2 difference inside both baryon and meson supermultiplets are relatively small as compared to the distances between different supermultiplets. On the other hand, the spin-orbit coupling predicts that A_2 and B, B and A_1, A_1 and δ, are equispaced in (mass)^2, in agreement with experimental values. The $\boldsymbol{L}\cdot\boldsymbol{S}$ coupling seems to be very small for baryons.

Obviously, the fact that the model allows only for p-wave π coupling constants which are diagonal in L is a serious desease. But it has also some nice properties and we will construct more realistic axial charges starting from it as a zeroth-order approximation [6]. We keep the $SU_6\times O_3$ classification, and write [7, 8, 9]

$$Q_5^\alpha = \exp[i\theta Z]\,A(\lambda_i\sigma_z)\exp[-i\theta Z]\,, \tag{5}$$

where Z is an SU_3 scalar Hermitean operator (in order to preserve the chiral algebra) with positive G-parity and normality (in order to keep both the G-parity and the normality of axial charges).

The choice of Z is suggested by the empirical observation that outside the $L=0$ supermultiplets the most important π coupling constants are those for the decays of $L=1$ states into $L=0$ ones. They connect $(\underline{35}+\underline{1})\leftrightarrow(\underline{35}+\underline{1})$ for mesons, and $\underline{56}\leftrightarrow\underline{70}$ for baryons (s- and d-wave transitions). That is what Z will be required to do and if, for simplicity, we choose it to be a member of a $\underline{35}$ under SU_6, it is determined to be [7]

$$Z=(\boldsymbol{W}\times\boldsymbol{M})_z\,. \tag{6}$$

Here $\boldsymbol{W}$ is a vector under $\boldsymbol{S}$ and connects only $\underline{56}\leftrightarrow\underline{70}$ for baryons. For

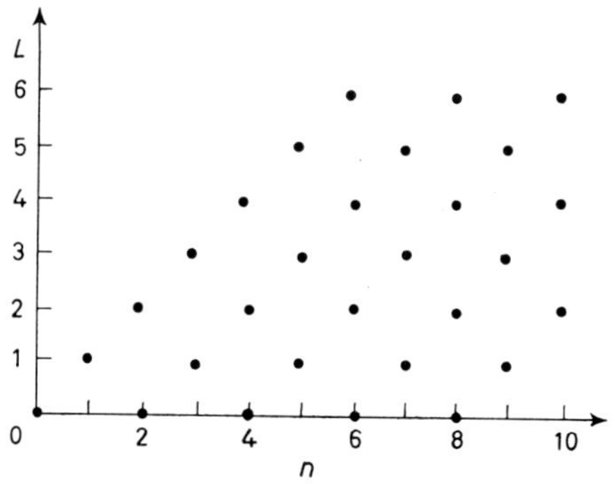

Fig. 1.

mesons it coincides essentially with the W-spin. The operator $\boldsymbol{M}$ is a vector under $\boldsymbol{L}$ and generates the orbital excitations. It gives rise to a new quantum number n, which counts the number of times one has to apply M to reach a state starting from the ground state ($L = 0$). Therefore, all the states that are connected to the $n = L = 0$ ground state can be labelled by $|nLL_z\rangle$ [10]. Because M is a $L = 1$ operator we obtain the set of states represented in the Figure.

As a first test to the model, a calculation of axial couplings has been performed to 2nd order in θ. For the $L = 1$ mesons we have only one parameter (the mixing angle θ) and we get the following results for the axial couplings (h = helicity) [7]:

	Theory	Experiments
$1\|\langle f^0\|Q_5^0\|\pi^0\rangle\|^2$	$\frac{1}{9}$ (normalization of θ)	0.11
$\sum_h \|\langle A_1^+\|Q_5^0\|\varrho^+\rangle_h\|^2$	$\frac{1}{2}$	$\leqslant 0.5$
$\|\langle\sigma^0\|Q_5^0\|\pi^0\rangle\|^2$	$\frac{2}{9}$	$\gg 0.1$
$\sum_h \|\langle B^0\|Q_5^0\|\omega\rangle_h\|^2$	$\frac{1}{3}$	0.33
$\sum_h \|\langle A_2^+\|Q_5^0\|\varrho^+\rangle_h\|^2$	$\frac{1}{6}$	0.13
$\langle A_1^+\|Q_5^0\|\varrho^+\rangle_{h=1}/\langle A_1^+\|Q_5^0\|\varrho^+\rangle_{h=0}$	$+\frac{1}{2}$	$\pm 0.48 \pm 0.13$
$\langle B^0\|Q_5^0\|\omega\rangle_{h=0}/\langle B^0\|Q_5^0\|\omega\rangle_{h=1}$	0	0.1 — 0.7

For the couplings of baryons to the nucleon we have one parameter for each supermultiplet (as we do not yet know the reduced matrix elements of the operator M). At $n = L = 0$ we predict that $F/D = \frac{2}{3}$ for the $\frac{1}{2}^+$ octet (experimentally, 0.67), and $\sqrt{2}\langle\Delta^+|Q_5^0|p\rangle = 4\langle p|Q_5^0|p\rangle = 4/5G_A$, in excellent agreement with experiments. The axial couplings are also well predicted for the $\underline{56}$ $n = L = 2$. The results are not so good for the $\underline{70}$ $n = L = 1$ but it is worth to remark that the worst predictions concern states which can be mixed with others belonging to the same supermultiplet, namely, states with the same total angular momentum J but different values for S ($\frac{1}{2}$ and $\frac{3}{2}$, respectively), since there is no reason but simplicity to suppose that the internal spin S is a good quantum number. Conversely, we predict a good result, $F/D = -\frac{1}{3}$ (experimentally, -0.13) for the $\mathcal{N}$ (1688) ($J^P = \frac{5}{2}^-$) octet, which cannot be mixed.

It is also possible to derive some results that are valid at any order in θ [8]. The B-meson should decay only transverserly into $\omega\pi$ and for the polarization in the $A_1 \to \rho\pi$ decay, one predicts

$$\frac{\langle A_1|Q_5|\varrho\rangle_{h=1}}{\langle A_1|Q_5|\varrho\rangle_{h=0}} = \frac{\langle A_2^+|Q_5^0|\varrho^+\rangle}{\langle f^0|Q_5^0|\pi^0\rangle} \times \frac{1}{\sqrt{3}} \text{(experimentally, 0.048 and 0.42, respectively)} .$$

Furthermore, any positive-parity octet trajectory should have $F/D = \frac{2}{3}$ and for the $\mathcal{N}$ (1688) ($J^P = \frac{5}{2}^-$) trajectory one also gets a constant F/D ratio with the value $-\frac{1}{3}$.

Up to now we only know the transformation properties of Z under $SU_6 \times O_3$. We still need the reduced matrix elements of the operator M. They have been studied for meson states by saturating the Weinberg condition, eq. (4), in a perturbative approach [9].

We restrict ourselves to nonstrange mesons and charges. First we multiply eq. (4) by $\exp[-i\theta Z]$ to the left and $\exp[+i\theta Z]$ to the right, and use the definition, eq. (5), to get

$$\left[A\left(\frac{\tau^+}{2}\sigma_z\right), \left[A\left(\frac{\tau^+}{2}\sigma_z\right), \exp[-i\theta z]\, m^2 \exp[i\theta z] = 0\right]\right] .$$

It means that $\exp[-i\theta Z]\, m^2 \exp[i\theta Z]$ has to behave like a $(0, 0) + (\frac{1}{2}, \frac{1}{2})$ representation (the analogous of $(1, 1)$ and $(3, \overline{3}) + (\overline{3}, 3)$, respectively) under the $SU_2 \times SU_2$ subalgebra of SU_4 generated by $A(\tau^i/2)$ and $A(\tau^i/2\sigma_z)$.

Then we look for a perturbative solution for the m^2-operator which we expand in a power series of θ^2:

$$m^2 = m_0^2 + \theta^2 m_2^2 + \theta^4 m_4^2 + \theta^6 m_6^2 + \dots .$$

The operator $\exp[-i\theta Z]\, m^2 \exp[i\theta Z]$ is also expanded in powers of θ:

$$\exp[-i\theta Z]\, m^2 \exp[i\theta Z] = m_0^2 - i\theta[Z, m_0^2] - \theta^2\{\tfrac{1}{2}[Z, [Z, m_0^2]] - m_2^2\} -$$
$$- i\theta^3\{\tfrac{1}{6}\big[Z, [Z, [Z, m_0^2]]\big] - [Z m_2^2]\} + \dots ,$$

and it is required to behave like a $(0, 0) + (\frac{1}{2}, \frac{1}{2})$ *at each order in* θ.

At 0-th order we get [5, 9]

$$m_0^2 = A(InL) + B(InL)(\boldsymbol{L}\cdot\boldsymbol{S}) ,$$

as in the unmixed case. At first order one obtains [9]

$$B(InL) = B(-)^{L+1} \qquad (B = \text{const}).$$

The 2nd order calculations are the crucial ones. First, they determine the m_0^2 operator to be [9]

$$m_0^2 = A_0 + \Delta n + B(-)^{L+1}(\boldsymbol{L}\cdot\boldsymbol{S} + \tfrac{1}{2}) \qquad (A_0, \Delta, B = \text{consts}).$$

We also get the $M_\pm$ matrix elements [9]. They can be put in a simple form if we define creation and anihilation operators, a_i and $a_i^\dagger$ ($i = +, 0, -$) corresponding to orbital angular-momentum excitation in each direction (up, collinear and down), which obey the commutation relations

$$[a_i^\dagger, a_j] = \delta_{ij} .$$

The operators a_i, $a_i^\dagger$ and $a_i a_i^\dagger$ generate the 3-dimensional harmonic oscillator group and we easily identify

$$n = \sum_i a_i a_i^\dagger ,$$

$$L_\pm = a_\pm a_0^\dagger + a_0 a_\mp^\dagger ,$$

$$L_z = a_+ a_+^\dagger + a_- a_-^\dagger .$$

Then, from the θ^2-calculations we obtain

$$M_\pm = a_\pm + a_\mp^\dagger ,$$

i.e. M_i is the *co-ordinate operator* of the 3-dimensional harmonic oscillator [9]. The set of all mesonic resonances should behave like an irreducible representation of that group.

Our m_0^2-operator consists of two pieces, the harmonic-oscillator energy and a spin-orbit coupling of the form $(-)^{L+1}(\boldsymbol{L}\cdot\boldsymbol{S} + \frac{1}{2})$ [9]. *The resulting spectrum consists of six families of equispaced linear trajectories.* The $I = 0$ and $I = 1$ states are degenerate and the π and A_1 trajectories coincide. Our spin-orbit term has some interesting properties: it satisfies eq. (4) at all orders, *i.e.* the operator $\exp[-i\theta Z]\,(-)^L(\boldsymbol{L}\cdot\boldsymbol{S} + \frac{1}{2})\exp[i\theta Z]$ behaves like a $(0, 0) + (\frac{1}{2}, \frac{1}{2})$; its alternating sign constitutes a crucial experimental test for the whole approach.

We get also some information on the m_2^2-operator, namely that it contains a piece which contributes with a constant splitting between $S = 0$ and $S = 1$ trajectories. It is completely normalized up to a factor θ^2 and to second order in θ we obtain the right sign and order of magnitude for the $\pi - \rho$ (mass)^2 difference. However, this new term spoils the nice equispacing of the $I = 1$, $L = 1$ states. One should therefore rather wait until one has the exact expression for m^2 before drawing a precise conclusion.

A similar program is in progress for baryons. An interesting point is that the coefficient of the spin-orbit coupling in the 0-th order (mass)^2 operator turns out to be zero. Indeed, it is quite small in nature. We should remark, however, that the SU_3 breaking seems to play an important role in the baryon (mass)^2 operator. Probably we should take it into account in order to get meaningful results but calculations would become very intrincate in this case.

A remark about the mixing operator $\exp[i\theta Z]$. We could replace it by $\exp[if(\theta Z)]$ where $f(\theta Z)$ is any odd function of θZ, without affecting the results quoted here. A $f(\theta Z) \neq \theta Z$ could be useful in order to improve the model. Anyway, this is an arbitrariness which still remains in our approach.

It is important to notice that once we know the spectrum of resonances and their coupling constants we can construct amplitudes for processes of the type $\pi\alpha \to \pi\beta$ in the narrow width approximation. Work has bean carried out in this direction.

REFERENCES

[1] The work reported here has been done by F. BUCCELLA, E. CELEGHINI, H. KLEINERT, F. NICOLÒ, A. PUGLIESE, E. SORACE and C. A. SAVOY.

[2] See, *e.g.*, S. L. ADLER and R. F. DASHEN: *Current Algebras and Application to Particle Physics* (New York, 1968).

[3] S. WEINBERG: *Phys. Rev.*, **177**, 2604 (1969).

[4] See, *e.g.*, R. H. DALITZ: *Proceedings of the Oxford Conference on Elementary Particles* (1965).

[5] C. BOLDRIGHINI, F. BUCCELLA, E. CELEGHINI, E. SORACE and L. TRIOLO: *Nucl. Phys.*, **22** B, 651 (1970).

[6] The idea of constructing the axial charges by the introduction of a mixing between $SU_6 \times O_3$ states has been first developped by R. GATTO, L. MAIANI and G. PREPARATA: *Phys. Rev. Lett.*, **16**, 377 (1966); *Physics*, **3**, 1 (1967); *Nuovo Cimento*, **44** A, 1279 (1966). The scheme proposed in ref. [7] is based on earlier work by F. BUCCELLA, E. CELEGHINI and E. SORACE: *Lett. Nuovo Cimento*, **1**, 558 (1969); **2**, 571 (1969).

[7] F. BUCCELLA, E. CELEGHINI, H. KLEINERT, C. A. SAVOY and E. SORACE: *Nuovo Cimento*, **69** A, 133 (1970).

[8] F. BUCCELLA, F. NICOLÒ and A. PUGLIESE: CERN preprint, TH 1331 (May 1971).

[9] F. BUCCELLA, E. CELEGHINI and C. A. SAVOY: *Nuovo Cimento*, **7** A, 281 (1972).

[10] As a matter of fact, each state labelled by $|nLL_z\rangle$ could be degenerate in nature but from the point of view of our sum rules it behaves like a single effective state.

Measuring the σ-Term in Pion-Nucleon Scattering.

G. ALTARELLI

Istituto di Fisica dell'Università - Roma

1. – Introduction.

The aim of this analysis, which was worked out by CABIBBO, MAIANI and ALTARELLI [1, 2], is to evaluate the σ-term in pion-nucleon scattering by relating its magnitude to the values of various low-energy parameters of the on-shell π-$\mathcal{N}$ amplitude. We start by recalling what the definition of the σ-term is, why it is important to know its magnitude, and by listing the previous attempts to evaluate this quantity.

We consider the amplitude for the process

$$\pi^a(q) + \mathcal{N}(p) \to \pi^b(q') + \mathcal{N}(p'),$$

which we call $T^{ab}_{\pi\mathcal{N}}$, the indices, a, $b = 1, 2, 3$ referring to isospin. The PCAC choice for the pion field [3]

$$\varphi^i = \frac{\sqrt{2}}{F_\pi \mu^2}\,\partial_\mu A^{\mu i} = \frac{\sqrt{2}}{F_\pi \mu^2} D^i \tag{1}$$

provides an off-mass–shell extrapolation for $T^{ab}_{\pi\mathcal{N}}$ and it is to this amplitude that the analysis of ADLER [4] and WEINBERG [5] can be applied. In eq. (1) A^i_μ is the axial vector current, D^i its divergence, $F_\pi \simeq 0{,}96\mu$ is the pion decay constant and μ is the pion mass. By using the standard technique of pulling the derivatives through the time-ordered product one obtains [3]

$$(2)\qquad (2\pi)^4\delta^4(p+q-p'-q')(\text{kinematic factors})\cdot T^{ab}_{\pi\mathcal{N}} =$$

$$= -\frac{2i(q^2-\mu^2)(q'^2-\mu^2)}{F^2_\pi \mu^4}\int \mathrm{d}^4x\,\mathrm{d}^4y \exp[iqx]\exp[-iq'y]\langle N|T(D^a(x)D^b(y))|N\rangle =$$

$$= \frac{2(q^2-\mu^2)(q'^2-\mu^2)}{F^2_\pi\mu^4}\int \mathrm{d}^4x\,\mathrm{d}^4y \exp[iqx]\exp[-iq'y]\langle N|\{-iq^\mu q'^\nu T(A^a_\mu(x)A^b_\nu(y)) -$$

$$- q^\mu\delta(x_0-y_0)[A^a_\mu(x), A^b_0(y)] + i\delta(x_0-y_0)[A^a_0(x), D^b(y)\}|N\rangle .$$

The first term contributes both to the symmetric and to the antisymmetric part of the amplitude with respect to the indices a, b, *i.e.* to isospin in the t channel. In the limit of $q_\mu \to 0$, $q'_\nu \to 0$ it reduces to the nucleon-pole term evaluated with the gradient coupling for the $\pi\mathcal{N}\mathcal{N}$ vertex (plus terms of second order in q_μ and q'_ν). The second term which is an equal-time commutator known from current algebra, is of first order in the pion momentum and antisymmetric in isospin indices. The last term is the one that corresponds to the σ-term. It is not a commutator which is given by current algebra and it derives its name from the σ-model where it simply reduces to the canonical σ-field. When $q_\mu \to 0$, $q'_\nu \to 0$, it reduces to

$$i\int \mathrm{d}^4x\,\mathrm{d}^4y \exp[iqx]\exp[-iq'y]\delta(x_0-y_0)\langle N|[A_0^a(x), D^b(y)]|N\rangle \to$$
$$\to (2\pi)^4\delta^4(p+q-p'-q')i\langle N|[Q_5^a, \dot{Q}_5^b]|N\rangle =$$
$$= -(2\pi)^4\delta^4(p+q-p'-q')\langle N|[Q_5^a, [Q_5^b, H_B]]|N\rangle\,,$$

Q_5^a is the axial-vector charge and H_B is that part of the Hamiltonian which breaks chiral $SU_2 \otimes SU_2$ invariance. By using the Jacobi identity for the double commutator and the fact that $[H_B, [Q_5^a, Q_5^b]] = 0$, we immediately obtain that it is symmetric in the isospin indices. Therefore we can also write

$$\sigma = \langle N|[Q_5^a, [Q_5^b, H_B]]|N\rangle = \tfrac{1}{3}\sum_{a=1}^{3}\langle N|[Q_5^a, [Q_5^a, H_B]]|N\rangle\,, \tag{3}$$

which is the definition of the σ-term.

We now specialize our amplitude to

$$T(\nu, t, q^2, q'^2) = A^+ + \nu B^+\,,$$

where A and B are the standard invariant amplitudes of $\pi\mathcal{N}$ scattering [6] and the plus index refers to the symmetric-isospin combination (*i.e.* isospin zero in the t-channel) which from now on we restrict ourselves to. The variables we will use are (m is the nucleon mass)

$$s = (p+q)^2\,, \qquad t = (p-p')^2\,, \qquad u = (p-q')^2\,,$$
$$\nu = \frac{s-u}{4m} = \frac{(p+p')\cdot(q+q')}{4m}\,,$$
$$\nu_B = \frac{t-q^2-q'^2}{4m} = -\frac{(qq')}{2m}\,.$$

We also introduce the amplitude F which is obtained from T by subtracting from it the corresponding pseudovector Born term:

$$F(\nu, t, q^2, q'^2) = A^+ + \nu B^+ - \frac{g(q^2)g(q'^2)}{m} \frac{\nu_B^2}{\nu_B^2 - \nu^2}$$

$g(q^2)$ reduces on the mass shell to the pion-nucleon coupling constant, while its off-mass-shell behaviour is dictated by the axial-vector form factors.

The σ-term is proportional to the value of F at the Weinberg point $\nu = t = q^2 = q'^2 = 0$:

$$\sigma = -\frac{F_\pi^2}{2} F(0, 0, 0, 0) . \tag{4}$$

The value of the σ-term is important because it is a measure of the amount of $SU_2 \otimes SU_2$ symmetry breaking in the Hamiltonian as it follows from eq. (3). In particular, in the elegant scheme where $SU_3 \otimes SU_3$ is broken by terms which transform according to a $(3, \overline{3}) \oplus (\overline{3}, 3)$ representation of $SU_3 \otimes SU_3$ [7, 8], $\frac{1}{3} \sum_a [Q_5^a, [Q_5^a, H_B]]$ simply reduces to H_B.

Thus, if

$$H = H_0 - \varepsilon_0 u_0 - \varepsilon_8 u_8 ,$$

where $\varepsilon_0 u_0$ breaks $SU_3 \otimes SU_3$ but not SU_3 and $\varepsilon_8 u_8$ is the SU_3 breaking, u_0 and u_8 being scalar densities of the $(3, \overline{3}) \oplus (\overline{3}, 3)$ representation, we have

$$\sigma = \langle N | H_B | N \rangle = -\frac{\sqrt{2}\varepsilon_0 + \varepsilon_8}{3} \langle N | (\sqrt{2} u_0 + u_8) | N \rangle .$$

From the value of the σ-term one can thus infer important consequences on the structure of the symmetry-breaking Hamiltonian.

A recent evaluation of the σ-term from dispersion relations by CHENG and DASHEN [9] led to a value ($\sigma \simeq 110$ MeV) much larger than that from a previous estimate by VON HIPPEL and KIM [10] ($\sigma \simeq 26$ MeV), therefore arising a considerable interest on this matter. In a work that went relatively unnoticed OSYPOWSKI [11] had also evaluated this quantity along a line which is similar to ours ($\sigma \simeq 60$ MeV). The determination of CHENG and DASHEN was in turn challenged by HOEHLER, JACOB and STRAUSS [12] who re-evaluated the σ-term, again from dispersion relations ($\sigma \simeq 40$ MeV) and by ERICSON and RHO [13] who instead extracted this quantity from a study of pion-nuclear interactions in π-mesic atoms ($\sigma \simeq 34$ MeV). The aim of our approach is to relate the σ-term to more familiar quantities such as the s- and p-wave scattering lengths of $\pi\mathcal{N}$ scattering, the $\pi\mathcal{N}\mathcal{N}$ and $\pi\mathcal{N}\mathcal{N}^*$ coupling constants and to a well-convergent integral over on shell total cross-sections—an integral whose value is precisely determined and uncontroversial. The main advantage of this method, besides its simplicity, is that it allows us to make use for the evaluation of the σ-term of the large amount of work which was done in the last decade on the determination of the low-energy parameters of $\pi\mathcal{N}$ scattering.

To conclude this introduction I want to recall that the size of the σ-term is also important from another point of view. The experimental smallness of the isospin even combination of the s-wave scattering lengths $a_1 + 2a_3$ was thought [3] to follow from the fact that both the σ-term and the regular part of the isospin even amplitude are of second order in the pion momenta (the Born term is also of that order at threshold) to be compared with the odd amplitude which is of first order. Actually, the values which follow from the various determinations of the σ-term are too large to support this statement and a cancellation between the σ-term and the regular part of the amplitude is necessary in order to reproduce the experimental value of $a_1 + 2a_3$.

2. – Determination of the σ-term.

As we have seen in the introduction, the σ-term is proportional to the amplitude F evaluated at the Weinberg point $\nu = t = q^2 = q'^2 = 0$. This is an unphysical point and in order to relate the σ-term to physical values of the amplitude some smoothness assumptions on F have to be made. We assume that F can be expanded in powers of the pion momenta q_μ and q'_ν. In the soft-pion region of interest here each component of q_μ and q'_ν is of order μ and, according to a philosophy originally put forward by WEINBERG [5], one expects the n-th term in the expansion to be of order $(\mu/M)^n$, M being a «typical» hadronic mass. Thus one would naturally be led to neglect all but the first few terms of the expansion. However, a closer analysis immediately shows that for the nearby-lying s-channel resonances the «typical» hadronic mass turns out to be equal to the difference of the resonance and the nucleon masses, which for the $\mathcal{N}^*$ (1233) is equal to only twice the pion mass.

As a matter of fact also nearby singularities in the t or q^2, q'^2 channels may introduce rapid variations in the amplitude, the typical mass being in this case the position of the singularity. However, from what we know from mass spectra in these channels, the mass of the singularities should be large enough in comparison with μ not to endanger the neglecting of high-order terms arising from them in the expansion of F.

It is then reasonable to separate the contributions to F of the low-lying s-channels resonances by writing down a decomposition for F of the form

$$F = \sum_{\substack{\text{nearby}\\ s\text{-channel}\\ \text{poles}}} F_{\text{Born}} + F_{\text{smooth}}\,.$$

The resonance Born terms have to be evaluated with the gradient coupling for the pion, in agreement with the off-mass-shell definition of F (Eq. (2)) which is provided by PCAC. It is however immaterial what definition we use for, say, the $\mathcal{N}^*$ propagator, because these different expressions for the resonance contributions only differ by nonpole terms which can be included in F_{smooth}.

We now expand F in powers of the pion momenta by taking crossing symmetry into account:

$$F(\nu, t, q^2, q'^2) = a\mu^2 + bt + c(q^2 + q'^2) + d\nu^2 + R(\nu, t, q^2, q'^2)\,. \tag{5}$$

Here a, b, c and d are constants. The function $R(\nu, t, q^2, q'^2)$ measures the deviations of F from linearity and it is defined in such a way as to vanish, together with its first derivatives $\partial R/\partial t$, $\partial R/\partial \nu^2$, $\partial R/\partial q^2$ and $\partial R/\partial q'^2$ at the Weinberg point.

Up to this point everything is rather general. But now we assume, according to our previous discussion, that F_{smooth} is well represented by its linear expansion, so that R will be identified with $\sum F_{\text{Born}}$ with its linear expansion near the Weinberg point subtracted away.

The σ-term is then given by

$$\sigma = -F_\pi^2 a\mu^2/2\,, \tag{6}$$

a is therefore the quantity we wish to evaluate. We first enforce on F the Adler-consistency condition [3], which is the statement that in the limit $q_\mu \to 0$, $q'^2 = \mu^2$, F vanishes, that is

$$F(0, \mu^2, \mu^2, 0) = 0\,. \tag{7}$$

As well known this is a general consequence of PCAC.

Equation (7) implies that

$$(a + b + c)\mu^2 + R(0, \mu^2, \mu^2, 0) = 0\,. \tag{8}$$

This equation is of fundamental importance in the evaluation of the σ-term, because once R is explicitly known, it allows to evaluate a from on-mass-shell $\pi\mathcal{N}$ scattering.

In fact, data on the mass shell can be used to determine $a + 2c$, b and d. Once these quantities are known we can disentangle a from c by eq. (8). Observe that it is also from eq. (8) that it can be seen why we have separated out a factor of μ^2 from a, *i.e.* why the σ-term is expected to be of order μ^2.

In order to evaluate a, b, c and d we have to add three more imputs to eq. (8).

The first equation of this set of three is obtained by recalling that F satisfies a substracted fixed $t = 0$ dispersion relation which reads ($q_L = (\nu^2 - \mu^2)^{\frac{1}{2}}$ is the pion laboratory momentum)

$$\operatorname{Re} F(\nu, 0, \mu^2, \mu^2) = F(0, 0, \mu^2, \mu^2) + \frac{\nu^2}{\pi} \int_{\mu}^{\infty} \frac{d\nu'}{\nu'} q'_L \frac{\sigma^{\pi^+ p}(\nu') + \sigma^{\pi^- p}(\nu')}{\nu'^2 - \nu^2}.$$

From this we can express the variation of F between threshold and the unphysical point $\nu = t = 0$ in terms of a well-convergent integral over the sum of π^+p and π^-p total cross-sections, which is very accurately known. We have

$$(9) \qquad F(\mu, 0, \mu^2, \mu^2) - F(0, 0, \mu^2, \mu^2) = d\mu^2 + R(\mu, 0, \mu^2, \mu^2) - R(0, 0, \mu^2, \mu^2) =$$

$$= \frac{\mu^2}{\pi} \int_{\mu}^{\infty} \frac{d\nu'}{\nu'} \frac{\sigma^{\pi^+ p} + \sigma^{\pi^- p}}{q'_L} \equiv A_0 = (1.45 \pm 0.02)\mu^{-1}.$$

All the determinations of this integral which we know [14, 15, 16] agree within the quoted error, which is negligible in the limits of this discussion. The second input is provided by the evaluation of F at threshold where it is given by the nucleon Born term plus the isospin even combination of the s-wave scattering lengths [6]:

$$(10) \qquad F(\mu, 0, \mu^2, \mu^2) = (a + 2c + d)\mu^2 + R(\mu, 0, \mu^2, \mu^2) =$$

$$= 4\pi \left(1 + \frac{\mu}{m}\right) \frac{a_1 + 2a_3}{3} + \frac{4\pi f^2}{m} = A_1 + \frac{4\pi f^2}{m}.$$

Here $f^2 = (g^2/4\pi)(\mu/2m)^2 = 0.081 \pm 0.004$ [17] is the pseudovector $\pi\mathcal{N}$ coupling constant.

Finally, we consider the expansion of F in terms of $\boldsymbol{q}^2$ and $\boldsymbol{q}^2 \cos\theta$ around threshold ($\boldsymbol{q}$ and θ being the centre-of-mass momentum and angle). The coefficient of $\boldsymbol{q}^2 \cos\theta$ equals a linear combination of s- and p-wave scattering lengths [6] plus the nucleon Born-term contribution. The relation we obtain is

$$(11) \qquad \mu^2\left(2b + \frac{\mu}{m} d\right) + 2\mu^2 \frac{\partial R}{\partial t}(\mu, 0, \mu^2, \mu^2) + \frac{\mu^3}{m} \frac{\partial R}{\partial \nu^2}(\mu, 0, \mu^2, \mu^2) =$$

$$= \frac{4\pi\mu^2}{3}\left[\frac{a_1 + 2a_3}{4m^2} + a_{11} + 2a_{31} + 2\left(1 + \frac{3\mu}{2m}\right)(a_{13} + 2a_{33})\right] -$$

$$- \frac{8\pi f^2}{m}\left(1 + \frac{\mu}{2m}\right) \equiv A_2 - \frac{8\pi f^2}{m}\left(1 + \frac{\mu}{2m}\right).$$

From eqs. (8), (9), (10), (11) one can solve for a, b, c and d in terms of A_0, A_1,

A_2, the nucleon Born term and a correction from R. This leads to

$$(12)\quad \begin{cases} -a\mu^2 = A_1 + A_2 - \left(1+\dfrac{\mu}{m}\right)\left(A_0 + \dfrac{4\pi f^2}{m}\right) + \varDelta_R\,, \\ b\mu^2 = \dfrac{1}{2}\left[A_2 - \dfrac{\mu}{m}A_0 - \dfrac{8\pi f^2}{m}\left(1+\dfrac{\mu}{2m}\right) + \varDelta_R + R(0,0) - 2R(0,\mu^2,\mu^2,0)\right], \\ d\mu^2 = A_0 + R(0,0) - R(\mu,0)\,, \\ c\mu^2 = -(a+b)\mu^2 - R(0,\mu^2,\mu^2,0)\,, \end{cases}$$

$\varDelta_R$ is the correction to the σ-term due to R:

$$(13)\quad \varDelta_R = \frac{\mu}{m}R(\mu,0) - \left(1+\frac{\mu}{m}\right)R(0,0) - \\ -2\mu^2\frac{\partial R}{\partial t}(\mu,0) - \frac{\mu^3}{m}\frac{\partial R}{\partial \nu^2}(\mu,0) - 2R(0,\mu^2,\mu^2,0)\,.$$

In eqs. (12) and (13) we have used the simplified notation $R(\nu,t)$ for $R(\nu,t,\mu^2,\mu^2)$.

The $\mathcal{N}^*$ (1233) contribution to R was evaluated from the standard coupling $(g^*/\mu)\bar{\psi}_\lambda\psi\partial^\lambda\varphi$ with the $\mathcal{N}^*$ propagator taken as follows:

$$(14)\qquad p_{\mu\nu} = \frac{p+M}{p^2-M^2}\left\{g_{\mu\nu} - \frac{1}{3}\gamma_\mu\gamma_\nu - \frac{1}{3M}(\gamma_\nu p_\mu - \gamma_\nu p_\mu) - \frac{2}{3M^2}p_\mu p_\nu\right\}.$$

We stress again at this point that any other choice for the propagator would only amount to a redefinition of F_{smooth} and would be immaterial for our analysis. The $\mathcal{N}^*$ contribution to R then turns out to be $[\nu_B^* = \nu_B + (M^2-m^2)/2m]$

$$(15)\quad R(\nu,t,q^2,q'^2) = \frac{8}{q}\frac{g^{*2}}{\mu^2}\left\{(m+M)\left(\nu_B + \frac{m}{2M^2}\nu^2\right)\left(\frac{\nu_B^*}{\nu_B^{*2}-\nu^2} - \frac{2m}{M^2-m^2}\right) + \right. \\ \left. + \frac{\nu^2}{\nu_B^{*2}-\nu^2}\left[\frac{m}{2M^2}\nu^2 + \frac{m+M}{8mM}(q^2+q'^2)\left(1+\frac{2m}{M}\right) + 3\nu_B\frac{M-m}{2M}\left(1+\frac{2m}{3M}\right)\right]\right\}.$$

Terms of higher order in the pion momenta in the numerator have been neglected.

The simplest way to fix g^* would be from the $\mathcal{N}^*$ width through the relation

$$\varGamma = \frac{g^{*2}}{12\pi\mu^2}\frac{E^*+m}{M}q^{*3}\,.$$

By this method one obtains $g^* = 4.67 \pm 0.43$, from $M = (1233 \pm 3)$ MeV and $\varGamma = (116 \pm 6)$ MeV. It is however possible to take into account finite width effects by fitting the Born-term expression to the amplitude evaluated through the dispersion relation with the immaginary part obtatained from the δ_{33} phase-shift values [18].

This method leads to $g^* \simeq 3.32$ and $M = 1219$ MeV as a best fit for the $\mathcal{N}^*$ contribution near threshold [17, 18].

This $\mathcal{N}^*$ contribution is accurate within 10 % around and below threshold. Details can be found in ref. [18]. When Δ_R is evaluated, one obtains

$$\Delta_R = -(0.55 \pm 0.05)\,\mu^{-1} \qquad \text{from } g^* \simeq 4.67\,, \tag{16 a}$$

$$\Delta_R = -(0.47 \pm 0.05)\,\mu^{-1} \qquad \text{from } g^* \simeq 3.32\,. \tag{16 b}$$

(We have affixed an error of 10 % to the second determination, according to the previous discussion.) The difference between the two determinations is rather small, since the lower value for g^* is partially compensated by the lower value for the «effective» mass $M = 1.219$ GeV. We are going to use the value b) for Δ_R in the following. The corresponding values for $R(0, 0)$ and $R(\mu, 0)$ are $[R(0, \mu^2, \mu^2, 0) \simeq 0]$

$$R(0, 0) \simeq -0.049\,\mu^{-1}\,, \qquad R(\mu, 0) \simeq -0.038\,\mu^{-1}\,, \tag{17}$$

Numerical estimates show that higher resonances such as the $\mathcal{N}^*$ (1470) and the $\mathcal{N}^*$ (1512) give negligible contributions to Δ_R (of a size which is smaller than the quoted error for Δ_R). This fact is due both to their larger mass and to their smaller coupling constants to the $\pi\mathcal{N}$ channel, since the increased phase space is not compensated by their elastic widths, which are of the same order as that of the $\mathcal{N}^*$ (1233).

We are now in a position to collect our results. The formula for $a\mu^2$ reads

$$\begin{aligned} -a\mu^2 = 4\pi\left[\left(1+\frac{\mu}{2m}\right)^2\frac{a_1+2a_3}{3}+\mu^2\,\frac{a_{11}+2a_{31}}{3}+2\mu^2\left(1+\frac{3\mu}{2m}\right)\frac{a_{13}+2a_{33}}{3}\right]- \\ -\left(1+\frac{\mu}{m}\right)\left(A_0+\frac{4\pi f^2}{m}\right)+\Delta_R = 14.51\,\frac{a_1+2a_3}{3}+12.57\,\frac{a_{11}+2a_{31}}{3}+ \\ +30.74\,\frac{a_{13}+2a_{33}}{3}-(1.84\pm 0.03)-(0.47\pm 0.05)\,. \end{aligned} \tag{18}$$

The present situation on the s- and p-wave $\pi\mathcal{N}$ isospin even scattering lengths is summarized in the following Table which is extracted from the compilation of ref. [17] to which we refer for further references.

Using the «recommended values» in this Table we obtain

$$-a\mu^2 = (1.26 \pm 0.48)\mu^{-1}\,, \tag{19}$$

which corresponds to

$$\sigma = (81 \pm 30)\ \text{MeV}\,. \tag{20}$$

TABLE I (*).

s-wave		*p*-wave		
a_1+2a_3	Authors	$a_{11}+2a_{31}$	$a_{13}+2a_{33}$	Authors
-0.002 ± 0.008	HAMILTON, 1966	-0.168	0.396	HOEHLER *et al.*, 1969
$+0.056\pm0.022$	LOVELACE, 1967	-0.160	0.431	COLLINS, SAMARANAYAKE and WOOLCOCK, 1969
$+0.045$	ZOVKO, 1969			
-0.021 ± 0.010	HOEHLER *et al.*, 1969			
-0.026 ± 0.008	SAMARANAYAKE and WOOLCOCK, 1969			
-0.010	ENGELS, 1970			
$0.000^{+0.045}_{-0.035}$	Recommended values	-0.164 ± 0.08	0.414 ± 0.021	Recommended values

We also have

$$b\mu^2 = (1.24 \pm 0.11)\mu^{-1}, \qquad c\mu^2 = (0.00 \pm 0.35)\mu^{-1},$$

$$d\mu^2 = (1.44 \pm 0.03)\mu^{-1}.$$

Of the total-error on the σ-term, 40 % is due to the unprecise knowledge of the *s*-wave scattering lengths, another 40 % comes from the *p*-wave scattering lengths, while the remaining 20 % is the uncertainty on Δ_R.

The «recommended values» from ref. [17] are based on an average of scattering lengths evaluated by different groups. It is therefore worth-while to report the results which follow by using the scattering lengths of these two groups, quoted in the same ref. [17], who gave values for both the *s* and *p* waves. From the Hoehler *et al.* scattering lengths we obtain (the value of *d* which does not

(*) Units of $\mu = 1$ are used for the scattering lengths.

depend on the scattering lengths is not changed)

$$-a\mu^2 \simeq 0.94\mu^{-1}\,(\sigma = 60\ \text{MeV})\,, \qquad b\mu^2 = 1.14\mu^{-1}\,, \qquad c\mu^2 = -0.20\mu^{-1}\,,$$

while from the Samaranayake *et al.* scattering lengths we obtain

$$-a\mu^2 = 1.31\mu^{-1}\,(\sigma = 85\ \text{MeV})\,, \qquad b\mu^2 = 1.33\mu^{-1}\,, \qquad c\mu^2 = -0.02\mu^{-1}\,.$$

Finally in a very recent paper, HOEHLER *et al.* [18] made a new determination of the *s*- and *p*-wave scattering lengths. The results were

$$a_1 + 2a_3 = -0.075\,, \qquad a_{11} + 2a_{31} = -0.166\,, \qquad a_{31} + 2a_{33} = 0.408\,.$$

The *s*-wave scattering lengths are much larger than from previous determinations. From these values one obtains

$$-a\mu^2 = 0.68\mu^{-1}\,(\sigma = 44\ \text{MeV})\,, \qquad b\mu^2 = 1.15\mu^{-1}\,, \qquad c\mu^2 = -0.47\mu^{-1}\,.$$

These results are particularly interesting because in their recent determination by dispersion relations of the σ-term HOEHLER *et al.* [12] made use of this large value for $a_1 + 2a_3$, and it is satisfactory to check that our approach reproduces exactly their result, although the two methods are quite different.

It is clear that the large discrepancies of the various determinations reported above reflect the unsatisfactory state of our knowledge of low-energy pion nucleon scattering.

As a final remark we recall that the determination of the σ-term by dispersion relations as performed by CHENG and DASHEN [9], and by HOEHLER *et al.* [12], is based upon the relation

$$F(0, 0, 0, 0) = -F(0, 2\mu^2, \mu^2, \mu^2)\,, \tag{21}$$

which relates the σ-term to the value of F at a point which is on the mass shell and within the Lehmann ellipse and therefore can be obtained from fixed t-dispersion relations. Equation (21) immediately follows, if R is neglected, from the expansion of F, eq. (5) and from the Adler-consistency condition, eq. (8). Our evaluation of the $\mathcal{N}^*$ (1233) contribution to R allows a determination of the corrections to eq. (21), due to the nonlinearity of F. These corrections turn out to be exceedingly small, of order $0.01\mu^{-1}$, thus confirming the validity of the dispersion approach.

REFERENCES

[1] G. Altarelli, N. Cabibbo and L. Maiani: *Phys. Lett.*, **35** B, 415 (1971).
[2] G. Altarelli, N. Cabibbo and L. Maiani: *Nucl. Phys.*, **34** B, 621 (1971).
[3] See, for example, S. Adler and R. Dashen: *Current Algebra and Applications to Particle Physics* (New York, 1968) and references therein.
[4] S. Adler: *Phys. Rev.*, **139** B, 1638 (1965).
[5] S. Weinberg: *Phys. Rev. Lett.*, **17**, 616 (1966).
[6] R. G. Moorhouse: *Ann. Rev. Nucl. Sci.*, **19**, 301 (1969).
[7] S. L. Glashow and S. Weinberg: *Phys. Rev. Lett.*, **20**, 224 (1968).
[8] M. Gell-Mann, R. J. Oakes and B. Renner: *Phys. Rev.*, **175**, 2195 (1968).
[9] T. P. Cheng and R. Dashen: *Phys. Rev. Lett.*, **26**, 594 (1971).
[10] F. Von Hippel and J. K. Kim: *Phys. Rev. D* **1**, 151 (1970).
[11] E. T. Osypowski: *Nucl. Phys.*, **21** B, 615 (1970).
[12] G. Hoehler, H. P. Jacob and R. Strauss: *Phys. Lett.*, **35** B, 445 (1971).
[13] M. Ericson and M. Rho: *Phys. Lett.*, **36** B, 93 (1971).
[14] G. Hoehler and R. Strauss: *Zeits. Phys.*, **232**, 205 (1970).
[15] W. K. Samaranayake and W. S. Woolcock: *Phys. Rev. Lett.*, **15**, 936 (1965).
[16] E. Ferrari: private communication.
[17] G. Ebel *et al.*: *Compilation of Coupling Constants and Low Energy Parameters, September* 1970, edition in *Springer Tracts of Modern Physics*, Vol. **55**, edited by G. Hoehler.
[18] G. Hoehler, H. P. Jacob and R. Strauss: *A critical test of models for the low-energy* $\pi\mathcal{N}$ *amplitude*, University of Karlsruhe preprint, March 1971.

An Introduction to the Light-Cone.

C. G. CALLAN Jr.

Institute for Advanced Study - Princeton, N. J.

1. – Introduction: the relevance of the light-cone.

The lesson of the SLAC electroproduction experiments is that hadronic currents behave in a particularly simple fashion at large energy and virtual mass. Since large energy corresponds to short distance, one may equivalently say that these currents (more accurately, products of these currents) have simple short-distance behaviour. A possible approach to the problem of scaling would therefore be to make simple hypotheses about this short-distance behavior.

The subject of these lectures will be an approach to the problem of electroproduction scaling, known as the method of the light-cone, based on the above remark: One tries to make reasonable hypotheses about the short-distance behavior of currents which not only reproduce the electroproduction results, but suggest new experimental situations in which related phenomena should occur.

To start, we must see precisely what short-distance behavior is involved in electroproduction and related experiments. Consider first of all electron-positron annihilation. Under the assumption that the process is dominated by single virtual-photon exchange, the total cross-section is determined by the tensor

$$W_{\mu\nu}(q) = \sum_n (2\pi)^4\delta(q-n)\langle 0|J_\mu|n\rangle\langle n|J_\nu|0\rangle = \int \mathrm{d}^4x \exp[iq\cdot x]\langle 0|J_\mu(x)J_\nu(0)|0\rangle \,. \tag{1}$$

In this expression J_μ is the hadronic electromagnetic current, q is the total four-momentum of the electron-positron pair (and of the virtual photon as well) and both q_0 and q^2 are positive. $W_{\mu\nu}$ may be rewritten as

$$W_{\mu\nu}(q^2) = \sum\{(2\pi)^4\delta(q-n)\langle 0|J_\mu|n\rangle\langle n|J_\nu|0\rangle - (2\pi)^4\delta(q+n)\langle 0|J_\nu|n\rangle\langle n|J_\mu|0\rangle\} = \int \mathrm{d}x \exp[iq\cdot x]\langle 0|[J_\mu(x), J_\nu(0)]|0\rangle \,.$$

The added term is zero: since both q_0 and n_0 are positive, $\delta(q_0+n_0)$ vanishes identically. Furthermore, since $q^2>0$, we may choose a frame in which q has only a time component. Then, because the commutator of two local operators vanishes for spacelike separation, we may write

$$(2)\qquad \left\{\begin{aligned} W_{\mu\nu}(q) &= \int_{-\infty}^{\infty} dx_0 \exp\left[i\sqrt{q^2}x_0\right] F_{\mu\nu}(x_0)\,, \\ F_{\mu\nu}(x_0) &= \int_{|\mathbf{x}|<x_0} d^3x \langle 0|[J_\mu(x), J_\nu(0)]|0\rangle\,. \end{aligned}\right.$$

The behavior of $W_{\mu\nu}$ for large q^2 evidently determines the asymptotic behavior of the annihilation cross-section. Elementary theorems on Fourier transforms tell us that

$$\lim_{\lambda\to\infty}\int_{-\infty}^{\infty} dx \exp[i\lambda x] f(x) = \sum i\,\frac{\exp[i\lambda x_0]}{\lambda}\,\big(f(x_0+\varepsilon)-f(x_0-\varepsilon)\big)_{\varepsilon=0}\,,$$

where the sum is over all discontinuities of $f(x)$. According to eq. (2), $F_{\mu\nu}(x_0)$ is the integral of a commutator, and hence an odd function of x_0. Since the commutator is a smooth function of x, the only possible discontinuity of $F_{\mu\nu}(x_0)$ is at $x_0=0$, and the asymptotic behavior of $W_{\mu\nu}$ as $q^2\to\infty$ is determined by $F_{\mu\nu}(x_0)$ at $x_0=0$. From the definition of $F_{\mu\nu}$ in terms of the current commutator we see that this in turn is determined by the behavior of $[J_\mu(x), J_\nu(0)]$ as $x_\mu\to 0$. Thus the behavior of current commutators at zero space-time separation has direct phenomenological significance.

Now consider the process of inelastic electroproduction. It is by now a familiar fact that in the one-photon exchange approximation, the cross-section for this process is determined by the tensor

$$\begin{aligned} W_{\mu\nu}(p,q) &= \sum_n (2\pi)^4\delta(p+q-n)\langle p|J_\mu|n\rangle\langle n|J_\nu|p\rangle = \\ &= \int dx \exp[iq\cdot x]\langle p|J_\mu(x)J_\nu(0)|p\rangle = \int dx \exp[iq\cdot x]\langle p|[J_\mu(x), J_\nu(0)]|p\rangle\,, \end{aligned}$$

where J_μ is the hadron electromagnetic current, q is the virtual photon four-momentum and p is the four-momentum of the target, typically a proton. The passage from product to commutator is justified, as in the previous paragraph, by spectral conditions.

We can always choose a frame in which

$$p = (m, 0, 0, 0)\,,$$

$$q = (q_0, 0, 0, q_3)\,,$$

so that the two standard scalar invariants have the form

$$\nu = \frac{q \cdot p}{m} = q_0 \qquad q^2 = q_0^2 - q_3^2 .$$

It is also convenient to introduce slightly different variables

$$x_\pm = x_0 \pm x_3 , \qquad \boldsymbol{x}_\perp = (x_1, x_2) ,$$

$$q_\pm = \frac{q_0 \pm q_3}{2} = \frac{\nu}{2}(1 \pm \sqrt{1 - q^2/\nu^2}) .$$

Then we may write

$$(3) \quad \begin{cases} W_{\mu\nu}(q^2, \nu) = \int_{-\infty}^{\infty} \mathrm{d}x_- \exp[iq_+ x_-] G_{\mu\nu}(x_-) , \\ G_{\mu\nu}(x_-) = \int_{-\infty}^{\infty} \mathrm{d}x_+ \exp[iq_- x_+] \int_{\boldsymbol{x}_\perp^2 \leqslant x_+ x_-} \langle p | [J_\mu(x), J_\nu(0)] | p \rangle , \end{cases}$$

where the limit on the $\boldsymbol{x}_\perp$-integration comes from the vanishing of the commutator for $x^2 = x_+ x_- - \boldsymbol{x}_\perp^2 < 0$.

BJORKEN [1] has suggested that an interesting, experimentally accessible limit in which to study this object is $q^2 \to -\infty$ with $\bar{\omega} = -q^2/2\nu$ fixed. In this limit

$$q_+ \to \nu \to \infty ,$$

$$q_- \to \frac{q^2}{4\nu} = -\frac{\bar{\omega}}{2} ,$$

and we observe that, according to eq. (3) and our discussion following eq. (2), the asymptotic behavior of $W_{\mu\nu}$ is given by the discontinuities of $G_{\mu\nu}(x_-)$. Since $G_{\mu\nu}$ is an integral over the commutator, it is an odd function of x_-, and discontinuities will appear only at $x_- = 0$. The causality restriction, $x_\perp^2 \leqslant x_+ x_-$, then guarantees that only the behavior of the current commutator at $x^2 = 0$ influences the behavior of $G_{\mu\nu}$ at $x_- = 0$. Consequently, the behavior of current commutators at lightlike separations has direct phenomenological significance. The connection between the Bjorken scaling limit and the light-cone seems first to have been remarked by IOFFE [2].

If we look at the above arguments in somewhat more detail, we find that what determines the asymptotic behavior of measurable cross-sections is the leading singularity in the current commutator $[J_\mu(x), J_\nu(0)]$ at $x = 0$ or $x^2 = 0$, according to the process under consideration. The theoretical problem is to find ways of specifying these leading singularities.

2. – The operator-product expansion.

The basic tool for studying the singularities whose phenomenological importance was demonstrated in the previous Section is Wilson's operator-product expansion [3]. The first part of this hypothesis is the assertion that a product of two local operators $A(x)$ and $B(y)$ may be written as

$$A(x)B(y) = \sum_n C^n_{AB}(x-y)\,O_n\left(\frac{x+y}{2}\right), \tag{4}$$

where $\{O_n(x)\}$ is a complete set of Hermitean local operators and $\{C^n_{AB}\}$ is a corresponding set of c-number functions. Such a relation is more or less trivially true in free field theory as a consequence of the Wick expansion followed by Taylor expansion of well-defined normal-ordered products. It is also, though not so trivially, true in perturbation theory. In any case, its practical utility is rather limited so long as we have no further information about the C^n_{AB}, which, since they depend upon $x-y$, determine the behavior of the product as $x \to y$.

Such information is contained in the second part of the hypothesis, the assumption that the *leading* behavior of the $C^n_{AB}(x)$ as $x \to 0$ is that which would obtain if scale invariance were an unbroken symmetry. In a scale-invariant world each operator, A, is assigned a dimension, d_A, and transforms under scale transformations as

$$U(\lambda)A(x)\,U^{-1}(\lambda) = \lambda^{d_A}A(\lambda x)\,.$$

When applied to eq. (4), this yields the result that in the scale-invariant limit, $C_n(x)$ is homogeneous of degree $d_n - d_A - d_B$ in x. Therefore, the fundamental assumption of the operator product expansion is that in the limit $x \to 0$ $C^{AB}_n(x)$ behaves like $x^{d_n - d_A - d_B}$.

The expansion now has some useful implications. For example, the most singular contribution to $A(x)B(y)$ as $x \to y$ is given by the operator O_n having the lowest dimension, d_n, *independent* of d_A and d_B. By comparing the most singular short-distance behavior of different operator-products, AB and $\bar{A}\bar{B}$, we learn about their dimensions, since the effect of the dimensions of the O_n factors out of the comparison. Since the operators of lowest dimension clearly play a special role in this way of looking at things, it seems not unreasonable that they be especially simple operators such as the energy-momentum tensor or the $SU_3 \times SU_3$ currents, in which case even more interesting comparisons of different processes would be possible.

In fact, it is rather more instructive to see how, following BRANDT and PREPARATA [4], this analysis can be extended to discuss light-cone singularities as well. It is apparent that terms in eq. (4) which are finite in the limit $x \to 0$

may be singular in the limit $x^2 \to 0$. Consider for example, a term of the form

$$C_n(x)\,O_n(0) = \frac{x_{\mu_1} \dots x_{\mu_n}}{(x^2)^{n/2}}\,O_n^{\mu_1 \dots \mu_n}(0)\,.$$

It evidently has a finite limit as $x \to 0$, but diverges when $x^2 \to 0$ with x_μ finite. Therefore, we must expect an infinity of terms to contribute to the leading singularity on the light-cone, in contrast to the situation at $x = 0$ where the leading behavior is given by one term.

Let us look in more detail at the expansion of a product of a Lorentz scalar operator with itself, keeping track of the spin structure as suggested by the above remarks. Let us choose each O_n to be an irreducible tensor of spin l_n (*i.e.* it has l_n indices and is symmetric and traceless in these indices). Then, since the product we are expanding is a Lorentz scalar, we have

$$J\left(\frac{x}{2}\right) J\left(-\frac{x}{2}\right) = \sum_n C_n^{\mu_1 \dots \mu_{l_n}}(x)\,O^n_{\mu_1 \dots \mu_{l_n}}(0) = \sum C_n(x^2)\,x_{\mu_1} \dots x_{\mu_{l_n}} O_n^{\mu_1 \dots \mu_{l_n}}(0)\,.$$

Possible Kronecker deltas in $C_n^{\mu_1 \dots \mu_{l_n}}$ are irrelavant because of the tracelessness of O_n. Expressed in this form, we see that the behavior of the expansion for small x^2 is determined by scalar functions of x^2, the $C_n(x^2)$, in the neighborhood of $x = 0$. Therefore, so long as there are no extra-divergences introduced by the summation over n, short-distance behavior and light-cone behavior are connected!

The rules of the operator-product expansion determine the leading behavior of $C_n(x^2)$ to be

$$C_n(x^2) \propto (x^2)^{(d_n - l_n - 2d_J)/2} + (\text{higher order in } x^2)\,.$$

Evidently, the degree of singularity on the light-cone is determined by $\tau_n = d_n - l_n$, a quantity which GROSS and TREIMAN [5] have baptized « twist ». The leading singularity on the light-cone comes from all those O_n having the lowest twist

$$J\left(\frac{x}{2}\right) J\left(-\frac{x}{2}\right) = (x^2)^{\tau/2 - d_J} \left(\sum_n x^{\mu_1} \dots x^{\mu_{l_n}} O^n_{\mu_1 \dots \mu_{l_n}}(0)\right)_{\tau_n = \tau = \text{minimum twist}} + (\text{less singular})\,.$$

Note that unless there are an infinite number of operators with the minimum twist, the dependence of the leading singularity on position on the light-cone will be rather trivial—just a polynomial. It is interesting in this respect to note that in free field theory there do in fact exist infinite strings of operators with fixed twist, since operating on a field with a derivative raises the spin and the dimension by one unit simultaneously. On the other hand, if dimensions are modified by the interactions, as suggested by WILSON [3], it is not so obvious why dimensions should be correlated in this fashion.

As a final point we note that the singularity in the above expansion is not completely defined. This problem is removed by the standard « $i\varepsilon$ » trick for putting singularities in the upper half x_0-plane, as is required by causality: $x^2 \to x^2 - i\varepsilon x_0$.

Let us now begin to see what the formalism we have been developping has to say about electroproduction. For purposes of orientation, let us consider the scalar current analogue of the electroproduction tensor defined in Sect. **1**:

$$W(q^2, q\cdot p) = \int \mathrm{d}x \exp[iq\cdot x]\langle p|J(x)J(0)|p\rangle \,. \tag{5}$$

In the Bjorken limit, $q^2 \to -\infty$ with $q^2/q\cdot p$ fixed, we know that the integral is determined by the singularity of the integrand in the neighborhood of $x^2 = 0$. That in turn we know to be given by a series of the form

$$\langle p|J(x)J(0)|p\rangle = (x^2)^{(\tau-2d_J)/2} \sum x^{\mu_1} \dots x^{\mu_n}\langle p|O^n_{\mu_1\dots\mu_n}(0)|p\rangle + \text{less singular}\,.$$

If, as is usually the case, we are considering spin-averaged matrix elements (that is, unpolarized targets) we also have

$$\langle p|O_{\mu_1\dots\mu_n}(0)|p\rangle = C_n p_{\mu_1} \dots p_{\mu_n} + \dots\,,$$

where ... represent terms having at least one Kronecker delta. They may be ignored because, when substituted into eq. (5), they give extra factors of x^2, and so are less singular. Therefore we have

$$\langle p|J(x)J(0)|p\rangle = (x^2)^{(\tau-2d_J)/2} \sum C_n(p\cdot x)^n + \text{less singular}\,, \tag{6}$$

where the sum is over all operators having the minimum twist, τ. In order to evaluate W, it is convenient to pass to the rest frame of p and define

$$\langle p|J(x)J(0)|p\rangle = (x^2 - i\varepsilon x_0)^\alpha f(x_0) + \text{less singular}\,,$$

$$f(x) = \sum C_n m^n x^n\,,$$

$$\alpha = \frac{\tau - 2d_J}{2}\,,$$

W is then a Fourier transform of f which can be shown to have the following behavior in the Bjorken limit:

$$W(q^2, \nu) \xrightarrow[\nu\to\infty]{} (\nu)^{-(\alpha+2)} \int_{-\infty}^{\infty} \mathrm{d}x \exp\left[i\frac{q^2}{\nu}x\right] f(x)(x - i\varepsilon)^{\alpha+1}\,, \qquad -q^2/\nu \text{ fixed}\,.$$

The scaling observed in electroproduction experiments corresponds to W having a finite limit as $\nu \to \infty$ with $-q^2/\nu$ fixed. Evidently, for this to happen, we must have $\alpha = -2$ or $2d_J = \tau + 4$. If this special relation obtains, then

$$W(q^2, \nu) \to F(-q^2/2\nu) ,$$

$$F(a) = \int_{-\infty}^{\infty} \mathrm{d}x \exp[-i\, max] \frac{f(x)}{x} .$$

Finally, we must ask for the significance of the relation $2d_J = \tau + 4$ which is necessary in order to reproduce the observed scaling. To discuss this question we must assign a value to the dimension of the various operators appearing in an operator-product expansion. The only place where we have *a priori* knowledge of the dimension of operators is in free field theory. In an interacting theory, insofar as the concept of dimension makes sense, the dimension of an operator is in general coupling constant-dependent. In free field theory (equivalently, making use of naive canonical manipulations) the dimension associated with scale transformations is the same as the dimension in the sense of dimensional analysis. Thus, a scalar field has dimension 1, a spin-$\frac{1}{2}$ field has dimension $\frac{3}{2}$ and $\partial/\partial x_\mu$ has dimension 1.

With this in mind, let us ask what happens in a theory of quarks plus scalar gluons. Both scalar currents $(\bar{\psi}\psi)$ and vector currents $(\bar{\psi}\gamma_\mu\psi)$ have dimension 3 since the spin-$\frac{1}{2}$ field, ψ, has dimension $\frac{3}{2}$. Thus d_J is 3 and the relation $2d_J = \tau + 4$ indicates that the minimum twist of operators appearing in the operator product expansion is 2. Now the available operator series with low values of τ which we can construct out the scalar field, $\varphi(d_\varphi = 1)$, and the spin-$\frac{1}{2}$ field, ψ $(d_\psi = \frac{3}{2})$, are easily seen to be

$$\begin{array}{ll}
\tau = 0 & \text{none} \\
\tau = 1 & \varphi,\ \partial_\mu\varphi,\ \partial_\mu\partial_\nu\varphi,\ \dots, \\
\tau = 2 & \bar{\psi}\gamma_\mu\psi,\ \bar{\psi}\gamma_\mu\overset{\leftrightarrow}{\partial}_\nu\psi,\ \dots, \\
 & \varphi\overset{\leftrightarrow}{\partial}_\mu\varphi,\ \varphi\overset{\leftrightarrow}{\partial}_\mu\overset{\leftrightarrow}{\partial}_\nu\varphi,\ \dots .
\end{array}$$

The series with $\tau = 1$ is of no interest for the electroproduction problem since it consists of operators whose diagonal matrix elements in momentum space vanish. Therefore, the lowest useful value of τ is 2, which is precisely what is needed to obtain the observed scaling. Consequently, the hypothesis of free field or canonical dimensions, when coupled with the operator-product expansion, suffices to account for the scaling behavior of electroproduction cross-sections.

3. – Canonical dimensions.

In the previous Section we saw that the assumption of canonical dimensions for operators leads to the desirable asymptotic behavior of the «electroproduction» cross-section for scalar currents. There are, however, special features present when we consider Lorentz vector currents (several independent structure functions, possible relations between them), and we now want to see what, if anything, the hypothesis of canonical dimensions has to say about these special features.

If we want to treat neutrino-induced and photon-induced reactions on the same footing, the object we want to discuss is

$$W_{\mu\nu}(p, q) = \int dx \exp[iq\cdot x]\langle p|J^+_\mu(x)J_\nu(0)|p\rangle ,$$

where J_μ may be the weak or electromagnetic current. If we assume that the target is unpolarized, then we may decompose $W_{\mu\nu}$ in terms of scalar structure functions

$$W_{\mu\nu}(p, q) = W_1(q^2, \nu)\left(-g_{\mu\nu} + \frac{q_\mu q_\nu}{q^2}\right) + \\ + W_2(q^2, \nu)\left(p_\mu - \frac{p\cdot q}{q^2}q_\mu\right)\left(p_\nu - \frac{p\cdot q}{q^2}q_\nu\right) + W_3(q^2, \nu)\, i\varepsilon_{\mu\nu\lambda\sigma}q^\lambda p^\sigma + \\ + W_4(q^2, \nu)(p_\mu q_\nu + p_\nu q_\mu) + W_5(q^2, \nu)q_\mu q_\nu .$$

In order to discuss the behavior of these structure functions in the Bjorken limit it is again necessary to consider the leading light-cone singularities of the current product.

The dominant contribution to $W_{\mu\nu}$ in the Bjorken limit comes from the operator series of minimum twist. We shall assume, in accord with the hypothesis of canonical dimensions, that the minimum twist is two, and that the dimension of J_μ is three. Then we have

$$(7)\quad \begin{cases} J^+_\mu(x)J_\nu(0) \to (x^2 - i\varepsilon x_0)^{-3}\sum C^n_{\mu\nu\lambda_1\ldots\lambda_n}(x)\,O^{\lambda_1\ldots\lambda_n}_n(0) , \\ C^n_{\mu\nu\lambda_1\ldots\lambda_n} = a_n x_\mu x_\nu x_{\mu_1}\ldots x_{\lambda_n} + b_n x^2(x_\mu g_{\nu\lambda_1}x_{\lambda_2}\ldots x_{\lambda_n} + (\mu\leftrightarrow\nu)) + \\ \quad + c_n x^2 g_{\mu\nu}x_{\lambda_1}\ldots x_{\lambda_n} + d_n x^4(g_{\mu\lambda_1}g_{\nu\lambda_2}x_{\lambda_3}\ldots x_{\lambda_n}) + e_n x^2(\varepsilon_{\mu\nu\lambda_1\sigma}x^\sigma x_{\lambda_2}\ldots x_{\lambda_n}) , \end{cases}$$

where the number of powers of x^2 in each of the terms of C^n is determined by the usual dimension-counting arguments. Naturally the number of independent tensors in C^n is the same as the number of independent structure functions in $W_{\mu\nu}$. We notice that the degree of singularity in x^2 is not the same for the different tensor structures in C^n. We cannot, however, claim that $W_{\mu\nu}$ is determined by a_n alone in the Bjorken limit (it corresponds to an $(x^2)^{-3}$ singularity, which is stronger than any of the other four terms): the different tensors associated with the various powers of x^2 can perfectly well compensate for differing degrees of singularity in x^2. What is certain is that for a given tensor structure the leading contribution comes from the strongest singularity on the light-cone, and this is unambiguously given by the operator series of lowest twist.

Further restrictions on the expansion come from considering the divergence of the current. Let us denote the divergence of J_μ by D and its dimension by δ. Then the contribution of the minimum twist series to the product of D and J_μ is

$$D^+(x)J_\nu(0) \to \frac{1}{(x^2 - i\varepsilon x_0)^{(\delta+2)/2}} \sum_n \bar{C}^n_{\nu\lambda_1\dots\lambda_n} O_n^{\lambda_1\dots\lambda_n}(0)\,,$$

$$\bar{C}^n_{\nu\lambda_1\dots\lambda_n}(x) = \bar{a}_n x_\nu x_{\lambda_1} \dots x_{\lambda_n} + \bar{b}_n x^2 g_{\nu\lambda_1} x_{\lambda_2} \dots x_{\lambda_n}\,.$$

Note that the leading light-cone singularity is of degree $(\delta+2)/2$, which is less than three if δ is less than four. If J_μ is one of the $SU_3 \times SU_3$ currents, then in typical Lagrangian models D is just a symmetry rotation of that piece of the Lagrangian which breaks this symmetry. It is quite natural to let the symmetry breaking be caused by « soft » operators, such as mass terms, in which case the canonical dimension of D is less than four. We shall assume this to be true in the real world.

The latter expansion may be computed directly by differentiating the expansion for $J_\mu(x)J_\nu(0)$ with respect to x_μ, obtaining

$$D^+(x)J_\nu(0) \to \frac{1}{(x^2 - i\varepsilon x_0)^3} \sum_n \tilde{C}^n_{\nu\lambda_1\dots\lambda_n} O_n^{\lambda_1\dots\lambda_n}(0)\,,$$

$$\tilde{C}^n_{\nu\lambda_1\dots\lambda_n}(x) = \tilde{a}_n x_\nu x_{\lambda_1} \dots x_{\lambda_n} + \tilde{b}_n x^2 g_{\nu\lambda_1} x_{\lambda_2} \dots x_{\lambda_n}\,,$$

$$\tilde{a}_n = (n-1)\,a_n - 4b_n - 4c_n\,,$$

$$\tilde{b}_n = nb_n + nc_n - 2d_n\,.$$

Evidently, if $\delta < 4$, the two evaluations of D^+J_ν can be consistent only if

$$(8) \qquad \begin{cases} (n-1)\,a_n - 4b_n - 4c_n = 0\,, \\ nb_n + nc_n - 2d_n = 0\,, \end{cases}$$

or, equivalently, if

$$(9) \qquad \frac{\partial}{\partial x_\mu}\left[\frac{1}{(x^2 - i\varepsilon x_0)^3} C^n_{\mu\nu\lambda_1\dots\lambda_n}(x)\, O_n^{\lambda_1\dots\lambda_n}(0)\right] =$$

$$= \frac{1}{(x^2 - i\varepsilon x_0)^\Delta}\left[\bar{a}_n x_\nu x_{\lambda_1} \dots x_{\lambda_n} + \bar{b}_n x^2 g_{\nu\lambda_1} x_{\lambda_2} \dots x_{\lambda_n}\right] O_n^{\lambda_1\dots\lambda_n}(0) +$$

$$+ \text{less singular}, \qquad \Delta < 3.$$

These conditions reduce the number of independent tensors in $W_{\mu\nu}$ from five to three. We shall see that the reduction is precisely such as to render the Bjorken limit of $W_{\mu\nu}$ a conserved tensor. Consequently, we may say that on the light-cone both scale and chiral invariance are reinstated so long as they are broken only by « soft » operators in the Lagrangian.

In order to compute $W_{\mu\nu}$ we first take the appropriate matrix element of eq. (7) to obtain

$$\langle p|J_\mu^+(x)\,J_\nu(0)|p\rangle \to (x^2-ix_0)^{-3}\Big\{x_\mu x_\nu a(p\cdot x) + x^2(x_\mu p_\nu + x_\nu p_\mu)b(p\cdot x) + \\ + x^2 g_{\mu\nu}c(p\cdot x) + x^4 p_\mu p_\nu \frac{d(p\cdot x)}{(p\cdot x)^2} + x^2\varepsilon_{\mu\nu\lambda\sigma}p^\lambda x^\sigma \frac{e(p\cdot x)}{p\cdot x}\Big\},$$

where

$$a(x) = \sum a_n x^n\,, \text{ etc.,}$$

and

$$xa'(x) - a(x) - 4b(x) - 4c(x) = 0\,,$$

$$xb'(x) + xc'(x) - 2d(x) = 0\,.$$

It is then a straightforward matter to evaluate $W_{\mu\nu}$. To illustrate the technique, we first compute

$$q^\mu W_{\mu\nu} = q_\nu(q^2 W_5 + q\cdot p W_4) + p_\nu q^2 W_4 = i\int dx \exp[iq\cdot x]\frac{\partial}{\partial x_\mu}\langle p|J_\mu^+(x)J_\nu(0)|p\rangle = \\ = i\int dx \exp[iq\cdot x](x^2 - i\varepsilon x_0)^{-\Delta}[x_\nu \bar a(p\cdot x) + x^2 p_\nu \bar b(p\cdot x)]\,,$$

where the last equality comes from eq. (9). Performing an integration by parts on the term involving $\bar a$ gives

$$q^\mu W_{\mu\nu} = i\int dx \exp[iq\cdot x](x^2 - i\varepsilon_0)^{-\Delta+1}\Big\{p_\nu(\bar b + \bar a'/2(\Delta-1) + iq_\nu \frac{\bar a(p\cdot x)}{2(\Delta-1)}\Big\}\,,$$

an integral which can be evaluated with the help of the calculations done in Sect. **2**. The crucial point to notice is that in the Bjorken limit everything scales as $(q^2)^{\Delta-3}$, and vanishes in the limit since $\Delta < 3$. Therefore, in terms of the dimension, δ, of the divergence of the current, we have the following scaling law for the two « nonconserved »-structure functions [6]:

$$(10)\qquad \left.\begin{matrix} \nu W_4(q^2,\nu) \\ \nu W_5(q^2,\nu)\end{matrix}\right\} \to (q^2)^{(\delta-4)/2}\left\{\begin{matrix} F_4(q^2/\nu) \\ F_5(q^2/\nu)\end{matrix}\right. \to 0\,.$$

As far as the conserved-structure functions are concerned, we can easily use the above technique of integrating by parts to obtain

$$\text{(11)} \qquad \left.\begin{array}{r} W_1(q^2/\nu) \\ \nu W_2(q^2/\nu) \\ \nu W_3(q^2/\nu) \end{array}\right\} \to \left\{\begin{array}{l} F_1(q^2/\nu) \\ F_2(q^2/\nu) \\ F_3(q^2/\nu) \end{array}\right. .$$

These, of course, are the celebrated Bjorken scaling laws for electroproduction which seem to be so nicely obeyed by the SLAC experiments (at least for W_1 and W_2: W_3 is measured in neutrino processes, and it is not yet known whether it scales). This derivation reveals that scaling is simply a consequence of canonical dimensionality.

Unfortunately, it does not seem possible to carry this scheme much further. For example, in the SLAC experiments, it is found that a linear combination of W_1 and W_2 corresponding to longitudinal polarization is very small, if not zero. As far as the canonical dimension approach is concerned, there is no way of understanding such a phenomenon since W_1 and W_2 are independent functions about which nothing is known, except that they scale. If we want to understand such special relations between structure functions, an even stronger hypothesis than canonical dimensions is clearly called for.

4. – Beyond canonical dimensions.

In the preceeding Sections we have shown that the hypothesis of canonical dimensions suffices to explain the broad facts of electroproduction scaling, but not more detailed features of the data such as the possible vanishing of longitudinal cross-sections. We would now like to explain a yet more radical hypothesis, due to FRITZSCH and GELL-MANN [7], which yields such details. The rationale for this scheme, which might be called light-cone current algebra is as follows: On one hand, we know that the scaling limit of cross-sections is dominated by the leading light-cone singularity of current products. On the other hand, we know that to obtain the required degree of singularity, via the operator-product expansion, it is necessary to assign free-field values to the dimensions of operators. Rather than performing this indirect argument it might appear reasonable to calculate the singularity structure of current commutators *directly* in free-field theory. One would then adopt the resulting formal operator structure directly as a hypothesis about the behaviour of currents. Such a hypothesis would automatically yield the scaling results which follow from the assumption of free-field dimensions for operators, and would be expected to yield further information.

In order to carry out such a program, we must make a specific assumption about how the currents are constructed out of fields. For purposes of illustration we shall assume that the underlying fields are standard quarks. Accordingly, the $SU_3 \times SU_3$ currents which appear in weak and electromagnetically induced inelastic scattering are

$$J^a_\mu(x) = :\bar{\psi}(x)\gamma_\mu \begin{pmatrix} 1 \\ \gamma_5 \end{pmatrix} \lambda^a \psi(x): \qquad a = 1, \dots, 8,$$

where we choose 1 or γ_5 according to the choice of J^a_μ as a vector or axial-vector current. The : operation is the usual normal-ordering instruction which renders matrix elements of J_μ finite.

If ψ is a free field, the Wick expansion gives a complete enumeration of the singularities of products of operators involving ψ. In particular, if we make the definition

$$J_M(x) = :\bar{\psi}(x) M \psi(x):,$$

where M is some numerical matrix, then the Wick expansion is

$$(12) \qquad J_M(x) J_N(y) = \overbracket{\bar{\psi}(x) M \overbracket{\psi(x)\, \bar{\psi}(y)} N \psi(y)} + :\bar{\psi}(x) M \overbracket{\psi(x)\, \bar{\psi}(y)} N \psi(y): +$$
$$+ :\overbracket{\bar{\psi}(x) M \psi(x)\, \bar{\psi}(y) N \psi(y)}: + :\bar{\psi}(x) M(x)\, \bar{\psi}(y) N \psi(y):,$$

where

$$(13) \qquad \begin{cases} \overbracket{\psi_\alpha(x)\, \bar{\psi}_\beta(y)} = (i\gamma\cdot\partial - m)_{\alpha\beta} \Delta(x-y, m), \\ \Delta(x, m) = \dfrac{1}{x^2 - i\varepsilon x_0} + \text{less singular.} \end{cases}$$

Since the normal-ordering instruction renders (at least in free-field theory) all matrix elements of an operator finite, the singularities of eq. (12) arise only from the terms containing contractions (the object appearing in eq. (13)). If we throw away pure C-number terms (they in general contribute only to disconnected graphs and cancel out of the S-matrix), we have

$$J_M(x) J_N(y) = i\partial^\lambda \Delta(x-y)\{:\bar{\psi}(x) M\gamma_\lambda N \psi(y): + :\bar{\psi}(y) N \gamma_\lambda M \psi(x):\} -$$
$$- m\Delta(x-y)\{:\bar{\psi}(x) MN \psi(y): + :\bar{\psi}(y) NM \psi(\ ):\} + \text{finite}.$$

If we retain only the leading light-cone singularity we have

$$J_M(x) J_N(y) = i\partial^\lambda \left(\frac{1}{(x-y)^2 - i\varepsilon(x_0 - y_0)}\right) \{:\bar{\psi}(x) M\gamma_\lambda N \psi(y): + :\bar{\psi}(y) N\gamma_\lambda M \psi(x):\}.$$

The coefficient of the singular function, since it depends on two space-time points, is a «bilocal» operator, a notion introduced by FRISHMAN [8], and therefore has more complicated matrix elements than a local operator. On the other hand, for a specific choice of M and N, it will have definite Lorentz tensor character and definite internal symmetry properties.

If we make the choice of M and N appropriate to $SU_3 \times SU_3$ currents and reduce the products of γ and λ matrices which appear we find, for the leading light-cone singularity

$$(14)\quad \begin{cases} V^a_\mu(x)\,V^b_\nu(y) = A^a_\mu(x)\,A^b_\nu(y) = i\partial^\lambda\left(\dfrac{1}{z^2 - i\varepsilon z_0}\right)\{(g_{\mu\lambda}g_{\nu\sigma} + g_{\nu\lambda} - g_{\mu\nu}g_{\lambda\sigma}g)\cdot \\ \quad\cdot(d_{abc}\mathscr{F}^\sigma_{c+}(x, y) + if_{abc}\mathscr{F}^\sigma_{c-}(x, y)) + i\varepsilon_{\mu\nu\lambda\sigma}(d_{abc}\mathscr{F}^{\sigma5}_{c-}(x, y) + if_{abc}\mathscr{F}^{\sigma5}_{c+}(x, y)\}\,, \\ V^a_\mu(x)\,A^b_\nu(y) = A^a_\mu(y)\,V^b_\nu(y) = \{\mathscr{F}^\sigma \leftrightarrow \mathscr{F}^{\sigma5}\}\,, \end{cases}$$

where $z = x - y$ and the bilocal operators $\mathscr{F}^\sigma_{c\pm}$ are given by

$$(15)\quad \begin{cases} \mathscr{F}^\sigma_{c\pm}(x, y) = :\bar{\psi}(x)\gamma^\sigma\lambda^c\psi(y): \pm :\bar{\psi}(x)\gamma^\sigma\lambda^c\psi(g):\,, \\ \mathscr{F}^\sigma_{c\pm}(x, y) = :\bar{\psi}(x)\gamma^\sigma\gamma_5\lambda^c\psi(y): \pm :\bar{\psi}(x)\gamma^\sigma\gamma_5\lambda^c\psi(y):\,. \end{cases}$$

It is possible to abstract from free-field theory the form of this relation between the singular part of a current product and vector bilocal operators, and assume it to be true for the physical currents [7]. One would then renounce any detailed information on the matrix elements of the bilocal operators, and would retain only the assumption that their matrix elements are finite and that they have the various symmetry properties indicated by eq. (15). This is in the spirit of Gell-Mann's dictum that Nature reads the book of free-field theory in order to establish algebraic relations between operators, though naturally not to determine the value of specific matrix elements of these same operators.

It is a simple matter to extract the consequences of eq. (14) for the scaling limit of the inelastic scattering structure functions. If, as usual, we consider scattering from an unpolarized target, everything is determined by the spin-averaged single-particle matrix elements of the $\mathscr{F}^\sigma_{c\pm}$. Translation invariance evidently implies that

$$\langle p|\mathscr{F}^\sigma_{c\pm}(x, y)|p\rangle = \langle p|\mathscr{F}^\sigma_{c\pm}(x - y, 0)|p\rangle\,,$$

while Lorentz invariance then implies that

$$\langle p|\mathscr{F}^\sigma_{c\pm}(x, 0)|p\rangle = F^c_\pm(p\cdot x), x^2)\,p^\sigma + G^c_\pm(p\cdot x), x^2)\,x^\sigma\,,$$
$$\langle p|\mathscr{F}^{\sigma5}_{c\pm}(x, 0)|p\rangle = 0\,,$$

(the matrix element $\mathscr{F}^{\sigma5}$ would be nonzero if we did not spin-average). Because we are interested in the leading light-cone behavior, the functions G and F

may be evaluated at $x^2=0$. Actually, one can see by direct calculation that the function G does not contribute to the leading light-cone behavior. The contribution of the functions $F^c_\pm$ to the leading singularity of single-particle matrix elements of current products is then

$$\langle p|V^a_\mu(x)V^b_\nu(0)|p\rangle = \frac{-2i}{(x^2-i\varepsilon x_0)^2}(p_\mu x_\nu + p_\nu x_\mu - g_{\mu\nu}x\cdot p)\big(d_{abc}f^c_+(p\cdot x)+if_{abc}f^c_-(p\cdot x)\big),$$

$$\langle p|V^a_\mu(x)A^b_\nu(0)|p\rangle = \frac{2}{(x^2-i\varepsilon x_0)^2}\varepsilon_{\mu\nu\lambda\sigma}x^\lambda p^\sigma\big(d_{abc}f^c_-(p\cdot x)+if_{abc}f^c_+(p\cdot x)\big),$$

where

$$f^c_\pm(p\cdot x) = F^c_\pm(p\cdot x, 0).$$

It is worth noting that to the leading degree of singularity these expressions are automatically conserved, without requiring any equations of motion for the bilocal operator.

When we use these expressions to calculate the scaling limit of the structure functions we obtain, first of all. the desired Bjorken scaling

$$\begin{aligned} W_1(q^2,\nu) &\to F_1(\omega),\\ \nu W_2(q^2,\nu) &\to F_2(\omega),\\ \nu W_3(q^2,\nu) &\to F_3(\omega),\\ \nu W_{4,5} &\to 0. \end{aligned}$$

(The vanishing of the scaling limit of the structure functions W_4 and W_5 is a consequence of asymptotic current conservation.) We also obtain various relations between the surviving structure functions

$$(16)\qquad \left\{\begin{aligned} &F^{ab}_1(\bar\omega) = \frac{1}{2\bar\omega}F^{ab}_2(\bar\omega),\\ &(F^{ab}_1(\bar\omega))_{VV} = (F^{ab}_1(\bar\omega))_{AA},\\ &F^{ab}_1(\bar\omega)\pm F^{ab}_3(\bar\omega) = (d_{abc}\pm if_{abc})F^c_\pm(\bar\omega), \end{aligned}\right.$$

where

$$F^c_\pm(\bar\omega) = f^c_+(\bar\omega)\pm f^c_-(\bar\omega).$$

The first relation is just the statement that the cross-section for absorbing longitudinally polarized photons is zero in the Bjorken limit. The second and third equations imply the existence of sum rules between the cross-sections for different reactions described by current products of the type under discussion (ep, en, νp, νn for example). These sum rules are the consequence of the specific Lorentz and $SU_3 \times SU_3$ structure assumed in eq. (14) and are independent of the values of the matrix elements of the bilocal operators. In fact, the set of relations given in eq. (16) are identical in content to the general quark parton model, and we may regard the light-cone algebra as an explicit operator realization of the essential physics of the parton model.

5. – Conclusions.

It is apparent from this discussion that the study of the physics of the light-cone is in a very unfinished state. In the first place, the experimental results bearing on the question of scaling are very limited: so far we know only that electron-proton and electron-neutron inelastic structure functions scale and that for the former the longitudinal cross-section is very small. For a serious check of the scheme we have discussed, we must also know the structure functions for the weak-interaction analogs of inelastic electron scattering. These processes will be measured sometime within the next few years, but for the moment we know nearly nothing about them. It should also be emphasized that an absolute prediction of this scheme is the vanishing of the longitudinal cross-section in the scaling region. This can be checked directly in electroproduction experiments and the evidence is not yet overwhelming that the longitudinal cross-section does vanish.

Another area where much remains to be done is the extension of scale invariance or light-cone arguments to processes other than those of the deep inelastic electroproduction type. One possibility is to study semi-inclusive reactions, such as $\gamma + p \to \pi + X$. Some suggestions have been made as to how the hypotheses of Sect. **4** might be extended to such reactions [9], but the subject has not yet received the attention it deserves. A particularly interesting possibility is that crucial tests of the basic hypotheses (vanishing of longitudinal cross-sections etc.) may well be experimentally more accessible in the framework of semi-inclusive processes than in totally inelastic processes.

Finally, we have the problem of achieving some deeper theoretical understanding. Although we see why the concept of dimension makes asymptotic sense, we have no way of seeing why these dimensions should be canonical. Even granting canonical dimensions, we do not know why the singularity structure of current products should be the same as in free-field theory (of course, we do not know for sure that experiment is consistent with this simple hypothesis). Indeed, it seems that in perturbation theory the simple free-field structure is drastically modified by interactions. Of course, nature may ignore the complications which arise in renormalized perturbation theory, but it must be borne in mind that in the question of PCAC anomalies [10], nature seems to do just the opposite. It is to be hoped that we will soon have some information on these questions.

REFERENCES

[1] J. D. BJORKEN: *Phys. Rev.*, **179**, 1547 (1969).
[2] B. L. IOFFE: *Proceedings of the International Seminar on Vector Mesons and Electromagnetic Interactions* (Dubna, 1969).
[3] K. WILSON: *Phys. Rev.*, **179**, 1499 (1969).
[4] R. BRANDT and G. PREPARATA: CERN preprint TH 1208 (1970).
[5] D. GROSS and S. B. TREIMAN: *Light-cone commutators in gluon models*, Princeton preprint, May 1971.
[6] J. MANDULA, A. SCHWIMMER, J. WEYERS and G. ZWEIG: *Measuring chiral symmetry breaking in neutrino scattering*, Caltech preprint CALT-68-302 (971).
[7] H. FRITZSCH and M. GELL-MANN: *Scale invariance and the light-cone*, Caltech preprint CALT 68-297 (1971).
[8] Y. FRISHMAN: *Phys. Rev. Lett.*, **25**, 966 (1970).
[9] J. ELLIS: CERN preprint TH 1681 (1971).
[10] S. L. ADLER: *Phys. Rev.*, **177**, 2426 (1969); J. S. BELL and R. JACKIW: *Nuovo Cimento*, **60** A, 47 (1969).

Scaling Anomalies.

S. COLEMAN (*)

Lyman Laboratory of Physics, Harvard University - Cambridge, Mass.

1. – Introduction.

This lecture forms an appendix to Callan's series of lectures. In his lectures, CALLAN developed the formal theory of broken scale-invariance; the aim of this lecture is to report on some investigations into how well this theory does in the theorist's laboratory of renormalized perturbation theory. We shall see that the theory encounters severe difficulties; the Ward identities that follow from the formal theory of broken scale-invariance are afflicted with anomalies. However, at least in some cases, it is possible to express these anomalies in simple and compact form; that is to say, it is possible to find the true equations that replace the false equations deduced from the formal theory. Also, in favorable circumstances, some results of the formal theory (in particular, statements about the asymptotic behaviour of Green's functions) can also be derived, in somewhat modified form, from the corrected theory.

Section **2** recapitules that part of the formal theory of broken scale-invariance that will be necessary for our investigation. Section **3** is a demonstration that the formal theory is wrong in renormalized perturbation theory, in particular, that the low-energy theorems of broken scale-invariance contain anomalies. Section **4** is a first look at the anomalies and an introduction to the concept of anomalous dimensions. Section **5** is a derivation of the true equations that replace the false low-energy theorems, the Callan-Symanzik equations and the renormalization group equations of GELL-MANN and LOW. Section **7** gives an argument due to WILSON that shows that under favorable circumstances, despite the Callan-Symanzik anomalies (indeed, because of them) simple scaling behaviour (with anomalous dimensions) is regained in an appropriate asymptotic limit. Section **8** is a summary of our main results.

(*) Work supported in part by the Air Force Office of Scientific Research Under Contract F44620-70-C-0030 and by the National Science Foundation Under Grant GP-30819X.

Most of what I know about this subject is the result of conversations with C. CALLAN, R. JACKIW, K. SYMANZIK and K. WILSON. I would like to express both my gratitude to them and my hope that I have not distorted their ideas too badly.

2. – Some consequences of the formal theory of broken scale-invariance [1].

I will restrict myself mainly to the simplest renormalizable field theory, the theory of a self-interacting meson field [2]

$$\mathscr{L} = \frac{1}{2}\partial_\mu\varphi\,\partial^\mu\varphi - \frac{1}{2}\mu_0^2\varphi^2 - \frac{\lambda_0}{4!}\varphi^4 \tag{2.1}$$

although occasionally I will say something about the slightly more complicated theory of a meson and a nucleon with Yukawa interaction

$$\mathscr{L} = \frac{1}{2}\partial_\mu\varphi\partial^\mu\varphi - \frac{1}{2}\mu_0^2\varphi^2 + \psi(i\partial_\mu\gamma^\mu - m)\psi + g_0\bar{\psi}\gamma_5\psi\varphi - \frac{\lambda_0}{4!}\varphi^4\,. \tag{2.2}$$

These restrictions are just for the sake of notational simplicity; nearly everything I will say can readily be extended to a general renormalizable field theory.

Scale transformations, or dilatations, are transformations of the form

$$\varphi(x) \to \exp[d\alpha]\varphi(\exp[\alpha]x)\,, \tag{2.3}$$

where α is a real number, and d is a constant called the scale dimension of the field. The associated infinitesimal transformation is

$$\left.\frac{\mathrm{d}\varphi}{\mathrm{d}\alpha}\right|_{\alpha=0} = (d + x^\lambda\partial_\lambda)\varphi\,. \tag{2.4}$$

If we set the bare-mass term in (2.1) equal to zero, then these transformations are invariances of the theory defined by (2.1), if d is chosen to be one. This implies that we can find a current associated with the infinitesimal transformation (2.4), which we call the scale current and denote by s_μ, such that

$$\partial^\mu s_\mu = 0\,. \tag{2.5}$$

Equivalently, if we define the scale charge, D, by

$$D(t) = \int \mathrm{d}^3\boldsymbol{x}\, s_0(\boldsymbol{x}, t)\,, \tag{2.6}$$

then

$$\mathrm{d}D/\mathrm{d}t = 0\,. \tag{2.7}$$

The scale charge is the generator of infinitesimal scale transformations in the usual sense of quantum mechanics

$$i[D(t), \varphi(\boldsymbol{x}, t)] = (d + x_\lambda \partial^\lambda)\varphi \,. \tag{2.8}$$

If the bare mass is not zero, (2.1) is no longer scale-invariant; however, the scale current may still be defined, and eq. (2.5) is replaced by

$$\partial_\mu s^\mu \equiv \Delta = \mu_0^2 \varphi^2 \,. \tag{2.9}$$

This is the local version of the statement that only the bare mass breaks scale invariance. The scale charge is, of course, no longer time-independent; however, eq. (2.8) remains valid, but only as an equal-time commutator. All of this is readily extendible to the meson-nucleon theory defined by (2.2). The nucleon field transforms in the same way as the meson field, except that its scale dimension is $\frac{3}{2}$. Likewise, eq. (2.9) is replaced by

$$\partial^\mu s_\mu \equiv \Delta = \mu_0^2 \varphi^2 + m_0 \bar{\psi}\psi \,. \tag{2.10}$$

You have probably noticed that the numbers $\frac{3}{2}$ and 1, which occur in the transformation laws for the fields, are just the dimensions of the fields, in the sense of ordinary dimensional analysis. You may be wondering, therefore, whether what we have been doing is just dimensional analysis, disguised by a fancy formalism. If this is the case, you are wrong: The transformations of dimensional analysis not only scale the dynamical variables of a physical theory (in our case, the fields), they also scale all nondimensionless numerical parameters (in our case, the masses). Phrased somewhat more abstractly, the transformations of dimensional analysis turn one physical theory into another, different theory (*e.g.*, with different masses), and are always exact symmetries—given the exact solutions to the first theory, they yield the exact solutions to the second. Scale transformations, as we have defined them, are very different animals: They do not change numerical parameters—that is to say, they stay within a given physical theory—and they are not exact symmetries, except in special cases (vanishing masses).

It is to emphasize this difference that we call the numbers that occur in field transformation laws like (2.3) and (2.4), « scale dimensions ». Since, at the current stage of our investigation, they appear to be identical with the dimensions of the fields in the sense of dimensional analysis, this may seem mere nitpicking. However, I assure you that the distinction will be important in the future.

This whole structure is very much like that which arises for field theories that are invariant (or almost invariant) under chiral transformations, such as the sigma model. There also we have currents (the axial currents) with known divergences (the pion fields) and whose charges have known equal-time commutators with the fields. In the chiral case, there is s standard procedure for turning operator statements like these into constraints on the Green's functions of the theory, Ward identities. Among the most useful consequences of these Ward identities are the so-called low-energy theorems, statements that relate the matrix element of a current divergence (pion field) at zero momentum transfer to a chirally transformed Green's function without the pion field. This whole machinery can be carried over without alteration to the case of broken scale-invariance; I will now write down the low-energy theorems that can be deduced in this way for the Lagrangian (2.2).

Let us denote the one-particle irreducible [3] renormalized Green's function with n external lines by

$$\Gamma^{(n)}(p_1 \dots p_n) ,$$

where the p's are the momenta carried by the external lines, oriented so that they all go inward. (The p's are, of course, not all independent; their sum must be zero.) Likewise, let us denote the one-particle irreducible Green's function with n external meson lines and one insertion of Δ, the divergence of the scale current (in this case, $\mu_0^2\varphi^2$), by

$$\Gamma_\Delta^{(n)}(k; p_1 \dots p_n) ,$$

where the p's are defined as before, and k is the momentum carried by the insertion. Then the low-energy theorems are

$$\left(\sum_{r=1}^{n-1} p_r \cdot \frac{\partial}{\partial p_r} + nd - 4\right) \Gamma^{(n)}(p_1 \dots p_n) = -i\Gamma_\Delta^{(n)}(0; p_1 \dots p_n) , \tag{2.11}$$

where d is the scale dimension of the field. Of course, d is one, but I would like to suppress this information momentarily, both to make it easy for you to see the generalization to Fermi fields, for which d is $\frac{3}{2}$, and for another reason, which will become clear shortly.

Equation (2.11) is fairly easy to understand. The first two terms on the left are just the transformation law of the fields, written in momentum space; the sum only runs over $n-1$ momenta because only $n-1$ of the momenta are independent. The four appears because in passing from the Fourier transform of a time-ordered product to the conventionally defined $\Gamma^{(n)}$, we must factor out a four-dimensional δ-function, which leaves the four behind as it passes through the differential operator. The right-hand side of the equation is just the matrix element of the current divergence at zero momentum transfer, the analog to the pion field in the chiral case.

We can write (2.11) in another form, which will be useful in the sequel. Let us trade the p's for a set of variables consisting of

$$s = \sum_{r=l}^{n} p_r^2 \tag{2.12}$$

and the dimensionless kinematic variables $p_i \cdot p_j/s$. Then ordinary dimensional analysis tells us that $\Gamma^{(n)}$ is of the form

$$\Gamma^n = s^{(4-n)/2} F^{(n)}\left(\frac{s}{\mu^2}, \lambda, \frac{p_i \cdot p_j}{s}\right), \tag{2.13}$$

where μ is the renormalized mass, and λ is the renormalized coupling constant. (Remember, λ is dimensionless.) From this it is easy to see that (2.11) can equivalently be written as

$$\left[\mu \frac{\partial}{\partial \mu} + n(1-d)\right] \Gamma^{(n)}(p_1 \dots p_n) = i\Gamma_\Delta^{(n)}(0; p_1 \dots p_n) . \tag{2.14}$$

3. – A disaster in the deep Euclidean region.

The Euclidean region is that region in multiparticle momentum space in which all four-momenta are Euclidean; that is to say, they all have real space parts and imaginary time parts. The deep Euclidean region is that part of the Euclidean region in which the magnitude of s (defined by eq. (2.12)) gets very large, while the dimensionless variables $p_i \cdot p_j/s$ stay fixed; furthermore, no partial sum of the p's is zero. The deep Euclidean region is the maximally unphysical limit in which to study Green's functions: all external lines are far off the mass shell, and, furthermore, no matter how the diagram is cut in two, the momentum transferred between the two halves is far off the mass shell.

There are famous bounds on the behavior of Feynman amplitudes in the deep Euclidean region, first established by WEINBERG [4]. For the functions that appear in eq. (2.14), these Weinberg bounds say that $\Gamma^{(n)}$ grows no faster than $s^{(4-n)/2}$, times a polynomial in $\ln(s/\mu^2)$, to any finite order in renormalized perturbation theory. (The coefficients in the polynomial are, in general, functions of the coupling constant and of the dimensionless variables $p_i \cdot p_j/s$. Also, the order of the polynomial grows with the order of perturbation theory.) Likewise, $\Gamma_\Delta^{(n)}$ grows no faster than $s^{(2-n)/2}$, again times a polynomial in $\ln(s/\mu^2)$. Crudely, the reason why $\Gamma_\Delta^{(n)}$ grows less rapidly than $\Gamma^{(n)}$ is that, in the deep Euclidean region, all internal momenta are getting large; adding a mass insertion adds an internal propagator, which knocks out one power of s.

We can now combine this with the broken scale-invariance low energy theorems, eq. (2.14), to get a much more powerful statement about the asymptotic behaviour of the Green's functions than is given by the Weinberg bounds alone. For the Weinberg bounds tell us that in the deep Euclidean region, the right-hand side of (2.14) is negligible compared to the individual terms on the left-hand side. Thus, we can neglect it, and obtain an equation for the asymptotic form of $\Gamma^{(n)}$

$$\left[\mu\frac{\partial}{\partial\mu}+n(1-d)\right]\Gamma^{(n)}_{\text{as}}=0\,, \tag{3.1}$$

where the subscript indicates the asymptotic form in the deep Euclidean region. Or, since d is one

$$\mu\frac{\partial}{\partial\mu}\Gamma^{(n)}_{\text{as}}=0\,. \tag{3.2}$$

That is to say, there are no logarithmic factors in every order of renormalized perturbation theory. This is indeed a powerful statement; unfortunately, it is also a false one; anyone who has ever done any Feynman calculation involving a closed loop knows that the logarithms are in fact present. Therefore, eq. (2.14), from which we deduced the false statement (3.2), must itself be false. In the current technical language the Ward identities of broken scale-invariance must contain anomalies. Phrased more straightforwardly, *the entire theoretical structure of Sect.* **2** *is a lie*!

I want to make the logic of this argument clear. I am not saying that the formal theory of broken scale-invariance is wrong because it makes asymptotic predictions that differ from the asymptotic behavior found in perturbation theory; only a madman would take the asymptotic behavior of perturbation theory so seriously [5]. I need only assume that perturbation theory properly gives the successive derivatives of the Green's functions with respect to the coupling constant at zero coupling constant. This seems to me to be a very weak and extremely reasonable assumption, at least for renormalizable field theories. Under this assumption if eq. (2.14) is generally valid, it must be true order by order in perturbation theory. If the right- and left-hand sides of this equation are equal in a fixed order in perturbation theory, they must be equal in the deep Euclidean limit. They are not.

The technical reason for the occurrence of the anomalies is not difficult to understand. The formal canonical manipulations required to prove Ward identities are justified only if we introduce a cut-off to remove the divergences from the theory. However, this does us no good unless the cut-off is chosen in such a way that the cut-off theory still obeys the Ward identities. For such familiar cases as quantum electrodynamics or the sigma model, for example, this condition presents no difficulties; it is easy to introduce a cut-off in such a way that the relevant equations (gauge invariance in one case and PCAC and current algebra in the other) remain true. For scale invariance, though, the situation is hopeless; any cut-off procedure necessarily involves a large mass, and a large mass necessarily breaks scale invariance in a large way. This argument does not show that the occurrence of anomalies is inevitable, but it does show that there is no reason to believe it is impossible.

4. – Anomalous dimensions and other anomalies.

Last year, R. JACKIW and I got interested in these anomalies and decided to get some information about them by the most simple-minded method imaginable. In meson-nucleon theory (the Lagrangian (2.12)), we simply computed separately the right- and left-hand sides of the low-energy theorem (2.14), to lowest nontrivial order in perturbation theory, order g^2, to see how they differed. We found [6] that, to this order, all anomalies could be absorbed in a change in the scale dimension of the fields. In particular, for the meson field, eq. (2.14) is changed from a falsehood to a truth if we replace the naive value of d (one) by

$$d = 1 + \frac{g^2}{8\pi^2} \,. \tag{4.1}$$

For the self-energy operator, for example, the solution to eq. (3.1) becomes

$$\Gamma^{(2)}_{\text{as}} \propto s(s/\mu^2)^{1-d} \,. \tag{4.2}$$

Expanding the exponent and discarding terms of order g^4 and higher, we find

$$\Gamma^{(2)}_{\text{as}} \propto s - \frac{g^2}{8\pi^2}\, s \ln(s/\mu^2) \,. \tag{4.3}$$

This is the correct asymptotic behavior, to this order in perturbation theory. Similar remarks apply to the nucleon field; here d must be changed from its naive value ($\frac{3}{2}$) to

$$d = 3/2 + \frac{g^2}{32\pi^2} \,. \tag{4.4}$$

(As it happens, these changes in d not only fix up (3.7), but also take care of all the anomalies in the Ward identities for arbitrary momentum transfers.)

This phenomenon is frequently described by saying that the fields acquire anomalous dimensions. This is a slightly misleading way of putting things, for its tempts us to confuse scale dimensions and dimensions in the sense of dimensional analysis, and to think that something counter to common sense has occurred. It is only the scale dimensions of the fields that have changed as a result of the interactions; the dimensions in the sense of dimensional analysis remain firmly at one and $\frac{3}{2}$. I emphasize again that there is no logical connection between these two sets of numbers; the (discredited) analysis of Sect. **2** is the only thing that ever led us to believe they were equal.

It would be very pleasant if this phenomenon persisted in higher orders of perturbation theory, if all anomalies could be absorbed into a redefinition of the scale dimension. If this were the case, then we would have a correct theory of broken scale-invariance with almost as much predictive power as the false theory of Sect. **2**; the only price would be the introduction of a new, dynamically determined parameter for each field in the theory, its anomalous dimension.

Unfortunately, this is not what happens. To see this, let us return to the theory of a self-interacting meson field, and assume that eq. (3.1) is valid in all orders of perturbation theory, except that d is not one. Then it is easy to see that

$$\mu \frac{\partial}{\partial \mu} [\Gamma^{(4)}_{\text{as}}/(\Gamma^{(2)}_{\text{as}})^2] = 0 . \tag{4.5}$$

That is to say, although both the numerator and denominator of this expression have factors of $\log(s/\mu^2)$ in their asymptotic expansions, the logarithms

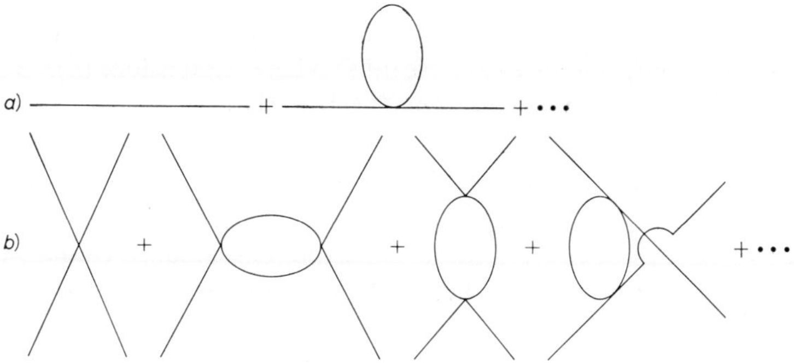

Fig. 7.

cancel in the ratio, order by order in perturbation theory. Let us check this prediction by calculating the ratio to order λ^2. The relevant diagrams for $\Gamma^{(2)}$ are shown in Fig. 1 *a*); the second diagram is just a constant, cancelled by mass renormalization. Thus, we obtain, in the deep Euclidean region

$$\Gamma^{(2)}_{\text{as}} \propto s + O(\lambda^2) . \tag{4.6}$$

Figure 1 *b*) shows the relevant diagrams for $\Gamma^{(4)}$. Every child knows that the three second-order diagrams grow logarithmically for high s; thus we obtain

$$\Gamma^{(4)}_{\text{as}} \propto \lambda + a\lambda^2 \ln(s/\mu^2) + b\lambda^2 , \tag{4.7}$$

where a and b are constants. (Actually, b is a function of the dimensionless kinematic variables, but, for our purposes, that's as good as being a constant.)

Putting this together we find

$$\Gamma^{(4)}_{\text{as}}/(\Gamma^{(2)}_{\text{as}})^2 \propto \lambda s^{-2} + a\lambda^2 s^{-2} \ln(s/\mu^2) + b\lambda^2 s^{-2} + O(\lambda^3)\,. \tag{4.8}$$

This is in contradiction to our prediction; thus there must be further anomalies in addition to anomalous dimensions.

5. – The last anomalies: the Callan-Symanzik equations.

We are lost in a dark wood. Whenever we look at a more complicated situation, investigate higher orders of perturbation theory, new anomalies appear. Happily, there exists a way to order this chaos, discovered by CALLAN and SYMANZIK [7], who independently found the true equations that replace the false low-energy theorems (2.14). These equations are surprisingly compact. For the theory of a self-interacting meson field, for example, they involve only two dynamically determined parameters. This is only one more than there would be if the only anomalies were anomalous dimensions [8]. I will now derive the Callan-Symanzik equations for this theory.

The simplest place to begin is in unrenormalized perturbation theory: here one computes the unrenormalized Green's functions (which I will distinguish by a subscript u) in a power series in the bare charge, with the bare mass held fixed. I remind you of the relation between the renormalized Green's functions, which we have been using until now, and the unrenormalized ones

$$\Gamma^{(n)}(p_1 \dots p_n) = (Z_3)^{n/2}\Gamma^{(n)}_u(p_1 \dots p_n)\,, \tag{5.1}$$

where Z_3 is the wave-function renormalization constant. Of course, unlike renormalized perturbation theory, unrenormalized perturbation theory is full of divergences. To control these, we will imagine that we have put a cut-off in the theory in some way, say by modifying the Feynman propagator in the standard manner. Once we re-express things in terms of the physical mass and coupling constant, and perform the multiplicative renormalization (3.18), everything is cut-off-independent in the limit of high cut-off; this is just the statement that (2.1) is a renormalizable Lagrangian. I will denote the unrenormalized Green's function with one mass ($\mu_0^2\varphi^2$) insertion by $\Gamma^{(n)}_{u\Delta}(k;p_1 \dots p_n)$. Then it is easy to see, just by looking at diagrams, that

$$i\Gamma^{(n)}_{u\Delta}(0;p_1 \dots p_n) = \mu_0 \frac{\partial}{\partial\mu_0}\Gamma^{(n)}_u(p_1 \dots p_n)\,. \tag{5.2}$$

This is the unrenormalized version of the low-energy theorem (2.14). As it stands, it is useless, since the differential operator cannot be turned into a scaling operator; the unrenormalized Green's functions depend on another parameter besides μ_0 with the dimensions of a mass, the cut-off.

Simple power counting shows that if we define

$$\Gamma^{(n)}_{\Delta}(k; p_1 \dots p_n) = Z(Z_3)^{n/2}\Gamma^{(n)}_{u\Delta}(k; p_1 \dots p_n), \tag{5.3}$$

then, for an appropriate (cut-off-dependent) choice of the constant Z, we can make the left-hand side of (5.3) cut-off-independent in the limit of high cut-off. This condition leaves Z undetermined up to a finite (cut-off-independent) multiplicative factor. We will choose this factor later, to make our final equations look as simple as possible.

Putting all this together, we find the consequence of (5.2) for the renormalized Green's functions

$$i\Gamma^{(n)}_{\Delta}(0; \dots) = Z\mu_0 \frac{\partial}{\partial\mu_0}\Gamma^{(n)}(\dots) - \frac{n}{2} Z\mu_0 \frac{\partial \ln Z_3}{\partial\mu_0}\Gamma^{(n)}(\dots). \tag{5.4}$$

We are not yet finished, since we have the derivative of a renormalized Green's function with respect to a bare mass, and a renormalized Green's function is cut-off-independent only when expressed in terms of renormalized masses and coupling constants. This is easily taken care of with the aid of the chain rule of differentiation

$$\left[\left(Z\mu_0 \frac{\partial\mu}{\partial\mu_0}\right)\frac{\partial}{\partial\mu} + \left(Z\mu_0 \frac{\partial\lambda}{\partial\mu_0}\right)\frac{\partial}{\partial\lambda} - \frac{n}{2}\left(Z\mu_0 \frac{\partial \ln Z_3}{\partial\mu_0}\right)\right]\cdot\Gamma^{(n)}(\dots) = i\Gamma^{(n)}_{\Delta}(0; \dots), \tag{5.5}$$

where all derivatives are taken with fixed cut-off. The only terms in these equations which retain a possible trace of cut-off-dependence are the three terms in parentheses. These, however, are constants, independent of n and the momenta. Therefore, it is easy to show, by evaluating the equations at three independent points and solving for these terms as functions of cut-off-independent quantities, thay thet are also cut-off-independent, in the limit of high cut-off. (An especially simple choice is the three points at which the renormalized mass, renormalized charge, and scale of the renormalized field are defined.)

We now choose Z (until now undetermined up to a finite multiplicative constant) such that

$$Z\mu_0 \frac{\partial\mu}{\partial\mu_0} = \mu, \tag{5.6}$$

and define

$$\beta = Z\mu_0 \frac{\partial\lambda}{\partial\mu_0} \tag{5.7}$$

and

$$\gamma = \frac{1}{2} Z\mu_0 \frac{\partial \ln Z_3}{\partial \mu_0} . \tag{5.8}$$

(Dimensional analysis shows that these functions can only depend on λ.) We thus obtain

$$\left[\mu \frac{\partial}{\partial \mu} + \beta(\lambda) \frac{\partial}{\partial \lambda} - n\gamma(\lambda)\right] \Gamma^{(n)}(...) = i\Gamma_{\Delta}^{(n)}(0; ...) . \tag{5.9}$$

These are the Callan-Symanzik equations.

Some remarks should be made about these equations:

1) As promised, these are the equations that replace the naive low-energy theorem (2.14). If β were zero (5.9) would be identical to (2.14), except that the scale dimension, d, would be anomalous:

$$d = 1 + \gamma . \tag{5.10}$$

Unfortunately, as we have already seen, β is not zero; there are further anomalies beyond anomalous dimensions. A direct computation [5] shows that

$$\beta = \frac{3\lambda^2}{16\pi^2} + O(\lambda^3) . \tag{5.11}$$

(It will be important in the sequel that this first nonzero term in the perturbation expansion of β is positive.)

2) The generalization to more complicated field theories is obvious: For every dimensionless coupling constant, there is a β-like term, and for every field, there is a γ-like term. Thus, for example, in meson-nucleon theory, if we denote a Green's function with n nucleon fields, n antinucleon fields, and r meson fields, by $\Gamma^{(n,n,r)}$, the Callan-Symanzik equations are

$$\left[\mu \frac{\partial}{\partial \mu} + m \frac{\partial}{\partial m} + \beta_1 \frac{\partial}{\partial \lambda} + \beta_2 \frac{\partial}{\partial g} - 2n\gamma_1 - r\gamma_2\right] \Gamma^{(n,n,r)}(...) = i\Gamma_{\Delta}^{(n,n,r)}(0; ...) ,$$

where the β's and the γ's are functions of λ, g and μ/m.

3) Likewise, if we attempt to study Green's functions for objects other than canonical fields (for example, conserved or partially-conserved currents), this will, in general, change the structure of the γ-terms, which refer to a particular field. The β-terms, though, which do not make reference to a particular choice of fields, but only to the underlying dynamics, will remain unchanged.

4) Unlike the low-energy theorems of current algebra, which they so closely resemble, the Callan-Symanzik equations are practically useless for

low-energy phenomenology [9]. It is the β-terms that make the difference. A current-algebra low-energy theorem is useful because it expresses one Green's function in terms of another in a way that does not depend on strong-interaction dynamics. The Callan-Symanzik equations, on the other hand, express one Green's function in terms of another *and* its derivatives with respect to coupling constants. If you know how to compute *these*, you have already solved the strong-interaction dynamics, and there is no reason for you to be piddling around with low-energy theorems.

5) Finally, there remains the possibility that β has a zero, and that, for some reason (the bootstrap?) the real value of λ is at this zero. (This is a speculation of WILSON, among others.) If this is the case, we would regain the naive theory of scale invariance (with anomalous dimensions). There is very little to be said for or against this possibility from a study of perturbation theory alone.

6. – The renormalization group equations and their solution.

By the arguments we used in Sect. **3**, in the deep Euclidean region, we can neglect the right-hand side of the Callan-Symanzik equations. Thus, for the case of a self-interacting meson field, we obtain the following differential equations for the asymptotic forms of the Green's functions:

$$\left[\mu\frac{\partial}{\partial\mu}+\beta(\lambda)\frac{\partial}{\partial\lambda}-n\gamma(\lambda)\right]\Gamma^{(n)}_{\text{as}}=0\,. \tag{6.1}$$

These equations are old friends; they are the equations associated with the so-called renormalization group, first devised by GELL-MANN and LOW to study quantum electrodynamics, and later extended to general renormalizable field theories [10].

The renormalization group equations (6.1) can be derived much more simply than the Callan-Symanzik equations. As I stated earlier, the deep Euclidean region is the maximally unphysical region: all external lines and all momentum transfers are far from the mass shell. One's first thought would be that in this regime all memory of the actual meson mass would be lost, and all Green's functions would be independent of the mass. However, this is too naive. The objects we are studying are Green's functions for renormalized fields, expressed as functions of the renormalized coupling constant. Both the normalization of the field and the value of the renormalized coupling constant are defined on the mass shell; these quantities remember the mass shell no matter how far we flee into the deep Euclidean region. Therefore, the correct statement is that all memory of the actual value of the mass is lost, except for that which is contained in the scale of the fields and the value of λ. In other words, in the deep Euclidean region, a small change in the mass can always be compensated for by an appropriate small change in λ and an appropriate small rescaling of the fields. The eqs. (6.1) are just the mathematical expression of this statement.

This argument seems to me to owe very little to asymptotic estimates derived from perturbation theory; therefore I will assume from now on that the renormalization group equations are valid independent of perturbation theory, and try to use them to get information about the asymptotic forms of Green's functions in a nonperturbative way.

The renormalization group equations are fairly easy to solve; they are linear partial differental equations in two variables, for which there exists a general method of solution [11]. The solution proceeds in two steps. First we define an auxillary function of two variables, $\lambda'(\lambda, t)$, which is the solution to the ordinary differential equation

$$\frac{d\lambda'(\lambda, t)}{dt} = \beta(\lambda') \tag{6.2}$$

with the boundary condition

$$\lambda'(\lambda, 0) = \lambda . \tag{6.3}$$

The general solution of (6.4) is then

$$\Gamma^{(n)}_{\text{as}} = s^{(4-n)/2} f^{(n)}\left(\lambda'\left(\lambda, \tfrac{1}{2}\ln\frac{s}{\mu^2}\right), \frac{p_i\cdot p_j}{s}\right) \exp\left[-n\int_0^{\frac{1}{2}\ln s/\mu^2} dt\gamma(\lambda'(\lambda, t))\right], \tag{6.4}$$

where $f^{(n)}$ is an arbitrary function. Here we have used dimensional analysis to explicitly restore the dependence on the variables that do not enter into (6.1), the large variable s and the dimensionless kinematic variables.

Just to check that we have not made a sign error, let us evaluate (6.4) under the assumption that we are actually at a zero of β, as conjectured at the end of Sect. **5**. In this case the solution of the ordinary differential equation is trivial, $\lambda' = \lambda$, and (6.4) reduces to

$$\Gamma^{(n)}_{\text{as}} = s^{(4-n)/2} \left(\frac{s}{\mu^2}\right)^{-n\gamma(\lambda)/2} f^{(n)}\left(\lambda, \frac{p_i\cdot p_j}{s}\right). \tag{6.5}$$

This is the correct answer in this special case: everything scales as it would naively, except that the scale dimension is anomalous.

7. – The return of scaling in the deep Euclidean region.

The final result of the last Section, eq. (6.5) evidently has a lot of content. It tells us that the asymptotic form of a Green's function in the deep Euclidean region, which *a priori* could depend in an arbitrary way on the dimensionless variables λ and $\ln s/\mu^2$, in fact, depends in an arbitrary way only on a certain function of these variables, λ', which is, in turn, completely determined in terms of the function $\beta(\lambda)$ by eqs. (6.3) and (6.4). This is useful information as it stands; for example, it has been used to enormously simplify the summation of leading logarithms in the deep Euclidean region. However, as was pointed out by WILSON [12], if we make some (apparently very mild) additional assumptions, we get a surprisingly greater amount of information.

These additional assumptions are: 1) The functions $f^{(n)}$ appearing in eq. (6.5) are continuous functions of the variable λ'. Likewise, β and γ are continuous functions of the variable λ. This is certainly mild and is trivially true in any finite order of perturbation theory, in which these functions are polynomials in these variables. 2) The function $\beta(\lambda)$ has a zero some place along the positive real axis. It is very difficult to say whether this is a plausible or implausible assumption from perturbation theory alone, since any finite order of perturbation theory merely tells us the behavior of β near the origin, and gives us no information about the presence of zeros away from the origin It is certainly much weaker than the conjecture we toyed with at the end of Sect. **5**, that β had a zero *and* that this zero was the physical coupling constant.

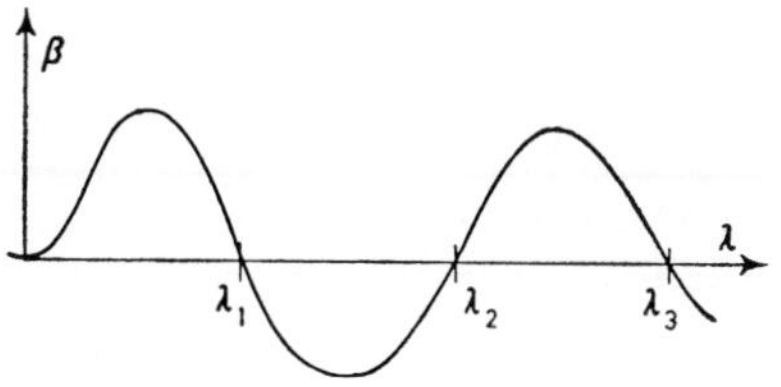

Fig. 2.

The assumed behavior of β is shown in Fig. 2. The quadratic zero at the origin is not part of the assumption; that can be established by direct perturbative computation (see eq. (5.11)). The zero at λ_1 is an assumption. The two subsequent zeroes at λ_2 and λ_3 are not necessary to the argument, but they lead to interesting results if they are assumed; I have put them in the Figure so I can talk about them when I have finished the main argument.

Now, let us assume that the physical value of λ lies between zero and λ_1. Then, when we integrate the differential equation (6.3), starting from the boundary condition (6.4), we will find that λ' is a monotonically increasing function of t, for β is positive in this region. Indeed, λ' stays monotonically increasing for all t, for its derivative can change sign only when λ' exceeds λ_1, which it cannot do because $\beta(\lambda_1)$ is zero. In other words, λ' asymptotically approaches λ_1 from below as t goes to infinity:

$$\lim_{t\to\infty} \lambda'(\lambda, t) = \lambda_1 . \tag{7.1}$$

This is the essential point. Under the stated assumptions, for a range of values of λ, λ' approaches a fixed limit as t goes to infinity. Furthermore, the value of this limit is independent of the value of λ; as t goes to infinity, λ' loses all memory of where it started.

Likewise, if λ is between λ_1 and λ_2, the next zero of β (although it is not essential to the argument such a second zero exists), λ' monotonically approaches λ_1 from above. Continuing along the real axis, λ_2 is an exceptional point; if λ equals λ_2, λ' also equals λ_2 for all t. If λ is between λ_2 and λ_3, λ' approaches λ_3 from below as t goes to infinity, etc.

Now let us return to the case $(0<\lambda<\lambda_2)$ in which the limiting value of λ is λ_1. Inserting this limit in (6.5), we obtain « the asymptotic form of the asymptotic form »

$$\lim_{s\to\infty} \Gamma^{(n)}_{\text{as}} = s^{(4-n)/2} f^{(n)}\left(\lambda_1, \frac{p_i\cdot p_j}{s}\right)\left(\frac{s}{\mu^2}\right)^{-n\gamma(\lambda_1)/2} K^n , \tag{7.2}$$

where the constant K is given by

$$K = \exp\left[\int_0^\infty \mathrm{d}t\left[\gamma(\lambda_1) - \gamma(\lambda'(\lambda, t))\right]\right] . \tag{7.3}$$

(We will later argue that this integral is convergent.) But this is just simple scaling behavior again! Indeed, it is even simpler scaling behavior than we thought we had back in Sect. **4** when we thought for awhile that the only anomalies were anomalous dimensions. For if that were the case, the anomalous dimensions, the exponents that would appear in the asymptotic form (7.2), would depend continuously on λ. In fact, as (7.2) shows, the anomalous dimensions that appear in the asymptotic form are *independent* of λ, for a range of λ $(0<\lambda<\lambda_2)$, as is the whole asymptotic form, aside from the constant K. The β anomalies, which complicate things terribly at low energies, simplify things enormously in the deep Euclidean region. What a wonderful reversal! (This praise of the β-terms is justified, of course, only if our critical assumption —the existence of a zero in β—is true. If it is not, the β-terms remain trouble-makers, at high energies as well as at low.)

If we are willing to assume that the relevant functions are differentiable as well as continuous, then we can obtain an estimate of the error in (7.2), for then we can see how rapidly λ approaches λ_1. In the neighborhood of λ_1, eq. (6.3) can be approximated by

$$\frac{\mathrm{d}\lambda'}{\mathrm{d}t} = a(\lambda_1 - \lambda') , \tag{7.4}$$

where a is a positive number

$$a = -\left.\frac{\mathrm{d}\beta}{\mathrm{d}\lambda}\right|_{\lambda_1} . \tag{7.5}$$

The solution to (4.12) is

$$\lambda' = \lambda_1 + ce^{-at}\,, \tag{7.6}$$

where c is a constant. This immediately implies the convergence of the integral (7.3). Also, after some straightforward differentiation, it leads to the statement that the error in (7.2) is of the order $(s/\mu^2)^{-a/2}$, compared to the terms retained. Without a knowledge of a, it is impossible to say how small this is, but at least it is a good solid power, not just some logarithms. (All of this presumes, of course, that the zero at λ_1 is a single zero, as shown in Fig. 2. It is were a double or higher-order zero, the terms neglected in (7.2) would just be down by logarithms compared to the terms retained, and the limit would be much less interesting.)

8 – Conclusions.

We have gone through a lot of complicated analysis, so perhaps I should end by summarizing our main results:

1) The formal theory of broken scale-invariance is a pack of lies, hopelessly afflicted with anomalies. In one case only (the low-energy theorems associated with broken scale-invariance), the true equations have been found that replace the false equations. These are the Callan-Symanzik equations. They are simple and elegant, but practically useless for doing low-energy phenomenology, because they involve derivatives with respect to coupling constants.

2) Under a very strong assumption (that the physical coupling constant is a zero of the function β that appears in the Callan-Symanzik equations), the Callan-Symanzik equations become equivalent to the naive low-energy theorems, except that the dimensions of fields are, in general, anomalous. Nothing is known about what happens, under this assumption, to the rest of the formal theory.

3) Under a much weaker assumption (the existence of a zero in β, not necessarily at the physical coupling constant), Green's functions show simple scaling behavior in the deep Euclidean region, with anomalous dimensions [13].

Obviously, there remains much that is not understood. In particular, the connection between the concepts we have been playing with the operator-product expansion on the light-cone (in less formal language, the observed scaling behavior of deep inelastic electroproduction) is almost completely obscure. Nevertheless, our understanding of scaling has increased at such a rapid rate in the last few years that I, for one, am optimistic about how much more we will learn in the near future.

REFERENCES

[1] This Section is a quick summary of results that will be useful in the rest of the lecture. For more details, and proofs, see: This Volume p. 264.

[2] Notation: Greek indices run from 0 to 3, with 0 denoting the time axis. Space-time points are labelled in three ways $x \equiv x^\mu \equiv (x^0, \boldsymbol{x})$. $\partial_\mu = \partial/\partial x^\mu$. The signature of the metric tensor is $(+ - - -)$.

[3] I remind you that a one-particle irreducible Green's function is the sum of all connected Feynman diagrams that cannot be cut in two by breaking a single internal line. By convention the diagrams are evaluated with no propagators on the external lines. Thus, $\Gamma^{(2)}$ is the inverse propagator; $\Gamma^{(3)}$ is the proper three-line vertex (zero in the theory (3.1), by parity), $\Gamma^{(4)}$ is the two-into-two off-mass-shell scattering amplitude, less the pole terms (which are absent anyway in (3.1)), etc.

[4] S. WEINBERG: *Phys. Rev.*, **118**, 838 (1960).

[5] An instructive example is a recent computation of T. APPELQUIST and J. PRIMACK: *Phys. Rev. D*, **4**, 2454 (1971). In vector-meson–nucleon theory, these authors isolate those diagrams which give the dominant asymptotic behavior for the nucleon electromagnetic form factor, in each order of perturbation theory (« the leading logarithms »). They then sum up these leading terms, and find an expression proportional to $\exp[-c \ln^2 s]$, where c is a positive constant, proportional to the square of the coupling constant. This goes to zero more rapidly than any powers of s; that is to say, the sum of the leading terms is highly damped, unlike the individual leading terms, which grow like powers of logarithms. Also, and even more discouraging, in the asymptotic region the sum of the leading terms is much less important than any one of the terms that were neglected (the nonleading terms) in the original summation.

[6] S. COLEMAN and R. JACKIW: *Ann. of Phys.*, **67**, 552 (1971).

[7] C. G. CALLAN: *Phys. Rev. D*, **2**, 1541 (1970); K. SYMANZIK: *Comm. Math. Phys.*, **18**, 227 (1970).

[8] Unfortunately, for the Ward identities at arbitrary momentum transfers the situation remains chaotic.

[9] One would attempt to apply the naive low-energy theorems (2.14) to low-energy phenomenology by saturating the matrix element of Δ with the contribution from a scalar, isoscalar meson. (The most popular candidate is the elusive ε, last seen somewhere between 700 and 1000 MeV.) See, for example, J. ELLIS: *Nucl. Phys.*, **22** B, 478 (1970).

[10] M. GELL-MANN and F. E. LOW: *Phys. Rev.*, **95**, 1300 (1954); N. N. BOGOLIUBOV and D. V. SHIRKOV: *Introduction to the Theory of Quantized Fields* (New York, 1959).

[11] See, *e.g.*, R. COURANT and D. HILBERT: *Methods of Mathematical Physics*, Vol. **2** (New York, 1937).

[12] K. WILSON: *Phys. Rev. D*, **3**, 1818 (1971). The immediate ancestors of Wilson's arguments can be found in the studies of quantum electrodynamics by M. BAKER, K. JOHNSON and R. WILLEY: *Phys. Rev.*, **136**, B 1111 (1964); **163**, 1699 (1967); **183**, 1292 (1969); *Phys. Rev. D*, **3**, 2516 (1971), and also in the original work on the renormalization group by GELL-MANN and LOW (ref. [10]); however, I believe that the application of these to strong-interaction scaling first appeared in Wilson's work.

[13] K. SYMANZIK: DESY Preprint, has recently shown that these same assumptions lead to scaling behaviour (with anomalous dimensions) for the coefficient functions in the short-diatance operator-product expansion.

The Light-Cone and Symmetry Breaking.

R. A. BRANDT (*)

CERN - Geneva
New York University, Physics Department - New York

1. – Introduction.

In this lecture I will review some work done in this past year by PREPARATA and myself [1,2] on an application of the canonical light-cone (LC) operator product-expansion formalism [3-6] we had previously developed. The question we addressed ourselves to was: what are the observable consequences of a given operator algebra? If one is given an exact operator algebra which does not correspond to an exact symmetry, then the physical states will not form representations of the algebra and so it is not straightforward to calculate measurable quantities of interest, such as matrix elements of the operators in question. I will try to show here how the use of LC techniques together with several dynamical assumptions enable one to deduce approximate relations between physical parameters from the exact operator relations embodied in current algebra and its generalization to the LC.

Consider, for example, current algebra. The (presumably) exact operator statements are the well-known Gell-Mann [7] equal-time $SU_3\times SU_3$ commutation relations among the vector V^a_μ and axial vector A^a_μ currents:

$$\delta(x_0)[V^a_0(x), V^b_0(0)] = if^{abc}V^c_0(0)\,\delta^4(x)\,, \tag{1.1a}$$

$$\delta(x_0)[V^a_0(x), A^b_0(0)] = if^{abc}A^c_0(0)\,\delta^4(x)\,, \tag{1.1b}$$

$$\delta(x_0)[A^a_0(x), A^b_0(0)] = if^{abc}V^c_0(0)\,\delta^4(x)\,. \tag{1.1c}$$

The consequences of this algebra are exact zero energy and mass theorems [8,9], the most famous being the Adler-Weisberger theorem

$$T^{(-)\prime}(\nu=0,\, q^2=0) = -\frac{2}{f_\pi^2}(g_A^2-1) \tag{1.2}$$

for the energy derivative of the isospin odd off-shell pion-nucleon scattering

(*) Alfred P. Sloan Foundation Fellow.

amplitude at zero energy ν and squared pion mass q^2. Thus current algebra gives $T^{(-)\prime}(\nu, q^2)$ at the unphysical mass $q^2 = 0$ and the unphysical energy $\nu = 0$. This would not be much of a problem if only the energy were unphysical since energy dispersion relations (DR's) can be used to calculate (in principle exactly) the amplitude at $\nu = 0$ from the physical region data. The fact that q^2 is unphysical has, on the contrary, always presented a much more serious problem since off-shell data do not exist at present. Our approach is a step in the direction of resolving this problem. We will use mass DR's to effect the q^2 extrapolation in the same way that energy DR's provide the ν extrapolation.

A second way to pose the problem under consideration is in terms of corrections to pole dominance. If the pole dominance were exact, then the off-shell extrapolation would be trivial and there would be no problem. Thus, if the pion pole dominance were exact in the above example, then $T(\nu, 0)$ would equal the on-shell amplitude $T(\nu, \mu^2)$. In this language, what we are proposing to do is to compute corrections to pole dominance. For example, $T(\nu, 0) = T(\nu, \mu^2) +$ (correction to pion pole dominance). We emphasize that our use of the word « correction » should not be taken to imply that the corrections are in any sense assumed to be small compared to the pole contributions. Indeed, in many examples the corrections will turn out to be more important than the pole contributions. In particular, our calculations of corrections will be strictly nonperturbative in character. We will discuss here corrections to vector-meson (ρ, ω) and particularly pseudoscalar-meson (π, K) pole dominance.

A third perspective on our problem is provided by an assumed underlying broken $SU_3 \times SU_3$ symmetry of the hadrons generated by the charges of V_0^a and A_0^a [7, 9-11]. If this symmetry were exact (*i.e.* exact in the multiplet sense for vector SU_3 and in the Goldstone sense for axial SU_3), then the pseudoscalar mesons would be massless and so $q^2 = 0$ would be a physical point. Since the symmetry is, in fact, broken, the mesons acquire mass (and mass differences) and the problem becomes that of calculating (perhaps large) symmetry-breaking effects. This approach to the problem makes it clear that a $SU_3 \times SU_3$ symmetry-breaking interaction should be specified. Given this interaction, and even given the magnitudes of the symmetry-breaking parameters, the problem is, however, still far from being solved. What is further needed is a method for relating the symmetry-breaking parameters in Lagrangians to the breaking effects in states. This is a difficult task, especially if large symmetry-breaking parameters and/or dynamical enhancements are involved, so that a perturbative approach is not useful. We will see that our LC techniques offer the possibility of connecting physical and Lagrangian symmetry breaking and provide a framework for a nonperturbative approach to the problem.

We assume here the simplest symmetry-breaking scheme in which $SU_3 \times SU_3$ is broken only by a mass term

$$\mathscr{M} = 2\alpha_0 S^0 + 2\alpha_8 S^8 , \tag{1.3}$$

where the scalar nonet S^a, together with a pseudoscalar P^a, constitute a $(3, \bar{3}) + (\bar{3}, 3)$ representation of $SU_3 \times SU_3$ [7, 12, 13]:

$$\delta(x_0)[V_0^a(x), S^b(0)] = \quad if^{abc} S^c(0)\, \delta^4(x)\,, \tag{1.4a}$$

$$\delta(x_0)[V_0^a(x), P^b(0)] = \quad if^{abc} P^c(0)\, \delta^4(x)\,, \tag{1.4b}$$

$$\delta(x_0)[A_0^a(x), S^b(0)] = \quad id^{abc} P^c(0)\, \delta^4(x)\,, \tag{1.4c}$$

$$\delta(x_0)[A_0^a(x), P^b(0)] = -id^{abc} S^c(0)\, \delta^4(x)\,. \tag{1.4d}$$

In this paper we shall, in fact, be more specific, and abstract additional algebraic relations from the gluon Lagrangian model [14]

$$\mathscr{H} = \psi^\dagger(-i\boldsymbol{\alpha}\cdot\boldsymbol{\nabla} + g\beta\gamma^\mu B_\mu)\psi + \mathscr{M} + \mathscr{H}_B\,, \tag{1.5}$$

in which the quark fields ψ interact via a neutral vector meson B_μ. The currents are then simply Dirac bilinears and we have the additional relations

$$\delta(x_0)[S^a(x), P^b(0)] = if^{abc}\, V_0^c(0)\, \delta^4(x)\,, \tag{1.6a}$$

$$\delta(x_0)[S^a(x), P^b(0)] = id^{abc}\, A_0^c(0)\, \delta^4(x)\,, \tag{1.6b}$$

$$\delta(x_0)[P^a(x), P^b(0)] = if^{abc}\, V_0^c(0)\, \delta^4(x)\,. \tag{1.6c}$$

The divergences of the vector and axial vector currents are respectively thus

$$\mathscr{D}^a \equiv \partial^\mu V_\mu^a = 2\alpha_8 f^{8ab} S^b\,, \tag{1.7a}$$

$$D^a \equiv \partial^\mu A_\mu^a = 2(\alpha_0 d^{0ab} + \alpha_8 d^{8ab}) P^b\,. \tag{1.7b}$$

The $SU_2 \times SU_2$ symmetry-breaking parameter is $\varepsilon_2 \equiv \sqrt{2}\alpha_0 + \alpha_8$ and the SU_3 one is $\varepsilon_3 \equiv \alpha_8$. The axial divergences of interest are

$$D^{\pi^-} \equiv \partial^\mu A_\mu^{\pi^-} = \frac{2}{\sqrt{3}}(\sqrt{2}\alpha_0 + \alpha_8) P^{\pi^-} \equiv \varepsilon_\pi P^{\pi^-}\,, \tag{1.8a}$$

$$D^{K^-} \equiv \partial^\mu A_\mu^{K^-} = \frac{2}{\sqrt{3}}\left(\sqrt{2}\alpha_0 - \frac{1}{2}\alpha_8\right) P^{K^-} \equiv \varepsilon_K P^{K^-}\,. \tag{1.8b}$$

In the limit of exact $SU_2 \times SU_2$, $\varepsilon_\pi = 0$, and in the limit of exact SU_3, $\varepsilon_\pi = \varepsilon_K$ or $\alpha_8 = 0$.

The nature of the symmetry-breaking term (1.3) is specified by the ratio

$$c \equiv \frac{\alpha_8}{\alpha_0}\,. \tag{1.9}$$

The popular values are

$$(1.10a)\qquad SU_2\times SU_2 \text{ much better than } SU_3 \Rightarrow \varepsilon_\pi \ll \varepsilon_K \Rightarrow \frac{c}{\sqrt{2}} \sim -1\,,$$

$$(1.10b)\qquad SU_3 \text{ much better than } SU_2\times SU_2 \Rightarrow \varepsilon_\pi \sim \varepsilon_K \Rightarrow \frac{c}{\sqrt{2}} \sim 0\,.$$

The numerical values in the two cases are

$$(1.11a)\qquad \text{meson spectrum [12]} \Rightarrow \frac{c}{\sqrt{2}} \simeq -0.9\,,$$

$$(1.11b)\qquad \text{baryon spectrum [15]} \Rightarrow \frac{c}{\sqrt{2}} \simeq -0.2\,,$$

We will consider each of these possibilities below.

2. – Canonical light-cone expansions.

The effectiveness of the use of energy DR's to compute the energy extrapolation is due to a comprehensive phenomenology, namely Regge theory, of high-energy behaviour, in addition to the detailed knowledge of the low-energy data. We have used canonical LC operator product expansions [3,6] to provide a similar description of high-mass behaviour in order to control the mass DR's. This canonical framework will be briefly reviewed in this Section.

To introduce the LC expansion, let us consider the scalar currents $j(x) = :\varphi(x)\varphi(x):$ in the $\lambda\varphi^4$-theory. We assume canonical (*i.e.* free-field) dimensions for all the currents. Thus, for example, $\dim j = 2$ and $\dim :\varphi\partial_{\alpha_1}\ldots\partial_{\alpha_n}\varphi: = n+2$. It then follows that the LC behaviour is [3]

$$(2.1)\qquad j(x)\,j(0) \xrightarrow[x^2\to 0]{} \frac{1}{x^2}\sum_n x^{\alpha_1}\ldots x^{\alpha_n} O_{\alpha_1\ldots\alpha_n}(0)\,,$$

where $\dim O_{\alpha_1\ldots\alpha_n} = n+2$. As discussed in detail in ref. [3], the specific form of the O's can be obtained from consideration of the short-distance ($x_\mu \to 0$) expansions of all the products $\partial_{\alpha_1}\ldots\partial_{\alpha_n} j(x)\,j(0)$ or, alternatively, from the highest-spin contributions to the equal-time commutators

$$[\partial_{\alpha_1}\ldots\partial_{\alpha_n} j(x),\, j(0)]\,\delta(x_0)\,.$$

These commutators can be formally evaluated by using the canonical commutation relations for the fields $\varphi(x)$ and the equation of motion $(\Box + \mu^2)\varphi(x) = \lambda :\varphi(x)^3:$ to eliminate higher time derivatives. The results for the leading LC singularity are obviously independent of the interaction term since, for a given dimension, the leading LC singularity is carried by fields with the most Lorentz indices and these come from the kinetic term $\varphi\partial_\alpha\partial^\alpha\varphi$ (*e.g.*, the interaction contribution $:\varphi\varphi:\lambda:\varphi\varphi:$ cannot carry a LC singularity whereas the free contribution $:\varphi\partial_\alpha\partial_\beta\varphi:$ of the same dimension (four) carries a LC singularity $x^\alpha x^\beta/x^2$) and so the free-field expansion remains formally valid. This free-field expansion follows simply from Wick's theorem

$$j(x)j(y) = 4\Delta_+(x-y;\mu^2):\varphi(x)\varphi(y): + c\text{-number} + \text{nonsingular}\,, \tag{2.2}$$

to be

$$j(x)j(y) \xrightarrow[\xi^2\to 0]{} \Delta_+(\xi)\sum_n \frac{1}{n!}\xi^{\alpha_1}\dots\xi^{\alpha_n}:\varphi(\eta)\overleftrightarrow{\partial}_{\alpha_1}\dots\overleftrightarrow{\partial}_{\alpha_n}\varphi(\eta):\,, \tag{2.3}$$

where

$$\xi = \frac{x-y}{2}\,,\qquad \eta = \frac{x+y}{2}\,,\qquad \overleftrightarrow{\partial} = \overleftarrow{\partial} - \overrightarrow{\partial}\,, \tag{2.4}$$

and

$$\Delta_+(\xi) = -\frac{1}{4\pi^2}\frac{1}{\xi^2 - i\varepsilon\xi_0} \tag{2.5}$$

is the massless free-field Wightman function. Expansions of the type (2.3), which uniquely follow from the assumptions of canonical commutators and field equations, will be referred to as canonical LC expansions. They have been extensively used in ref. [1-6, 16, 17].

Comparison of (2.1) and (2.3) gives

$$O_{\alpha_1\dots\alpha_m} = -\frac{1}{4\pi^2}\frac{1}{n!}:\varphi\overleftrightarrow{\partial}_{\alpha_1}\dots\overleftrightarrow{\partial}_{\alpha_m}\varphi:\,.$$

The highest-spin component of the equal-time commutators immediately follows from (2.3). Form example

$$\delta(x_0-y_0)[\partial_0 j(x), j(y)] = j(\eta)\,\delta(\boldsymbol{\xi})\,.$$

For actual physical applications we will use the canonical gluon model (1.5). The interesting currents are the vector, axial-vector, scalar and pseudoscalar ones:

$$V^a_\mu = :\bar\psi\gamma_\mu\frac{\lambda^a}{2}\psi:\,, \tag{2.6a}$$

$$A^a_\mu = :\bar\psi\gamma_\mu\gamma_5\frac{\lambda^a}{2}\psi:\,, \tag{2.6b}$$

$$S^a = :\bar\psi\frac{\lambda^a}{2}\psi:\,, \tag{2.6c}$$

$$P^a = :\bar\psi\gamma_5\frac{\lambda^a}{2}\psi:\,, \tag{2.6d}$$

of dimension three. The free-field expansions, for example

(2.7) $$V^0_\mu(x)\,V^0_\nu(y)\xrightarrow[\xi^2\to 0]{}\frac{\partial}{\partial\xi_\alpha}\,\Delta_+(\xi)\,[g_{\mu\alpha}O^{[-]}_\mu(\xi,\eta)+g_{\nu\alpha}O^{[-]}_\mu(\xi,\eta)-\\ -g_{\mu\nu}O^{[-]}_\alpha(\xi,\eta)+i\varepsilon_{\mu\nu\alpha\beta}O^{[+]}_{5\beta}(\xi,\eta)]\,,$$

(2.8) $$O^{[\pm]}_\mu(\xi,\eta)=\\ =\sum_{ij}\frac{1}{i!\,j!}\,\xi^{\alpha_1}\ldots\xi^{\alpha_i}\xi^{\beta_1}\ldots\xi^{\beta_j}:\bar\psi(\eta)\gamma_\mu\overleftarrow{\partial}_{\alpha_1}\ldots\overleftarrow{\partial}_{\alpha_i}\overrightarrow{\partial}_{\beta_1}\ldots\overrightarrow{\partial}_{\beta_j}\psi(\eta):\pm(x\leftrightarrow y)\,,$$

are now, however, altered by the interaction term since, for example $:\bar\psi\partial_\alpha\partial_\beta\psi:$ and $g^2:\bar\psi B_\alpha B_\beta\psi:$ can carry the same (leading) LC singularity. The effect of the interaction can, however, be simply accounted for by invoking the invariance of the theory under the gauge transformation [3]

(2.9) $$\begin{cases}\psi(x)\to\exp[ig\Lambda(x)]\,\psi(x)\,, & B_\mu(x)\to B_\mu(x)+\partial_\mu\Lambda(x)\,,\\ (\Box+\mu^2)\Lambda(x)=0\,.\end{cases}$$

The result [3] is simply to replace the derivatives ∂_ν in (2.8) by the gauge-invariant derivatives

(2.10) $$\Delta_\nu=\partial_\nu-igB_\nu\,.$$

The resulting expansions are then the unique ones which follow from the assumptions of canonical commutators and field equations.

An example of a canonical expansion, to be used below, is [1]

(2.11) $$P^a\left(\frac{x}{2}\right)S^b\left(-\frac{x}{2}\right)\xrightarrow[x^2\to 0]{}\partial_\mu\left(\frac{1}{x^2}\right)\sum_n d^{abc}O^{\mu,c}_{\alpha_1\ldots\alpha_n}(0)\,x^{\alpha_1}\ldots x^{\alpha_n}+\\ +\partial_\mu\partial_\nu(\ln x^2)\sum_n f^{abc}O^{\mu\nu,c}_{\alpha_1\ldots\alpha_n}(0)\,x^{\alpha_1}\ldots x^{\alpha_n}+\left(\frac{1}{x^2}\right)\sum_n d^{abc}O^{c}_{\alpha_1\ldots\alpha_n}(0)\,x^{\alpha_1}\ldots x^{\alpha_n}\,.$$

The O's can be obtained from the canonical evaluation of the equal-time commutators or, more simply, from the free-field expansions with the covariant derivatives (2.10). In practice, we will only make use of the first few O's in the expansion.

Another expansion of interest is for the product of (conserved) electromagnetic currents

(2.12) $$J_\mu=V^3_\mu+\frac{1}{\sqrt3}\,V^8_\mu\,.$$

The manifestly conserved form of the expansion is [3]

(2.13) $$J_\mu(x)J_\nu(0)\xrightarrow[x^2\to 0]{}(\partial_\mu\partial_\nu-g_{\mu\nu}\Box)\,x^{-2}\sum_n x^{\alpha_1}\ldots x^{\alpha_n}\mathscr{R}_{(0)\alpha_1\ldots\alpha_n}(0)+\\ +i\varepsilon_{\mu\nu\alpha\beta}\,\partial^\alpha x^{-2}\sum_n x^{\alpha_1}\ldots x^{\alpha_n}\mathscr{R}^\beta_{(1)\alpha_1\ldots\alpha_n}(0)+\\ +[g_{\mu\nu}\partial_\alpha\partial_\beta-g_{\alpha\nu}\partial_\beta\partial_\mu-g_{\alpha\mu}\partial_\beta\partial_\nu+g_{\alpha\mu}g_{\beta\nu}\Box](\log x^2)\sum_n x^{\alpha_1}\ldots x^{\alpha_n}\mathscr{R}^{\alpha\beta}_{(2)\alpha_1\ldots\alpha_n}(0)\,.$$

3. – Large mass behaviour.

In this Section we will recall how the LC expansions of Sect. **2** can be used to determine the behaviour of scattering amplitudes in the limit of large virtual external mass. Knowledge of this behaviour will be crucial for our use of mass dispersion relations in the following Sections. We consider first the scaling limit and then show how, in composite-particle theories, the large-mass limit of interest arises as a special case. The treatment here is an expanded version of the discussion of ref. [2, 4, 6, 16].

The main points can be illustrated by the consideration of the vertex function

$$A(p^2, q^2) = -i\int dx \exp[i(p+q)\cdot x/2]\langle 0|T\left[j\left(\frac{x}{2}\right)j\left(-\frac{x}{2}\right)\right]|k\rangle \tag{3.1}$$

between two scalar currents and the one scalar particle state $|k\rangle$. We take the scalar currents to have dimension two, as in (2.1). The currents carry momentum p and q and the scalar (on-shell) particle has momentum $k = p - q$ and mass $k^2 = m^2$. In addition to the (virtual) mass variables p^2 and q^2, we will use the energy variable

$$\nu = p\cdot k \tag{3.2}$$

and the scaling variable

$$\omega = \frac{p^2}{2\nu} = \frac{p^2}{p^2 - q^2 + m^2}. \tag{3.3}$$

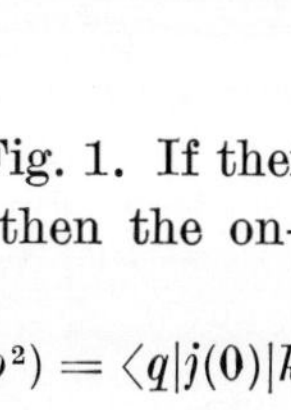

Fig. 1. – The scalar vertex function $A(p^2, q^2)$. The momentum $k = p - q$ labels an (on-shell) scalar particle, and p and q label (off-shell) scalar currents.

The kinematics are illustrated in Fig. 1. If there is a particle in the spectrum of $j(x)$ with squared mass $q^2 = m^2$, then the on-shell vertex function

$$A(p^2) = \langle q|j(0)|k\rangle \tag{3.4}$$

is given by

$$A(p^2) = \lim_{q^2\to m^2}(q^2 - m^2)A(p^2, q^2)\,; \tag{3.5}$$

i.e. $A(p^2)$ is the residue of the particle pole. We will always assume that the particles $\langle q|$ and $|k\rangle$ are composite ones that the form factor (3.4) decreases rapidly for $p^2 \to \infty$:

$$A(p^2) \xrightarrow[p^2\to\infty]{} 0 \text{ fast}. \tag{3.6}$$

This is the observed behaviour of all the measured hadronic form factors, presumably corresponding to the composite nature of the low-lying hadrons (*e.g.*, they lie on Regge trajectories). This rapid fall-off occurs in all composite models [18]. This is to be contrasted with the form factors of elementary particles (as in finite orders of perturbation theory for renormalizable field theories), which approach constants (within log *s*) for $p^2 \to \infty$. It will not be necessary for our purposes to specify the actual rate of decrease of on-shell form factors. We need only to assume a faster decrease than $1/p^2$. Recall that the nucleon electromagnetic form factors decrease like $1/(p^2)^2$, possibly corresponding to the three-quark structure of the nucleon, or perhaps even decrease exponentially, as would be expected in an « infinitely composite » picture of the nucleon.

By the usual arguments [3, 4], the behaviour of $A(p^2, q^2)$ in the scaling limit $p^2 \to \infty$, ω fixed is determined by the leading LC singularity (2.1). The matrix element behaviour

$$\langle 0|\, T\left[j\left(\frac{x}{2}\right) j\left(-\frac{x}{2}\right)\right] |k\rangle \xrightarrow[x^2\to 0]{} \frac{1}{x^2 - i\varepsilon} f(x\cdot k) \tag{3.7}$$

is determined by the matrix elements

$$\langle 0|O_{\alpha_1\ldots\alpha_n}(0)|k\rangle = c_n k_{\alpha_1}\ldots k_{\alpha_n} + g_{\alpha_i\alpha_j}\text{— terms} \tag{3.8}$$

via the formal power series

$$f(\lambda) = \sum_n c_n \lambda^n . \tag{3.9}$$

The series (3.9) is assumed to be everywhere convergent so that $f(\lambda)$ is entire. (This is true in perturbation theory and follows if the only singularities are on the LC.) Substitution of (3.7) in (3.1) gives

$$A(p^2, q^2) \xrightarrow[\substack{p^2\to\infty \\ \omega \text{ fixed}}]{} \frac{1}{p^2} G(\omega) , \tag{3.10}$$

where

$$G(\omega) = (4\pi^2)\,\omega \int_{-1}^{1} d\eta \frac{F(\eta)}{\omega - \eta} . \tag{3.11}$$

Here $F(\eta)$ is the Fourier transform

$$F(\eta) = \frac{1}{2\pi}\int d\lambda \exp[i\lambda\eta] f(\lambda) , \tag{3.12}$$

and the support property exhibited in (3.11) is a consequence of the usual analyticity.

A special case is the old Bjorken [19] limit $p_0 \to \infty$ with $\boldsymbol{p}$ fixed, in which

$$A \to \frac{1}{p_0^2} \int dx\, \delta(x_0) \langle 0| \left[\partial_0 j\left(\frac{x}{2}\right), j\left(-\frac{x}{2}\right)\right] |k\rangle = \frac{1}{p_0^2} 4 \langle 0| j(0) |k\rangle . \tag{3.13}$$

In this limit, $p^2 \to p_0^2$ and $\omega \to p_0/2k_0 \to \infty$, so that (3.10) and (3.11) give

$$A \to \frac{1}{p_0^2} (4\pi^2) \int_{-1}^{1} d\eta\, F(\eta) . \tag{3.14}$$

Comparison of (3.13) and (3.14) gives

$$4\pi^2 \int_{-1}^{1} d\eta F(\eta) = 4\langle 0|j(0)|k\rangle . \tag{3.15}$$

This result could have been immediately derived from (3.13) and (3.7), which give respectively

$$\int d\eta F(\eta) = f(0) \tag{3.16}$$

and

$$4\pi^2 f(0) = 4\langle 0|j(0)|k\rangle . \tag{3.17}$$

The actual limit of interest to us here is the fixed mass limit $p^2 \to \infty$ with q^2 fixed. This is a special case of (3.10) in which, according to (3.3), $\omega \to +1$. Let us for the moment assume that $G(1)$ exists and that our fixed mass limit can be obtained by first taking the scaling limit (3.10) and then letting $\omega \to 1$. We will afterwards verify this assumption for composite particles. We thus obtain

$$A(p^2, q^2) \xrightarrow[q^2 \text{ fixed}]{p^2 \to \infty} \frac{1}{p^2} G(1) . \tag{3.18}$$

The same result can be obtained directly from (3.1) and (3.7) provided the LC dominates this fixed-mass limit. It follows immediately from (3.18), since the right-hand is independent of q^2, that the on-shell form factor (3.5) decreases faster than $1/p^2$ for $p^2 \to \infty$. This is precisely the behaviour we expect for composite particles and the behaviour we, of course, want to have. We emphasize, however, that we have not derived this good result from the LC behaviour. We have essentially assumed it via our assumption that $G(1)$ exists. The point is that we want on-shell form factors to decrease rapidly and we incorporate this requirement into our LC formalism by assuming the existence of $G(1)$.

Of course, in models containing elementary particles, the form factors involving these particles need not decrease and, correspondingly, the appropriate $G(1)$ will not exist. Examples of this are provided by the amplitudes given by finite-order Feynman diagrams. Consider, for example, the low-order

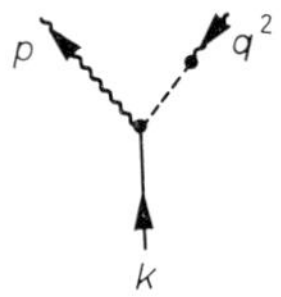

Fig. 2. – A pole (in q^2) contribution to the scalar vertex function.

diagram of Fig. 2, in which the current of momentum q couples to an elementary scalar meson of mass m^2. Taking nonderivative couplings, the amplitude is essentially

$$A_0(p^2, q^2) = \frac{1}{q^2 - m^2} = \frac{1}{p^2}\frac{\omega}{\omega - 1}\,. \tag{3.19}$$

Comparison with (3.10) gives the corresponding scaling function to be

$$G_0(\omega) = \frac{\omega}{\omega - 1}\,. \tag{3.20}$$

Also, by (3.11),

$$F_0(\eta) = \frac{1}{4\pi^2}\,\delta(\eta - 1)\,. \tag{3.21}$$

Finally, the corresponding on-shell form factor (3.5) is

$$A_0(p^2) = 1\,. \tag{3.22}$$

The first thing we notice is that $G_0(\omega)$ has a pole at $\omega = 1$ so that $G_0(1)$ does not exist and eq. (3.18) is meaningless. The correct behaviour in the limit (3.18) is $A_0(p^2, q^2) \to 1/(q^2 - m^2)$. Roughly speaking, the linear vanishing of (3.18) for $p^2 \to \infty$ is made up for by the linear divergence of $G(1)$ so that the resultant asymptotic limit is a constant in p^2. For the same reason, the form factor behaviour (3.6) is not obtained, but rather (3.22) gives $A_0(p^2) \to 1$. (The diagram for the form factor is shown in Fig. 3.) We must conclude that the leading LC singularity does not dominate the fixed mass limit in this case. This is accompained, as it must be, by the bad asymptotic behaviour of the form factor. The culprit is, of course, the elementary particle in Fig. 2 which simultaneously ruins LC dominance of the fixed-mass off-shell limit *and* fast decrease of the on-shell form factor. As we have stressed above, we rule out this unphysical situation by fiat. We assume composite particles and we get LC dominance and rapid decrease of form factors. If the elementary particle

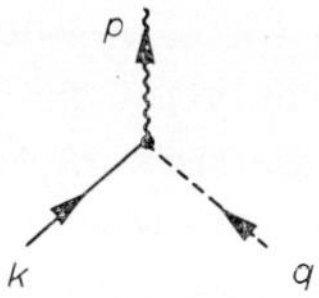

Fig. 3.

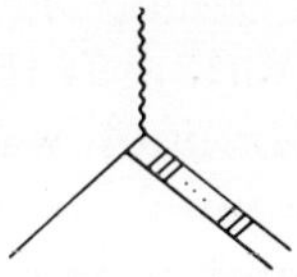

Fig. 4.

Fig. 3. – The scalar form factor for an elementary particle. The momenta k and q label (on-shell) scalar particles and p labels the (off-shell) scalar current.

Fig. 4. – The scalar form factor for a composite particle.

in Fig. 2 is replaced by a composite particle, as in Fig. 4 (and likewise for the particle of momentum k), we then obtain a falling form factor and can show that the LC dominates the fixed-mass limit.

Let us now return to the general case of composite-particle theories where (3.6) and (3.18) are valid. Reference to the representation (3.11) provides an alternative description of our composite-particle assumption. Since

$$G(1) = (4\pi^2)\int_{-1}^{1} d\eta \frac{F(\eta)}{1-\eta}, \tag{3.23}$$

the existence of $G(1)$ is essentially equivalent to the vanishing of $F(1)$. In composite models, the vanishing of $F(\eta)$ at the threshold point $\eta = 1$ is expected since the only contribution to $F(\eta)$ at threshold is from the single-particle intermediate state so that $F(1)$ is controlled by the elastic form factor. Thus, a rapidly decreasing form factor ensures the vanishing of $F(1)$. This is another way of seeing that a rapidly decreasing form factor requires the existence of $G(1)$. From either point of view, we conclude that $F(1) = 0$. $F(\eta)$ is, in fact, expected to vanish rapidly near $\eta = 1$. In composite models, the rate of decrease of $A(p^2)$ for $p^2 \to \infty$ is directly correlated to the rate of vanishing of $F(\eta)$ near $\eta = 1$.

The rapid vanishing of $F(\eta)$ near $\eta = 1$ and, similarly, near $\eta = -1$, does more than guarantee the existence of $G(1)$ and the LC dominance of the fixed-mass limit. It also provides a means of estimating $G(1)$ and hence of using the limit (3.18) in a quantitative way. If $F(\eta)$ is strongly peaked near $\eta = 0$, then from (3.23) and (3.15) we obtain the approximate relation

$$G(1) \simeq (4\pi^2)\int_{-1}^{1} d\eta F(\eta) = 4\langle 0|j(0)|k\rangle . \tag{3.24}$$

(The same approximation works quite well for the SLAC-MIT structure function $F_2(\eta)$.) Since decay constants of the form $\langle 0|j(0)|k\rangle$ are in general known, (3.24) gives an estimate of $G(1)$ and hence of (3.18). Such estimates will be useful to us in later Sections.

We emphasize that, in spite of its appearance, (3.18) with (3.24) does not in any way constitute an assumption that the fixed-mass limit (3.18) is the same as the Bjorken limit (3.13). If this were exactly true, then only the first term $(1/x^2 O(0) = 1/x^2(-1/4\pi^2)j(0))$ would be present in (2.1), so that (3.9) would become $f(\lambda) = c_0 = f(0) = (1/4\pi^2)4\langle 0|j|k\rangle$ and (3.12) would become $F(\eta) = c_0\,\delta(\eta)$. This form of $F(\eta)$ is unacceptable both physically and theoretically (*i.e.*, it violates causality [3]). Our assumption is rather that $F(\eta)$ is peaked near $\eta = 0$ so that $G(1)$ is of the same order as $(4\pi^2)f(0)$.

We conclude this Section by giving a more complete discussion of the on-shell form factor and a more exact statement of the condition for LC dominance of the fixed-mass limit. To do this, we return to the scaling limit (3.10). The form (3.10) is the asymptotic form of the contribution of the leading LC singularity (3.7). Taking into account the contributions of the leading and nonleading contributions of the leading and nonleading LC singularities, assuming always canonical dimensions, we obtain an expansion of the form

$$A(p^2, q^2) \xrightarrow[\omega\ \text{fixed}]{p^2\to\infty} \frac{1}{p^2}G_1(\omega) + \frac{1}{(p^2)^2}G_2(\omega) + \ldots + \frac{1}{(p^2)^r}G_r(\omega) + \ldots\,. \tag{3.25}$$

Here $G_1(\omega) = G(\omega)$ gives the leading contribution of the leading LC singularity, $G_2(\omega)$ gives the leading contribution of the next-leading LC singularity ($\log x^2$) as well as the next-leading contribution of the leading LC singularity, etc. For example

$$A(p^2, q^2) \to \tag{3.26}$$

$$\to \int_{-1}^{1} \mathrm{d}\eta\,[F_1(\eta)(p^2 - 2\eta\nu + \eta^2 m^2)^{-1} + F_2(\eta)(p^2 - 2\eta\nu + \eta^2 m^2)^{-2} + \ldots] =$$

$$= \frac{\omega}{p^2}\int \mathrm{d}\eta\,F_1(\eta)\left(\omega - \eta + \frac{\eta^2 m^2\omega}{p^2}\right)^{-1} + \frac{\omega^2}{p^4}\int \mathrm{d}\eta\,F_2(\eta)\left(\omega - \eta + \frac{\eta^2 m^2\omega^2}{p^2}\right)^{-2} + \ldots =$$

$$= \frac{1}{p^2}\left[\omega\int \mathrm{d}\eta\,F_1(\eta)\frac{1}{\omega-\eta}\right] + \frac{1}{p^4}\left[m^2\omega^2\int \mathrm{d}\eta\,F_1(\eta)\frac{\eta^2}{\omega-\eta} + \omega^2\int \mathrm{d}\eta\,F_2(\eta)\frac{1}{(\omega-\eta)^2}\right] + \ldots\,.$$

If each $G_r(\omega)$ exists at $\omega = 1$, then

$$A(p^2, q^2) \xrightarrow[q^2\ \text{fixed}]{p^2\to\infty} \frac{1}{p^2}G_1(1) + \frac{1}{p^4}G_2(1) + \ldots\,, \tag{3.27}$$

and the on-shell form factor (3.4) decreases faster than any power for $p^2 \to \infty$. Suppose, on the contrary, that $G_l(\omega)$ is the first G_r to have a pole at $\omega = 1$. (The usual analyticity implies that the only possible singularities of the G_r's are simple poles.) Thus

$$G_l(\omega) \sim \frac{a}{\omega - 1} = a\left(\frac{p^2}{q^2 - m^2} - 1\right), \tag{3.28}$$

and we can only conclude from (3.25) that

$$A(p^2, q^2) \xrightarrow[q^2 \text{ fixed}]{p^2 \to \infty} \frac{1}{p^2} G_1(1) + \ldots + \frac{1}{(p^2)^{l-1}} G_{l-1}(1) + \frac{1}{(p^2)^{l-1}} \frac{a}{q^2 - m^2} + \ldots . \tag{3.29}$$

Furthermore, in this case, we have

$$A(p^2) \xrightarrow[p^2 \to \infty]{} \frac{a}{(p^2)^{l-1}} . \tag{3.30}$$

As long as $l > 2$, we have a reasonable rate of decrease for the form factor and LC dominance of the fixed-mass limit. This, in fact, is the precise condition for LC dominance of the fixed-mass limit in our framework.

The above discussion makes it clear that the LC formalism can accommodate any rate of decrease (3.30) of the form factor, but that it cannot predict what this rate is. From an ascetic point of view, perhaps the exponential decrease implied by the existence of each $G_r(1)$ is to be preferred.

4. – Mass dispersion relations.

In this Section we continue to work with the scalar vertex function (3.1). For subsequent applications, however, we allow a fixed-mass asymptotic behaviour more general than (3.18). We take

$$A(p^2, q^2) \xrightarrow[q^2 \text{ fixed}]{p^2 \to \infty} (p^2)^{r-2} F_r \tag{4.1}$$

with r unspecified. The behaviour (4.1) corresponds to an LC singularity $(1/x^2)^r$ (ref. [2, 4]). At this point we introduce an assumption which we have abstracted from the SLAC-MIT deep inelastic electroproduction experiments. Namely, we assume that the leading asymptotic term (4.1) completely dominates A already for p^2 as low as 2.5 $(\text{GeV})^2$. The experimental support for this assumption, which we call « precocious asymptopia », is reviewed in ref. [4].

We now wish to exploit the fact that, for fixed q^2, $A(p^2, q^2)$ is an analytic function of p^2 in the complex p^2-plane cut along the positive p^2-axis. This enables us to write the relations

$$(4.2)\qquad A(p^2, q^2) = \frac{1}{\pi}\int_0^{\Lambda} \mathrm{d}z\,\frac{a(z, q^2)}{z - p^2} + \frac{1}{2\pi i}\oint_{C_\Lambda} \mathrm{d}z\,\frac{A(z, q^2)}{z - p^2}\,,$$

$$(4.3)\qquad 0 = \frac{1}{\pi}\int_0^{\Lambda} \mathrm{d}z\,a(z, q^2) + \frac{1}{2\pi i}\oint_{C_\Lambda} \mathrm{d}z\,A(z, q^2)\,,$$

where $a = \text{abs } A$ and C_Λ is the circular contour $|z| = \Lambda$. Choosing $\Lambda = 2.5\ (\text{GeV})^2$, (4.1) can be used to evaluate the contour integrals ($p^2 \leqslant \Lambda$):

$$(4.4)\qquad \frac{F_r}{2\pi i}\oint_{C_\Lambda} \mathrm{d}z\,\frac{z^{r-2}}{z - p^2} = (p^2)^{r-2} A_r\,,$$

$$(4.5)\qquad \frac{F_r}{2\pi i}\oint_{C_\Lambda} \mathrm{d}z\,z^{r-2} = B_r\,.$$

Suppose there is a single low-lying particle P of mass μ with the quantum numbers of $A(x)$ so that

$$(4.6)\qquad a(z, q^2) = \pi\delta(z - \mu^2)\,a_p(q^2) + a_N(z, q^2)\,.$$

Then $a_N(z, q^2)$ should not oscillate in the short integration range (this is the great virtue of precocious asymptopia) so that the mean-value theorem can be used to conclude that for small p^2

$$(4.7)\qquad \int_0^{\Lambda} \mathrm{d}z\,\frac{a_N(z, q^2)}{z - q^2} = \frac{1}{M^2 - p^2}\int_0^{\Lambda} \mathrm{d}z\,a_N(z, q^2)\,, \qquad 0 \leqslant M^2 \leqslant \Lambda\,.$$

We have always found that $1.5 \leqslant M^2 \leqslant 2$. This intermediate value is expected since $a_N(z)$ should be very small for $z < 1$, where the particle contribution dominates, and for $z > 2.5$, where (4.1) dominates so that $a_N \to 0$ since F is real. Putting all this into (4.2) and (4.3), we obtain [2, 4, 16]

$$(4.8)\qquad A(p^2, q^2) \simeq a_P(q^2)\left(\frac{1}{\mu^2 - p^2} - \frac{1}{M^2 - p^2}\right) - \frac{B_r}{M^2 - p^2} + (p^2)^{r-2} A_r\,.$$

Equation (4.8) is our master equation which we shall use repeatedly in the following Sections. It contains essentially no unknowns since A_r and B_r can be approximately determined from decay amplitudes as in (3.24) and $a_P(q^2)$

is determined from on-shell data. It, therefore, provides an approximate description of the off-shell behaviour of the amplitude. As we discussed in Sect. **1**, this information is what we have been seeking.

Although (4.8) has been derived and discussed for scalar vertex functions, it should be clear that analogous results can be obtained when spin is included and for scattering amplitudes. Examples will be given in subsequent Sections.

5. – Deviations from vector meson dominance.

Before applying the relation (4.8) to discuss the off-shell extrapolations relevant to symmetry-breaking calculations, in this Section we will review some of the applications we have made of it in the realm of vector-meson dominance (VMD) [16]. Since in these applications the left-hand side of (4.8) will be evaluated at a physical (photon) point, the effect of the off-shell extrapolation predicted by (4.8) can be directly compared with experiment. These applications thus serve as an important check of (4.8) and of the assumptions invoked in its derivation.

In this Section we always take $p^2 = 0$ in (4.8) and take $j(x)$ to be the electromagnetic current $S_\mu(x)$ in (3.1). Then the low-lying particle P is the appropriate vector meson. Equation (4.8) then relates the photon amplitude $A(0)$ to the vector-meson amplitude a_P and the decay amplitudes B_r and A_r: The ordinary VMD result is the $M^2 \to \infty$, $A_r \to 0$ limit of (4.8):

$$A(0) = a_P/\mu^2 \qquad \text{(VMD)}\,. \tag{5.1}$$

Our relation (4.8) is thus seen to supply corrections to this VMD result.

Let us first use (4.8) to compute the $\pi^0 \to \gamma\gamma$ decay amplitude. The off-shell amplitude for $\pi^0(k) \to \gamma^3(p) + \gamma^8(q)$

$$(2\sqrt{3})\,F(p, q) = \varepsilon_{\mu\nu\alpha\beta}\varepsilon^\mu(p)\,\varepsilon^\nu(q)\,p^\alpha q^\beta A(p^2, q^2) \tag{5.2}$$

is given by (3.1) with $j = eJ_\mu$, eJ_ν and $|k\rangle = |\pi^0(k)\rangle$. The relevant operator product expansion is (2.13), where only the second sum contributes. The fixed-mass asymptotic behaviour is

$$A(p^2, q^2) \xrightarrow[q^2\ \text{fixed}]{p^2 \to \infty} \frac{1}{p^2}\, G(1) \simeq \frac{1}{p^2}\, \frac{\sqrt{2} f_\pi e^2}{3}\,, \tag{5.3}$$

where we have used the approximation analogous to (3.24). The relevant equal-time commutation relation is

$$[J_i^3(0, \boldsymbol{x}), J_j^8(0)] = i\,\frac{1}{\sqrt{3}}\,\varepsilon_{ijk} A_k^3(0)\,\delta(\boldsymbol{x})\,, \tag{5.4}$$

and we have used the definition $\langle 0|A^3_\mu(0)|k\rangle = (1/\sqrt{2})k_\mu f_\pi$ of the pion decay constant f_π. Thus, in (4.4) and (4.5), $A_r = 0$ and $B_r = \sqrt{2}f_\pi e^2/3$. The residue of the ω pole in $A(p^2, 0)$ is essentially the $\omega \to \pi\gamma$ decay amplitude:

$$a_\omega = (em_\omega^2/2\gamma_\omega)A(\omega \to \pi\gamma) \tag{5.5}$$

in the usual notation. Equation (4.8), evaluated at $p^2 = q^2 = 0$, thus becomes [16]

$$A(0, 0) \simeq \frac{e}{\gamma_\omega}\left(1 - \frac{m_\omega^2}{M^2}\right)A(\omega \to \pi\gamma) + \sqrt{2}f_\pi e^2/3M^2 . \tag{5.6}$$

It is even simpler to use our methods to estimate $A(0) \equiv A(\omega \to \pi\gamma)$ itself. As we have already said, $A(0)$ is essentially the residue of the ω pole fo $A(p^2, 0)$. The off-shell invariant amplitude $A(p^2)$ is the « on-shell » form factor

$$\langle \omega(q)|J_\mu(0)|\pi(k)\rangle \sim A(p^2)\varepsilon_{\mu\nu\alpha\beta}\varepsilon^\nu(q)q^\alpha p^\beta$$

and, therefore, decreases rapidly (faster than $1/p^2$) for $p^2 \to \infty$. Therefore, applying our formalism, we have a relation of the form (4.8) for $A(p^2)$ but with $A_r = B_r = 0$, as given by (4.4) and (4.5). The residue of the ρ pole in $A(p^2)$ is essentially the $\omega \to \rho\pi$ amplitude: $a_\rho = g_{\omega\rho\pi}(e/2\gamma_\rho)$ in the usual notation. Thus (4.8) gives [16]

$$A(0) \simeq g_{\omega\rho\pi}\frac{e}{2\gamma_\rho}\left(1 - \frac{m_\rho^2}{M^2}\right) . \tag{5.7}$$

Let us now compare our predictions (5.6) and (5.7) with experiment. We begin with (5.6). The latest data [20] indicate that the usual VMD result $A(0) \sim g_{\omega\rho\pi}(e/2\gamma_\rho)$ is wrong by about a factor of two in the rates. With $M^2 \simeq 2$ (GeV)^2, the correction factor $(1 - m_\rho^2/M^2)$ implied by our analysis nicely resolves this discrepancy. Consider next (5.6). Here the usual VMD result $A(0, 0) \sim (e/\gamma_\omega)A(0)$ is wrong by about a factor of three in the rates [20]. Our result (5.6), with $M^2 \simeq 2$, is again in agreement with the data. Our formalism is thus seen to supply important and numerically accurate corrections to VMD.

For our present purposes, the precise numerical results (5.6) and (5.7) need not be taken too seriously. What is important is that the comparison of (5.6) and (5.7) with experiment has revealed no gross failures of our assumptions and calculations. On the contrary, the reliability of our methods is strongly suggested [21]. We, therefore, proceed with some confidence to apply (4.8) to discuss symmetry-breaking effects where, unlike in the present situation, the left-hand side will not be directly known.

6. – Symmetry breaking for vertex functions.

The formalism of Section 4 can be directly applied in order to relate symmetry-breaking parameters in Lagrangians to breaking effects in states and deduce approximate relations between physical S-matrix elements from operator algebras. An example of this is our analysis [1] in which some $SU_3 \times SU_3$ symmetry-breaking effects were estimated. We used the mass DR's to provide algebraic relations between the values of amplitudes at zero mass (given by equal-time commutation relations), at physical points (given by experiment), and at mass $\simeq 2.5\,(\text{GeV})^2$ (given by the LC expansions). Since use of the smooth threshold assumption relates the LC behaviour back to the equal-time behaviour, we end up with coupled equations for various physical parameters which can be solved simultaneously. This is how our algebraic equations for operators lead to algebraic equations for physical parameters and how the (exact) operator symmetry embodied in eq. (1.1)-(1.6) leads to (broken) symmetry for the physical states. This is, in fact, a different way of saying that we are calculating corrections to pseudoscalar meson pole dominance. In the exact $SU_3 \times SU_3$ limit ($\alpha_0 = \alpha_8 = 0$), these mesons are massless and so current algebra gives on-shell SU_3-symmetric predictions. In the real world with $\alpha_8 \neq 0$ and $\alpha_0 \neq 0$, the effects of both off-shell extrapolations and SU_3 violations must be taken into account, and that is what our formalism attempts to accomplish. This, incidentally, puts pion PCAC and kaon PCAC on the same footing.

The vertex functions of interest are the (vacuum)–(one-pseudoscalar meson) matrix elements of A^a_μ and P^a. We define the usual pion and kaon decay constants by

$$\langle 0|A^{\pi^-}_\mu|\pi^+\rangle = i\varrho_\mu f_\pi\,, \qquad \langle 0|A^{K^-}_\mu|K^+\rangle = ik_\mu f_K\,, \tag{6.1}$$

and « renormalization » constants by

$$\langle 0|P^{\pi^-}|\pi^+\rangle = Z_\pi\,, \qquad \langle 0|P^{K^-}|K^+\rangle = Z_K\,. \tag{6.2}$$

We always label (on or off-shell) pions (kaons) with momentum $p_\mu(k_\mu)$. On-shell, $p^2 = m^2_\pi = \mu^2$, $k^2 = m^2_K = m^2$. The divergence equations (1.8) immediately give the relations

$$\mu^2 f_\pi = \frac{2}{\sqrt{3}}(\sqrt{2}\,\alpha_0 + \alpha_8)\,Z_\pi\,, \qquad m^2 f_K = \frac{2}{\sqrt{3}}\left(\sqrt{2}\,\alpha_0 - \frac{1}{2}\,\alpha_8\right) Z_K\,. \tag{6.3}$$

The analysis of ref. [1] goes as follows. We define the two vertex functions ($k = p + q$)

$$\int \mathrm{d}x \exp[i(p-q)\cdot x/2]\langle 0|TD^a\left(\frac{x}{2}\right)\mathscr{D}^K\left(-\frac{x}{2}\right)|b\rangle\,, \qquad a = \pi, K \tag{6.4}$$

Each of these satisfies two exact zero-mass theorems, one at $q=0$ ($p^2=m^2$ or $k^2=\mu^2$) and one at $p=0$ ($q^2=m^2$) or $k=0$ ($q^2=\mu^2$). We define two more vertex functions:

(6.5) $$\int dx \exp[i(p-q)\cdot x/2]\langle 0|\, TD^a\left(\frac{x}{2}\right) V^{\mathrm{K}}_\mu\left(-\frac{x}{2}\right)|b\rangle\,, \qquad a=\pi,\,\mathrm{K}\,.$$

For $a=\pi$ we get a zero-mass theorem at $p=0$ ($q^2=m^2$) and for $a=\mathrm{K}$ we get one at $k=0$ ($q^2=\mu^2$). We next write mass DR's for these vertex functions and obtain equations of the form (4.8). The left-hand sides are known in terms of the zero-mass theorems, the pole contributions are known in terms of the on-shell K_{l3} amplitudes defined by

(6.6) $$\langle\pi^0(p)|\,V^{\mathrm{K}}_\mu(0)\,|K^+(k)\rangle=\frac{1}{\sqrt{2}}\,[(k+p)_\mu f_+(q^2)+(k-p)_\mu f_-(q^2)]\,,$$

and the LC contributions are known from the operator-product expansions (if the leading contribution does not contribute, we must keep the next leading one). Equation (2.11) is a typical LC expansion used. In this way, using the relations (6.3), we end up with six equations (given in ref. [1]) in the six unknowns:

$$f_+(0)\,,\qquad \xi(0)\equiv f_-(0)/f_+(0)\,,\qquad f_+(m^2)+f_-(m^2)\,,\qquad \alpha_8\,,\qquad A\,,\qquad B\,.$$

A and B are uninteresting parameters defined in ref. [1].

The solution to these equations depends on the choice of the parameter C defined in eq. (1.9). Let us first take the nonchiral value (1.11b). The solution is then

(6.7a) $$f_+(0)\simeq\quad 0.94\,,$$

(6.7b) $$\xi(0)\simeq-0.7\,,$$

(6.7c) $$f_+(m^2)+f_-(m^2)\simeq\frac{f_{\mathrm{K}}}{f_\pi}\,(0.7)\,,$$

(6.7d) $$\alpha_8\simeq-140\ \mathrm{MeV}\,.$$

These values were obtained using the usual value $M^2=1.5$ [22] but they are quite insensitive to M^2 in the range $1.5\leqslant M^2\leqslant 2$.

Let us comment on these results [23]. Equation (6.7) is in good agreement with the Ademollo-Gatto theorem. Equation (6.7) is in good agreement with the recent K_{l3} data [24]. Note that is was obtained without making any assumptions about the q^2-dependence of the form factors. Equation (6.7) represents a 30 % violation of the Callan-Treiman relation. Equation (6.7) is, as expected, a small SU_3-violating parameter. The chiral $SU_2\times SU_2$ symmetry-breaking parameter is thus

(6.8) $$\varepsilon_? = \sqrt{2}\,\alpha_0 + \alpha_g \simeq 580 \text{ MeV} .$$

We see that this formalism leads to a completely consistent picture of $SU_3 \times SU_3$ symmetry breaking.

With the value (1.11*b*), eq. (6.3) gives

(6.9) $$\frac{\mu^2}{m^2}\frac{f_\pi}{f_K} \simeq (0.77)\frac{Z_\pi}{Z_K} .$$

Thus, since $f_\pi/f_K \sim 1$, we have $Z_\pi/Z_K \sim 1/12$. The above formalism clearly exhibits the mechanism which is responsible for the large difference between Z_K/Z_π and f_K/f_π in the nonchiral framework. The $M^2 \to \infty$ limits of four of the above-mentioned six equations are

(6.10*a*) $$Z_K \sim Z_\pi f_+(0) \;-\varepsilon_3(f_K + A) ,$$

(6.10*b*) $$Z_\pi \sim Z_K f_+(0) \;+\varepsilon_3(f_\pi - A) ,$$

(6.11*a*) $$f_K \sim f_\pi[f_+(m^2) + f_-(m^2)] ,$$

(6.11*b*) $$f_\pi \sim f_K[f_+(\mu^2) - f_-(\mu^2)] .$$

The important difference between (6.10) and (6.11) is the presence of the $\mp\,\varepsilon_3 f_{K,\pi}$ terms in (6.10) which come from the LC. These terms are numerically important and their occurrence with opposite signs in (6.10*a*) and (6.10*b*) gives rise to the difference between Z_K and Z_π. We thus see that this difference can be completely understood as a dynamical consequence of a formalism with a small symmetry-breaking parameter at the Lagrangian level. We conclude that a large value of Z_K/Z_π is, contrary to what might be naively expected, completely consistent with the usual picture of a small symmetry-breaking parameter. It is clearly very dangerous to assume that there is a simple connection between Lagrangian and state symmetry-breaking, especially when the pseudoscalar octet is involved.

As is clear from the solution (6.7), the contributions from the order $1/M^2$ terms are also important. These terms constitute the corrections to pion and kaon pole dominance.

Let us now solve our six equations using the chiral value (1.11*a*) of C. We then find that $\xi(0)$ is small and $f_+(m_K^2) + f_-(m_K^2)$ is near f_K/f_π. These are the familiar chiral results [9] obtained by assuming that the off-shell K_{l3} amplitudes are pion-pole dominated or, equivalently, that the chiral symmetry-breaking parameter ε_π is small enough to justify a perturbative treatment. Actually, we have learned more since it is possible to obtain a large (negative) $\xi(0)$ in the chiral framework provided the slope parameter λ_+, defined by

$$f_+(t) = f_+(0)[1 + \lambda_+ t/\mu^2] , \tag{6.12}$$

is much larger than the value $\lambda_+ \simeq 0.023$ predicted by K* dominance [25]. The most recent experiment [26], which has achieved the highest degree of accuracy to date, gives $\lambda_+ = 0.023 \pm 0.005$. The latest X2-collaboration result is $\lambda_+ = 0.027 \pm 0.010$ [27]. Thus it appears that the chiral prediction for $\xi(0)$ is too small. Further experimental clarification is, of course, needed and should come in the near future. It seems fair to say, however, that the present experimental information favours (1.11*b*) over (1.11*a*).

We conclude this Section with a brief discussion of the chiral symmetry-breaking effects expected for on-shell form factors. Consider, as a well-known example, the nucleon matrix element of (1.8*a*)

$$\langle N|O^\pi(0)|N\rangle = \bar{u}\gamma_5 u\, \mathrm{d}(p^2) . \tag{6.13}$$

Keeping only the pion pole in the p^2 dispersion relation gives the Goldberger-Treiman relation

$$2m_{\mathcal{N}} g_A = \mathrm{d}(0) \simeq \sqrt{2} f_\pi g_r , \tag{6.14}$$

in conventional notation. It should be clear that our formalism gives instead

$$2m_{\mathcal{N}} g_A \simeq \sqrt{2} f_\pi g_r \left(1 - \frac{\mu^2}{M^2}\right) . \tag{6.15}$$

Since the correction factor $(1 - \mu^2/M^2)$ is very near to unity in this case, we predict that (6.14) should be satisfied to within a few percent. This is consistent with experiment. The deviation from pion-pole dominance should be similarly small for all form factors. The general statement is [16]

$$\langle \alpha|D^a|\beta\rangle|_{p^2=0} \simeq \frac{\sqrt{2}}{f_\pi} \langle \alpha\pi^a|\beta\rangle \left(1 - \frac{\mu^2}{M^2}\right) \tag{6.16}$$

for arbitrary low-lying states $\langle\alpha|$, $|\beta\rangle$.

7. – Symmetry breaking for scattering amplitudes.

The formalism which we used in previous Sections for vertex functions can be easily extended to scattering amplitudes [2]. Consider an off-shell forward scattering amplityde $T(\varkappa, \nu)$ for

$$\sigma(p) + j(q) \to \sigma(p) + j(q) ,$$

where $\sigma(p)$ is a scalar particle of momentum p and mass $p^2 = 1$ and $j(q)$ is a scalar current of dimension two, momentum q, and mass $q^2 = \varkappa$. $T(\varkappa, \nu)$ is defined to have its $\varkappa$ poles removed so that the on-shell amplitude is $T(\mu^2, \nu)$. We assume the usual Regge behaviour

$$T(\varkappa, \nu) \xrightarrow[\varkappa \text{ fixed}]{\nu \to \infty} \beta(x)\nu^{\alpha}, \tag{7.1}$$

the usual scaling behaviour

$$T(\varkappa, \nu) \xrightarrow[\omega = \varkappa/2\nu \text{ fixed}]{\nu \to \infty} \nu^{-1}F(\omega), \tag{7.2}$$

and the usual connection between these limits [4]

$$\beta(\varkappa) \xrightarrow[\varkappa \to \infty]{} \beta\varkappa^{-\alpha-1}, \tag{7.3}$$

$$F(\omega) \xrightarrow[\omega \to 0]{} \beta\cdot(2\omega)^{-\alpha-1}. \tag{7.4}$$

We cannot apply the analysis of Sect. **4** directly to the mass DR's satisfied by $T(\varkappa, \nu)$ since, for finite ν, the $\varkappa$ discontinuities come from intermediate states in both the current and the current-particle channels. In the limit $\nu \to \infty$, however, only the pure current discontinuities remain. It is the absorptive part corresponding to these pure discontinuities which should not oscillate in the short $\varkappa$-integration range so that the mean-value theorem can be applied. The $\nu \to \infty$ limit of the mass DR for $T(\varkappa, \nu)$ gives, via (7.1), a mass DR for $\beta(\varkappa)$. This leads to a relation of the form (4.8) for $\beta(\varkappa)$. The constants A_r and B_r in (4.4) and (4.5) can be calculated in terms of β by use of (7.3). With this then explicit representation for $\beta(\varkappa)$, we can calculate the effect

$$\Delta\beta \equiv \beta(\mu^2) - \beta(0) \tag{7.5}$$

of the off-shell extrapolation.

The next step is to write the finite-energy sum rule

$$T(\varkappa, \nu) = \frac{1}{\pi}\int_0^N d\nu' \frac{\operatorname{Im} T(\varkappa, \nu')}{\nu' - \nu} + \frac{1}{2\pi i}\oint_{C_N} d\nu' \frac{T(\varkappa, \nu')}{\nu' - \nu}. \tag{7.6}$$

We choose N large enough (say, $N \geqslant 2.5\,(\text{GeV})^2$) so that the leading Regge term (7.1) dominates the contour integral I_N. We obtain for $\nu \ll N$

$$I_N = \beta(\varkappa)c_{\alpha}N^{\alpha}, \tag{7.7}$$

where c_{α} is a constant. (Care must be taken in using the correct phase implicit in (7.1).) From the 0-N integration in (7.6), we separate out the Born and low-lying resonance contributions and treat them explicitly. The remainder $\operatorname{Im}\bar{T}(\varkappa, \nu')$ can be expressed as a sum over low-lying (since $\nu' \lesssim N = 2.5\ (\text{GeV})^2$) intermediate states, for each term of which a relation of the form (6.16) is valid. Putting all this into (7.6) and evaluating at $\nu = 0$, we obtain [2].

$$(7.8) \qquad T(\mu^2, 0) - T(0, 0) = \left(\frac{2\mu^2}{M^2}\right)\bar{T}(\mu^2, 0) + (\Delta\beta)c_\alpha N^\alpha + \\ + (\text{Born and resonance contribution})\,.$$

The relation (7.8) expresses the off-shell amplitude $T(0, 0)$ in terms of quantities which can be calculated from on-shell data. The new ingredient which we have supplied is the contribution of the high-energy part of T to the off-shell extrapolation. We emphasize again that neither pole dominance nor small symmetry-breaking assumptions were involved in the derivation of (7.8). Our mass DR's have enabled us to approximately *compute* the effects of the off-shell extrapolation.

We now follow ref. [2] and apply (7.8) to π-$\mathcal{N}$ scattering. The off-shell amplitudes $T^{(\pm)}(\varkappa, \nu)$ satisfy the exact zero-energy theorems [8]

$$(7.9) \qquad T^{(-)\prime}(0, 0) = -\frac{2}{f_\pi^2}(g_A^2 - 1)\frac{1}{2m_\mathcal{N}}\,,$$

$$(7.10) \qquad T^{(+)}(0, 0) = -\frac{2}{f_\pi^2}\sigma_\mathcal{N}\,,$$

where

$$(7.11) \qquad \sigma_\mathcal{N} \equiv \langle N|\sigma(0)|N\rangle$$

is the nucleon expectation value of the σ-term

$$(7.12) \qquad \sigma(0) = \tfrac{2}{3}\varepsilon_2[\sqrt{2}\,S^0(0) + S^8(0)]\,.$$

The analogues of eq. (7.8) can be used to check (7.9) and (7.10), by relating them to on-shell quantities. For this purpose, the generalizations $\beta^{(\pm)}$ of the constant in (7.3) must be estimated. This is done in ref. [2] by relating $T^{(\pm)}$ to the Compton-scattering amplitudes in the deep inelastic limit. (To obtain this relation, the canonical formalism of Sect. 2 must be used strongly.)

From the relation for the isospin odd amplitude $T^{(-)}$, using the nonchiral value (1.11*b*) of C, we obtain $g_A \sim 1.23$ [2]. The chiral value (1.11*a*) of C gives $g_A \sim 1.30$. The most recent experimental determinations of g_A give about 1.24 [28]. The nonchiral value for C is thus favoured, although the theoretical uncertainties in the calculations are sufficiently large in the present case that this must be taken as just a suggestion. Since completely ignoring off-shell effects gives only $g_A \sim 1.16$, however, these effects are rather important and indicate that chiral symmetry is not too good.

From the relation for the isospin even amplitude $T^{(+)}$, the nonchiral value (1.11*b*) gives a huge σ-term $\sigma_{\mathcal{N}} \sim (430 \pm 150)$ MeV [2, 29]. All we can say for the result of using the chiral value (1.11*a*) is that $\sigma_{\mathcal{N}} \sim (100 \pm 100)$ MeV. There is, unfortunately, no independent nonperturbative determination of $\sigma_{\mathcal{N}}$ to check which result is better. It is clear from (7.12), however, that, unless $\langle N|S^0|N\rangle$ is anomalously large, $\sigma_{\mathcal{N}}$ should be quite small if ε_2 is small (chiral value) and (1.3) is the correct symmetry-breaking term. Some recent determinations [30, 31] of $\sigma_{\mathcal{N}}$, based on the *assumption* that ε_2 is small, give too large values to be consistent with this. Therefore, if these determinations are correct and if ε_2 is really small, then one must either give up (1.3) or invent some mechanism (*e.g.*, a new low-lying meson [32, 33]) for making $\langle N|S_0|N\rangle$ very big.

It is unfortunate that, at the present time at least, we cannot tell for certain from π-$\mathcal{N}$ scattering which, if either, value of C is correct. Nor can consideration of other scattering processes give us this information at present. All of the other current algebra, PCAC results usually taken as indicating the smallness of ε_2 (*e.g.*, Adler consistency condition, Weinberg-Tomozawa scattering lengths, Fubini photoproduction sum rules, $K_{3\pi}$ decay) can also be derived with our methods without this assumption [15]. We conclude that these scattering processes at present give us little information on the magnitude of ε_2.

8. – Discussion.

In the above Sections we have shown how LC-controlled mass DR's can be used to accurately calculate the deviations from pion (and kaon) pole dominance for both vertex functions and scattering amplitudes. Our mass DR's can be looked upon as a means of deducing observable (dynamical) consequences from an underlying operator (algebraic) structure. This operator structure has been extended from the original $SU_3 \times SU_3$ equal-time algebra [7], to the $(3, \bar{3}) + (\bar{3}, 3)$ symmetry-breaking relations [7], to the additional equal-time relations [14] and full LC expansions [3] abstracted from the canonical gluon model. The equal-time commutation relations imply exact zero-energy and -mass theorems which are at unphysical (zero) masses for the pseudoscalar at these points to their values at the mass-shell points of the pseudoscalar mesons or vector mesons and to their values at mass Λ which, by precocious asymptopia, are given by the LC. The smooth threshold assumption relates this LC behaviour to either equal-time relations or to other measurable quantities. We are thus provided with algebraic relations among physical quantities which are the observable reflections of the basic underlying operator algebra.

The present experimental situation is unfortunately not yet accurate enough for us to unequivocally conclude whether $C/\sqrt{2}$ is near zero (nonchiral world) or near -1 (chiral world). The K_{l3} decay provides the most decisive test. Here the present results definitely favour a small C, but further clarification is needed. Pion-nucleon scattering perhaps also indicates a slight preference for small C since off-shell effects seem important in the Adler-Weisberger relation and rather large σ-terms may be present.

Some further support for a small C comes from consideration of the $\pi^0 \to \gamma\gamma$ and $\eta \to 3\pi$ decays, both of which vanish in the limit $\varepsilon_2 \to 0$ [8]. The first problem can be got around if perturbative anomalies [34] are present in D^{π^0} in the presence of electromagnetism (although the rate or sign comes out wrong in the usual perturbative models [15]), but more drastic remedies, such as the presence of nonelectromagnetic isospin-violating interactions or additions to the usual electromagnetic current (2.12), are needed to get the η to decay in the chiral framework [35-37]. Similarly, the Dashen [11] electromagnetic mass shift sum rule, valid to order ε_2 but in terrible agreement with experiment, suggests that conventional electromagnetism should be changed if ε_2 is really small.

It is clear that difficulties are encountered if one wants a small ε_2. Each difficulty can, of course, be argued away. Thus, the present K_{l3} experiments may be wrong, the $(3, \bar{3}) + (\bar{3}, 3)$ symmetry-breaking model may be wrong or a new scalar meson may be around to enhance $\langle N|S^0|N\rangle$, an anomaly may be present in D^{π^0}, and our present understanding of electromagnetism and isospin violation may be wrong. These are all the possibilities which may turn out to be correct. It seems far simpler, however, to simply give up the idea that ε_2 is small. Indeed, it seems that every time one goes beyond the good current algebraic results which are independent of the assumption that ε_2 is small and tries to do perturbation theory in ε_2, one encounters an enormous disagreement with experiment unless one simultaneously invents a new particle or a new interaction. The *new* predictive power of good chiral symmetry has, in fact, been nonexistent. The beauty of the chiral framework of course makes these attempts to save it understandable. Experiment, however, must remain the final judge.

* * *

I thank the CERN Theoretical Study Division, where these notes were written, for its hospitality. I am grateful to B. Renner for a careful and critical reading of the manuscript.

REFERENCES

[1] R. Brandt and G. Preparata: *Phys. Rev. Lett.*, **26**, 1605 (1971); **29**, 244 (1972).
[2] R. Brandt and G. Preparata: *Phys. Rev. D*, **7** (1973), to be published.
[3] R. Brandt and G. Preparata: *Nucl. Phys.*, **27** B, 541 (1971).

[4] For a review of the theory and a summary of applications, see: R. BRANDT and G. PREPARATA: *Broken Locale Invariance and the Light Cone* (New York, 1971), p. 43.
[5] A more complete review is contained in: R. BRANDT and G. PREPARATA: *Proceedings of the Summer Institute in Theoretical Physics* (Hamburg, 1971).
[6] See also the lectures by: R. BRANDT and G. PREPARATA: *1970 Brandeis, Erice and Heidelberg Summer Schools.*
[7] M. GELL-MANN: *Phys. Rev.*, **125**, 1067 (1962).
[8] See, for example: S. ADLER and R. DASHEN: *Current Algebra and Applications to Particle Physics* (New York, 1968).
[9] S. WEINBERG: Rapporteur talk at the *XIV International Conference on High-Energy Physics* (Vienna, 1968).
[10] Y. NAMBU and G. JONA-LASINIO: *Phys. Rev.*, **122**, 345 (1961); **124**, 246 (1961).
[11] R. DASHEN: *Phys. Rev.*, **183**, 1245 (1968).
[12] M. GELL-MANN, R. OAKES and B. RENNER: *Phys. Rev.*, **175**, 2195 (1968).
[13] S. GLASHOW and S. WEINBERG: *Phys. Rev. Lett.*, **20**, 224 (1968).
[14] See: R. BRANDT and G. PREPARATA: *Phys. Rev. D*, **1**, 2577 (1970), for a previous application of this extended algebra.
[15] R. BRANDT and G. PREPARATA: *Ann. of Phys.*, **61**, 119 (1970).
[16] R. BRANDT and G. PREPARATA: *Phys. Rev. Lett.*, **25**, 1530 (1970).
[17] G. ALTARELLI, R. BRANDT and G. PREPARATA: *Phys. Rev. Lett.*, **26**, 42 (1970).
[18] See, for example: D. AMATI, R. JENGO, H. R. RUBINSTEIN, G. VENEZIANO and M. A. VIRASORO: *Phys. Lett.*, **27** B, 38 (1968).
[19] J. D. BJORKEN: *Phys. Rev.*, **148**, 1469 (1966).
[20] J. E. AUGUSTIN, D. BENAKSAS, J. C. BIZOT, J. BUON, B. DELCOURT, V. GRACCO, J. HAISSINSKI, J. JEANJEAN, D. LALANNE, F. LAPLANCHE, J. LEFRANÇOIS, P. LEHMANN, P. MARIN, H. NGUYEN NGOC, J. PEREZ-Y-JORBA, F. RICHARD, F. RUMPF, E. SILVA, S. TAVERNIER and D. TREILLE: *Phys. Lett.*, **28** B, 503 (1969).
[21] For further application to VMD problems, see ref. [4] and: R. BRANDT, W. C. NG, P. VINCIARELLI and G. PREPARATA: *Lett. Nuovo Cimento*, **2**, 937 (1971); G. PREPARATA: *Phys. Lett.*, **36** B, 53 (1971).
[22] The smaller value $M^2 = 1.5$ is expected for the π case since $m_\pi^2 = 0.02$ whereas $m_\rho^2 = 0.5$. These mesons are expected to dominate near their masses so that the effective continuum is from 0 to 2.5 in the π case and from 1 to 2.5 in the ρ case.
[23] See ref. [9] for a discussion of some previous attempts to calculate these quantities.
[24] For a review, see: M. K. GAILLARD and L. CHOUNET: CERN 70-14 (1970).
[25] R. DASHEN and M. WEINSTEIN: *Phys. Rev. Lett.*, **22**, 1337 (1969).
[26] AACHEN-CERN-TORINO COLLABORATION, V. BISI *et al.*: *Phys. Lett.*, to be published.
[27] X2-COLLABORATION, H. J. STEINER *et al.*: *Phys. Lett.*, to be published.
[28] C. J. CHRISTENSON, V. E. KROHN and C. R. RINGO: *Phys. Lett.*, **28** B, 411 (1969), from polarized neutron β decay, find a radiatively corrected $(g_A/g_V)_0 = 1.26 \pm 0.02$. From the ratio of the neutron β decay rate of C. J. CHRISTENSON, A. NIELSEN, A. BAHNSEN, W. K. BROWN and B. M. RUSTAD *et al.* (*Phys. Lett.*, **26** B, 11 (1968)) and the rate of ^{26}Al one gets $(g_A/g_V)_0 = 1.226 \pm 0.011$, and from the rate of ^{14}O one obtains $(g_A/g_V)_0 = 1.232 \pm 0.012$.
[29] See also: W. C. NG: *Phys. Rev. D*, **4**, 2079 (1971).
[30] T. CHENG and R. DASHEN: *Phys. Rev. Lett.*, **26**, 594 (1971).
[31] G. HÖHLER, H. JAKOB and R. STRAUSE: *Phys. Lett.*, **35** B, 4451 (1971); the value here is somewhat smaller than that of ref. [30].

[32] R. CREWTHER: *Phys. Rev. D*, **3**, 3152 (1971).
[33] G. ALTARELLI, N. CABIBBO and L. MAIANI: *Phys. Lett.*, **35** B, 4 (1971).
[34] S. ADLER: *Phys. Rev.*, **177**, 2426 (1969).
[35] M. WEINSTEIN: *Phys. Rev. D*, **4**, 2544 (1971).
[36] There is, moreover, experimental evidence against these remedies. See: T. CHENG and R. DASHEN: *Phys. Rev. D*, **4**, 1561 (1971).
[37] In the nonchiral framework, on the other hand, not only is there no problem about the magnitude of the decay rate, but the good current algebraic prediction of the slope is maintained. See: R. BRANDT, M. GOLDHABER, C. ORZALESI and G. PREPARATA: *Phys. Rev. Lett.*, **24**, 1517 (1970).

Properties of Noncanonical Scaling.

P. Menotti

Scuola Normale Superiore - Pisa
Istituto Nazionale di Fisica Nucleare - Sezione di Pisa

In this seminar we shall examine some general properties of the Compton amplitude and scaling functions which are obtained in the presence of noncanonical singularities of the commutator forward matrix element on the light-cone. The work I shall describe has been done in collaboration with Gatto [1], and makes extensive use of semi-Fourier transforms of light-cone singularities. The interest in examining such a situation is to see whether in case of noncanonical singularities [2] some pathological feature obtains, which would enable one to reject the noncanonical case, as it seems from experiment that the canonical case is the one realized in Nature. As we shall see, no definite clash with commonly accepted principles emerges even if the noncanonical case shows up rather peculiar features. The new features with respect to the canonical case are:

1) the scaling functions $F(\omega)$ have unbounded support;

2) fixed Regge poles at noninteger values of the angular momentum with complex residues, or multiple poles occur in the Compton amplitude [3].

As byproducts we shall obtain a very simple derivation of the Bjorken and Regge limits for canonical and noncanonical light-cone singularities and a straightforward derivation of the Cornwall-Corrigan-Norton sum rule [4] for the total Compton cross-section.

We start from the retarded forward Compton amplitude for off-shell photons, which for $q_0 > 0$ coincides with the T^*-amplitude:

$$T_R^{\mu\nu}(q, p) = -(g^{\mu\nu} q^2 - q^\mu q^\nu) T_1(q^2, \nu) - \\ - [p^\mu p^\nu q^2 - (p^\mu q^\nu + q^\mu p^\nu)(p\cdot q) + g^{\mu\nu}(q\cdot p)^2] T_2(q^2, \nu) , \quad (1)$$

where q and p are the photon and nucleon momenta.

The invariant amplitudes T_1 and T_2 are assumed to be Fourier transforms of two retarded functions $r_i(x^2, x\cdot p)$ $(i = 1, 2)$:

$$T_i(q^2, \nu) = \int d^4x \exp[iq\cdot x]\, r_i(x^2, x\cdot p) , \quad (2)$$

with the following behavior near the light-cone:

$$r_i(x^2, x\cdot p) \simeq \theta(x\cdot p)f_i(x\cdot p)[\Gamma(\lambda_i+1)]^{-1}(x^2)^{\lambda_i}\theta(x^2)\,, \tag{3.1}$$

or

$$r_i(x^2, x\cdot p) \simeq \theta(x\cdot p)f_i(x\cdot p)[\Gamma(\lambda_i+1)]^{-1}(x^2)^{\lambda_i}\theta(x^2)\ln(x^2)\,. \tag{3.2}$$

λ_i are real numbers and $f_i(x\cdot p)$ must be real for Hermitean currents. More general behaviors containing $(x^2)^{\lambda_i}\ln^n(x^2)$ with n integer can be similarly discussed.

One goes over to the Fourier transform of the connected commutator of the two electromagnetic currents by taking the imaginary part of $T_R^{\mu\nu}$:

In fact, we have

$$\begin{aligned}\text{Im}\, T_R^{\mu\nu}(q, p) &= \tfrac{1}{2}\int \exp\,[iq\cdot x]\,\langle p|[J^\mu(x), J^\nu(0)]|p\rangle\, \mathrm{d}^4x = \\ &= -(g^{\mu\nu}q^2 - q^\mu q^\nu)V_1(q^2, \nu) - (p^\mu p^\nu q^2 - (p^\mu q^\nu + q^\mu p^\nu)(p\cdot q) + g^{\mu\nu}(p\cdot q)^2]V_2(q^2, \nu)\,.\end{aligned} \tag{4}$$

$V_1(q^2, \nu)$ and $V_2(q^2, \nu)$ have well-defined supports as can be seen by considering the laboratory frame, $p = (m, 0)$ and by introducing a complete set of intermediate states in the integral in (4)

$$\begin{aligned}&\int \exp\,[iq\cdot x]\langle p|[J^\mu(x), J^\nu(0)]|p\rangle\, \mathrm{d}^4x = \\ &= (2\pi)^4\sum_n \langle p|J^\mu(0)|n\rangle\langle n|J^\nu(0)|p\rangle\delta^4(q+p-p_n) - \\ &\qquad - (2\pi)^4\sum_{n'}\langle p|J^\nu(0)|n'\rangle\langle n'|J^\mu(0)|p\rangle\delta^4(q-p+p_{n'})\,.\end{aligned} \tag{5}$$

For $q_0 > 0$ $(\nu > 0)$ only the first sum contributes, and one must have

$$(q+p)^2 = q^2 + 2q\cdot p + p^2 = p_n^2 \qquad \text{with} \qquad p_n^2 = m^2\,,$$

for the single-nucleon intermediate state or otherwise

$$p_n^2 \geqslant (m+\mu)^2\,,$$

i.e.

$$\omega = -\frac{q^2}{2\nu} = 1 - \frac{p_n^2 - p^2}{2\nu} \leqslant 1\,, \qquad \nu = p\cdot q\,. \tag{6}$$

Moreover, one has

$$q^2 = q_0^2 - \mathbf{q}^2 < q_0^2 = \frac{(q\cdot p)^2}{m^2} = \frac{\nu^2}{m^2}\,. \tag{7}$$

For $q_0 < 0$ the second sum contributes and the new support is obtained by reflecting in ν.

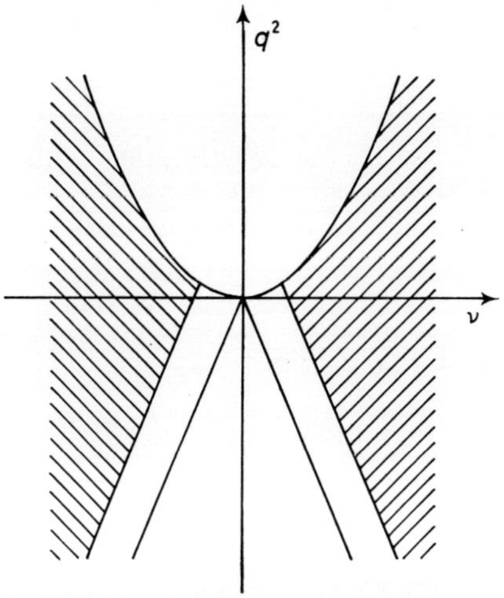

Fig. 1. – Spectral support for the commutator matrix element.

In the first sum one can distinguish three types of contributions, the connected part (C), the semi-disconnected part (D) and the pair part (P) according to the possible intermediate states as shown in Fig. 2.

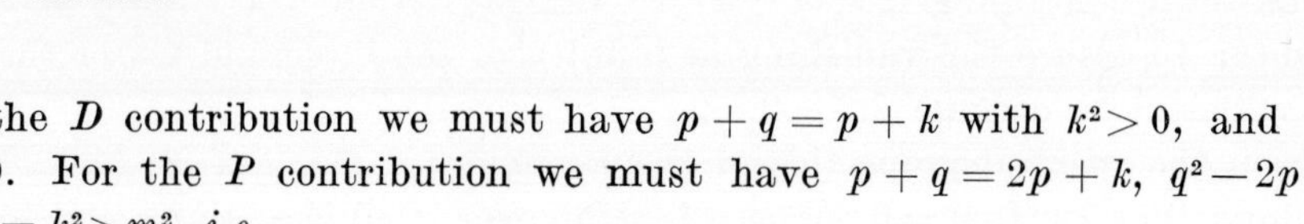

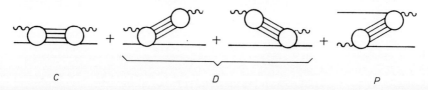

Fig. 2. – Connected, semi-disconnected and pair contributions to the commutator matrix element.

For the D contribution we must have $p + q = p + k$ with $k^2 > 0$, and thus $q^2 > 0$. For the P contribution we must have $p + q = 2p + k$, $q^2 - 2p \cdot q + p^2 = k^2 \geqslant m^2$, *i.e.*

$$\omega = -\frac{q^2}{2\nu} = -1 - \frac{k^2 - p^2}{2\nu} \leqslant -1\,.$$

Thus we have that for $\nu > 0$, C contributes for $\omega < 1$, D for $\omega < 0$ and P for $\omega < -1$.

A factorized form for the singularity near the light cone as considered in eqs. (3) gives rise, as we shall see, to a generalized scaling behavior. The canonical case obtains for $\lambda_1 = -1$ and $\lambda_2 = 0$. To evaluate the integrals in eq. (2) we introduce the Fourier transform of the product $f(x\cdot p)\theta(x\cdot p)$; in this way we will be able to exploit the analyticity properties implied by the insertion of the step function.

Putting

$$t(\xi) = \frac{1}{2\pi}\int \exp\left[-ip\cdot x\xi\right] f(x\cdot p)\theta(x\cdot p)\,\mathrm{d}(x\cdot p)\,, \tag{8}$$

one obtains from eq. (2) with $r_i(x^2, x\cdot p)$ as given by eq. (3.1)

$$T(q^2, \nu) \simeq 2\pi 4^{\lambda+1}\Gamma(\lambda+2)\int \mathrm{d}\xi\, t(\xi)\cdot \cdot\{\exp\left[i\pi\lambda\right]\left[(q+p\xi)^2+i\varepsilon\right]^{-\lambda-2} + \exp\left[-i\pi\lambda\right]\left[(q+p\xi)^2-i\varepsilon\right]^{-\lambda-2}\}\,. \tag{9}$$

The function $t(\xi)$ has the following important properties:

i) it is analytic in the lower half-plane;

ii) $t^*(\xi) = t(-\xi)$ as it follows from the reality of $f(x\cdot p)$.

Moreover, in the following we shall need the asymptotic behavior of $t(\xi)$ for $\xi\to\infty$. Assuming $f(x\cdot p)$ to be regular and at infinity polynomially bounded, one has, integrating by parts that $t(\xi) = 0(\xi^{-1})$ for $|\xi|\to\infty$.

We now perform the Bjorken limit [5] in eq. (9) by taking $\nu\to+\infty$ with $\omega = -q^2/2\nu$ fixed:

$$T(q^2, \nu) \underset{B}{\to} 2\pi 4^{\lambda+1}\Gamma(\lambda+2) I(\omega)(2\nu)^{-\lambda-2}\,, \tag{10}$$

with

$$I(\omega) = \exp\left[i\pi\lambda\right]\int \mathrm{d}\xi\, t(\xi)(\xi-\omega+i\varepsilon)^{-\lambda-2}\,. \tag{11}$$

The term containing $(\xi-\omega-i\varepsilon)^{-\lambda-2}$ vanishes from usual analyticity arguments; to assure the vanishing at infinity of the integrand faster than ξ^{-1} it is sufficient to assume $\lambda > -2$.

From the spectral properties of the commutator (see eq. (6) and Fig. 1) it follows that $\mathrm{Im}\, I(\omega) = 0$ for $\omega > 1$. Moreover, $I(\omega)$ is analytic and regular in the lower half-plane and from ii) above we have

$$I^*(\omega) = \exp\left[-i\pi\lambda\right] I(-\omega)\,. \tag{12}$$

The Bjorken limit of νW_2 is obtained from the imaginary part of $T_2(q^2, \nu)$:

$$\nu W_2 \to (2\nu)^{-\lambda} F_2(\omega)\,, \qquad F_2(\omega) = 2\pi 4^{\lambda+1}\Gamma(\lambda+2)\omega\,\mathrm{Im}\, I(\omega)\,. \tag{13}$$

To understand the structure of this generalized scaling function $F_2(\omega)$ we put

$$I(\omega) = (\omega - i\varepsilon)^{-\lambda} f(\omega) . \tag{14}$$

This combined with eq. (12) gives

$$f^*(\omega) = f(-\omega) , \tag{15}$$

and in particular $\operatorname{Im} f(-\omega) = -\operatorname{Im} f(\omega)$.

Thus for $\lambda =$ integer $F_2(\omega) \propto \omega^{-\lambda+1} \operatorname{Im} f(\omega)$ and it is even in ω for $\lambda =$ even and odd for $\lambda =$ odd. Moreover, for integer λ, $F_2(\omega)$ has compact support because it has definite parity and vanishes for $\omega > 1$. On the other hand, for $\lambda \neq$ integer one has for $\omega < -1$, $F_2(\omega) \propto (-\omega)^{-\lambda+1} \sin \pi\lambda \operatorname{Re} f(\omega)$, since $\operatorname{Im} f(\omega)$ vanishes identically for $\omega < -1$. Thus for $\lambda \neq$ integer $F_2(\omega)$ cannot vanish for $\omega < -1$ except at isolated points, and thus the support of $F_2(\omega)$ stretches down to $-\infty$. It is important to note that we take the Bjorken limit with $\nu \to +\infty$ independently of the sign of ω. In the canonical case a well defined relationship has to hold among the contributions $C(\omega)$, $D(\omega)$ and $P(\omega)$ (respectively of the connected, semi-disconnected and pair contributions to the commutator) to the scaling function $F_2(\omega)$ in order to make it even between -1 and 1 and vanishing for $\omega < -1$.

Actually, assuming the analyticity of $C(\omega) + D(\omega)$ around the point $\omega = -1$, say for $-1 - a < \omega < -1 + a$, one can give a connection [6] between $P(\omega)$, which apart for a sign is the scaling function in the annihilation region, and the analytic continuation of $C(\omega)$ above $\omega = 1$. In fact, for $-1 < \omega < 0$ we must have

$$C(\omega) + D(\omega) = C(-\omega) ,$$

which ensures the analytic continuability of $C(\omega)$ above 1 for $\omega < 1 + a$, while from

$$C(\omega) + D(\omega) + P(\omega) = 0 \qquad \text{for } \omega < -1,$$

we get

$$P(\omega) = - C(-\omega) \qquad \text{for } -a - 1 < \omega < -1.$$

We come now to the evaluation of the Regge limit of $T(q^2, \nu)$, *i.e.* $\nu \to +\infty$, $q^2 = \text{const}$. In order to perform this limit one has to separate in $f(x \cdot p)$ the contribution of the higher Regge terms.

We write

$$f(x \cdot p) = \tilde{f}(x \cdot p) + \sum_j b_j (x \cdot p)^{\delta_j} , \tag{16}$$

where the summation is extended to the asymptotic terms with $\delta_j > -\lambda - 2$. We exclude the accidental degeneracy $\delta_j = -\lambda - 2$. Then we can write

$$t(\xi) = \tilde{t}(\xi) + t_R(\xi)\,, \tag{17}$$

with

$$\tilde{t}(\xi) = \frac{1}{2\pi}\int \exp\left[-ip\cdot x\xi\right]\tilde{f}(x\cdot p)\theta(x\cdot p)\,\mathrm{d}(x\cdot p) \tag{18}$$

and

$$t_R(\xi) = -\frac{i}{2\pi}\sum_j b_j \exp\left[-\frac{i\pi\delta_j}{2}\right]\Gamma(\delta_j + 1)(\xi - i\varepsilon)^{-\delta_j - 1}\,. \tag{19}$$

The contribution of $\tilde{t}(\xi)$ and $t_R(\xi)$ to the integral in eq. (9) have to be treated differently in the limit $\nu \to +\infty$. In the term with $\tilde{t}(\xi)$ the Regge limit $\nu \to +\infty$ is performed by taking $(q + p\xi)^2 + i\varepsilon \to 2\nu(\xi + i\varepsilon)$. The limit of the term with $t_R(\xi)$ is obtained by introducing $z = 2\nu\xi$ and taking

$$(q + p\xi)^2 + i\varepsilon \to q^2 + z + i\varepsilon\,.$$

The result of the Regge limit on the amplitudes as given by eq. (2) is then given by

$$T(q^2, \nu) \underset{R}{\to} 2\pi 4^{\lambda+1}(2\nu)^{-\lambda-2}\sum_j b_j \exp\left[-\frac{i\pi\delta_i}{2}\right]\Gamma(\lambda + \delta_j + 2)\cdot \\ \cdot\left(-\frac{q^2}{2\nu}\right)^{-\lambda-\delta_j-2} + 2\pi 4^{\lambda+1}\Gamma(\lambda+2)(2\nu)^{-\lambda-2}\tilde{I}(0) + \text{less dominant terms}\,. \tag{20}$$

Again in the evaluation, the two integrals with the factors $(q^2 + z - i\varepsilon)^{-\lambda-2}$ and $(\xi - \omega - i\varepsilon)^{-\lambda-2}$ vanish identically for analyticity reasons.
$\tilde{I}(\omega)$ is given by

$$\tilde{I}(\omega) = \exp\left[i\pi\lambda\right]\int \mathrm{d}\xi\,\tilde{t}(\xi)(\xi - \omega + i\varepsilon)^{-\lambda-2}\,, \tag{21}$$

i.e. it is nothing else than $I(\omega)$ with $t(\xi)$ replaced by $\tilde{t}(\xi)$. Thus we see that in addition to the usual Regge-poles contributions as given by the first term in eq. (20) we have a fixed Regge-pole contribution $\propto (2\nu)^{-\lambda-2}$. This pole is called fixed because its position depends only on the nature of the light-cone singularity which, as follows from the Wilson [2] expansion of products of currents as generalized to the light-cone [7], should be independent of the momentum transfer between the protons [8].

It is important to note that the residue of this fixed pole is independent of q^2 and this feature will allow us to connect it to the physical Compton cross-section; the other Regge-poles have a residue that depends on q^2 in such a way as to ensure scaling [9]. We recall that in deriving the above written Regge-limit we have excluded the accidental degeneracy of a moving Regge-pole with $\delta_j(0) = -\lambda - 2$. Nondominant light-cone singularities should not alter the residue of the fixed pole as each of them would contribute with a different λ.

$\tilde{I}(\omega)$ has the same reality properties as $I(\omega)$; in particular we have $\tilde{I}(0) = \exp[i\pi\lambda/2] \times$ (real quantity). Thus a noncanonical singularity would give rise to a fixed pole in the Compton amplitude at a noninteger value of the angular momentum with a corresponding complex residue. Such a pole therefore shows up both in the real and in the imaginary part of the amplitude.

The logarithmic singularity on the light-cone as in eq. (3.2) can be obtained by simply applying the operation $[\Gamma(\lambda+1)]^{-1}(\mathrm{d}/\mathrm{d}\lambda)\Gamma(\lambda+1)$ to eq. (3.1). Thus we have in the Regge-limit in addition to the higher poles contribution the asymptotic behavior

$$(2\nu)^{-\lambda-2}\left[-\ln 2\nu + \frac{i\pi}{2}\right]\tilde{I}(0),$$

which is typical of a double Regge-pole.

$\tilde{I}(\omega)$ is obtained by replacing in $I(\omega)$, $t(\omega)$ with $\tilde{t}(\omega)$. Thus we can express the residue of the fixed Regge pole in terms of the Regge subtracted scaling function $\tilde{F}_2(\omega)$. By this we mean the function obtained by subtracting from $F_2(\omega)$ the Regge-terms behaving as $\omega^{-\lambda-1-\delta_j}$ with $\delta_j > -\lambda - 2$. $\tilde{F}_2(\omega)$ satisfies eq. (13) with $I(\omega)$ replaced by $\tilde{I}(\omega)$. Using the analyticity of $\tilde{I}(\omega)$ in the lower half-plane one can then express, by writing a dispersion relation, the residue of the fixed Regge-pole as follows:

$$R = -\frac{1}{\pi}\int_{-\infty}^{+\infty}\frac{\tilde{F}_2(\omega)\,\mathrm{d}\omega}{\omega(\omega+i\varepsilon)}. \tag{22}$$

As we have discussed above, this residue is real in the canonical case $\lambda_2 = 0$. As the residue is independent of q^2 we can express it in terms of the physical Compton amplitude. One starts from the Regge-subtracted dispersion relation for the amplitude $A_+(\nu) = -(\nu^2/2)T_2(\nu)$

$$\tilde{A}_+(\nu) = 1 - \frac{\nu^2}{2\pi^2\alpha}\int_0^\infty \frac{\tilde{\sigma}_\gamma(\nu')\,\mathrm{d}\nu}{\nu'^2 - \nu^2}.$$

By taking the limit $\nu \to +\infty$, and using eq. (22) for the residue, we get the sum rule for Compton scattering

$$1 + \frac{1}{2\pi^2\alpha}\int_0^\infty \tilde{\sigma}_\gamma(\nu)\,\mathrm{d}\nu = \frac{1}{4\pi}\int_0^\infty \frac{\mathrm{d}\omega}{\omega^2}\tilde{F}_2(\omega),$$

which was obtained by CORNWALL, CORRIGAN and NORTON [4], by using the DESER-GILBERT-SUDARSHAN [10] representation. The possibility of writing a sum rule for the residue in terms of the Compton cross-section is bound to the case $\lambda_2 =$ even integer; in such a situation in fact the pole shows up only in the real part of the amplitude.

REFERENCES

[1] R. GATTO and P. MENOTTI: *Phys. Rev. D*, **5**, 1493 (1972).

[2] K. G. WILSON: *Phys. Rev.*, **179**, 1499 (1969); C. G. CALLAN: *Phys. Rev. D*, **2**, 1541 (1970); K. SYMANZIK: *Comm. Math. Phys.*, **18**, 227 (1970); K. G. WILSON: SLAC-PUB. 807; S. COLEMAN and R. JACKIW: MIT preprint CTP 172 (1970).

[3] J. B. BRONZAN, I. S. GERNSTEIN, B. W. LEE and F. E. LOW: *Phys. Rev.*, **157**, 1448 (1967); T. P. CHENG and WU-KI TUNG: *Phys. Rev. Lett.*, **35** B, 234 (1971); V. F. MUELLER and W. RUEHL: *Nucl. Phys.*, **30** B, 513 (1971); G. MACK: *Phys. Lett.* **35** B, 234 (1971).

[4] J. M. CORNWALL, D. CORRIGAN and E. R. NORTON: *Phys. Rev. Lett.*, **24**, 1141 (1970); *Phys. Rev. D*, **3**, 536 (1971).

[5] J. D. BJORKEN: *Phys. Rev.*, **179**, 1547 (1969).

[6] S. D. DRELL, D. LEVY and T. M. YAN: *Phys. Rev.*, **187**, 2159 (1969); J. PESTIEAU and P. ROY: *Phys. Lett.*, **30** B, 483 (1969).

[7] R. BRANDT and G. PREPARATA: *Nucl. Phys.*, **27** B, 541 (1971); Y. FRISHMAN: *Phys. Rev. Lett.*, **25**, 966 (1970).

[8] This pole occurs in the canonical case at $J = 0$. Fixed poles in weak amplitudes at $J = 1$ on the other hand are related to the Dashen–Fubini–Gell-Mann sum rules; R. F. DASHEN and M. GELL-MANN: *Proceedings of the III Coral Gables Conference*; S. FUBINI: *Nuovo Cimento*, **43** A, 475 (1966).

[9] H. D. ABARBANEL, M. L. GOLDBERGER and S. B. TREIMAN: *Phys. Rev. Lett.*, **22**, 500 (1969).

[10] S. DESER, W. GILBERT and E. C. G. SUDARSHAN: *Phys. Rev.*, **115**, 731 (1959); N. NAKANISHI: *Progr. Theor. Phys. (Kyoto)*, **23**, 1151 (1960).

Introduction to General Relativity.

R. SEXL

Institut für Theoretische Physik der Universität Wien - Wien

1. – Introduction to general relativity.

1'1. *Why is gravitation different?* – For several decades the theory of gravitation had been disconnected from the rest of physics: Einstein's general relativity, admittedly one of the most admirable and logically perfect physical theories, was not part of the mainstream of physics. The main reason for this was, of course, the lack of progress of the experimental tests of general relativity. This situation has changed, and since 1960 « experimental relativity » has become a reality. This is due to the progress of high-precision measurement techniques (Mössbauer effect, Radar technology etc.), to space exploration (Mariner etc.) and to new astronomical discoveries (pulsars, quasars).

There is, however, a second reason for the isolation of gravitation within physics. The mathematical concepts used in general relativity are rather different from those employed in other field theories (electrodynamics, weak and strong interactions). The gravitational field is explained in terms of the curvature of a Riemannian space-time, while other field theories are based on the concept of Poincaré invariance.

The reason for this difference is the following: The *principle of equivalence* postulates that a freely falling small system (satellite etc.) constitutes a local

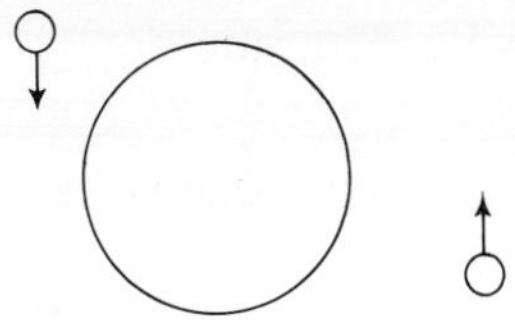

Fig. 1. – Two satellites as local internal system.

inertial system. Effects of gravity are absent in such a system, if it is sufficiently small. The relation between these local inertial systems at different points in space-time is obviously rather complicated, as Fig. 1 shows. This relation is described by the curvature of space-time, and the curvature in turn is determined by the mass distribution, which governs the relation between freely falling systems at different points.

There is thus no *global* inertial system and the concepts of Poincaré-invariant field theories have to be abandoned.

1·2. *Christoffel symbols.* – In a local inertial system the line element is given by

$$\mathrm{d}s^2 = \eta_{ik}\,\mathrm{d}y^i\,\mathrm{d}y^k\,, \tag{1.1}$$

where $\eta_{ik} = \mathrm{diag}\,(1, -1, -1, -1)$ (we put $c = 1$) is the metric of flat space-time. This line element is valid, however, only in a small neighborhood of $y = 0$.

We have now to find an expression for $\mathrm{d}s^2$ which is valid globally (*), *i.e.* in all of space-time. For this purpose we introduce some global co-ordinate

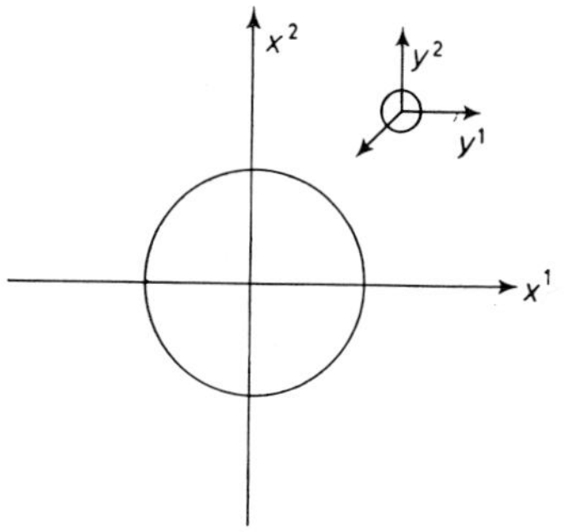

Fig. 2. – Earth with one local inertial system (y^i) and global co-ordinate system (x^i).

system x^i. The relation between y^i and x^i is in general rather complicated, since the x^i-system can be any curvilinear co-ordinate system, and the y^i-system is accelerated with respect to it. The functions

$$y^i = f^i(x^k) \tag{1.2}$$

are thus generally complicated, but can be determined experimentally (in principle) very easily by reading off the x and y co-ordinates simultaneously,

(*) The term « global » is used here in a loose way; no problems of « global differential geometry » will be dealt with in this chapter.

while the local inertial system is falling. Inserting (1.2) into (1.1) we obtain

$$\text{(1.3)} \qquad \mathrm{d}s^2 = \eta_{ik}\mathrm{d}y^i\mathrm{d}y^k = \eta_{ik}\frac{\partial y^i}{\partial x^l}\frac{\partial y^k}{\partial x^m}\mathrm{d}x^l\mathrm{d}x^m = :g_{lm}(x)\,\mathrm{d}x^l\,\mathrm{d}x^m\,,$$

where g_{lm} is the (covariant) *metric tensor* of space-time.

A large number of local inertial systems (satellites) are obviously necessary to determine g_{lm} in the manner described here, since (1.3) is valid only along the path of the local inertial system, and only by combining the information from these local systems the global metric $g_{ik}(x)$ results. When only a limited class of local systems is used, an incomplete description of space-time is obtained, as we shall see in an important example in Sect. **5**'3.

The basic task of any theory of gravitation is to determine the metric tensor g_{ik} as a function of the mass-energy distribution and to calculate the equations of motion of given masses in the gravitational field. The latter task is simple in the case of structureless (spin 0, uncharged etc.) test masses.

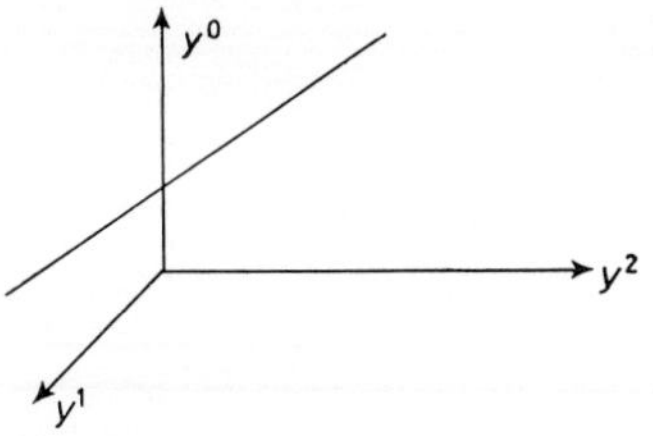

Fig. 3. – Motion of particles through local internal systems is in straight lines in space-time.

Particles move uniformly along straight lines in the local inertial system, *i.e.* along a straight line through space time (Fig. 3)

$$\text{(1.4)} \qquad y^i = v^i s\,.$$

Now we have to translate this motion into arbitrary curvilinear co-ordinates. For this purpose we start from the variational principle

$$\text{(1.5)} \qquad \delta\int \mathrm{d}s = 0$$

fulfilled by (1.4). Inserting (1.3) into (1.5) we have

$$\text{(1.6)} \qquad \delta\int\sqrt{g_{ik}\dot{x}^i\dot{x}^k}\,\mathrm{d}\lambda = :\delta\int L\,\mathrm{d}\lambda = 0\,,$$

where λ is an arbitrary parameter along the world line of the particle and $\dot{x}^i = \mathrm{d}x^i/\mathrm{d}\lambda$.

The Euler equations of the variational principle are

$$\frac{\partial L}{\partial x^i} = \frac{\mathrm{d}}{\mathrm{d}\lambda}\frac{\partial L}{\partial \dot{x}^i}\,, \tag{1.7}$$

i.e.

$$\frac{1}{L}\dot{x}^l\dot{x}^k\frac{\partial g_{ik}}{\partial x^l} - \frac{1}{L^2}g_{ik}\dot{x}^k\frac{\mathrm{d}L}{\mathrm{d}\lambda} + \frac{1}{L}\ddot{x}^k g_{ik} = \frac{1}{2L}\frac{\partial g_{kl}}{\partial x^i}\dot{x}^k\dot{x}^l\,. \tag{1.8}$$

Defining the Christoffel symbols of first kind by

$$\Gamma_{ikl} = \frac{1}{2}\left(\frac{\partial g_{ik}}{\partial x^l} + \frac{\partial g_{li}}{\partial x^k} - \frac{\partial g_{kl}}{\partial x^i}\right), \tag{1.9}$$

and chosing a parameter λ such that $\mathrm{d}L/\mathrm{d}\lambda = 0$ (*e.g.* $\lambda = s$), we obtain the geodesic equations

$$g_{ik}\ddot{x}^k + \Gamma_{ikl}\dot{x}^k\dot{x}^l = 0 \tag{1.10}$$

as the equation of motion of the particle.

To simplify (1.10) we define the contravariant metric tensor g^{kl} by

$$g_{ik}g^{kl} = \delta_i{}^l\,. \tag{1.11}$$

Multiplying (1.10) by g^{im} and summing over i we have

$$\ddot{x}^m + \Gamma^m{}_{kl}\dot{x}^k\dot{x}^l = 0\,, \tag{1.12}$$

where the Christoffel symbols of second kind have been defined by

$$\Gamma^m{}_{kl} = g^{mi}\Gamma_{ikl}\,. \tag{1.13}$$

1·3. *The Newtonian approximation.* – In the previous Section we have calculated the motion of a test mass in a gravitational field. Since we know, *e.g.*, the motion of the planets in Newtonian approximation, we are able to calculate the metric g_{ik} corresponding to the sun's gravity field in this approximation. From experiment we know that special relativity is an excellent approximation in the solar system and we can therefore put

$$g_{ik}(x) = \eta_{ik} + 2\psi_{ik}(x)\,, \tag{1.14}$$

where $|\psi|^2 \ll 1$. All planetary velocities in the solar system are nonrelativistic, so that we can approximate (1.12) by

$$\frac{\mathrm{d}^2x^i}{\mathrm{d}s^2} \simeq \frac{\mathrm{d}^2x^i}{\mathrm{d}t^2} = -\Gamma^i_{kl}\dot{x}^k\dot{x}^l \simeq -\Gamma^i_{00}\,. \tag{1.15}$$

For the spatial part of the acceleration vector we obtain thus

$$\frac{\mathrm{d}^2x^\alpha}{\mathrm{d}t^2} \simeq -\Gamma^\alpha_{00} = -g^{\alpha i}\Gamma_{i00} \simeq -\eta^{\alpha i}\Gamma_{i00} = +\Gamma_{\alpha 00} = \\ = \frac{\partial g_{0\alpha}}{\partial x^0} - \frac{1}{2}\frac{\partial g_{00}}{\partial x^\alpha} = -\frac{1}{2}\frac{\partial g_{00}}{\partial x^\alpha} = -\frac{\partial \psi_{00}}{\partial x^\alpha} = -\frac{\partial U}{\partial x^\alpha}\,. \tag{1.16}$$

Thus $\psi_{00} = U$, where U is the gravitational potential. The line element becomes therefore in *Newtonian approximation*

$$\mathrm{d}s^2 = (1 + 2U)\mathrm{d}t^2 - \mathrm{d}\boldsymbol{x}^2\,. \tag{1.17}$$

Specializing to a spherically-symmetric mass distribution (mass M) we have (putting the Newtonian gravitational constant $G = 1$) $U = -M/r$ and therefore

$$\mathrm{d}s^2 = \left(1 - \frac{2M}{r}\right)\mathrm{d}t^2 - \mathrm{d}\boldsymbol{x}^2\,, \tag{1.18}$$

$2M$ (in conventional units $2GM/c^2$) is thus a length, the *Schwarzschild radius* of the mass M. For the sun $2M = 3$ km, for the earth $2M = 0.9$ cm.

The line element (1.17) enables us to calculate all effects of Newtonian gravitation and also the red shift of spectral lines.

In order to go beyond this approximation and to find the full field equation, more formal developments are necessary which will be given in the next Section.

1·4. *Tensor algebra and tensor analysis.* – The starting point of tensor algebra is the behaviour of scalars, vectors and tensors under co-ordinate transformations $x^i = f^i(x^{\bar{k}})$, where $x^{\bar{k}}$ is an arbitrary new co-ordinate system.

For a *scalar field* T we have

$$T(x) = \overline{T}(\overline{x})\,. \tag{1.19}$$

A *contravariant vector field* v^i transforms in the same way as the co-ordinate differentials

$$\mathrm{d}x^{\bar{i}} = \frac{\partial x^{\bar{i}}}{\partial x^k}\mathrm{d}x^k =: p^{\bar{i}}_{\ k}\mathrm{d}x^k\,, \tag{1.20}$$

$$v^{\bar{i}}(\overline{x}) = p^{\bar{i}}_{\ k}(x)v^k(x)\,. \tag{1.21}$$

A *covariant vector field* is exemplified by the gradient of a scalar

$$w_{\bar{i}}(\bar{x}) = \frac{\partial \bar{T}(\bar{x})}{\partial x^{\bar{i}}} = \frac{\partial x^k}{\partial x^{\bar{i}}} \frac{\partial T(x)}{\partial x^k} =: p^k{}_{\bar{i}}\, w_k(x)\,. \tag{1.22}$$

Tensors (*e.g.*, products of vectors $C^{ik} = A^i B^k$ etc.) transform according to

$$T^{\bar{i}\bar{k}\cdot\cdot}{}_{\bar{l}\bar{m}\dots} = p^{\bar{i}}{}_a p^{\bar{k}}{}_b p^c{}_{\bar{l}} p^d{}_{\bar{m}} T^{ab\cdot\cdot}{}_{cd\dots}\,. \tag{1.23}$$

Tensors can be added, multiplied with complex numbers and contracted as in special relativity.

The *metric tensor* g_{ik} is a nonsingular ($\det g_{ik} =: g \neq 0$) symmetric tensor

$$g_{ik} = g_{ki} \tag{1.24}$$

defined on space-time. It enables us to define the length A of a vector A^i and to convert co- and contravariant vectors into one another:

$$A_i = g_{ik} A^k\,, \qquad A^k = g^{ik} A_i\,, \tag{1.25}$$

$$A^2 := A_i A^i = g_{ik} A^i A^k\,. \tag{1.26}$$

The distance of neighboring points in space-time is given by

$$\mathrm{d}s^2 = g_{ik}\,\mathrm{d}x^i \mathrm{d}x^k\,. \tag{1.27}$$

The *Christoffel symbols* Γ_{ikl} formed from g_{ik} do not form a tensor. They transform under co-ordinate transformations according to

$$\Gamma_{\bar{i}\bar{k}\bar{l}} = \Gamma_{abc}\, p^a{}_{\bar{i}} p^b{}_{\bar{k}} p^c{}_{\bar{l}} + g_{ab} p^a{}_{\bar{i}} p^b{}_{\bar{k}\bar{l}}\,, \tag{1.28}$$

where

$$p^b{}_{\bar{k}\bar{l}} = \frac{\partial^2 x^b}{\partial x^{\bar{k}} \partial x^{\bar{l}}}\,. \tag{1.29}$$

Now we turn to *tensor analysis*. In special relativity a new tensor $D^{ik\dots}{}_{\cdot\cdot lmn}$ can be formed from a tensor $D^{ik\dots}{}_{\cdot\cdot lm}$ by differentiating D:

$$D^{ik\dots}{}_{\dots lmn} = \frac{\partial}{\partial x^n} D^{ik\dots}{}_{\dots lm}\,. \tag{1.30}$$

This is no longer possible in curvilinear co-ordinates as the example of a vector field shows

$$v^{\bar{m}}{}_{,\bar{k}} := \frac{\partial v^{\bar{m}}}{\partial x^{\bar{k}}} = p^l{}_{\bar{k}} \frac{\partial}{\partial x^l} p^{\bar{m}}{}_i v^i(x) = p^l{}_{\bar{k}} p^{\bar{m}}{}_i v^i{}_{,l} + p^l{}_{\bar{k}} p^{\bar{m}}{}_{li} v^i\,. \tag{1.31}$$

No tensor transformation law results, due to the presence of the last term in (1.31). This term cancels, however, when the *covariant derivative*

$$v^i_{;k} = v^i_{,k} + \Gamma^i_{km} v^m \tag{1.32}$$

is used. $v^i_{;k}$ transforms as a tensor as can be seen by straightforward computation:

$$v^{\bar{i}}_{;\bar{k}} = p^{\bar{i}}_{m} p_{\bar{k}}^{n} v^m_{;n} \,. \tag{1.33}$$

Similarly, the covariant derivative of an arbitrary tensor $T^{ik\dots}_{lm\dots}$ is defined by

$$T^{ik\dots}_{lm\dots;r} = T^{ik\dots}_{lm\dots,r} + \Gamma^i_{rs} T^{sk\dots}_{lm\dots} + \Gamma^k_{rs} T^{is\dots}_{lm\dots} - \\ - \Gamma^s_{rl} T^{ik\dots}_{sm\dots} - \Gamma^s_{rm} T^{lk\dots}_{ls\dots} + \dots \,. \tag{1.34}$$

The tensor property of $T^{ik\dots}_{m\dots;r}$ can again be established by straightforward computation.

The following properties of covariant derivatives will be used frequently:

$$(A_i B_k)_{;l} = A_{i;l} B_k + A_i B_{k;l} \,, \tag{1.35}$$

$$g_{ik;l} = \delta^k_{i;l} = g^{ik}_{;l} = 0 \,. \tag{1.36}$$

From (1.36) we obtain

$$A^i_{;k} = (g^{il} A_l)_{;k} = g^{il} A_{l;k} \tag{1.37}$$

for an arbitrary vector field A^i. Indices can thus be raised and lowered within covariant derivatives.

1·5. *The Riemann tensor.* – Covariant derivatives do not commute, except in the case of a scalar field where

$$\Phi_{;i;j} = \Phi_{;j;i} = \Phi_{,i,j} - \Phi_{,m} \Gamma^m_{ij} \tag{1.38}$$

is obviously symmetric in i and j.

In the case of a vector we have

$$A_{i;k;j} = (A_{i;k})_{,j} - A_{m;k} \Gamma^m_{ji} - A_{i;m} \Gamma^m_{jk} = \\ = A_{i,jk} - A_{m,j} \Gamma^m_{ik} - A_m \Gamma^m_{ik,j} - A_{i,m} \Gamma^m_{kj} + \\ + A_r \Gamma^r_{mi} \Gamma^m_{kj} - A_{m,k} \Gamma^m_{ji} + A_r \Gamma^m_{ji} \Gamma^r_{mk} \,. \tag{1.39}$$

Commuting i and j we obtain

$$A_{i;k;j} - A_{i;j;k} = : A_r R^r{}_{ikj} , \tag{1.40}$$

where

$$R^r{}_{ikj} = \Gamma^r{}_{ij,k} - \Gamma^r{}_{ik,j} + \Gamma^r{}_{mk}\Gamma^m{}_{ji} - \Gamma^r{}_{mj}\Gamma^m{}_{ki} \tag{1.41}$$

is the Riemann tensor.

Lowering the first index we obtain the covariant curvature tensor

$$R_{hijk} = R^r{}_{ijk} g_{hr} , \tag{1.42}$$

which can be written in the equivalent forms

$$R_{hijk} = \Gamma_{ikh,j} - \Gamma_{ijh,k} + \Gamma^l{}_{ij}\Gamma_{hkl} - \Gamma^l{}_{ik}\Gamma_{hjl} , \tag{1.43}$$

$$R_{hijk} = \tfrac{1}{2}(g_{hk,ij} + g_{ij,hk} - g_{ik,jh} - g_{jh,ik}) + g^{ml}(\Gamma_{ijm}\Gamma_{hkl} - \Gamma_{ikm}\Gamma_{hjl}) . \tag{1.44}$$

The last form of the curvature tensor displays the symmetry properties explicitly:

$$R_{hijk} = -R_{ihjk} , \tag{1.45a}$$

$$R_{hijk} = -R_{hikj} , \tag{1.45b}$$

$$R_{hijk} = R_{jkhi} , \tag{1.45c}$$

$$R_{hijk} + R_{hjki} + R_{hkij} = 0 . \tag{1.45d}$$

One finds that R_{hijk} has 20 independent components.

Contracting R_{hijk} we obtain the Ricci tensor

$$R_{ik} = R^h{}_{ikh} = R_{ki} , \tag{1.46}$$

and the curvature scalar

$$R = g^{ik} R_{ik} . \tag{1.47}$$

The *Bianchi identities*

$$R^h{}_{ijk;l} + R^h{}_{ikl;j} + R^h{}_{ilj;k} = 0 \tag{1.48}$$

can be proved from (1.41) by explicit—but lengthy—calculations. Contracting (1.48) over h and j we have

$$R_{ik;l} - R^h{}_{ikl;h} - R_{il;k} = 0 . \tag{1.49}$$

Multiplication with g^{ik} leads to

$$R^h{}_{k;h} = \tfrac{1}{2} R_{;k} . \tag{1.50}$$

This result can also be written as

$$G^i{}_k = R^i{}_k - \tfrac{1}{2}\delta^i{}_k R , \qquad G^i{}_{k;i} = 0 . \tag{1.51}$$

(1.51) shows that the divergence of the *Einstein tensor* G_{ik} vanishes.

2. – Field equations and experimental tests.

2·1. *The field equations.* – The field equations determining the metric g_{ik} are generalizations of the Poisson equation

$$\Delta U = 4\pi\varrho \tag{2.1}$$

of the Newtonian theory. The basic equations of Einstein's theory are

$$R_{ik} - \tfrac{1}{2} g_{ik} R = -\varkappa T_{ik} , \tag{2.2}$$

where

$$\varkappa = 8\pi G/c^2 = 1.86 \cdot 10^{-27}\ \text{cm/g} \mathrel{\hat=} 8\pi . \tag{2.3}$$

T_{ik} is the energy-momentum tensor, which is conserved as a consequence of the Bianchi identities

$$(R_{ik} - \tfrac{1}{2} g_{ik} R)^{;k} = 0 , \qquad T_{ik}{}^{;k} = 0 . \tag{2.4}$$

The Bianchi identities are 4 relations between the 10 field equations (2.2). Therefore only 6 indepenent field equations are available to determine g_{ik}. The remaining 4 degrees of freedom correspond to the co-ordinate transformations

$$x^i = f^i(x^{\bar{k}}) . \tag{2.5}$$

Remark. It is customary to include in (2.2) an additional term, the *cosmological constant.* This term takes into account the energy-momentum density of the vacuum. In this case the source T_{ik} which enters into (2.2) is split into two parts

$$T_{ik} = T_{ik}^{\text{matter}} + T_{ik}^{\text{vac}} . \tag{2.6}$$

Because of invariance reasons T^{vac}_{ik} has to be of the form

$$T^{\text{vac}}_{ik} = \lambda/\varkappa g_{ik}\,,$$

λ is the cosmological constant, $\lambda/\varkappa$ is the (unknown) energy-momentum density of the vacuum. We shall disregard this term in the sequel, since it is of interest only in cosmology.

2·2. *The linear approximation.* – In this Section we shall show that Einstein's field equations (2.2) reduce to Newtonian theory in the limit. For this purpose we put

$$g_{ik} = \eta_{ik} + 2\psi_{ik} \tag{2.7}$$

with $\psi^2_{ik} \simeq 0$. In this approximation R_{km} becomes

$$R_{km} = \psi_{km,\,\,i}^{\;\;\;\;i} + \psi^i_{\;i,km} - \psi^i_{\;m,ik} - \psi^i_{\;k,mi}\,. \tag{2.8}$$

Inserting this into (2.2) we obtain the linearized form of Einstein's equations

$$\psi_{km,i}^{\;\;\;\;\;i} + \psi^i_{\;i,km} - \psi^i_{\;m,ik} - \psi^i_{\;k,im} + \eta_{km}\psi^{ij} - \eta_{km}\psi^i_{\;i,j}{}^{j} = -\varkappa T_{km}\,. \tag{2.9}$$

This system of differential equations is linear. In order to decouple it and solve it by means of Green's function techniques, we use the freedom (2.5) in the choice of co-ordinates. A co-ordinate transformation

$$x^i = x^{\bar{i}} + 2\Lambda^i\,, \qquad (\Lambda^i)^2 \simeq 0\,, \tag{2.10}$$

leaves the field equations invariant and changes ψ_{ik} according to

$$\psi_{ik} = \bar{\psi}_{ik} + \Lambda_{i,k} + \Lambda_{k,i}\,. \tag{2.11}$$

This « gauge transformation » is analogous to the one in electrodynamics and can be used to decouple the equations (2.9). Imposing the co-ordinate conditions

$$\psi_{ik,}^{\;\;\;\;k} = \tfrac{1}{2}\psi^k_{\;k,i} \tag{2.12}$$

on ψ_{ik}, (2.8) becomes

$$R_{ik} = \Box\psi_{ik}\,, \qquad R = \Box\psi_i^{\;i}\,. \tag{2.13}$$

The linear approximation thus becomes

$$\Box(\psi_{km} - \tfrac{1}{2}\eta_{km}\psi_l^{\;l}) = -\varkappa T_{km}\,, \tag{2.14}$$

or

$$\square\psi_{km} = -\varkappa(T_{km} - \tfrac{1}{2}\eta_{km}T_l{}^l)\,. \tag{2.15}$$

Approximating T_{km} by

$$T_{km} = \left(\begin{array}{c|c} \varrho & 0 \\ \hline 0 & 0 \end{array}\right), \tag{2.16}$$

we find

$$\square\psi_{00} = -4\pi\varrho \tag{2.17}$$

and

$$\psi_{00} = \psi_{11} = \psi_{22} = \psi_{33}\,, \tag{2.18}$$

all nondiagonal ψ_{ik} vanishing. (2.17) shows that ψ_{00} is identical with the Newtonian gravitational potential U. Inserting this and (2.18) into the line element, we find

$$\mathrm{d}s = \mathrm{c}^2\,\mathrm{d}t^2\left(1 + \frac{2U}{c^2}\right) - \mathrm{d}\boldsymbol{x}^2\left(1 - \frac{2U}{c^2}\right). \tag{2.19}$$

The factors c^2 have been reinserted into (2.19) in order to show the orders of magnitude.

For a slowly-*rotating mass* we can approximate T_{ik} by

$$T_{ik} = \varrho\begin{pmatrix} 1 & \boldsymbol{v} \\ \boldsymbol{v} & 0 \end{pmatrix} \tag{2.20}$$

(we neglect pressure effects, which is possible only for nonrelativistic rotation). For *stationary rotation* we have for the nondiagonal parts of ψ_{ik}

$$-\square\psi_{0\alpha} = \Delta\psi_{0\alpha} = \varkappa T_{0\alpha}\,, \tag{2.21}$$

and thus ($r = |\boldsymbol{x}|$)

$$\psi_{0\alpha}(\boldsymbol{x}) = -\frac{\varkappa}{4\pi}\int\frac{\mathrm{d}^3x'}{|\boldsymbol{x}-\boldsymbol{x}'|}\,T_{0\alpha}(\boldsymbol{x}') = -\frac{2G}{r}\int\mathrm{d}^3x'\,T_{0\alpha}(x') - \frac{2G}{r^3}\int\mathrm{d}^3x'\,x^\beta x'^\beta\,T_{0\alpha}(x') + \dots\,, \tag{2.22}$$

$P_\alpha = \int\mathrm{d}^3x'\,T_{0\alpha}(x') = 0$ is the linear momentum of the mass which we shall assume to be at rest at the origin of the co-ordinate system. (2.22) then simplifies to

$$\psi_{0\alpha}(\boldsymbol{x}) = -\frac{2G}{r^3}x^\beta \int d^3x' \varrho x'^\beta v_\alpha . \tag{2.23}$$

Assuming rigid rotation with angular frequency ω^μ

$$-v_\alpha = v^\alpha = \varepsilon^{\alpha\mu\nu}\omega^\mu x^\nu , \tag{2.24}$$

we obtain

$$\psi_{0\alpha}(\boldsymbol{x}) = \frac{2G}{r^3}\varepsilon^{\alpha\mu\nu}\omega^\mu x^\beta \int d^3x' \varrho x'^\beta x'^\nu .$$

For spherically symmetric $\varrho(\boldsymbol{x}) = \varrho(r)$

$$\int d^3x' \varrho x'^\beta x'^\mu = \tfrac{1}{2} I \delta^{\beta\mu} , \tag{2.25}$$

where I is the moment of inertia of the mass. Thus

$$\psi_{0\alpha}(\boldsymbol{x}) = \frac{GI}{r^3}\varepsilon^{\alpha\beta\nu}\omega^\beta x^\nu . \tag{2.26}$$

The line element in the exterior of uniformly and rigidly rotating mass becomes in the linear approximation

$$ds^2 = (1 + 2U)dt^2 - (1 - 2U)d\boldsymbol{x}^2 + \frac{4GI}{r^3}\varepsilon^{\alpha\beta\gamma}dx^\alpha \omega^\beta x^\gamma dt . \tag{2.27}$$

The presence of the term $\propto dx^\alpha dt$ shows that the gravitational field is *stationary* but not static (no time inversion).

(2.27) can be used to compute all present experimental tests of general relativity, except for perihelion advance. For this we need the nonlinear terms in the Schwarzschild solution, which will be given in the next Section.

2·3. *The Schwarzschild solution.* – The Schwarzschild solution is the only spherically-symmetric solution of the Einstein vacuum-field equations

$$R_{ik} = 0 , \tag{2.38}$$

and is therefore the exterior solution for an arbitrary spherical mass distribution.

Symmetry arguments lead to the general form

$$ds^2 = e^\nu dt^2 - e^\lambda dr^2 - r^2 d\Omega^2 , \qquad d\Omega^2 = d\theta^2 + \sin^2\theta \, d\phi^2 , \tag{2.29}$$

of the line element, where $\nu = \nu(r, t)$, $\lambda = \lambda(r, t)$.

A lengthy but straightforward calculation leads to the following nonvanishing components of the Einstein tensor:

$$G^1{}_1 = e^{-\lambda}\left(\frac{\nu'}{r} + \frac{1}{r^2}\right) - \frac{1}{r^2}, \tag{2.30a}$$

$$G^2{}_2 = G^3{}_3 = \frac{1}{2} e^{-\lambda}\left(\nu'' + \frac{\nu'^2}{2} + \frac{\nu' - \lambda'}{r} - \frac{\nu'\lambda'}{2}\right) - \frac{1}{4} e^{-\nu}(2\bar{\bar{\lambda}} + 2\bar{\lambda}^2 - \bar{\lambda}\bar{\nu}), \tag{2.30b}$$

$$G^0{}_0 = \frac{1}{r^2} e^{-\lambda}(1 - \lambda' r) - \frac{1}{r^2}, \tag{2.30c}$$

$$G^0{}_1 = e^{-\lambda}\bar{\lambda}/r, \tag{2.30d}$$

where $^- = \partial/\partial t$, $' = \mathrm{d}/\mathrm{d}r$. (2.30*d*) leads to $\bar{\lambda} = 0$, *i.e.* λ is time-independent. Adding (2.30*a*) and (2.30*c*) we obtain

$$\lambda' + \nu' = 0, \qquad \lambda + \nu = q(t). \tag{2.31}$$

Introduction of a new time variable $\tilde{t}$ by

$$\tilde{t} = \int \mathrm{d}t \exp[-q(t)/2] \tag{2.23}$$

has the consequence $\nu \to \nu - q$, and thus

$$\lambda + \nu = 0. \tag{2.33}$$

λ is determined by (2.30*c*):

$$r\lambda' = 1 - e^{\lambda}, \qquad \int \frac{\mathrm{d}\lambda}{1 - e^{\lambda}} = \int \frac{\mathrm{d}r}{r}, \qquad \ln r = \ln\left(\frac{e^{\lambda}}{e^{\lambda} - 1}\right) + C. \tag{2.34}$$

Putting the variable of integration $C = \ln 2M$, we have $e^{-\lambda} = 1 - 2M/r$ and the *Schwarzschild line element* results

$$\mathrm{d}s^2 = \left(1 - \frac{2M}{r}\right)\mathrm{d}t^2 - \left(1 - \frac{2M}{r}\right)^{-1}\mathrm{d}r^2 - r^2\,\mathrm{d}\Omega^2. \tag{2.35}$$

Comparison of (2.35) with the linear approximation (2.19) proves that the mass parameter M has been identified correctly.

The transformation

$$r = \left(1 + \frac{M}{2\bar{r}}\right)^2 \bar{r} \tag{2.36}$$

leads to the isotropic form

$$(2.37)\qquad \mathrm{d}s^2 = \left(\frac{1-M/2\bar r}{1+M/2\bar r}\right)^2 \mathrm{d}t^2 - (1+M/2\bar r)^4 \mathrm{d}\boldsymbol{x}^2\,, \qquad \mathrm{d}\boldsymbol{x}^2 = \mathrm{d}\bar r^2 + \bar r^2 \mathrm{d}\Omega^2\,,$$

of the Schwarzschild line element, which we shall use in the next Section.

2·4. *Geodesics in the Schwarzschild metric.* – In order to calculate light deflection and perihelion advance we have to find the geodesics in the Schwarzschild metric.

Consider a line element of the form

$$(2.38)\qquad \mathrm{d}s^2 = e^{\nu}\mathrm{d}t^2 - e^{\lambda}\mathrm{d}r^2 - e^{\mu} r^2 \mathrm{d}\theta^2 - e^{\mu} r^2 \sin^2\theta\, \mathrm{d}\phi^2\,,$$

where ν, λ, μ are functions of r.

The variational principle

$$(2.39)\qquad \delta\int K \mathrm{d}\lambda = 0\,, \qquad K = e^{\nu}\dot t^2 - e^{\lambda}\dot r^2 - e^{\mu} r^2 \dot\theta^2 - e^{\mu} r^2 \sin^2\theta\, \dot\phi^2$$

is equivalent to (1.6).

t and ϕ are cyclic variables in (2.39), so that two conservation laws

$$(2.40)\qquad \frac{\partial K}{\partial\dot\phi} = 2e^{\mu} r^2 \sin^2\theta\dot\phi =: 2l = \text{const}\,,$$

$$(2.41)\qquad \frac{\partial K}{\partial\dot t} = 2e^{\nu}\dot t =: 2a = \text{const}\,,$$

are obtained (angular momentum, energy conservation). One shows easily that $\theta = \pi/2$, $\dot\theta = 0$ is a solution of the equation of motion for θ.

(2.40) then simplifies to

$$(2.42)\qquad e^{\mu} r^2 \dot\phi = l\,.$$

The equation of motion for r can be replaced by the equation $K = \text{const}$, where $K = 1$ for timelike, $K = 0$ for null-geodesics. The remaining geodesic equations are thus

$$(2.43a)\qquad e^{\mu} r^2 \dot\phi = l\,,$$

$$(2.43b)\qquad e^{\nu}\dot t = a\,,$$

$$(2.43c)\qquad e^{\nu}\dot t^2 - e^{\lambda}\dot r^2 - e^{\mu} r^2 \dot\phi^2 = K \qquad (K = 0, 1)\,.$$

These equations cannot be solved in terms of elementary functions, but can be very easily discussed graphically. Using the standard form (2.35) of the

Schwarzschild metric $\exp[\nu] = \exp[-\lambda] = 1 - 2M/r$, $\mu = 0$, (2.43) becomes

$$\dot{t}(1 - 2M/r) = a\,, \qquad r^2\dot{\phi} = l\,, \tag{2.44}$$

$$\frac{\dot{r}^2}{2} - \frac{MK}{r} + \frac{l^2}{2r^2} - \frac{Ml^2}{r^3} = \frac{a^2 - K}{2}\,. \tag{2.45}$$

(2.45) expresses the law of energy conservation for a particle of mass 1 moving in a potential

$$V_{\text{eff}} = -\frac{M}{r} + \frac{l^2}{2r^2} - \frac{Ml^2}{r^3}\,, \tag{2.46}$$

where $E = (a^2 - 1)/2$ is the energy of the particle.

General relativity therefore leads but to an additional term $-Ml^2/r^3$ in the potential.

For photons ($K = 0$) the effective potential becomes

$$V_{\text{eff}} = \frac{l^2}{2r^2} - \frac{Ml^2}{r^3}\,. \tag{2.47}$$

The potential (2.46) is shown for various values of l/M in Fig. 4.

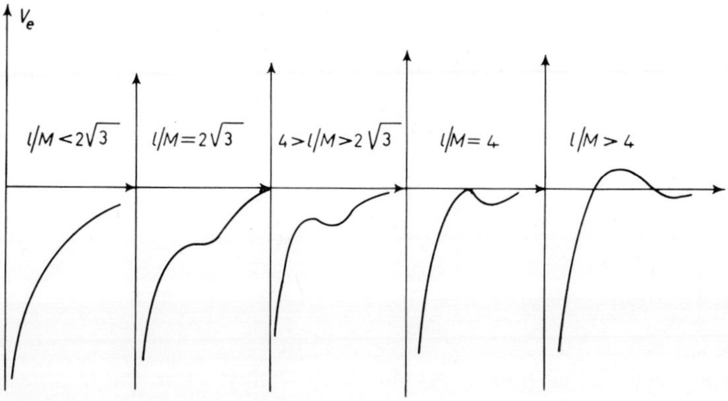

Fig. 4. – Effective potential for $K = 1$.

The potential is strongly attractive for small r, and particles released with $l < 2M$ from infinity fall into the singularity $r = 0$ (in Newtonian theory only for $l = 0 \to r \to 0$).

For $l > 2\sqrt{3}\,M$ there is a maximum and minimum of V_{eff} for

$$r = l^2\left[1 \mp \sqrt{1 - 12M^2/l^2}\right]/2M\,, \tag{2.48}$$

and bound orbits are possible.

The smallest circular orbit is $r = 6M$, its binding energy $E = -\frac{1}{18} = -5.5\%$. This will be relevant in our discussion of gravitational radiation.

The calculation of light deflection and perihelion advance is a trivial task; it is best to start out from (2.43) and use isotropic co-ordinates (2.37). The results are

$$\Delta\phi = \frac{6\pi M}{A(1-\varepsilon^2)} \text{ perihelion advance}, \tag{2.49}$$

(where A is the major semi-axis and ε the eccentricity of the orbit) and

$$\delta = \frac{2M}{R} \text{ light deflection}, \tag{2.50}$$

where R is the distance of closest approach.

The latest experimental results are (all experimental results will be expressed as fractions of the corresponding effects predicted by general relativity) (SHAPIRO [1])

$\Delta\phi$:	0.98	± 0.03	(planetary radar),
	1.00	± 0.03	(optical),
	1.00	± 0.01	(planetary + optical);
δ:	0.99	± 0.12	(short base line interferometry, radio-wave deflection).
	0.90	± 0.05	
	1.04	$\begin{cases} +0.15 \\ -0.10 \end{cases}$	

The *Shapiro experiment* measures the time delay of a radar signal reflected by Venus. It can be shown that the Maxwell equations can be written in the Schwarzschild metric in the same form as in flat space-time, except for a dielectric constant $\varepsilon = (1-2M/r)^{-1}$ and a permeability $\mu = (1-2M/r)^{-1}$. The velocity of light is thus

$$c = \frac{1}{1-2M/r}, \tag{2.51}$$

where the flat space-time picture is used. These dielectric properties lead to an additional time delay, which corresponds to an apparent increase of the Earth-Venus distance of about 30 km. This effect has been measured recently by SHAPIRO. The results are

$$1.02 \pm 0.05 \qquad \text{(planetary radar)},$$

$$1.00 \pm 0.04 \qquad \text{(spacecraft)}.$$

Red shift can be calculated either from energy-conservation arguments or by considering the propagation of radiation in the Schwarzschild metric. When light emitted with wavelength λ in the Schwarzschild metric is observed at infinity its wavelength is given by

$$\frac{\lambda_\infty}{\lambda} =: 1 + z = \frac{1}{\sqrt{1 - 2M/r}}\,.$$

3. – Gravitational waves.

The literature on gravitational waves is very extensive, and only the simplest approach, the linearized approximation will be dealt with.

3·1. *The energy of gravitational waves.* – One of the main problems in the theory of gravitational radiation is the definition of the energy of the waves. In the linearized approximation this is accomplished by deriving the field equation of general relativity from a *variational principle*

$$W = \frac{1}{2\varkappa}\int R\sqrt{-g}\,\mathrm{d}^4x + \int \Lambda\sqrt{-g}\,\mathrm{d}^4x\,, \tag{3.1}$$

where Λ is the Lagrangian of the particles moving in the gravitational field. For a scalar field ϕ, Λ is given by

$$\Lambda = \tfrac{1}{2}(\phi_{,i}\phi_{,k}g^{ik} - m^2\phi^2)\,. \tag{3.2}$$

(3.1) is no standard Lagrangian, since R contains the second derivatives of the field variables g_{ik}. These second derivatives can be removed, however, by partial integration, since

$$\sqrt{-g}R = \sqrt{-g}\,g^{ik}R_{ik} = \sqrt{-g}\,g^{ik}\Gamma^m{}_{im,k} - g^{ik}\Gamma^m{}_{ik,m} + {} + g^{ik}\Gamma^m{}_{ir}\Gamma^r{}_{km} - g^{ik}\Gamma^m{}_{ik}\Gamma^r{}_{mr} \tag{3.3}$$

can be rewritten by means of

$$\sqrt{-g}\,g^{ik}\Gamma^m{}_{ik,m} = (\sqrt{-g}\,g^{ik}\Gamma^m{}_{ik})_{,m} - \Gamma^m{}_{ik}(\sqrt{-g}\,g^{ik})_{,m} \tag{3.4}$$

into

$$\sqrt{-g}\,R = \sqrt{-g}\,G - (\sqrt{-g}\,g^{ik}\Gamma^m{}_{ik} - \sqrt{-g}\,g^{im}\Gamma^r{}_{ir})_{,m}\,, \tag{3.5}$$

where

$$G = g^{ik}(\Gamma^m{}_{ir}\Gamma^r{}_{km} - \Gamma^m{}_{ik}\Gamma^r{}_{mr})\,. \tag{3.6}$$

$G\sqrt{-g}$ can be used as a Lagrangian for the gravitational field, since

$$\delta\int\sqrt{-g}\,R\,\mathrm{d}^4x = \delta\int\sqrt{-g}\,G\,\mathrm{d}^4x\,, \tag{3.7}$$

and a lengthy calculation shows that the correct field equations actually result.

Putting

$$g_{ik} = \eta_{ik} + 2\psi_{ik}\,, \qquad |\psi_{ik}|^2 \ll 1\,, \tag{3.8}$$

we obtain for the Lagrangian of the free gravitational field in the linearized approximation

$$L = \frac{1}{\varkappa}\left(\frac{1}{2}\psi^{ik,j}\psi_{ik,j} - \psi^{ik,j}\psi_{jk,i} - \frac{1}{2}\psi_{,i}\psi^{,i} + \psi^{ik}{}_{,k}\psi_{,i}\right). \tag{3.9}$$

This is the Pauli-Fierz Lagrangian for a massless particle of spin 2.

The canonical energy-momentum tensor can then be calculated from (3.9):

$$t_{jl} = \frac{1}{\varkappa}(\psi^{ik,j} - 2\psi^{jk,i} - \eta^{ik}\psi^m{}_m{}^{,j} + \eta^{jk}\psi^m{}_m{}^{,i} + \eta^{ik}\psi^{jm}{}_{,m})\,\psi_{ik,l} - \eta_{jl}L\,. \tag{3.10}$$

The calculation of the radiated energy is then completely analogous to the corresponding calculation in electrodynamics. The linearized field equation

$$\Box(\psi_{ik} - \tfrac{1}{2}\eta_{ik}\psi) = -\varkappa T_{ik} \tag{3.11}$$

is solved by means of Green's functions and the result is inserted into (3.10). The energy flux is then integrated over a sphere at infinity.

The result is

$$-\frac{\mathrm{d}E}{\mathrm{d}t} = \frac{G}{5c^5}\dddot{Q}^{\mathrm{ret}}_{\alpha\beta}\dddot{Q}^{\mathrm{ret}}_{\alpha\beta}\,, \tag{3.12}$$

where

$$Q_{\alpha\beta} = \int\mathrm{d}^3x(x_\alpha x_\beta - \tfrac{1}{3}\delta_{\alpha\beta}r^2)\,\varrho \tag{3.13}$$

is the *quadrupole tensor* of the source of the waves.

3'2. *Radiation from particles in the Schwarzschild metric.* – The radiation from a test mass m falling freely in the Schwarzschild metric of M can be calculated in the linear approximation. The quadrupole-moment tensor is

$$Q_{\alpha\beta} = M(y_\alpha y_\beta - \tfrac{1}{3} y^2 \delta_{\alpha\beta}) + m(x_\alpha x_\beta - \tfrac{1}{3}\delta_{\alpha\beta} x^2)\,, \tag{3.14}$$

where x and y are the distances of m and M from the origin. Putting $r_\alpha = x_\alpha - y_\alpha$ this becomes

$$Q_{\alpha\beta} = m\gamma^3(x_\alpha x_\beta - \tfrac{1}{3} r^2 \delta_{\alpha\beta})\,, \tag{3.15}$$

where $\gamma := 1 + m/M$.

For a circular orbit $r = \text{const}$ in the x-y–plane

$$\dddot{Q}^2_{\alpha\beta} = 32\gamma^6 m^2 r^4 \omega^6\,,$$

where ω is given by $\omega^2 r^3 = G(m+M)$. The energy radiated becomes

$$\frac{\mathrm{d}E}{\mathrm{d}t} = -\frac{32}{5c} G^4 m^2 M^3 \gamma / r^5 = -\frac{\gamma c^5}{5G}\left(\frac{2M}{r}\right)^3 \left(\frac{2m}{r}\right)^2. \tag{3.17}$$

The two factors in brackets are dimensionless and

$$c^5/5G = 7.28 \cdot 10^{58}\ \text{erg/s} = 4 \cdot 10^4\, M_0 c^2/\text{s}\,, \tag{3.18}$$

which is $2 \cdot 10^{25}$ solar luminosities.

The energy loss will lead to a slow decrease of the radius of the circular orbit, which can be calculated to be

$$r^4 = r_0^4 - \frac{256}{5} m M^2 \gamma c t\,. \tag{3.19}$$

The *frequency* of the radiation follows from (3.15) and (3.10)

$$\nu = c\sqrt{\gamma M / r^3}/\pi\,. \tag{3.20}$$

Angular momentum is radiated according to

$$\frac{\mathrm{d}L_\alpha}{\mathrm{d}t} = -\frac{2G}{5c^5} \varepsilon_{\alpha\beta\gamma} \dddot{Q}_{\beta\delta} \dddot{Q}_{\gamma\delta}\,. \tag{3.21}$$

The following picture emerges: Radius and eccentricity of an elliptic orbit decrease slowly due to gravitational radiation and almost circular orbits result.

The radius of these circular orbits continues to shrink until the last stable circular orbit $r = 6M$ is reached. At this point the energy is $E = -m/18$ and 5.5% of the rest mass of the body have been radiated in the form of gravitational waves. Most of this radiation (0.03 m) is radiated in the transition from $r = 20\,M$ to $r = 6\,M$. The duration of this intense radiation is

$$t = \frac{(20M)^4 - (6M)^4}{mM^2\gamma c}\,\frac{5}{256} \simeq 10^{-2}\,\text{s}\,\frac{M}{\gamma m}\,\frac{M}{M_0}, \tag{3.22}$$

and its frequency

$$\nu = \frac{c}{M\pi}\sqrt{\gamma}\,\frac{1}{\sqrt{1000}} \simeq 2.10^3\,\text{Hz}\,\frac{M}{M_0}. \tag{3.23}$$

Very little radiation is emitted during the final phase of the orbit $6M \geqslant r > 2M$:

$$E = mc^2\,\frac{2\gamma^5}{105}\,\frac{m}{M}. \tag{3.24}$$

This radiation lasts for about

$$t = 1{,}5\,\frac{M\gamma}{c} = 7.5\cdot 10^{-6}\,\text{s}\,\frac{\gamma M}{M_0}. \tag{3.25}$$

Linearized theory is, however, not very suitable at radii $6r \leqslant M$ and tends to overpredict the amount of radiation emitted.

4. – The use of exterior forms in Riemannian geometry.

4·1. *Tangent space.* – Let us define a vector in space-time by

$$v = v^i\,\frac{\partial}{\partial x^i}. \tag{4.1}$$

This differs from our previous definition, which defined only the components v^i, but not v itself. The vector components can be recovered from (4.1) by applying the vector (which is now a differential operator) to the function x^k:

$$v(x^k) = v^i\,\frac{\partial x^k}{\partial x^i} = v^k. \tag{4.2}$$

For an arbitrary function $f(x)$

$$v(f) = v^i\,\frac{\partial f}{\partial x^i}. \tag{4.3}$$

(4.1) can also be written in the form

$$v = v^i e_i$$

with the *basis vectors*

$$e_i = \frac{\partial}{\partial x^i}, \qquad i = 0, \ldots, 3 . \tag{4.4}$$

These basis vectors are *natural basis vectors*, which can be visualized as tangent vectors to the co-ordinate lines on space time (Fig. 5).

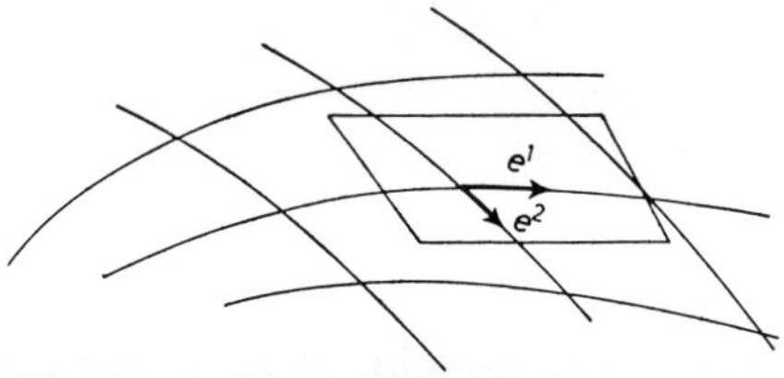

Fig. 5. – Tangent vectors and tangent space.

These vectors span the 4-dimensional *tangent space T* at each point of space-time. Any arbitrary base can obviously be introduced in tangent space, and while classical differential geometry worked only with natural bases (tangent to some co-ordinate lines), it has turned out recently to be of advantage to consider more general bases.

Tensors are defined by, *e.g.*,

$$T = T^{ik} e_i \times e_k . \tag{4.5}$$

A *scalar product* is defined in tangent space by

$$\begin{cases} (v, w) = g_{ik} v^i w^k = v^i w^k (e_i, e_k) , \\ (e_i, e_k) = g_{ik} , \end{cases} \tag{4.6}$$

where $g_{ik} = g_{ki}$ is the metric tensor.

It is possible to choose $g_{ik} = \eta_{ik}$ at each point of space-time, when the restriction to natural bases is abandoned.

The *cotangent space* T^* is the dual space to T, *i.e.* the space of linear functionals over T. (4.3) is not only linear in v, but also in f, so that it can be considered as defining a scalar product $\langle\,\rangle$ by

$$v(f) =: \langle \mathrm{d}f, v \rangle = v^i \frac{\partial f}{\partial x^i} , \tag{4.7}$$

where

$$df = \frac{\partial f}{\partial x^i} dx^i \tag{4.8}$$

is an element of T^*. A basis of T^* is given by

$$e^i = dx^i . \tag{4.9}$$

Its duality to e_i is expressed by

$$\langle e^i, e_j \rangle = \delta^i{}_j . \tag{4.10}$$

Covectors, the elements of T^*, are

$$v = v_i e^i . \tag{4.11}$$

The vectors e^i can again be thought of as being tangent to co-ordinate lines and more general bases, ω^i, can be introduced in cotagent space

$$v = v_i \omega^i , \tag{4.12}$$

ω^i are differential forms, which can be written in terms of the natural basis $dx^{\bar{i}}$ as

$$\omega^i = h^i{}_{\bar{j}} dx^{\bar{j}} . \tag{4.13}$$

The coefficients $h^i{}_{\bar{j}}$ are called vierbein components.

A scalar product is defined in cotangent space by

$$\begin{cases} (v, w) = v_i w_k g^{ik} , \\ (\omega^i, \omega^k) = g^{ik} . \end{cases} \tag{4.14}$$

An orthonormal base fulfills

$$(\omega^i, \omega^k) = \eta^{ik} . \tag{4.15}$$

Connections define isomorphic mapping between the tangent vector spaces in neighboring points (Fig. 6).

When one compares two vectors at different points P, Q of an Euclidian space, one transports one vector parallely from P to Q and compares the vectors there; *i.e.* one maps the vector space attached to P in some way onto the vector space attached to Q. If one takes two neighboring points P (co-ordi-

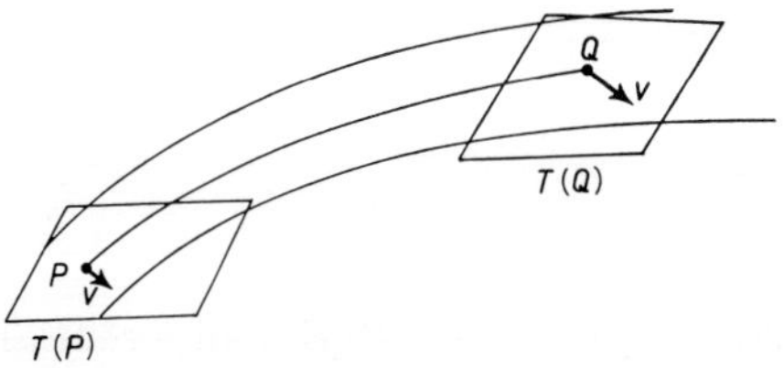

Fig. 6. – Connections.

nates x^k) and $Q(x^k + dx^k)$ of a manifold the basis vectors e_i of $T(P)$ will in general not be mapped into the basis vectors at Q but differ from these by de_i (Fig. 7).

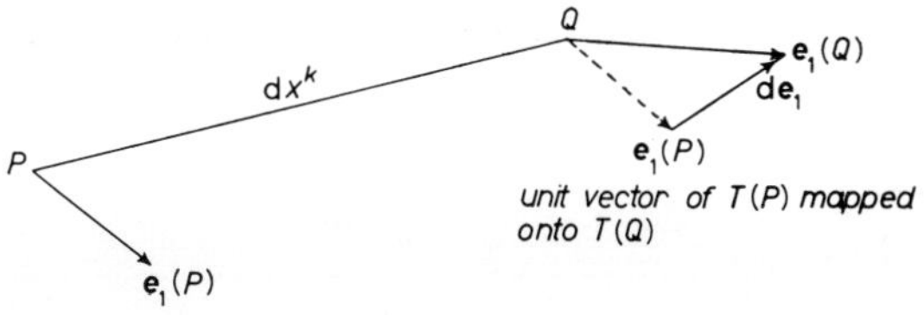

Fig. 7. – Mapping of unit vectors $P \to Q$.

Since the e_j form a basis of the vector space $T(Q)$, de_i can be expressed in the general form

$$de_i = \omega^j{}_i e_j \,, \tag{4.16}$$

where $\omega^j{}_i$ is a set of differential coefficients expressing the mutual relations of unit vectors at different points.

In a *Finsler space* the $\omega^j{}_i$ are homogeneous of first degree in the dx^n, in an *affine space* the $\omega^j{}_i$ are differential forms linear in dx^k:

$$\omega^j{}_i = L^j{}_{ik} dx^k \,. \tag{4.17}$$

The functions $L^j{}_{ik}$ are the *coefficients of the affine connection* (for a natural base they become the Christoffel symbols).

The differential of an arbitrary vector v can be expressed in the form

$$\begin{aligned} dv = d(v^i e_i) = dv^i e_i + v^i de_i = (dv^j + L^j{}_{km} v^k dx^m)\, e_j = \\ =: (Dv^j)\, e_j =: v^j{}_{;k} dx^k e_j \,. \end{aligned} \tag{4.18}$$

In (4.18) the *absolute differential* Dv^j of the vector components and the covariant *derivative*:

$$v^j_{;k} = v^j_{,k} + L^j_{mk} v^m \tag{4.19}$$

have been defined.

4·2. *Exterior differential forms.* – Exterior differential forms have the general structure

$$\omega = \sum_H a_H(x^1, \ldots, x^n) \mathrm{d}x^H , \tag{4.20}$$

where

$$\mathrm{d}x^H = \mathrm{d}x^{h_1} \wedge \mathrm{d}x^{h_2} \wedge \ldots \wedge \mathrm{d}x^{h_r} , \qquad h_1 < h_2 \ldots < h_r . \tag{4.21}$$

The sum over H is extended over all combinations of the h_r fulfilling (4.21).

The exterior derivative of ω is given by

$$\mathrm{d}\omega = \sum_H \mathrm{d}a_H \wedge \mathrm{d}x^H , \tag{4.22}$$

$$\mathrm{d}(\mathrm{d}\omega) = 0 . \tag{4.23}$$

Equation (4.23) contains many of the well-known rules (*e.g.*, div rot $\boldsymbol{v} = 0$) of vector analysis.

Let us apply this now to space time. The distance $\mathrm{d}P$ between two neighboring points on a manifold can be written heuristically as

$$\mathrm{d}P = \mathrm{d}x^i e_i , \tag{4.24}$$

since P and $P + \mathrm{d}P$ lie both in the tangent space of P. The condition

$$\mathrm{d}\,\mathrm{d}P = 0 \tag{4.25}$$

is no identity (since $\mathrm{d}P$ is not an exact differential of a quantity « P ») but expresses the fact that space time is *torsion free*. Writing (4.25) in the form

$$\begin{aligned} \mathrm{d}\,\mathrm{d}P &= \Omega^i e_i = \mathrm{d}(\mathrm{d}x^i e_i) = \mathrm{d}x^i \mathrm{d}e_i = \mathrm{d}x^i \wedge \omega^j{}_i e_j = \\ &= \mathrm{d}x^i \wedge L^j{}_{ik} \mathrm{d}x^k e_j = \tfrac{1}{2}(L^j{}_{ik} - L^j{}_{ki}) \mathrm{d}x^i \wedge \mathrm{d}x^k e_j =: T^j{}_{ik} \mathrm{d}x^i \wedge \mathrm{d}x^k e_j , \end{aligned} \tag{4.26}$$

introduces the *torsion forms* Ω^i and the torsion tensor $T^j{}_{ik}$. Heuristically speaking, $\Omega^i = 0$ means that one returns to the same point P after completing and infinitesimal « circle ». (Spaces with torsions can be visualized as crystals with screw-type dislocations.) $\mathrm{d}e^i$ is no exact differential either, since the result of the parallel displacement will depend, in general, on the infinitesimal path taken from P to Q. (Fig. 8).

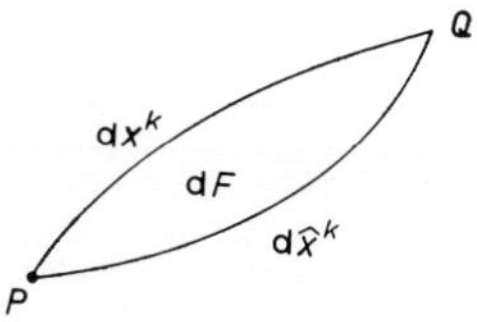

Fig. 8. – Two different paths leading from P to Q.

The difference between the resulting $\mathrm{d}e_k$ will be proportional to the area $\mathrm{d}F$ between the two paths, *i.e.* of second order in $\mathrm{d}x^k$. This is expressed by the fact that

$$\mathrm{d}\,\mathrm{d}e_i \neq 0 .$$

We put

$$\mathrm{d}\,\mathrm{d}e_i =: \Omega^j{}_i e_j , \tag{4.27}$$

where $\Omega^j{}_i$ is the *curvature form.*

The connection coefficients $L^i{}_{jk} = L_{jk}{}^i$ are determined by the postulate that the space is torsion free (which implies that $\Gamma_{jk}{}^i = \Gamma_{kj}{}^i$; see (4.26)) and by the postulate that (4.6) is preserved during parallel transport (*i.e.* the angle between two vectors remains unchanged). Differentiating (4.6) gives

$$(\mathrm{d}e_i, e_k) + (e_i, \mathrm{d}e_k) = \mathrm{d}g_{ik} . \tag{4.28}$$

Inserting

$$\mathrm{d}e_i = \omega^j{}_i e_j = \Gamma_{ik}{}^j \mathrm{d}x^k e_j ,$$

we obtain

$$\omega_{ik} + \omega_{ki} = \mathrm{d}g_{ik} \tag{4.29}$$

with

$$\omega_{ik} = g_{il}\omega^l{}_k , \tag{4.30}$$

or

$$\Gamma_{ik}{}^j \mathrm{d}x^k g_{jk} + \Gamma_{kl}{}^j \mathrm{d}x^l g_{ji} = \mathrm{d}g_{ik} , \tag{4.31}$$

which is equivalent with the standard definition of the Christoffel symbols.

In a *general* (it need neither be natural nor be orthonormal) *basis* e_i and

$e^i = \omega^i$, respectively, the relevant equations become

$$dP = \omega^i e_i \,, \tag{4.32}$$

and $ddP =: (D\omega^i)e_i = \Omega^i e_i = 0$ (this defines the absolute differential $D\omega^i$ of ω^i). The torsion form Ω^i is now given by

$$\Omega^i = D\omega^i = d\omega^i + \omega^i{}_k \wedge \omega^k = 0 \,. \tag{4.33}$$

Inserting $\omega^i{}_k = L^i{}_{kl}\omega^l$, this becomes

$$d\omega^i + L^i{}_{kl}\omega^l \wedge \omega^k = 0 \,. \tag{4.35}$$

Because of $d\omega^i \neq 0$ (except for a natural basis), we obtain $L^i{}_{kl} \neq L^i{}_{lk}$, *i.e.* the connection coefficients $L^i{}_{lk}$ are symmetric only when referred to a natural basis, in which case they become the Christoffel symbols. When we specialize to an orthonormal base (4.28) becomes

$$\omega_{ij} + \omega_{ji} = d\eta_{ij} = 0 \,, \tag{4.36}$$

i.e. the forms ω_{ij} are antisymmetric. In this case the $L^i{}_{jk}$ are called the Ricci-rotation coefficients (the most general relation between two orthonormal bases at different points is a rotation).

Finally we have to find an explicit expression for the curvature form $\Omega^i{}_j$ in a general base. We have

$$d(de_i) = \Omega^j{}_i e_j = d(e_j\omega^j{}_i) = de_j \wedge \omega^j{}_i + e_j d\omega^j{}_i = \omega^k{}_j \wedge \omega^j{}_i e_k + e_j d\omega^j{}_i \,, \tag{4.37}$$

or

$$\Omega^j{}_i = d\omega^j{}_i + \omega^j{}_l \wedge \omega^l{}_i \,. \tag{4.38}$$

We define the curvature tensor by

$$\Omega^j{}_j = \tfrac{1}{2} R^j{}_{ikl}\omega^k \wedge \omega^l \,. \tag{4.39}$$

By specializing to a natural base $\omega^k = dx^k$ one finds immediately that $R^j{}_{ikl}$ agrees with the curvature tensor

$$\begin{aligned}\Omega^j{}_i &= d(\Gamma_{ik}{}^j dx^k) + \Gamma_{lm}{}^j \Gamma_{ik}{}^l dx^m \wedge dx^k = (\Gamma_{ik}{}^j{}_{,m} + \Gamma_{lm}{}^j \Gamma_{ik}{}^l)\, dx^m \wedge dx^k = \\ &= \tfrac{1}{2}(\Gamma_{ik}{}^j{}_{,m} - \Gamma_{im}{}^j{}_{,k} + \Gamma_{lm}{}^j \Gamma_{ik}{}^l - \Gamma_{lk}{}^j \Gamma_{im}{}^l)\, dx^m \wedge dx^k \,.\end{aligned} \tag{4.40}$$

Collecting our results we obtain the fundamental equations of Riemannian geometry in an arbitrary base:

$$(4.41)\quad \begin{cases} \mathrm{d}P = e_i\omega^i\,, & \omega_{ij} = g_{ik}\omega^k{}_j\,,\\ (e_i, e_j) = g_{ij}\,, & \omega_{ij} + \omega_{ji} = \mathrm{d}g_{ij}\,,\\ \mathrm{d}s^2 = (\mathrm{d}P, \mathrm{d}P) = g_{ij}\omega^i \otimes \omega^j\,, & \mathrm{d}\omega^i + \omega^i{}_j\wedge\omega^j = 0\,,\\ \mathrm{d}e_i = \omega^j{}_i\wedge e^j\,, & \Omega^i{}_j = \mathrm{d}\omega^i{}_j + \omega^i{}_k\wedge\omega^k{}_j\,,\\ & \Omega^i{}_j = \tfrac{1}{2}R^i{}_{jkl}\omega^k\wedge\omega^l\,. \end{cases}$$

4·3. *Application: Spherically-symmetric solutions of the Einstein field equations.* – The preceding equations (4.41) can be applied to the explicit calculation of curvature tensors and turn out to be superior to the classical techniques. As an example let us apply these equations to a metric of the form

$$(4.42)\qquad \mathrm{d}s^2 = e^{2\nu}\,\mathrm{d}t^{2\nu} - e^{2\lambda}\,\mathrm{d}r^2 - r^2\,(\mathrm{d}\theta^2 + \sin^2\theta\,\mathrm{d}\phi^2)\,,$$

where

$$(4.43)\qquad \nu = \nu(r, t)\,, \qquad \lambda = \lambda(r, t)\,.$$

It can be shown that every spherically-symmetric line element can be written in this form, (4.42) will therefore be the metric to be used in the theory of spherically-symmetric gravitational collapse.

We can introduce an orthonormal base in cotangent space by putting

$$(4.44)\quad \begin{cases} \omega^0 = e^\nu dt\,,\\ \omega^1 = e^\lambda\,\mathrm{d}r\,,\\ \omega^2 = r\,\mathrm{d}\theta\,.\\ \omega^3 = r\sin\theta\,\mathrm{d}\phi\,. \end{cases}$$

Denoting derivatives with respect to t and r by a bar and prime, respectively, we obtain

$$(4.45)\quad \begin{cases} \mathrm{d}\omega^0 = e^\nu\nu'\,\mathrm{d}r\wedge\mathrm{d}t\,,\\ \mathrm{d}\omega^1 = e^\lambda\bar{\lambda}\,\mathrm{d}t\wedge\mathrm{d}r\,,\\ \mathrm{d}\omega^2 = \mathrm{d}r\wedge\mathrm{d}\theta\,,\\ \mathrm{d}\omega^3 = \sin\theta\,\mathrm{d}r\wedge\mathrm{d}\phi + r\cos\theta\,\mathrm{d}\theta\wedge\mathrm{d}\phi\,. \end{cases}$$

Comparing this with

$$\mathrm{d}\omega^i + \omega^i{}_j \wedge \omega^j = 0 \,, \tag{4.46}$$

and taking into account

$$\omega_{ij} + \omega_{ji} = 0 \,, \tag{4.47}$$

one obtains immediately

$$\left\{\begin{aligned} &\omega^0{}_1 \equiv \exp[\nu - \lambda]\,\nu'\,\mathrm{d}t + \exp[\lambda - \nu]\,\dot{\lambda}\,\mathrm{d}r = \omega^1{}_0 \,, \\ &\omega^2{}_1 = \exp[-\lambda]\,\mathrm{d}\theta = -\,\omega^1{}_2 \,, \\ &\omega^3{}_1 = \exp[-\lambda]\sin\theta\,\mathrm{d}\phi = -\,\omega^1{}_3 \,, \\ &\omega^3{}_2 = \cos\theta\,\mathrm{d}\phi = -\,\omega^2{}_3 \,. \end{aligned}\right. \tag{4.48}$$

All other $\omega^i{}_k$ vanish. The exterior derivatives $\mathrm{d}\omega^i{}_k$ become

$$\left\{\begin{aligned} \mathrm{d}\omega^0{}_1 = {} & \big[-\exp[-2\lambda]\,(\nu'^2 - \nu'\lambda' + \nu'') + \\ & + \exp[-2\nu](-\dot{\nu}\dot{\lambda} + \dot{\lambda}^2 + \ddot{\lambda})\big]\,\omega^0\wedge\omega^1 \equiv B\omega^0\wedge\omega^1 \,, \\ \mathrm{d}\omega^2{}_1 = {} & -\exp[-\lambda]\,\dot{\lambda}\,\mathrm{d}t\wedge\mathrm{d}\theta - \exp[-\lambda]\,\lambda'\,\mathrm{d}r\wedge\mathrm{d}\theta \,, \\ \mathrm{d}\omega^3{}_2 = {} & -\sin\theta\,\mathrm{d}\theta\wedge\mathrm{d}\phi \,. \end{aligned}\right. \tag{4.49}$$

We were able to omit $\mathrm{d}\omega^3{}_1$ from this list, because of symmetry arguments. The 2 and 3 directions are the θ and ϕ directions on the sphere. These directions are completely equivalent geometrically and the curvature tensor has to be symmetric when the 2 and 3 directions are interchanged. Note, however, that this argument is valid only in an orthonormal base. In the natural base $(\mathrm{d}\theta, \mathrm{d}\phi)$ the basis vectors in these directions are of unequal length and the symmetry is destroyed. The fact that symmetries are preserved by the use of orthonormal basis vectors is an additional advantage of the method used here.

For the curvature form $\Omega^i{}_j$ we obtain

$$\Omega^0{}_1 = B\omega^0\wedge\omega^1 \,,$$

thus

$$R^0{}_{101} = B = -\,R^0{}_{110} \,,$$

all other

$$R^0{}_{1ik} = 0 \,.$$

Furthermore

$$
(4.50)\quad \begin{cases}
\Omega^0{}_2 = \omega^0{}_1\wedge\omega^1{}_2 = -(\exp[-\lambda]\nu'\omega^0 + \exp[-\nu]\bar{\lambda}\omega^1)\wedge\exp[-\lambda]\omega^2/r\,,\\
R^0{}_{202} = -\exp[-2\lambda]\nu'/r = R^0{}_{303}\,,\\
R^0{}_{212} = -\exp[-\nu-\lambda]\bar{\lambda}/r = R^0{}_{313}\,,\\
\Omega^1{}_2 = \exp[-\nu-\lambda]\bar{\lambda}\omega^0\wedge\omega^2/r + \exp[-2\lambda]\lambda'\omega^1\wedge\omega^2/r\,,\\
R^1{}_{202} = \exp[-\nu-\lambda]\bar{\lambda}/r = R^1{}_{303}\,,\\
R^1{}_{212} = \exp[-2\lambda]\lambda'/r = R^1{}_{313}\,,\\
\Omega^2{}_3 = \omega^2\wedge\omega^3/r^2 - \exp[-\lambda]\omega^2\wedge\exp[-\lambda]\omega^3/r^2{}_r\,,
\end{cases}
$$

$$
(4.51)\qquad R^2{}_{323} = \frac{1}{r^2}(1-\exp[-2\lambda])\,.
$$

This completes the list of nonvanishing $R^i{}_{klm}$. The components of the Ricci tensor and of $G_{ik} = R_{ik} - \frac{1}{2}\eta_{ik}R$ can be easily obtained from these equations:

$$
(4.52)\qquad G_{00} = -\frac{1}{r^2} - \exp[-2\lambda]\left(\frac{2\lambda'}{r} - \frac{1}{r^2}\right),
$$

$$
(4.53)\qquad G_{11} = \frac{1}{r^2} - \exp[-2\lambda]\left(\frac{2\nu'}{r} + \frac{1}{r^2}\right),
$$

$$
(4.54)\qquad G_{22} = G_{33} = -\exp[-2\lambda][\nu'' + \nu'^2 + (\nu'-\lambda')/r - \nu'\lambda'] + \exp[-2\nu](\bar{\bar{\lambda}} + \bar{\lambda}^2 - \bar{\lambda}\bar{\nu})\,,
$$

$$
(4.55)\qquad G_{01} = -2\bar{\lambda}\exp[-\nu-\lambda]/r\,.
$$

These components of G_{ik} refer to the orthonormal base ω^i, the standard components $G_{\bar{i}\bar{k}}$ referring to dx^i can be calculated from G_{ik} with the help of

$$
(4.56)\qquad G_{ik}\omega^i\otimes\omega^k = G_{\bar{i}\bar{k}}dx^{\bar{i}}\otimes dx^{\bar{k}}\,.
$$

5. – Gravitational collapse, Kruscal space and Kerr metric.

5'1. *The Tolman-Oppenheimer-Volkoff equation.* – We shall begin with a discussion of the physics of static mass configuration of very high density.

In this case all time derivatives can be omitted in (4.52)-(4.55) and the energy-momentum tensor of matter becomes in an orthonormal frame

$$T_{ik} = \begin{pmatrix} \varrho & & & \\ & p & & \\ & & p & \\ & & & p \end{pmatrix}, \tag{5.1}$$

where ϱ is the density and p the pressure within the mass distribution. Inserting (5.1), (4.52) and (4.53) into the Einstein equations, we obtain

$$\exp[-2\lambda]\left(\frac{1}{r^2} - \frac{2\lambda'}{r}\right) - \frac{1}{r^2} = -\varkappa\varrho\,, \tag{5.2}$$

$$\exp[-2\lambda]\left(\frac{1}{r^2} + \frac{2\nu'}{r}\right) - \frac{1}{r^2} = \varkappa p\,. \tag{5.3}$$

These two equations are obviously sufficient to determine ν and λ; their solution is

$$\mathrm{d}s^2 = e^{2\nu}\mathrm{d}t^2 - \mathrm{d}\sigma^2\,, \tag{5.4}$$

$$\mathrm{d}\sigma^2 = (1 - 2m(r)/r)^{-1}\mathrm{d}r^2 + r^2\mathrm{d}\Omega^2\,, \tag{5.5}$$

$$2m(r) = \varkappa\int_0^r \varrho r^2\,\mathrm{d}r\,, \tag{5.6}$$

$$\nu = -\lambda + \int_\infty^r \mathrm{d}r\, e^{2\lambda} r(p + \varrho)\cdot 4\pi\,. \tag{5.7}$$

The geometry of the spatial part ($\mathrm{d}\sigma^2$) of the line element is therefore determined by ϱ only, p does not enter.

These equations determine the metric completely if p and ϱ are known as functions of r.

We need therefore two additional equations. One is given by the *equation of state*

$$p = p(\varrho)\,. \tag{5.8}$$

The derivation of a reliable equation of state at high densities is a problem which is still unsolved, we shall come back to it later. The second equation which we need can be derived from (4.54) (we have not used the angular part of the Einstein equations yet) or simpler from

$$T_{1\ ;k}^{\ k} = 0\,. \tag{5.9}$$

Inserting (5.1) and (5.4)-(5.7) into (5.9) we obtain the Tolman-Oppenheimer-

Volkoff (TOV) equation

$$-\frac{dp}{dr}=\frac{(p+\varrho)[m(r)+4\pi Gpr^3]}{r[r-2m(r)]}. \tag{5.10}$$

This is the relativistic generalization of the standard equation for hydrostatic equilibrium, which can be obtained from (5.10) by omitting the p-terms in the numerator and the $2m(r)$-term in the denominator. (5.10) has to be solved subject to the boundary condition that $p(R)=0$ at the boundary of the star ($r=R$).

Note that the pressure occurs twice at the r.h.s. of (5.10), this leads to the relativistic self-generation of pressure. At $r=R$ the line element (5.4)-(5.7) can be joined smoothly to the exterior Schwarzschild solution, the mass M being given by

$$M=4\pi\int_0^R dr\, r^2\varrho\,. \tag{5.11}$$

This equation for M looks like the standard definition of a mass in terms of density; it has, however, a number of remarkable properties:

a) M is positive if ϱ is positive. If one compares this to the Newtonian equation

$$M=M_0-GM_0^2/Rc^2\,,$$

for the mass one sees that this fact is not trivial.

b) In (5.11) the gravitational self-energy appears to be missing and no distinction seems to exist between the pre-assembly mass M_0 and the actual mass of the body. The error in this argument is that $4\pi r^2 dr$ is not the volume element dV of curved space; dV follows from (5.5) to be

$$dV=4\pi r^2 dr\left(1-\frac{2m(r)}{r}\right)^{-\frac{1}{2}}. \tag{5.12}$$

We can write (5.11) therefore in the form

$$M=\int dV\varrho\left(1-\frac{2m(r)}{r}\right)^{+\frac{1}{2}}. \tag{5.13}$$

The fact that the integrand

$$\varrho\left(1-\frac{2m(r)}{r}\right)^{\frac{1}{2}}$$

is smaller than ϱ takes into account the effects of the gravitational self-energy, as one can show in detail. The pre-assembly mass M_0 (which equals the baryon number A of the star if the unit of mass is suitably chosen) is given by

$$M_0 = A = \int \mathrm{d}V\, n(r)\,, \tag{5.14}$$

where $n(r)$ is the baryon density in the star. The connection between n, p and ϱ is given by

$$p = -\frac{\mathrm{d(energy/baryon)}}{\mathrm{d(volume/baryon)}} = \frac{\mathrm{d}(\varrho/n)}{\mathrm{d}(1/n)} = n\frac{\mathrm{d}\varrho}{\mathrm{d}n} - \varrho\,, \tag{5.15}$$

or

$$p + \varrho = n\frac{\mathrm{d}\varrho}{\mathrm{d}n}\,. \tag{5.16}$$

If the equation of state $p(\varrho)$ is known, $n(\varrho)$ can be determined from

$$\int \frac{\mathrm{d}\varrho}{\varrho + p} = \int \frac{\mathrm{d}n}{n} = \log n\,. \tag{5.17}$$

5·2. *Incompressible matter.* – The only case in which the TOV equation can be integrated analytically is for incompressible matter

$$T^0{}_0 = \varrho = \text{const}\,. \tag{5.18}$$

In this case $m(r)$ becomes

$$m = \frac{4\pi}{3} G\varrho r^3\,. \tag{5.19}$$

Inserting this into (5.10) we obtain for the pressure p

$$p = \varrho\,\frac{\sqrt{1-x^2} - \sqrt{1-X^2}}{3\sqrt{1-X^2} - \sqrt{1-x^2}}\,, \tag{5.20}$$

where x is a dimensionless radial variable defined by

$$x = (8\pi G\varrho/3)^{\frac{1}{2}} r\,, \tag{5.21}$$

and X is the value of x at the surface of the star ($r = R$). For $X \ll 1$ (5.20) agrees with the ordinary nonrelativistic pressure distribution in a star. For

$X = \sqrt{\frac{8}{9}}$ the pressure becomes infinite at $x = 0$, *i.e.* all matter is crushed and collapse sets in. Stable-mass configurations can therefore exist only for $X < \sqrt{\frac{8}{9}}$.

From (5.19) we have that the Schwarzschild radius $2M$ of the star is given by

$$2M = 8\pi G\varrho/3 \cdot R^3 , \tag{5.22}$$

X^2 can thus be written as

$$X^2 = 2M/R = \frac{\text{Schwarzschild radius}}{\text{radius}} . \tag{5.23}$$

No star with a radius $R < \frac{9}{8}(2M)$ can therefore be stable.

For incompressible matter the baryon density n agrees with the mass density, $n = \varrho = \text{const}$. The pre-assembly mass of the star is thus ($\mathrm{d}V$ is given in (5.12))

$$A = n\int \mathrm{d}V = \frac{4\pi\varrho G}{3} R^3 f(X) = Mf(X) , \tag{5.24}$$

where

$$f(X) = \tfrac{3}{2}\left(\arcsin X - X\sqrt{1-X^2}\right)/X^3 , \tag{5.25}$$

$$f(X) = \begin{cases} 1 + \dfrac{3}{10}X^2 , & X \ll 1 , \\ 3\pi/4 , & X = 1 , \\ 1.374 , & X = \sqrt{8/9} . \end{cases} \tag{5.26}$$

The mass defect of the star becomes

$$\frac{\delta M}{A} = \frac{A - M}{A} = \frac{f(X) - 1}{f(X)} . \tag{5.27}$$

At the limit of stability ($X = \sqrt{\frac{8}{9}}$) we obtain $\delta M/A \simeq 30\,\%$, *i.e.* $30\,\%$ of the pre-assembly mass of the star is radiated during the evolution of the star leading to this configuration.

The constant X, which characterizes the degree of compaction of the star can be determined in a rather direct way by observing the red shift of light emanating from the stellar surface. The ratio of wavelength λ at emission to the one observed at infinity, λ_∞, is given by

$$\frac{\lambda_\infty}{\lambda} \equiv 1 + z = \frac{\mathrm{d}t}{\mathrm{d}s} = \left(1 - \frac{2M}{R}\right)^{-\frac{1}{2}} = (1 - X^2)^{-\frac{1}{2}} . \tag{5.28}$$

For $X = \sqrt{\frac{8}{9}}$ we obtain $z = 2$, which is therefore the maximum red shift which can be explained by gravitational effects. Note that for $X=1$, $z=\infty$, *i.e.* light emanating from a source with $2M=R$ will be red-shifted completely.

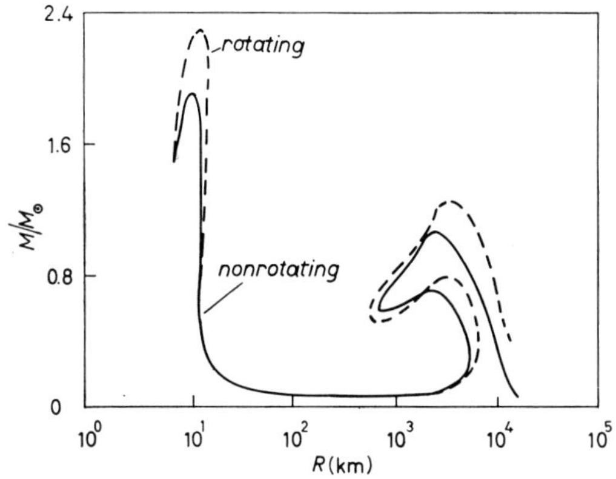

Fig. 9.

Inserting realistic equations of state into (5.10) and integrating (5.4)-(5.7), (5.10) numerically, one is led to the 2 families of stable high-density objects, *i.e.* white dwarfs and neutron stars shown in Fig. 9.

5·3. *The geometry of the Schwarzschild solution.* – In order to visualize the geometry of the Schwarzschild solution we have to restrict ourselves to a two-dimensional surface which can then be embedded in 3-space.

We shall study here the geometry on a plane through the center of the star, *i.e.* $\theta=\pi/2$, $t=\text{const}$, $\mathrm{d}\theta=\mathrm{d}t=0$. The line element on this surface is given by (see (5.5))

$$\mathrm{d}\sigma^2=\left(1-\frac{2M}{r}\right)^{-1}\mathrm{d}r^2+r^2\mathrm{d}\phi^2 \tag{5.29}$$

for $r>R$, *i.e.* in the exterior of the star. It is easy to show that a parabola (z is an embedding co-ordinate)

$$z=\sqrt{8M}\sqrt{r-2M} \tag{5.30}$$

rotated around the axis $r=0$. Figure 10 gives a surface of the same intrinsic geometry as (5.29).

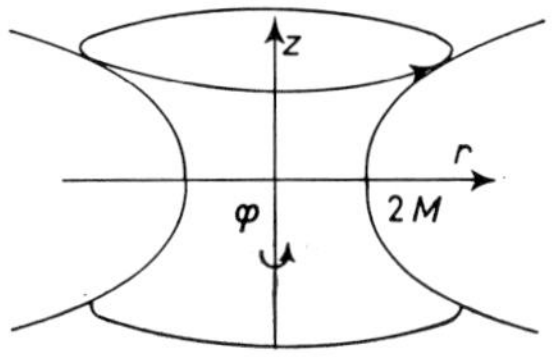

Fig. 10. – Surface of revolution corresponding to Schwarzschild metric.

In the interior of an incompressible-mass distribution $d\sigma^2$ can be derived from (5.5) and (5.6) to be

$$d\sigma^2 = \frac{dr^2}{1-((8\pi G\varrho)/3)\,r^2} + r^2 d\Omega^2 . \tag{5.31}$$

This is the geometry on a space of constant curvature, *i.e.* of a 3-dimensional hypersphere with radius

$$a = \left(\frac{3}{8\pi G\varrho}\right)^{\frac{1}{2}} . \tag{5.32}$$

(5.21) can therefore be written as $x = r/a$, $X = R/a$. The mass distribution on the hypersphere extends only to the maximum radius R. One can show that the hypersphere joins the exterior metric smoothly in $r = R$ (Fig. 11).

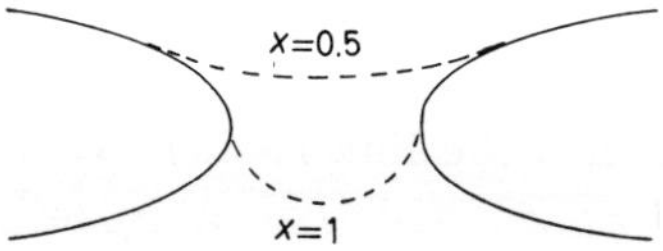

Fig. 11. – The complete geometry of the Schwarzschild metric.

5·4. *Kruskal space and gravitational collapse.* – In this Section we shall continue the study of the geometry of the Schwarzschild metric. It will turn out that the simple discussion presented before leaves many questions to be answered: Is there no space at all for $r < 2M$? Intuitively one feels that there should be a singularity at $r = 0$, whereas there seems to be one at $r = 2M$. If a test particle is dropped towards a Schwarzschild singularity where does it go once it reaches $r = 2M$? Can an object collapse beyond its own Schwarzschild radius? In this case the interior and exterior metrics could not be joint together in the simple manner shown in Fig. 11.

The key step towards the understanding of the full geometry of the Schwarzschild solution is a co-ordinate transformation $(t, r, \theta, \phi) \to (u, v, \theta, \phi)$ introduced by KRUSKAL which is defined in an implicit way by

$$ds^2 = f^2(u, v)(dv^2 - du^2) - r^2(u, v)\, d\Omega^2 , \tag{5.33}$$

$$u^2 - v^2 = ((r/2M) - 1)\exp[r/2M] , \tag{5.34}$$

$$2uv/(u^2 + v^2) = \operatorname{tgh}(t/2M) , \tag{5.35}$$

$$f^2(u, v) = \frac{32M^3}{r(u, v)} \exp[-r/2M] . \tag{5.36}$$

The geometrical significance of the transformation becomes most transparent in a (u, v) diagram, from which one can read off the space-time behaviour of radial trajectories (see Fig 12). The radial variables θ, ϕ are omitted in this diagram.

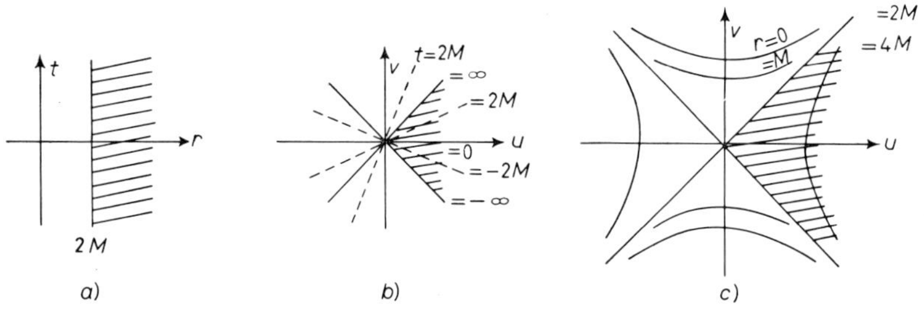

Fig. 12. – Kruskal space: space-time diagram.

The physical region $r > 2M$ of the Schwarzschild space is mapped onto one quadrant of the Kruskal space by the transformation (5.33)-(5.36). These equations show, however, that the line element is regular even in the other 3 quadrants up to the line $r = 0$, which is a hyperbola in the (u, v)-plane. This means that (t, r, θ, ϕ) is a co-ordinate patch (chart) which can be used only on one part of the whole manifold.

In (5.33) v is obviously a timelike co-ordinate, proportional to the proper time element of an observer located at a point $u = \text{const}$.

This observer starts at a finite radius $r > 2M$ for $v = 0$, begins to fall towards smaller values of r later on and reaches the singularity $r = 0$ at a finite proper time.

A special advantage of the Kruskal metric is that radial light rays are simply given by

$$du = \pm dv . \tag{5.37}$$

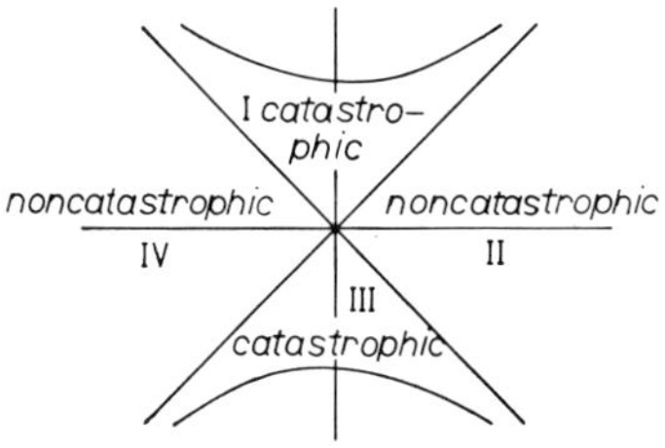

Fig. 13. – Kruskal space.

The light-cone has therefore the same form, as in special relativity, which simplifies the discussion of causality problems. The light-cone $u = \pm v$ (Fig. 13) separates Kruskal space into four regions, the catastrophic ones (I and III,

$r < 2M$) and the noncatastrophic ones (II and IV, $r > 2M$). Both branches of all light-cones emanating from the points within I reach the singularity $r = 0$; every observer within the catastrophic region I is therefore bound to fall into the singularity. It is possible to send light rays into I, but no signal can come back from there; similarly, it is possible to send signals from III to the noncatastrophic regions, but no signal can be sent from II or IV to III.

Figure 12 *c*) shows, furthermore, that the lines $r = \text{const} < 2M$ are not timelike. Only a small part of Kruskal space (the regions II + IV) can thus be explored with the help of static ($r = \text{const}$) observers.

In Subsect. **5**˙3 we have studied the geometry on a (spacelike) cross-section through the center of the star and have constructed an embedding diagram for this surface. Figure 12 *b*) shows, however, that the cross-sections $t = \text{const}$, $\theta = \pi/2$ do not give a complete picture of the manifold, since the catastrofic regions I and III cannot be obtained in this way. We have to choose, therefore, a different family of cross-sections through the center of the mass distribution; the only restrictions being that the surfaces should actually be spacelike (slope $< 45°$ in the u-v diagram) and that taken together they should give a complete picture of the manifold. The simplest choice fulfilling these conditions are the surfaces $v = \text{const}$, $\theta = \pi/2$.

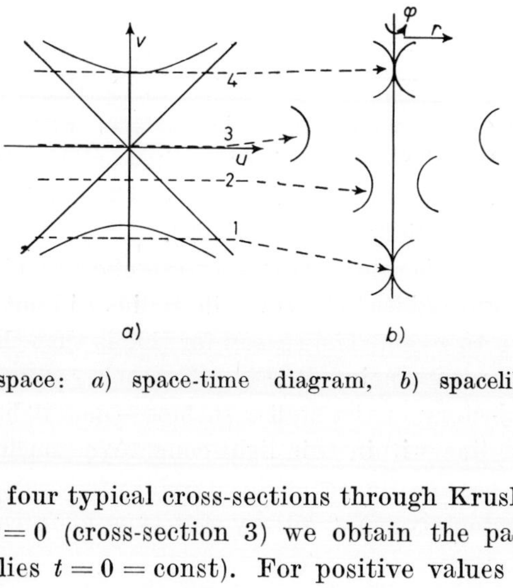

Fig. 14. – Kruskal space: *a*) space-time diagram, *b*) spacelike slices along the geometry.

Figure 14 shows four typical cross-sections through Kruskal space, obtained in this way. For $v = 0$ (cross-section 3) we obtain the paraboloid shown in Fig. 10 ($v = 0$ implies $t = 0 = \text{const}$). For positive values of v the Schwarzschild throat begins to close slowly and in $v = 1$ the two surfaces begin to touch. It seems slightly surprising at first that the Schwarzschild metric shows a dynamical behaviour; one has to bear in mind, however, that the static observers ($r = \text{const}$) cannot see the whole manifold and dynamic ones (*e.g.*, $u = \text{const}$) have to be used to explore all of space-time

We are now in a position to discuss the *dynamics of spherically symmetric gravitational collapse*.

As the initial configuration of the collapsing object we choose a static, homogeneous mass distribution of radius R, density ϱ at $t = v = 0$. For $t = 0$ the cross-section through the center of the star has one of the forms shown in Fig. 11, *i.e.* space is given by the Schwarzschild metric outside and a (momentarily static) space of constant curvature on the inside. These two spaces are joined together smoothly at the surface of the star $r = R$.

The subsequent dynamical evolution of the exterior metric can be read off Fig. 15. The stellar surface (in Fig. 15 the atom which is at $r = R$ for $t = 0$) begins to collapse along the world line shown in Fig. 15.

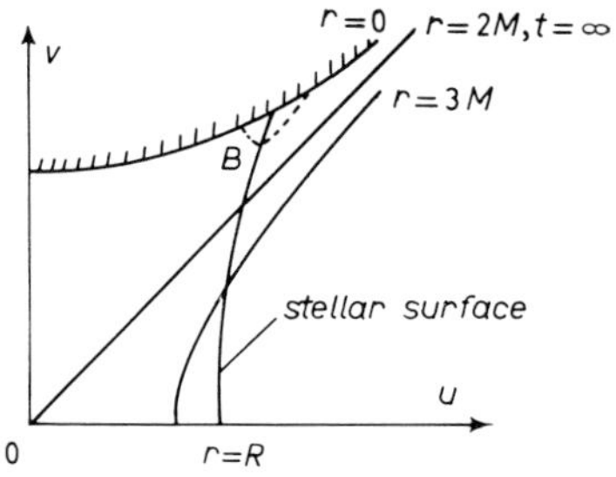

Fig. 15. – Gravitational collapse: exterior metric.

The Figure shows that the stellar surface crosses the Schwarzschild radius $r = 2M$ at a finite proper time and reaches $r = 0$ also within a finite time, as measured by an observer sitting on the surface. For an exterior observer, located at $r > 2M$ the collapse takes infinite time and the Schwarzschild radius is crossed by the collapsing object only at $t = \infty$. The object will lose its luminosity, however, quite rapidly (exponentially).

Figure 15 shows, furthermore, that pressure effects which become relevant after the object has crossed its Schwarzschild radius will not be able in principle to stop the collapse to $r = 0$. The reason for this is that the both parts of the light-cone emanating from point B (Fig. 15) reach $r = 0$ and pressure effects can change the trajectory of the stellar surface—once it has reached B—only to any other world line within this light-cone. We conclude, therefore, that once an object has crossed its own Schwarzschild radius it cannot be stopped from collapsing completely (to $r = 0$) by any type of force law.

5'5. *The Kerr metric.* – The Kerr solution has the form

$$(5.38) \qquad \mathrm{d}s^2 = \mathrm{d}t^2 - \frac{2Mr}{r^2 + a^2\cos^2\theta}(\mathrm{d}t + a\sin^2\theta\,\mathrm{d}\varphi)^2 - \\ - (r^2 + a^2)\sin^2\theta\,\mathrm{d}\varphi^2 - (r^2 + a^2\cos^2\theta)\left[\mathrm{d}\theta^2 + \frac{\mathrm{d}t^2}{r^2 - 2Mc + a^2}\right].$$

Repeating the calculation of Subsect. **1**'3 for $r \gg a$, one sees that M is actually the mass parameter. The significance of a can be seen by comparing (5.38) with the metric (2.27) of a rotating mass. For $\omega^s = (0, 0, \omega)$ (2.27) can be rewritten as

$$ds^2 = (1 + 2u)\,dt^2 - (1 - 2u)\,d\boldsymbol{x}^2 + \frac{4I}{r^3}\,\omega \sin^2\theta\, d\varphi\, dt\,. \tag{5.39}$$

The angular momentum $L = I\omega$ is then given by

$$L = -Ma\,, \qquad a = -L/M\,. \tag{5.40}$$

The parameter a is thus the angular momentum per unit mass. As in the case of the Schwarzschild metric there is a *surface of infinite red shift* $g_{00} = 0$ given by

$$r_+ = M + (M^2 - a^2\cos^2\theta)^{\frac{1}{2}}\,, \tag{5.41}$$

and an *event horizon* (*one way membrane*) where $g_{rr} = \infty$

$$r_0 = M + (M^2 - a^2)^{\frac{1}{2}}\,. \tag{5.42}$$

These two surfaces coincide ($r_+ = r = 2M$) for the Schwarzschild solution.

The region between the surface of infinite red shift and the event horizo is the *Ergosphere* of the Kerr black hole (Fig. 16).

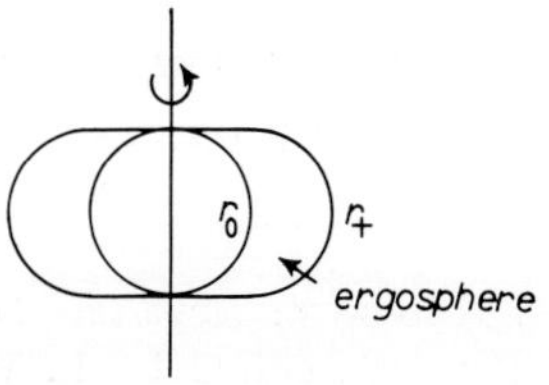

Fig. 16. – Ergosphere of the Kerr black hole.

Outside r_+ the world-line given by $dt \neq 0$, $dr = d\theta > d\varphi = 0$ is timelike, *i.e.* a static observer is possible there.

In the ergosphere $dt \neq 0$, $dr = d\theta = d\varphi = 0$ is no longer timelike, because of $g_{00} < 0$. However, $dt \neq 0$, $d\varphi \neq 0$, $d\theta = dt = 0$ can be timelike if dt and $d\varphi$ are suitably chosen. In the ergosphere one has to keep moving, therefore, but does not necessarily fall into the center.

For $r < r_0$ finally, both $g_{00} < 0$ and $g_{rr} < 0$ so that the roles of t and φ have been exchanged. Timelike world-lines are possible only for $dr \neq 0$, *i.e.* fall into the center is inevitable once r_0 has been crossed.

The importance of the ergosphere is due to the following: When a particle enters the ergosphere with energy E_1 and decays then into two pieces, one of which falls into the interior of the one way membrane, the other piece can come out with energy $E_2 > E_1$.

The Kerr-Newmann metric can be obtained from the Kerr metric by replacing $M \to M - e^2/2r$, *i.e.*

$$(5.43)\qquad ds^2 = dt - \frac{2Mr - e^2}{r^2 + a^2\cos^2\theta}(dt + a\sin^2\theta\, d\varphi)^2 - (r^2 + a^2)\sin^2\theta\, d\varphi^2 - -(r^2 + a^2\cos^2\theta)\left[d\theta^2 + \frac{dr^2}{r^2 - 2Mr + a^2 + e^2}\right].$$

It is the metric of a charged, rotating mass. Putting $a = 0$ in (5.43) leads to the Reisner-Nordström metric of a charged mass.

Iti s almost certain that (5.43) is the most general black hole. All other features (multiple momenta, etc.) are radiated during collapse and only the conserved quantities (e, a, M) remain.

BIBLIOGRAPHY

Section **1** and **2**:

J. L. ANDERSON: *Principles of Relativity Physics* (New York, 1967).

R. ADLER, M. BAZIN and M. SCHIFFER: *Introduction to General Relativity,* (New York, 1965).

J. L. SYNGE: *Relativity: The General Theory* (Amsterdam, 1960).

L. P. EISENHART: *Riemannian Geometry* (Princeton, 1926).

Section **2**:

B. BERTOTTI, D. BRILL and R. KROTKOV: *Experiments on gravitation,* in *Gravitation: An Introduction to Current Research,* edited by L. WITTEN (New York, 1962).

I. SHAPIRO: *Testing general relativity: progress, problems and prospects,* Talk given at the *VI International Conference on Gravitation and Relativity* (Copenhagen, 1971), to be published in GRG.

Section **3**:

J. WEBER: *General Relativity and Gravitational Waves* (New York, 1961).

F. A. E. PIRANI: *Gravitational radiation,* in *Gravitation: An Introduction to Current Research,* edited by L. WITTEN (New York, 1962).

D. W. SCIAMA: *Recent developments in the theory of gravitational radiation,* Talk given at the *VI International Conference on Gravitation and Relativity* (Copenhagen, 1971), to be published in GRG.

Section **4**:

T. J. WILLMORE: *An Introduction to Differential Geometry* (Oxford, 1964).

W. ISRAEL: *Differential forms in general relativity,* Communications of the Dublin Institute for Advanced Studies, Series A, No. 19.

C. W. MISNER: *Gravitational collapse*, in *Astrophysics and General Relativity*, edited by M. CHRETIEN, S. DESER and J. GOLDSTEIN (London, 1969).

Section **5**:

M. D. KRUSKAL: *Phys. Rev.*, **119**, 1743 (1960).
R. P. KERR: *Proceedings of the First Texas conference on Gravitational Collapse* (Dallas, 1963).
B. K. HARRISON, K. THORNE, M. WAKANO and J. A. WHEELER: *Gravitation Theory and Gravitational Collapse* (Chicago, 1965).
C. M. MISNER: *Gravitational Collapse*, edited by M. CHRETIEN, S. DESER and J. GOLDSTEIN (New York 1969).
B. CARTER: *Phys. Rev.*, **174**, 1559 (1968).

Equation of State of Ultradense Relativistic Matter.

S. A. BLUDMAN

Department of Physics, Tel Aviv University - Tel Aviv
University of Pennsylvania - Philadelphia, Pa. (*)

1. – Introduction.

In these lectures we will consider a question of principle and a practical question having to do with the equation of state $p = p(\varrho)$ of cold matter, of pressure p and mass density ϱ. In the many-body problem, one calculates the ground-state energy $E(V)$ of a system of N particles in a box of volume V. Then at zero temperature or under adiabatic conditions the pressure

$$p = -\frac{dE}{dV} = n\frac{d\mathscr{E}}{dn} - \mathscr{E}\,, \tag{1.1}$$

in terms of the energy density $\mathscr{E} = E/V$ and number density $n = N/V$. The speed of long-wavelength compressional sound waves is given by

$$c_s^2 = dp/d\varrho = c^2\, dp/d\mathscr{E}\,, \tag{1.2}$$

since relativistically $\mathscr{E} = \varrho c^2$. In terms of the chemical potential per particle

$$\mu = \frac{pV + E}{N} = \frac{p + \mathscr{E}}{n}\,, \qquad (c_s/c)^2 = \frac{n\, d\mu}{\mu\, dn}\,.$$

We will always consider a system of particles interacting in a Lorentz-invariant way. The question of principle which we investigate in the next section is whether, in such a relativistic system, the sound speed c_s can exceed the speed of light c. In the last Section we study the role of the equation of state in determining the critical parameters, mass M and radius R, of neutron stars.

(*) Assisted in part by the U.S. Atomic Energy Commission.

2. – Possibility of noncausality and of instability in relativistic matter.

2·1. *Lorentz electron.* – That noncausality is possible in a classical relativistic field theory is illustrated already in the seventy-year old example of the Lorentz classical electron. (By classical we mean that particle production is not considered.) By considering the action of one part of an extended electron on the rest, Lorentz derived the equation of motion

$$ma - \frac{2}{3}\frac{e^2}{c^3}\dot{a} = F_{\text{ext}}\,, \tag{2.1}$$

for the acceleration $a(t)$ of the electron in the external field $F_{\text{ext}}(t)$. In this equation, we have gone to the limit of point electron and the self-mass, which diverges, has been replaced by the finite (observable or renormalized) mass m. The radiation reaction term $m\tau\dot{a}$, $\tau = \frac{2}{3}(e^2/mc^3)$ is what distinguishes this equation of motion from $F = ma$ in Newtonian mechanics. The third-order equation of motion (2.1) has the integral

$$a(t) = \exp[t/\tau]\left[\text{constant} + \int_t^\infty \exp[-t'/\tau]\frac{F_{\text{ext}}(t')}{m\tau}\,\mathrm{d}t'\right].$$

LORENTZ and DIRAC imposed the asymptotic condition $a(\infty) < \infty$, so that

$$a(t) = \int_t^\infty \frac{\exp[-(t'-t)/\tau]}{m\tau}\cdot F_{\text{ext}}(t')\,\mathrm{d}t'\,; \tag{2.2}$$

otherwise, the solutions would be run-away: the electron would be unstable or self-accelerated even in the absence of external forces. In the Lorentz-Dirac solution (2.2), however, the force at time t' determines the acceleration at earlier times $t < t'$. This pre-acceleration is inevitable if the electron is to be stable. Notice, however that the time $\tau = \frac{2}{3}(e^2/mc^3)$ characterizing the pre-acceleration phenomena is nonzero only because the (finite) renormalized mass m was introduced in eq. (2.1).

The microscopic noncausality in this familiar electrodynamic example cannot be made macroscopic. Because of the infinite range of electromagnetic forces, in an N-electron system the pre-acceleration time is decreased to $\tau/N^{\frac{1}{3}}$.

2·2. *Instability vs. noncausality.* – In the dispersion analysis of eq. (2.1) there appears a pole at $\omega = i/\tau$ in the upper half ot the complex ω-plane. Any failure of the Kramers-Kronig relation can be indicative of *noncausality* or of *instability*, depending on whether one chooses a contour along the real axis or above the upper half-plane singularities. In the first case, the solutions though bounded are noncausal and, because some wave numbers k corresponding to the singularities $\omega(k)$ are excluded, nonlocal in space. In the second case all normal modes are included so that the solutions are local in space and time; the singularities lead, however, to solutions growing for $t > 0$. In either case, noncausality or instability, a stable Cauchy problem does not exist.

2·3. *Classical baryonic lattice.* – BLUDMAN and RUDERMAN [1] considered a model in which classical baryons, separated by distance a, interacted relativistically through the medium of a neutral vector-meson field of range $1/\mu$. Normal mode analysis gave [2], for longitudinal compressional oscillations, and for $\mu a \gg 1$, the dispersion relation

$$\left(\frac{\omega}{c}\right)^2 [mc^2 + (\mu a)\varphi \cos ka] + \frac{g^2}{4\mu^2}\left(\frac{\omega}{c}\right)^4 = 2\mu^2\varphi(1 - \cos ka) , \tag{2.3}$$

where $\varphi \equiv g^2 \exp[-\mu a]/a$.

$\omega(k)$ has branch cuts in the upper half of the complex plane. This lattice of baryons is not truly at minimum energy; it is unstable against self-accerelating run-away. If, as for the classical electron, the run-away solution is excluded then the baryon lattice becomes noncausal. While in the elementary particle problem one insists on stability even at the price of causality, in the many-body problem causality conditions are usually imposed even if this leads to instability.

The long-wavelength limit of eq. (2.3) is

$$\left(\frac{\omega}{kc}\right)^2 = \frac{(\mu a)^2 \varphi}{mc^2 + (\mu a)\varphi} ,$$

which is also obtained [1] by applying eq. (1.2) to the ground-state energy. If the particle spacing a is now decreased (keeping $\mu a \gg 1$), c_s = phase velocity = group velocity becomes superluminal when $g^2 \exp[-\mu a] > mc^2/\mu^2 a$.

In our model, which is a fairly realistic classical picture of nuclear matter, the finite range of interaction allows Lorentz's microscopic noncausality to become macroscopically manifest.

2·4. *Role played by correlations* [3]. – We return to the expressions (1.1) and (1.2) for the pressure and speed of low-frequency sound in terms of the ground-state energy density $\mathscr{E}(n)$. This latter formula always agrees with the results of a dynamical calculation of $\omega(k)$ in the limit $\omega, k \to 0$. The ground-state energy density may be split into three terms

$$\mathscr{E} = \mathscr{E}_0 + \mathscr{E}_H + \mathscr{E}_c , \tag{2.4}$$

where $\mathscr{E}_0$ is the rest or kinetic energy density, $\mathscr{E}_H$ is the Hartree energy density due to the interaction of a particle with the averaged field due to all the other $N-1$ particles, and $\mathscr{E}_c$ is the correlation energy or correction to the Hartree approximation. $\mathscr{E}_0 \sim n$ for a classical system of baryons at rest, and $\sim n^{\frac{5}{3}}$ and $n^{\frac{4}{3}}$ for a degenerate fluid of nonrelativistic or relativistic fermions. $\mathscr{E}_0 \sim n^2$ and $\mathscr{E}_c \sim -n^\gamma$ is always negative, because the Hartree approximation is an upper bound to the true energy density. In the model considered above, the baryonic lattice with nearest neighbor interactions ($\mu a \gg 1$)

$$\mathscr{E}_H = \frac{2\pi g^2}{\mu^2} n^2 , \qquad \mathscr{E}_c = -2\pi g^2 \beta n^{\frac{4}{3}} , \tag{2.5}$$

where $\beta = \frac{1}{12}$ for a simple cubic lattice. Generally, since each of N particles interacts with fewer than $N-1$ others, $-E_e/N$ increases with n slower than linearly ($\gamma < 2$) but—unless phase transitions take place— $-E_c/N$ still increases with N ($\gamma > 1$). Defining the correlation part of the pressure

$$p_c = \frac{n\,\mathrm{d}\mathscr{E}_c}{\mathrm{d}n} - \mathscr{E}_c = [\gamma(n) - 1]\,\mathscr{E}_c(n) , \qquad \mathscr{E}_c < p_c \tag{2.6}$$

and

$$\mu_c \equiv \frac{\mathrm{d}E_c}{\mathrm{d}n} = \frac{\mathscr{E}_c}{n} + \frac{p_c}{n} , \tag{2.7}$$

is expected to be $< \mathscr{E}_c/n$ if, as expected, $\gamma > 1$ and $p_c < 0$. From eqs. (1.2) and (1.1)

$$mc^2 < \frac{\mathrm{d}}{\mathrm{d}n}(p_c - \mathscr{E}_c) \text{ makes matter superluminal } (c_s > c) , \tag{2.8}$$

$$mc^2 < \frac{p_c - \mathscr{E}_c}{n} \text{ makes matter ultrabaric } (p > \mathscr{E}) . \tag{2.9}$$

Because $\mathscr{E}_c$ is negative and increasing in magnitude slower than n^2,

$$p_c - \mathscr{E}_c = n^3 \frac{\mathrm{d}}{\mathrm{d}n} \frac{\mathscr{E}_c}{n^2} > 0$$

and, if $p_c - \mathscr{E}_c$ or $-\mathscr{E}_c$ increases faster than n, matter will become superluminal and ultrabaric at high enough density.

In our classical baryonic lattice where $\gamma = \frac{4}{3}$, $c_s > c$ when $mc^2 < \frac{8}{9}(2\pi g^2 \beta) n^{\frac{1}{3}}$. In a quantum system, as long as particle production is neglected, $(-\mathscr{E}_c)$ does indeed increase faster than linearly $\gamma > 1$. Thus, for a relativistic Fermi fluid in Hartree-Fock approximation

$$\mathscr{E} = \left(\frac{9\pi}{8}\right)^{\frac{2}{3}} \hbar c n^{\frac{4}{3}} + \frac{4\pi g_0^2 n^2}{\mu^2} - \left(\frac{81}{8\pi}\right)^{\frac{1}{3}} g_0^2 n^{\frac{4}{3}} ,$$

and, provided $g_0^2/\hbar c > \pi/2$, the statistical anticorrelation by itself suffices to make such a fluid superluminal and ultrabaric. For a high-density Bose fluid with repulsive static Coulomb interaction $\mathscr{E}_c \sim -n^{\frac{5}{4}}$ nonrelativistically or $-n^{\frac{4}{3}}$ relativistically.

2'5. *Effect of virtual-particle production* [3]. – In neutral vector-meson theory the current of baryons is conserved. It was this that enabled us to identify the system by the number of baryons. Ward's Identity then makes the radiative corrections vanish in the ground state, a state of zero four-momentum. Quantum-mechanically, the Hartree energy (2.5) remains unchanged with g, μ the unrenormalized meson coupling and mass.

The baryon lattice becomes superluminal and ultrabaric at high density classically and quantum-mechanically, so long as statistical anticorrelation and radiative corrections are considered, but not real particle production.

2'6. *Effects of real baryon-antibaryon production*: *Instability* [3]. – We now consider the stability of an N-baryon system against the production of baryon-antibaryon pairs. To produce one such pair increases the energy by

$$E^{\text{pair}} = mc^2 + E_{\text{H}} + mc^2 - E_{\text{H}} + E_c^{\text{pair}},$$

where the Hartree energies of particle and antiparticle cancel and

$$E_c^{\text{pair}} < 2\mu_c,$$

because the baryon-antibaryon attraction makes the correlation energy of a pair less than the energy of separate baryons and antibaryons. Stability against spontaneous pair production requires $0 < E^{\text{pair}}$ or $-\mu_c < mc^2$. Stability is compatible with the conditions (2.8) or (2.9) for superluminal or for ultrabaric bevaviour only provided $0 < (\mathrm{d}/\mathrm{d}n)p_c$ or $0 < p_c$. But in eq. (2.7), μ_c is the work done against correlation forces in order to add one more particle and

$$p_c/n = p_c V/N$$

is that part of the work that must be done to prepare a space for the additional particle. Because the baryon-antibaryon forces are attractive, this space-preparing work is negative or $p_c < 0$. Thus stability is incompatible with ultrabaric or superluminal behaviour.

If baryon-antibaryon interactions were repulsive enough to make matter highly incompressible, then baryon-antibaryon interactions would be attractive enough to make the vacuum unstable and to give matter no lowest energy.

3. – Simple calculation of critical parameters of neutron stars.

3'1. *Introduction.* – Laboratory experiments in high-energy physics are limited by the particle energies and densities available. In astrophysical situations, however, baryons attain relativistic energies when the temperature is high (thermal energy $= mc^2$ at 10^{12} °K) or the density is high (Fermi momentum

$$p_F = \hbar(3\pi^2 n)^{\frac{1}{3}} = mc \tag{3.1}$$

when the number density $n = 3.6 \cdot 10^{39}$ nucleons/cm³ or the mass density $nm = 6 \cdot 10^{15}$ g/cm³). These conditions obtain generally in late stages of stellar evolution. In supernovae, high temperatures are realized and neutrino processes become significant. At the end point of stellar evolution, nuclear energy sources are depleted or inaccessible; the temperature is relatively low and the white dwarf or neutron star is supported against gravity by electron or baryon degeneracy pressure. For both electron stars (white dwarfs, typically of density $(10^4 \div 10^9)$ g/cm³) and neutron stars (density $(10^{14} \div 10^{15})$ g/cm³) there exists a critical mass above which the star will collapse. Our main purpose will be to compute the critical mass M_{CR}, critical density ϱ_{CR}, and associated radius R.

The condition for hydrostatic equilibrium is

$$-\frac{dp}{dr} = G\frac{(m + 4\pi r^3(p/c^2))(\varrho + (p/c^2))}{r^2(1 - (2G/c^2)(m/r))}, \tag{3.2}$$

$$m(r) \equiv \int \varrho 4\pi r^2 \, dr \tag{3.3}$$

in general relativity (TOV equation). In the nonrelativistic limit, the right-hand side of eq. (3.2) reduces to the Newtonian gravitational force

$$-\frac{dp}{dr} = G\frac{m\varrho}{r^2}. \tag{3.4}$$

We neglect thermal structure, chemical structure, and any possible stellar rotation and superfluid mantle, crystalline crust and plasma magnetosphere. That any star observed is not at zero temperature and is radiating, can be handled in a straightforward way. In pulsars, however, rotation is the outstanding property, and the signals received are determined by the superfluid, crystalline and plasma layers surrounding the neutronic core.

Many detailed calculations of the equation of state of neutronic matter and numerical integrations of the TOV equations including variable chemical structure have been done. Before these results can be compared with pulsar observations, a great deal of work has to be done on the effects of rotation, superfluidity, seismology, magnetohydrodynamics, etc. This is why, in this introductory review, we concern ourselves with a simple picture serving to illuminate the roles of relativity and of equation of state in determining the density, mass and radius of simple nonrotating neutron stars.

3·2. *General features of electron and neutron stars.* – Figure 1 shows the results of typical detailed calculations of the total mass M as a function of

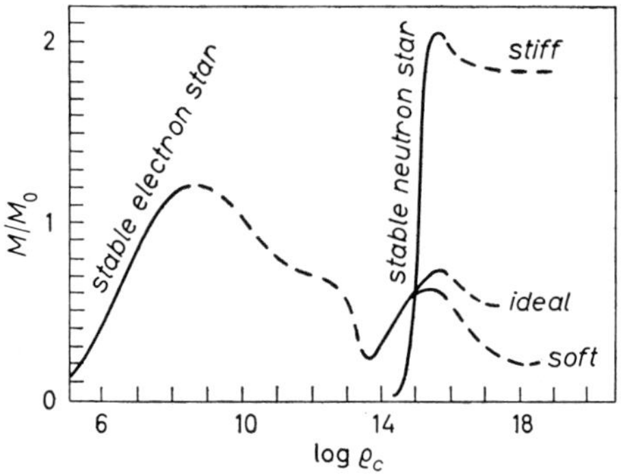

Fig. 1. – Mass (in solar units $M_\odot$) of electron and neutron stars, stable (heavy line) and unstable (dashed line), as function of central radius ϱ_c. The three calculated curves on the right are for stiff equation of state for neutronic matter (CAMERON [6]), ideal gas (OPPENHEIMER and VOLKOFF [7]), and soft equation of state (AMBURTSUMYAN and SAAKYAN [8]).

central density ϱ_e. The main features to be noted which we will try to explain are:

1) Electron stars (white dwarfs) and neutron stars are stable up to a critical central density $\varrho_{CR} = (10^9 \div 10^{10})$ g/cm³ and $\varrho_{CR} = (2 \div 4) \cdot 10^{15}$ g/cm³, respectively. In these two regions of stability, M increases roughly as $\varrho_c^{\frac{1}{2}}$.

2) Above ϱ_{CR} electron stars and neutron stars become unstable, for different reasons as we shall see. Neutron stars become unstable when $\varrho_c \approx \approx 3 \cdot 10^{15}$ g/cm³. Above this density, M decreases roughly as $\varrho_c^{-\frac{1}{2}}$. (Above $\varrho_c = 10^{18}$ g/cm³, M is insensitive to ϱ_c, approaching a constant value. This is an effect of general relativity, due to its nonlinearities. We shall, however, not concern ourselves with such very unstable stars.)

3) For electron stars the results of all calculations agree that the maximum mass is about $M_{CR} = 1.4\, M_\odot$. For neutron stars, different calculations give results ranging from $M_{CR}/M_\odot = 0.2$ to 2. In the absence of neutron interactions (Oppenheimer Volkoff results for ideal Fermi fluid), $M_{CR}/M_\odot = = 0.72$. If the neutron interactions are predominantly repulsive, a stiffer equation of state is obtained leading, for example to $M_{CR}/M_\odot = 2$. If the neutron interactions are predominantly attractive or if many statistically

independent baryons can be produced, a softer equation of state is obtained leading, for example, to $M_{CR}/M_\odot = 0.2$.

4) If we plotted a density profile, $\varrho(r)$ as a function of r, we would find that especially for large ϱ_c, $\varrho(r)$ is fairly constant, *i.e.* the neutronic matter is ultimately incompressible.

The gross features of the M *vs.* ϱ_c curves will follow from dimensional analysis or from the virial theorem, which asserts nonrelativistically

$$-3\int p\,\mathrm{d}V = -\int \frac{Gm(r)\,\mathrm{d}m(r)}{r} \sim -G\frac{M^2}{R}, \tag{3.5}$$

or the average pressure

$$\bar{p} \sim G\left(\frac{M}{R^2}\right)^2 \sim G(R\bar{\varrho})^2, \tag{3.6}$$

in terms of the average density $\bar{\varrho} = M/((4/3\pi)R^3)$. Thus

$$M \sim (\bar{p}/G\bar{\varrho}^{\frac{4}{3}})^{\frac{3}{2}}, \qquad R = (\bar{p}/G\bar{\varrho}^2)^{\frac{1}{2}}. \tag{3.7}$$

The first of eqs. (3.7) shows that a star will be stable ($\mathrm{d}M/\mathrm{d}p_c > 0$) so long as $\bar{\varrho}\,\mathrm{d}\bar{p}/\bar{p}\,\mathrm{d}\bar{\varrho} > \frac{4}{3}$, or if the equation of state is $p \sim \varrho^\Gamma$ so long as

$$\Gamma > \Gamma_{CR} = \frac{4}{3}. \tag{3.8}$$

The above analysis, which is familiar in Newtonian theory, still applies qualitatively in general relativity provided that over most of the star $p \ll \varrho c^2$ or $p = k\varrho$. The TOV equation is invariant under the scaling law

$$r \to Ar, \qquad p, \varrho \to p/A^2, \qquad \varrho/A^2,$$

so that if $p = k\varrho$, the solutions depend only on the scaling invariants $4\pi r^2 p$ $4\pi r^2\varrho$. So long as gravity is treated only linearly ($2Gm/c^2 r \ll 1$), relation (3.6) and therefore (3.7) must follow. (In even the most massive neutron stars considered $GM/c^2R < 0.3$.)

In Subsect. **3·4** below, we shall see that the features 1) and 2) of Fig. 1 are direct consequences of eq. (3.7) together with a reasonable assumption about the equation of state $p(\varrho)$.

3·3. *Effects of general relativity.* – General relativity enhances the gravitational force on the right side of eq. (3.2) over the Newtonian force in eq. (3.4) in two ways. The factor $(1-(2G/c^2)(m/r))$ in the denominator is a *kinematic* effect of the curvature of space; in the most massive neutron stars considered

$GM/c^2R \leqslant 0.3$ and $(1-(2G/c^2)(m/r))^{-1} \leqslant 3$. The pressure terms in the numerator are *dynamical* effects saying that, because the entire stress energy tensor $T_{\mu\nu}$ is a source of the general relativistic gravitational field, pressure as well as energy density acts upon and is acted upon by the gravitational field. Because for most of the equations of state that have been considered $p \leqslant \frac{1}{3}\varrho c^2$, these dynamical effects enhance the gravitational force by factors

$$\left(1+\frac{3p}{\varrho c^2}\right)\left(1+\frac{p}{\varrho c^2}\right) \leqslant 2\cdot\frac{4}{3}\,.$$

The qualitative effect of both kinematic and dynamical factors is to enhance the gravitational force as compared with the Newtonian force. This requires a steeper pressure gradient and makes it harder for hydrostatic equilibrium to be satisfied. Since $\varrho(r)$ increases inwards, a stiffer equation of state $p \sim \varrho^{\Gamma}$ with

$$\Gamma > \Gamma_{CR} > \frac{4}{3} \tag{3.9}$$

is needed for stability.

In dense white dwarfs, where the electrons become relativistic $\Gamma \simeq \frac{4}{3}$, so that general relativity would destabilize white dwarfs; in fact, except in C or He white dwarfs, electron capture

$$e^- + P \to \bar{\nu} + \mathcal{N}\,, \tag{3.10}$$

which also sets in when the electrons have MeV energies, already destabilizes electron stars by reducing the source of degeneracy pressure. This is the actual reason for electron star instability above $\varrho_c = 10^{9.5}$ g/cm³.

The reason for neutron star instability is entirely different and does not depend upon general relativity at all. We saw that above $6\cdot 10^{15}$ g/cm³ the neutrons became special relativistic; we shall soon see that this onset of special relativity softens the neutronic equation of state to where $\Gamma < \frac{4}{3}$, the threshold instability even on Newtonian gravitational theory. Although general relativity makes the ultimate collapse of supermassive stars unavoidable, neutron stars collapse already at $\varrho_c = 6\cdot 10^{15}$ g/cm³ for essentially Newtonian reasons.

Qualitatively, one may think of neutron stars in a Newtonian way. Quantitatively, the effects of Einsteinian gravity are to reduce the star mass by $3\cdot\frac{8}{3} = 8$, in the case of the densest stars.

3·4. *Ideal Fermi-fluid approximation.* – For an ideal Fermi fluid, the total energy density

$$\mathscr{E} = \int\sqrt{(p_F c)^2 + (mc^2)^2}\,\mathrm{d}n = \begin{cases} nmc^2 + \dfrac{\hbar^2(3\pi^2)^{\frac{2}{3}}}{2m}\dfrac{3}{5}n^{\frac{5}{3}}, & \text{(nonrelativistic)}\,, \\ \hbar c(3\pi^2)^{\frac{1}{3}}\dfrac{3}{4}n^{\frac{4}{3}}, & \text{(relativistic)} \end{cases} \tag{3.11}$$

and the pressure

$$p = n\frac{\mathrm{d}\mathscr{E}}{\mathrm{d}n} - \mathscr{E} = \begin{cases} \frac{2}{3}(\mathscr{E} - nmc^2) & \text{(nonrelativistic)}, \\ \frac{1}{3}\mathscr{E} & \text{(relativistic)}. \end{cases} \tag{3.12}$$

The source of mass density is, in any case, the baryons, of which there are A/Z per electron. Therefore, for an electron gas, relativistic or not

$$\varrho = nm\frac{A}{Z}, \tag{3.13}$$

where m is the nucleon mass.

For a neutron gas

$$\varrho = \begin{cases} nm & \text{(nonrelativistic)}, \\ 3p/c^2 & \text{(relativistic)}. \end{cases} \tag{3.14}$$

Thus

1) Nonrelativistic electron or neutron stars have

$$p \sim n^{\frac{5}{3}} \sim \varrho^{\frac{5}{3}}, \qquad M \sim \bar{\varrho}^{\frac{1}{2}}. \tag{3.15}$$

This explains the rising or stable portions of Fig. 1.

2) Relativistic electron stars have

$$p \sim n^{\frac{4}{3}} \sim \varrho^{\frac{4}{3}}, \qquad M \to \text{constant}. \tag{3.16}$$

As explained, such stars are neutrally stable in Newtonian theory and would therefore be especially sensitive to general relativity effects. In Newtonian theory, M would approach a limiting value, the Chandresekhar limit, were it nor for reaction (3.10), which makes $M(\varrho_c)$ decrease below the Chandresekhar limit.

3) Relativistic neutron stars have

$$p \sim \mathscr{E} \sim \varrho, \qquad M \sim \bar{\varrho}^{-\frac{1}{2}}. \tag{3.17}$$

Neutron stars therefore become unstable when the neutron becomes special relativistic or when the equation of state approaches the causality limit $p = \varrho c^2$. This explains feature 2) of Fig. 1, specifically why, on all models, the critical average density $\bar{\varrho}$ is about $6 \cdot 10^{15}$ g/cm³; the precise relation between $\bar{\varrho}$ and the central density ϱ_c is, of course, model-dependent. Detailed model calculations give critical values for ϱ_c from 1.6 to $4.0 \cdot 10^{15}$ g/cm³.

In a neutron star, the critical baryon number A_{CR} is that at which, on the addition of one more baryon, the decrease in gravitational potential energy

$$-\frac{GmA}{R}\cdot m\,, \tag{3.18}$$

would exceed the statistical energy gained by addition of one more baryon

$$\frac{\mathrm{d}\mathscr{E}}{\mathrm{d}n}=\hbar c(3\pi^2)^{\frac{1}{3}}n^{\frac{1}{3}}\,. \tag{3.19}$$

Since $n=A/\frac{4}{3}\pi R^3$ and

$$\tilde{m}\equiv\left(\frac{\hbar c}{G}\right)^{\frac{1}{2}}=2.2\cdot10^{-5}\text{ g}\,, \tag{3.20}$$

(3.18) and (3.19) are equal at

$$A_{CR}=3\left(\frac{\pi}{4}\right)^{\frac{1}{2}}\left(\frac{\tilde{m}}{m}\right)^3=7\cdot10^{57}\text{ nucleons}\,. \tag{3.21}$$

This number of nucleons has a (proper) mass

$$mA_{CR}=6M_{\odot}\text{ (Newtonian gravity)}\,. \tag{3.22}$$

In eq. (3.19) we have used the relativistic form of the energy density eq. (3.11), which is more correct for electrons than for neutrons. In an electron star, es. (3.18) is multiplied by A/Z the number of baryons per electron. Thus

$$mA_{CR}=\left(\frac{Z}{A}\right)^2 6M_{\odot}=1.4M_{\odot}\,,\qquad\text{(electron star)}\,, \tag{3.23}$$

the Chandresekhar limit.

When eq. (3.22) is corrected for general relativity effects as in Subsection **3˙3**, we find for the (gravitational) mass

$$mA_{CR}\approx\frac{1}{8}\cdot6M_{\odot}=0.75M_{\odot}\,,\qquad\text{(neutron star)} \tag{3.24}$$

the Oppenheimer-Volkoff critical mass. Note that, except for comparable $(Z/A)^2$ and general relativity factors, the critical baryon number (3.21) is the same for electron stars; it is for entirely different reasons that the sun, a much younger star, has a mass close to these maximum masses (3.23), (3.24).

The remainder of this paper is concerned only with neutron star model calculations.

3·5. *Realistic nuclear forces and equations of state.* – The ideal gas calculations in the preceding Subsection have determined ϱ_{CR} within a factor $1.4 \div 4.0$ and M_{CR} to order of magnitude. We would now like to consider realistic nuclear forces and the equation of state to which they lead.

1) For $\varrho < 6 \cdot 10^{14}$ g/cm³ the nuclear forces are fairly well determined and Brueckner theory is presumed to apply. For these densities, all nuclear matter calculations should agree irrespective of assumptions concerning the core or intermediate range nuclear forces, but do not. Suffice it to say, that at $\varrho = 6 \cdot 10^{14}$ g/cm³, $M/M_\odot = 0.3$ and is still rising. This makes it most doubtful that M_{CR} can be much less than say $M_\odot$.

2) Above $6 \cdot 10^{14}$ g/cm³, hyperons (whose forces are essentially unknown) are produced and Brueckner theory does not converge.

3·6. *Limiting equation of state.* – We need to know the equation of state up to $\varrho = 6 \cdot 10^{15}$ g/cm³, which is also the density at which the nucleon cores begin to overlap. Any extrapolation up to these densities must be optimistic. For this reason, and in the spirit of simplicity, we assume a limiting equation of state

$$p = \tfrac{1}{3}\mathscr{E} \,. \tag{3.25}$$

This is the equation of state that would apply if the baryons were noninteracting and relativistic. Baryon forces at close range are spin- and isospin-dependent, but are on average repulsive (especially if due to vector meson exchange). When many baryonic channels are opened up, it will be energetically favourable for baryons to transform into states in which the predominantly repulsive forces are inoperative. This is the rationale for regarding the dense baryonic gas to be noninteracting with some effective mass m_{eff}. Because the baryons in massive neutron stars are nearly relativistic, we are as to order of magnitude correct in neglecting m_{eff}. The $\frac{1}{3}$ in eq. (3.24) could also be 1, consistent with causality and with Hartree approximation applied to repulsive interactions at high density.

The TOV equation has already been integrated in scale-invariant units by Bondi [4] in calculating the star of maximum M/R or gravitational red-shift (*). A star terminates at that radius R, where p vanishes with $\mathscr{E}$ or ϱ finite. Since this is impossible with the equation of state (3.24), Bondi applies this equation of state only out to a radius $t(\frac{4}{3}\pi\varrho_c)^{\frac{1}{2}} = 0.909$, where $\varrho/\varrho_c = 0.126$ and $m(r)/r = 0.231$. Then he surrounds this core region with an envelope region of constant density $\varrho_e/\varrho_c = 0.126$ in which p falls to zero at $R(\frac{4}{3}\pi\varrho_c)^{\frac{1}{2}} = 1.37$, where $M/R = 0.319$. A profile of the Bondi model is shown in Fig. 2. The

(*) Units $G/c^2 = 1$.

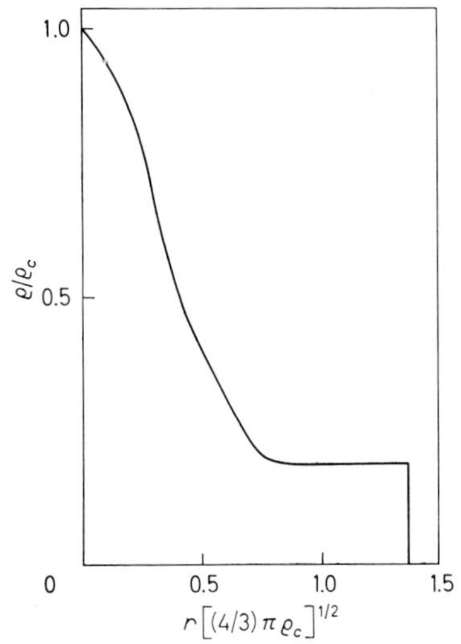

Fig. 2. – Density profile of Bondi's star with maximum M/R for $p \leqslant \frac{1}{3}\mathscr{E}$. The equation of state is $p = \frac{1}{3}\mathscr{E}$ in the core and ϱ = constant in the envelope. Core and envelope join at $r(\frac{4}{3}\pi\varrho_c)^{\frac{1}{2}} = 0.909$ and star terminates at $R(\frac{4}{3}\pi\varrho_c)^{\frac{1}{2}} = 1.37$. 40% of the mass is in the core, 60% in the envelope. The central, average and envelope densities are in the ratio $\varrho_c : \bar{\varrho} : \varrho_e = 1 : \frac{1}{6} : \frac{1}{8}$. This star shows central peaking of density typical of general relativity. The mass and radius of star are much more sensitive to $\bar{\varrho}$ than to ϱ_c.

average density $\bar{\varrho} = \frac{1}{6}\varrho_c = \frac{4}{3}\varrho_e$. Choosing $\bar{\varrho} = 6 \cdot 10^{15}$ g/cm³, the results in the first column of Table I are obtained.

A very similar model which is almost maximal in M/R is that of incompressible matter $\varrho = \bar{\varrho}$ = constant. This model, Schwarzschild's interior solution, is analytically solvable giving, for $p_c = \frac{1}{3}\bar{\varrho}$, $M/R = \frac{4}{3}\pi\varrho R^2 = \frac{5}{18} = 0.278$, so that $R(\frac{4}{3}\pi\bar{\varrho})^{\frac{1}{2}} = 0.55$. Choosing $\bar{\varrho} = 6 \cdot 10^{15}$ g/cm³, the second column in Table I is obtained.

TABLE I. – *Radius and mass of neutron stars on three models*: 1) Bondi's maximal model: core with $p = \frac{1}{3}\varrho c^2$, surrounded by constant density envelope, invented to maximize M/R. 2) Schwarzschild interior solution: ϱ = constant $p_c = \frac{1}{3}\varrho c^2$. 3) Results of COHEN *et al.*, ref. [5]. In models (1) and (2), the central density is chosen so that the average density $\bar{\varrho} = 6 \cdot 10^{15}$ g/cm³. Column (3) is that star in Cohen *et al.* Table III which gives largest M.

	Maximal model	Incompressible matter	COHEN *et al.*
ϱ_c (10^5 g/cm³)	36	6	1
R (km)	4.1	3.65	11.9
M (km)	1.34	1.03	3.6
$M/M_\odot$	0.94	0.69	2.43
M/R	0.319	0.278	0.302

In the last column, the results of the very detailed calculations by COHEN *et al.* [5] are presented. The results of all three models are quite comparable for M/R. Because our estimate of $\bar{\varrho}$ is high, our scale of M and R separately is too small by a factor 2-3. The comparison shows that, in massive neutron stars, nuclear matter is approximately incompressible and that Cohen's stiff equation of state is close to the limiting form (3.25).

REFERENCES

[1] S. A. BLUDMAN and M. A. RUDERMAN: *Phys. Rev.*, **170**, 1176 (1968).
[2] M. A. RUDERMAN: *Phys. Rev.*, **172**, 1286 (1968).
[3] S. A. BLUDMAN and M. A. RUDERMAN: *Phys. Rev. D*, **1**, 3243 (1970).
[4] H. BONDI: *Proc. Roy. Soc.*, A **282**, 303 (1964).
[5] J. M. COHEN, W. D. LANGER, L. C. ROSEN and A. G. W. CAMERON: *Astr. Space Sci.*, **6**, 228 (1970), Table III.
[6] A. G. W. CAMERON: *Astrophys. Journ.*, **130**, 884 (1959).
[7] J. R. OPPENHEIMER and G. M. VOLKOFF: *Phys. Rev.*, **55**, 374 (1939).
[8] V. A. AMBURTSUMYAN and G. S. SAAKYAN: *Astr. Zh.*, **38**, 785 (1961), English translation: *Sov. Astr.-AJ*, **5**, 601 (1961).

Nonpolynomial Lagrangians, Infinities and Gravity Theory.

A. SALAM

International Centre for Theoretical Physics - Trieste
Imperial College - London

1. – Introduction.

Field-theoretic infinities—first encountered in Lorentz's computation of electron self-mass—have persisted in classical electrodynamics for seventy and in quantum electrodynamics for some thirty-five years. These long years of frustration have left in the subject a curious affection for the infinities and a passionate belief that they are an inevitable part of nature. Most field theorists feel they can live with renormalizable theories.

They are thus content in never being able to compute g_A/g_V in weak interaction physics or $m_{\pi^+} - m_{\pi^0}$.

A new argument for dissatisfaction with *renormalizable* theories has recently, however, emerged from scaling behaviour of such theories. It appears from the work of GATTO, MENOTTI and others that a renormalizable theory like ϕ^4 does not scale in Bjorken's sense, while a superrenormalizable theory like ϕ^3 may.

It seems that it would be even better if we could work with theories which are « softer » and with even fewer infinities than ϕ^3 so far as scaling is concerned.

As is well known, the infinities result from a lack of proper definition of singular distributions which occur in field theory. One of the major obstacles to progress in the subject has been the uncertainty of whether these singularities have their origin in the circumstance that a perturbation expansion is being made or whether it is the form of the Lagrangian—assumed to be polynomial in the field variables—which is at fault. An important suggestive advance in resolving this uncertainty has come from the work of JAFFE and GLIMM (proceedings of Varenna school, 1969) who, using exact and mathematically well-defined solutions of *polynomial* Lagrangian-field theories (in two and three space-time dimensions), have shown that infinities persist even in exact solutions. If their conclusions may be extrapolated to physical four-dimensional space-time, it would seem that the origin of the infinities is not so much in the *bad mathematics* of the perturbation solution. Rather, the fault lies with the *bad physics* of the assumed polynomiality of the electromagnetic interaction.

Now nonpolynomial Lagrangian theories have been studied since 1954 (in fact they date back to the Born-Infeld nonlinear electrodynamics of the 1930's) and it is well known that a variety of these do indeed possess perturbation solutions free of infinities. However, in modifying electrodynamics to a nonpolynomial version one has been presented with two dilemmas:

1) There are a million nonpolynomial ways of «completing» the conventional polynomial version. Which represents physics?

2) Since the methods developed for solving nonpolynomial theories are radically different from those for polynomial theories—for example, they involve analytic continuation procedures in an essential manner—one would wish to be sure that the field-theory solutions thus defined do satisfy the conventional canons of *good* field theories, like appropriate analyticity, unitarity, positive-definiteness and Froissart boundedness.

In respect of the first problem, *i.e.* that of discovering the missing (nonpolynomial) physics, which should complete conventional lepton electrodynamics for example, in a series of papers my collaborators STRATHDEE, ISHAM and DELBOURGO and I have revived [15] the conjecture of LANDAU, KLEIN, PAULI, DEWITT, DESER and others (S. DESER, Texas symposium 1970, unpublished) which suggested that it may be the neglect (of the intrinsic nonpolynomiality) of tensor gravity—and the associated curvature of space-time produced by an electron or a photon in the space surrounding it—which may be the direct cause of the electron's and photon's self-mass and self-charge infinities. We have made other suggestions regarding weak and strong interactions which I wish to discuss in these lectures.

In respect of the second problem, an advance has just recently been made by LEHMANN and POHLMEYER [5] and TAYLOR [11] who have shown rigorously that the analytic procedures developed in earlier papers by VOLKOV, FILIPPOV, SALAM, STRATHDEE and others [5] do indeed define *good* field theories, *good* in the perturbational sense, provided the associated nonpolynomial theory falls into the *localizable* class, satisfying the principle of microcausality.

To anticipate what will follow: we shall see that entire nonpolynomial functions of the field variable like $\exp[\varkappa\phi]$ give *finite localizable* and microcausal theories. Rational functions with singularities in the $\varkappa\phi$-plane, like $(1+\varkappa\phi)^{-1}$ give rise to finite but *nonlocalizable* theories (see Appendix for details).]

My lectures will be divided into three parts:

Part I: Analytic regularization methods, their applicability to nonpolynomial Lagrangians and finite computation of renormalization constants in such theories.

Part II: Effects of including quantized *tensor* gravity with the electrodynamics of leptons—*i.e. the finite and gauge-invariant computation of electron's self-charge and self-mass* in the *quantum theorist's* version of curved space and time.

Part III: The speculative suggestion that F-mesons couple to the hadronic stress tensor in the same (nonpolynomial) manner as Einstein's gravitons do to leptons (*i.e.* the postulate of the two-tensor theory of gravity), and the possibility of using this nonpolynomiality to regularize renormalization constants in strong-interaction physics.

Before I proceed it may be useful to make a list of some of the nonpolynomial Lagrangians important in physics.

Nonpolynomial Lagrangians of physical interest include the following:

A) *Chiral $SU_2 \times SU_2$ Lagrangian for strong interactions.* A typical example is the π-meson Lagrangian in its different parametric versions

$$\mathscr{L} = \mathrm{Tr}\, \partial_\mu S \partial_\mu S^+ ,$$

where

$$S = \frac{1 + i\lambda_{\mathrm{W}} \boldsymbol{\tau}\cdot\boldsymbol{\pi}}{1 - i\lambda_{\mathrm{W}} \boldsymbol{\tau}\cdot\boldsymbol{\pi}}, \qquad \text{(Weinberg-Schwinger co-ordinates)},$$

or

$$S = \exp\left[i\lambda_{\mathrm{G}} \boldsymbol{\tau}\cdot\boldsymbol{\pi}\right] \qquad \text{(Gürsey co-ordinates)}.$$

Here λ, (λ_{W} or λ_{G}), which we call the minor coupling constant, has dimensions of inverse length (empirically $\lambda \approx m^{-1}$). An important open question for field theory in general is this: Are on-shell S-matrices for these two versions equal; particularly as the Gürsey (*exponential* from defines a localizable chiral theory of π-mesons and the Weinberg (*rational*) form a nonlocalizable one. It is perhaps worth remarking that recently Lehmann and Trute have shown that the Gürsey form defines the *only* localizable chiral theory.

B) *Intermediate-boson mediated weak Lagrangian.* A typical example is a neutral W-meson of mass m interacting with quarks (Q) of mass M with

$$\mathscr{L}_{\mathrm{int}} = f\bar{Q}\gamma_\mu(1+\gamma_5)QW_\mu \quad \text{and} \quad m^{-2}f^2 \approx G_{\mathrm{F}} \qquad \text{(the Fermi constant)}.$$

The essential nonpolynomiality of the theory is concealed in the derivative coupling of the spin-zero daughter of the physical spin-one particle which is described by the four-vector field W_μ. To make this manifest, write $W_\mu = A_\mu + (1/m)\partial_\mu B$ in the well-known Stückelberg form and transform the quark-field $Q' = \exp[if\gamma_5 B/m]Q$. The transformed $\mathscr{L}_{\text{int}}$ equals

$$\mathscr{L}_{\text{int}} = f\bar{Q}'\gamma_\mu(1+\gamma_5)Q'A_\mu + M\bar{Q}'\left(\exp\left[\frac{2if\gamma_5 B}{m}\right]-1\right)Q' .$$

The constant $f/m \approx \sqrt{G_F}$ plays the role of the minor coupling constant in the second term of this Lagrangian. The important point I wish to stress is this: *A derivative coupling of the daughter field B can look deceptively polynomial in form*; by suitable field transformations its essential nonpolynomiality (with the characteristic property that, in Feynman's language, a whole host of lines emanate from a single vertex) can be made manifest.

C) *Einstein's tensor gravity and gravity-modified matter fields.* The conventional Lagrangian for gravity is

$$L_{\text{Einstein}} = \varkappa^{-2} g^{\mu\nu}(\Gamma^\lambda_{\mu\varrho}\Gamma^\varrho_{\nu\lambda} - \Gamma^\lambda_{\mu\nu}\Gamma^\varrho_{\lambda\varrho})/\sqrt{-\det(g^{\mu\nu})} ,$$

where

$$\Gamma^\lambda_{\mu\nu} = \tfrac{1}{2} g^{\lambda\varrho}(\partial_\mu g_{\nu\varrho} + \partial_\nu g_{\mu\varrho} - \partial_\varrho g_{\mu\nu}) .$$

If $g^{\mu\nu}$ is the fundamental field, the covariant quantity $g_{\mu\nu}$ is intrinsically nonpolynomial and *vice versa*. The simplest example for matter field in curved space-time is the spin-zero field

$$L_{\text{matter}} = \frac{g^{\mu\nu}\partial_\mu\phi\partial_\nu\phi}{\sqrt{-\det g^{\mu\nu}}} .$$

L_{matter} is also nonpolynomial. Now the quantity $g^{\mu\nu}$ (the metric tensor of classical physics) is conventionally parametrized (when space-time at infinity is Minkowskian) in the form

$$g^{\mu\nu} = \eta^{\mu\nu} + \varkappa h^{\mu\nu}, \qquad \eta^{\mu\nu} = \begin{pmatrix} 1 & & & \\ & -1 & & \\ & & -1 & \\ & & & -1 \end{pmatrix} .$$

Here $h^{\mu\nu}$ is the physical field (which describes gravitons) and $\varkappa^2$, the coupling constant of the theory, equals $8\pi G_N$ (G_N is the Newtonian constant) $(10^{-44}/m_e^2)$, *i.e.* $\varkappa^{-1} \approx 10^{18}$ GeV. An alternative (and by the mathematicians the more favoured) parametrization is given by

$$g^{\mu\nu} = [\exp[\varkappa\gamma_{ab}h^{ab}]]^{\mu\nu} ,$$

where γ_{ab} are 4×4 pseudosymmetric matrices. (Note that for this «exponential» parametrization $\det g^{\mu\nu} = \exp[\varkappa h^a_a]$). The great merit of the exponential parametrization (pointed out to me by DeWitt) is that det g does not change sign in this parametrization and the signature of the metric $g^{\mu\nu}$ is preserved. There is one further complication for quantized gravity which we must remember. When we come to consider spin-$\frac{1}{2}$ particles, Einstein's tensor cannot be treated as the fundamental field. Instead one must work with vierbein gravity $L^{\mu a}$ whose relation to $g^{\mu\nu}$ is given by $g^{\mu\nu} = L^{\mu a} L^{\nu}_{a}$ (see Sexl's notes in this Volume). To summarize, the basic gravitational structure will be taken as the vierbein field $L^{\mu a}$ which we shall parametrize exponentially in the form $L^{\mu a} = [\exp[(\varkappa/2)\gamma_{ab}\phi^{ab}]]^{\mu a}$. The field ϕ describes gravitons. The analogy of the structure $g^{\mu\nu}$ is to the structure $S = \exp[i\lambda\pi\cdot\tau]$ in chiral theories where π is the field describing pions. One may conjecture (following Lehmann and Trute that the exponential parametrization of gravity defines the *only* localizable gravity theory. All other parametrizations, including the one used by Einstein ($g^{\mu\nu} = \eta^{\mu\nu} + \varkappa h^{\mu\nu}$) give rise to nonlocalizable theories.

PART I

The first point to note about nonpolynomial Lagrangians like $g(e^{\varkappa\phi} - 1)$ or $g[(1 + \varkappa\phi)^{-1} - 1]$ is that it is a mistake to think of these as an infinite sum of nonrenormalizable polynomial Lagrangians. These theories are really different, needing their own procedures to work them out. To expect that we shall see nothing in working them out except things we have encountered before in polynomial cases is simply wrong. To give one example of the new type of phenomenon which one had not encountered before, we shall see that the simplest computation of a matrix element in such theories (up to a given order g^n in the major coupling constant g and to *all* orders in the minor constant $\varkappa$) gives rise to terms like $\log(\varkappa^2 p^2)$. This dependence of matrix elements on the logarithm of the minor constant $\varkappa^2$, is something *we* at least had never budgeted for.

The major technical difference in working with nonpolynomial theories is the Euclidicity hypothesis which one makes for such theories. I do not know how essential this hypothesis is, but it is certainly convenient in such theories to start the computation of a matrix element for spacelike external momenta (*i.e.* in the Symanzik region). For this regime one can consistently transform $x_0 \to ix_4$, *i.e.* go to a Euclidean-field theory. One then continues the result to the physical region of momenta. It appears like a miracle to me that after this continuation one can show that the continued matrix elements possess the right analyticity structure consistent with unitarity. (This is one of the things I was referring to when I spoke of LEHMANN, POHLMEYER and TAYLOR having shown that the field theories defined after this continuation are *good* field theories.)

In the next Subsections we review the various concepts needed in dealing with these theories. The impatient reader may wish to turn to Sect. **2** where a model calculation is exhibited (self-mass of the electron in scalar-gravity-modified electrodynamics) making plain the appearance of the factor $\log(\varkappa^2 p^2)$.

1. – Gel'fand-Shilov method [1] and infinity suppression in localizable theories.

1'1. *The problems.* – Given a localizable Lagrangian like

$$L_I = g(\exp[\varkappa\phi] - 1)\,, \tag{1}$$

we wish to compute the superpropagator

$$S(x) = \big(TL_I(\phi_{\text{in}}(x))\, L_I(\phi_{\text{in}}(0))\big)\,. \tag{2}$$

Formally

$$S(x) = \sum_{n=1}^{\infty} \frac{(\varkappa^2)^n}{n!} \langle T\phi^n_{\text{in}}(x)\phi^n_{\text{in}}(0)\rangle\,. \tag{3}$$

We specialize to zero-mass particles where

$$(T\phi_{\text{in}}(x)\phi_{\text{in}}(0)) = D(x) = -\frac{1}{x^2} \tag{4}$$

with the Fourier transform (FT) proportional to $1/p^2$. In evaluating a term like $(T\phi^2(x)\phi^2(0))$, the first problem is the meaning to be ascribed to $\phi^2(x)$. The conventional procedure uses (4) to define a *normal product* $:\phi^2:$ from the relation $:\phi^2: = \phi^2 + D(0)$. Here $D(0)$ is the infinite renormalization constant $\lim_{x\to 0} 1/x^2$. One shows that

$$(T:\phi^2(x)::\phi^2(0):) = \frac{2!}{(x^2)^2} \tag{4'}$$

up to a distribution-theoretic ambiguity of the form $b\delta(x)$ (b-ambiguity). This simplest of situations already poses the three problems which lie at the heart of our discussion:

a) *Normal ordering*: is there any physics concealed in $D(0)$ and being discarded with it by the normal-ordering procedure?

b) $(1/x^2)^2$ is a product of singular distributions $1/x^2 \otimes 1/x^2$. Is there a natural definition for its Fourier transform?

c) The role of the « ambiguity constant » b.

Conventional renormalization theory treats problems *b*) and *c*) as parts of one problem; in Fourier space, a faltung is used to write

$$\int \frac{1}{x^2} \otimes \frac{1}{x^2} \exp[ipx]\, d^4x = \frac{1}{(2\pi)^4} \int \frac{d^4k}{(p-k)^2\, k^2}\,.$$

The integral on the right-hand side exhibits a logarithmic infinity. A subtraction procedure is devised to separate this from the integral and the constant *b* is adjusted to compensate this infinity.

This faltung method and infinity separation become prohibitively complicated when we consider objects like

$$(T{:}\phi^n_{\text{in}}(x){:}\,{:}\phi^n_{\text{in}}(0){:})\,,$$

represented by a cocoonlike graph with *n*-lines

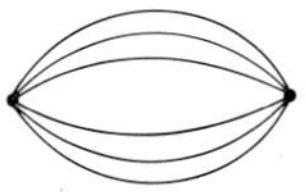

with its $(n-1)$ divergent subintegrations in momentum space. This was one reason why nonrenormalizable theories with *polynomial* Lagrangians (*e.g.*, $\mathscr{L}_{\text{int}} = g\phi^6$, etc.) were soon abandoned. Even a subtraction procedure was hard to define. (In fact, we shall see later that the use of momentum-space methods and the faltung are perhaps the greatest disasters which could happen to field theories, so far as the infinity problem is concerned. In momentum space we completely lose contact with the underlying distribution theory of *x*-space singular functions.)

1·2. *Gel'fand method.* – Nonpolynomial Lagrangian theories, on the other hand, offer, through the Gel'fand-Shilov procedures [1], a different approach, *where we separate problems b*) and *c*). (Basically this happens because a superpropagator in such theories is a sum of a series of singular functions

$$S(x) = \sum \frac{1}{n!} \left(\frac{-\varkappa^2}{x^2}\right)^n.$$

This sum is far less singular, when $x \to 0$ from an appropriate direction, than each single term of the series. Roughly speaking, $\exp[-\varkappa^2/x^2] \to 0$ when x^2 is spacelike and $\varkappa^2$ is negative. Analytic continuation then *fills in for other directions* and other $\varkappa^2$.)

To be more precise let us return to (4'). We wish to compute the FT of $(1/x^2)^2$; more generally $D^n(x) = (1/x^2)^n$. GEL'FAND and SHILOV remark that since the FT of $(1/x^2)^z$ is a well-defined classical-mathematics object with no ambiguities whatever whenever $0 < \mathrm{Re}\, z < 2$, and is proportional to $(\Gamma(2-z))/\Gamma(z) \times (1/p^2)^{2-z}$, the FT of $(1/x^2)^n$ with n lying outside this region, may be defined by an appropriate *analytic continuation of this function in the variable z*. (Contrast the elegance of this definition with the clumsiness of the conventional faltung procedure with its multiple divergent loop subintegrations. We make the word « appropriate » more precise in a minute.)

The Gel'fand-Shilov method was discussed in physics literature by GÜTTINGER [2] as early as 1966 and, in an equivalent formulation, by GUSTAFSON [3] even earlier. BOLLINI and GIAMBIAGI [4] were perhaps the first to use it purposefully for rewriting conventional renormalization theory. Its power and value, however, become apparent particularly when we use it together with nonpolynomial Lagrangians, because here the somewhat vague concept of « appropriate analytic continuation » in the variable z becomes dove-tailed with the analyticity properties of the superpropagator $S(x)$.

To give the bare bones of the method, consider the superpropagator for the Lagrangian $L_I = (:g^2 e^{\varkappa\phi}: - 1)$. This is an entire function of $\varkappa^2 D(x)$:

$$S(x) = g^2 \sum_{n=1}^{\infty} \frac{1}{n!} [\varkappa^2 D(x)]^n . \tag{5}$$

First write its Sommerfeld-Watson transform

$$S(x) = \frac{g^2}{2\pi i} \int \frac{\mathrm{d}z}{\Gamma(z+1)} \frac{1}{\mathrm{tg}\, \pi z} [\varkappa^2 D(x)]^z . \tag{6}$$

The conditions under which transition from (5) to (6) is justified are stated in the papers of VOLKOV, SALAM, STRATHDEE, LEHMANN and POHLMEYER [5]. There are b-ambiguities in writing (6) which are discussed below. The contour as usual encloses the positive real axis from $\mathrm{Re}\, z < 1$ to infinity.

Second, rotate the contour to lie parallel to the imaginary axis—in this particular case along $0 < \mathrm{Re}\, z < 1$. The Gel'fand-Shilov condition for « classical » Fourier-transforming is met, and we write

$$\tilde{S}(p) = \int\limits_{0<\mathrm{Re}\, z<1} (\varkappa^2)^z \left(\frac{1}{p^2}\right)^{2-z} \frac{\Gamma(2-z)\,\mathrm{d}z}{\Gamma(z)\Gamma(z+1)\,\mathrm{tg}\, \pi z} . \tag{7}$$

The integral has a single pole at $z = 1$ (corresponding to the $-\varkappa^2/x^2$ term in $S(x)$) and double poles at $z = 2, 3, 4, \ldots$. These give rise to characteristic terms proportional to $(\varkappa^2 p^2)^r \log (\varkappa^2 p^2)$. *This logarithmic dependence of the Green's function on the minor coupling constant is a hallmark of nonpolynomial Lagrangian theories.*

1'3. *Infinity suppression.* – To see that this logarithmic dependence represents infinity suppression, consider a mixed theory with $L_{\text{int}} = g\chi^3 \exp[\varkappa\phi]$. The superpropagator equals

$$S(x) = g^2 \sum_{n=3} \frac{D^{n+3}(\varkappa^2)^n}{n!}$$

with the Sommerfeld-Watson transform

$$\frac{g^2}{2\pi i} \int\limits_{\mathrm{Re}\, z<3} \frac{(\varkappa^2)^z D^{z+3}}{\mathrm{tg}\, \pi z \Gamma(z+1)} \, \mathrm{d}z \,. \tag{8}$$

For the Gel'fand condition to be met, the contour must be shifted to the left of $\mathrm{Re}\, z + 3 < 2$. This is perfectly possible since the integrand is not singular at $z = -1$. The FT $\tilde{S}(p)$ is easily evaluated and contains terms proportional to $g^2/\varkappa^2$, $g^2 \log(\varkappa^2 p^2)$, $g^2\varkappa^2 p^2 \log(p^2\varkappa^2)$, etc. Clearly, $g^2/\varkappa^2$ is the relic of the quadratic infinity in a polynomial theory given by the Lagrangian $g\chi^3 = \mathrm{lt}_{\varkappa\to 0}\, (g\chi^3)e^{\varkappa\phi}$; likewise, $\log(\varkappa^2 p^2)$ is the relic of the logarithmic infinity. We recover these infinities in the limit $\varkappa \to 0$. *To put it another way,* $\varkappa^{-1}$ *(more precisely* $4\pi/\varkappa$*) is the inbuilt, realistic, regularizing cut-off in the nonpolynomial theory* $(g\chi^3 \exp[\varkappa\phi])$.

2. – Computation of self-mass in « scalar-gravity » modified electrodynamics.

The infinity suppression mechanism of localizable nonpolynomial theories is so transparent and so easily exhibited that I feel the following illustrative calculation should form part of first-year courses in quantum field-theory.

Consider a *model* electrodynamic interaction $\mathscr{L} = e\bar{\psi}\psi A \exp[\varkappa\phi]$, where ψ is the electron field, A is the (*scalar*) photon and ϕ is the (scalar) gravitational field

$$\left(\alpha = e^2/4\pi = \frac{1}{137}\,, \qquad m_e^2\varkappa^2 = (16\pi)^{-1} G_N m_e^2 \approx 10^{-44}\right),$$

G_N is the Newtonian constant.

We wish to exhibit the realistic regularization of the otherwise infinite electron self-mass through gravity. Writing the most singular parts of the relevant propagators in the form (we have dropped some factors of 4π)

$$\begin{aligned}
\langle\psi\bar{\psi}\rangle &= -(i\gamma\cdot\partial + m)\frac{1}{x^2} + \text{singular terms}\,,\\
\langle AA\rangle &= -\frac{1}{x^2}\,,\\
\langle\phi\phi\rangle &= -\frac{1}{x^2}\,,
\end{aligned}$$

and noting that $\langle \exp[\varkappa\phi(x)] \exp[\varkappa\phi(0)]\rangle = \exp[-\varkappa^2/x^2]$, the contribution to the electron self-mass from the sum of the chain of graphs in Fig. 2 (with $y = 0$) is given by

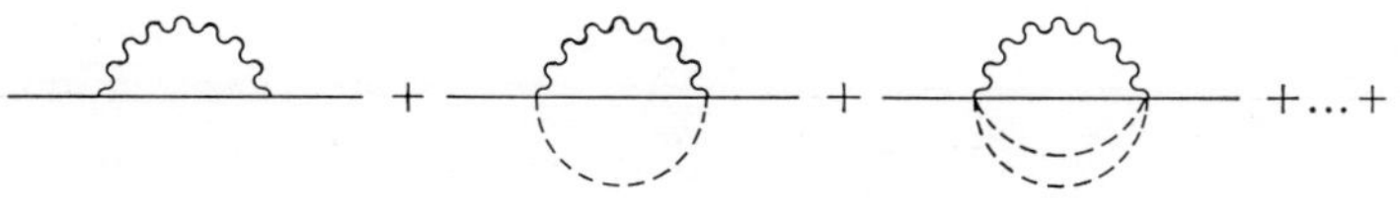

Fig. 2.

$$F(x) = \alpha \sum_{n=0}^{\infty} \frac{1}{n!} \left[(i\gamma\cdot\partial + m)\left(-\frac{1}{x^2}\right)\right]\left(-\frac{\varkappa^2}{x^2}\right)^n \left(-\frac{1}{x^2}\right).$$

Using a Sommerfeld-Watson transform, this equals

$$F(x) = \frac{\alpha}{2\pi i} \int \mathrm{d}z \, \frac{(\varkappa^2)^z(-\lambda)^z}{\Gamma(z+1)\sin \pi z} \left(-\frac{1}{x^2}\right)^{z+1} (i\gamma\cdot\partial + m)\left(-\frac{1}{x^2}\right),$$

where the contour lies round the positive real z-axis. This contour may be rotated to lie parallel to the imaginary axis to give

$$F(x) = \frac{\alpha}{2\pi i} \int_{\mathrm{Re}\, z < 0} \mathrm{d}z \, \frac{(\varkappa^2)^z(-\lambda)^z}{\Gamma(z+1)\sin \pi z} \left(\frac{i\gamma\cdot\partial}{z+2} + m\right)\left(-\frac{1}{x^2}\right)^{z+2}.$$

The advantage of doing this is that now we can use the unambiguous expression for the Fourier transform of the (classical) function $(-1/x^2)^z$ valid in the range $0 < \mathrm{Re}\, z < 2$, given by $-1/(4\pi)^2(-p^2/4\pi^2)(\Gamma(z-2))/(\Gamma(z))$. Thus

$$\tilde{F}(p) = -\frac{\alpha}{(2\pi i)(4\pi)^2} \int_{\mathrm{Re}\, z < 0} \frac{\mathrm{d}z(\varkappa^2)^z(-\lambda)^z}{(\sin \pi z)\Gamma(z+1)} \left\{-\frac{\gamma\cdot p}{z+2} + m\right\} \left[-\frac{p^2}{(4\pi)^2}\right]^z \frac{\Gamma(-z)}{\Gamma(z+2)}.$$

This is a Meijer G-function, well known in mathematical literature.

To estimate its behaviour as series in $\varkappa^2$, we may rotate the contour back to the real axis, the *double pole* in z-space at $z = 0$, gives regularized contributions to self-mass $\tilde{F}(p)|_{\gamma\cdot p - m = 0}$ of the form $\alpha m \log(\varkappa m/4\pi)^2$. The double pole at $z = 1$ gives a contribution $\alpha\varkappa^2 p^2 \log \varkappa^2 p^2$ and so on. This simple idealized example illustrates the basic technique used in this and earlier papers. The conventional infinity of $\delta m/m$ is instantly recovered by taking the limit $\varkappa \to 0$. Since numerically $\log(G_N m_e^2) \approx \log(\varkappa^2 m_e^2) \approx 100 \approx \alpha^{-1}$, the magnitude of $\delta m/m$ is not outrageously different from unity—lending support to the Lorentz view that all electron self-mass may be (gravity-modified) electrodynamic in the origin.

3. – Ambiguities.

Let us now turn to the ambiguity problem [6]. Even the *finite* theories in the sense defined in the last Subsection suffer from these. Their origin is *distribution-theoretic*; the distribution $(T{:}\phi^n(x){:}\,{:}\phi^n(0){:})$ is ambiguous up to terms of the type $\sum b_n(\partial^2)^{n-2}\delta(x)$. Alternatively, one can see these b-ambiguities in the Sommerfeld-Watson formulation. In passing from (5) to (6) and (7) we have written down an extrapolated function merely from a knowledge of its value at integer points. To be more accurate we should have written, in (6), the factor $[(1/(\operatorname{tg}\pi z)) + b(z)]$ rather than just $1/\operatorname{tg}\pi z$.

As I said earlier, Lehmann and Pohlmeyer's crucial contribution is to show that there exists a simple physical criterion—and this applies to all orders in the major coupling constant—using which one can eliminate these distribution-theoretic ambiguities from localizable theories. The basic idea is that for *finite* localizable theories the b-dependent (and ambiguous) contributions in (7) can be distinguished from those with $b=0$ through their behaviour in p-space for large $|p^2|$. The b-dependent terms do not fall in any direction in the $|p^2|$-plane, while the b-independent terms do and thus define a minimal-singularity superpropagator $\tilde{S}(p)$. The b-dependent ambiguous terms are thus sharply distinguished from the Volkov-Filippov solution—they are as clearly separated as oil and water. The same criterion was also used earlier by Filippov [7] for the second-order superpropagator to eliminate ambiguities and used by all workers in this field. Lehmann and Pohlmeyer give a general formulation valid for all higher-order superpropagators. They show further that this distinction between b-dependent and b-independent terms cannot be made for polynomial theories; *the nonpolynomials are far superior to polynomials in this regard.*

4. – Nonlocalizable Lagrangians of rational variety.

Let us now turn to the case of *normally ordered* nonlocalizable theories, *e.g.*, $L_{III} = {:}g/(1+\varkappa\phi){:}$. Here

$$S_{III}(x) = g^2 \sum_{n=0}^{\infty} \Gamma(n+1)\big(\varkappa^2 D(x)\big)^n . \tag{9}$$

Contrast this with the superpropagator $S_I(x)$. Considered as a series in $\varkappa^2 D(x)$, (9) in contrast to (5) has *zero* radius of convergence. Efimov and Fradkin [8] suggested *defining* the superpropagator as a Borel sum of (9).

Formally write

$$S_{III}(x) = \sum_{n=0}^{\infty} \int_0^{\infty} \exp[-\zeta](\varkappa^2 D(x)\zeta)^n \, d\zeta = \tag{10}$$

$$= \int_0^{\infty} \frac{1}{1 - \varkappa^2 D(x)\zeta} \exp[-\zeta] \, d\zeta \, . \tag{11}$$

For the definition of the physical superpropagator, EFIMOV and FRADKIN adopted a principal-value definition of the integral (11) in the ζ-plane. From this definition (or equivalently from (10)) one can go on to use Gel'fand and Shilov's methods for Fourier-transforming, precisely as in (7).

Quite clearly, the Borel ansatz introduces here a new and additional source of ambiguity, *the Borel ambiguity which just does not exist for localizable theories*. Another unfortunate aspect of nonlocalizable theories is that the Lehmann-Pohlmeyer ansatz for dealing with b-ambiguities is useless; both the b-ambiguity terms as well as the principal value terms of the FT of (11) fall equally fast in the $|p^2|$-plane for large $|p^2|$. A further difficulty has been noted by FELS [9] who shows that there is no guarantee that the requisite field-theoretic positive-definiteness of the principal-value propagator (11) can be guaranteed.

NOW STRATHDEE, DELBOURGO and myself have proved a theorem which might save the situation. Sparked by some remarks of FELS, what we have shown is this. *The trouble for nonlocalizable theories lies in normal ordering.* If $\mathscr{L}_{III}$ is not normally ordered—*i.e.* we leave $D(0)$ in, formally as an undetermined parameter in the theory—then the high-energy behaviour of the *momentum space* superpropagator is drastically altered. To be precise, we show that for nonlocalizable theories of the rational variety (with Lagrangians of the type $\sum a_i(1/(1+\varkappa_i\phi))$) the nonnormal-ordered series expansion for a superpropagator converges, when considered as a series in $D(x)$. This radius of convergence is given by $D(0)$. If $D(0)=\infty$, the radius of convergence is infinitely large, no principal value ansatz is needed to define (11) and the superpropagators visibly satisfy positive definiteness criteria. This is nice; unfortunately, as FRIED noted [10], if we do set $D(0) = \operatorname{lt}_{x^2\to 0}(-1/x^2) = \infty$ nearly all matrix elements will vanish. Clearly $D(0)$ must be renormalized. If we have physical reasons for believing that $D(0)$ should be *finite* (for example, if a scalar field in the theory possesses a nonzero expectation value $\langle\phi\rangle \neq 0$, which can be expressed as a function of $D(0)$), then the superpropagators would possess *finite* radii of convergence. In p-space this would *mean that for nonnormally-ordered rational theories*, the two-point spectral function $\varrho(p^2)$ would behave like $\exp[\sqrt{|p^2|}]$—*i.e.*, we would have a *just localizable* situation. The proof of our results is simple and can easily be constructed. A useful remark for its construction is the following: given a Lagrangian $L(\phi)$, one obtains the corresponding normally-ordered Lagrangian from it by a simple operator identity

$$:L(\phi): = \exp\left[+\frac{D(0)}{2}\frac{\partial^2}{\partial\varphi^2}\right] L(\phi)\,.$$

Our conclusion is that normal ordering is a crime for nonlocalizable theories. *Rational* Lagrangians are *just localizable* provided $D(0)$ is renormalized to a finite nonzero value; they become *nonlocalizable* only when they are normally ordered [11] (equivalently stated when $D(0)$ is renormalized to zero).

We have not completed the investigation of whether the Lehmann ansatz for isolating b-ambiguities applies to *just-localizable* theories, nor can we say whether the inbuilt cut-off for these theories is still $(\varkappa)^{-1}$ or if it is modified through the appearance of the new constant $D(0)$. *Clearly there is a certain amount of physics concealed in $D(0)$ whenever we work with nonlocalizable theories.*

5. – Finite *vs.* renormalizable Lagrangians.

Now it is not always the case that every infinity can be regularized. Consider the Lagrangian $L_{II} = (\exp[\varkappa\phi] - 1 - \varkappa\phi - (\varkappa^2\phi^2)/2!)$. Here the superpropagator $S_{II}(x)$ has the same expression as in (6); the contour of integration, however, lies along $2 < \operatorname{Re} z < 3$. We cannot interchange z-integration with the FT since the Gel'fand-Shilov condition $\operatorname{Re} z < 2$ is not met. We must write $S_{II}(x) = S_I(x) - (\varkappa^2 D^2)/2!:$ before doing so. While $S_I(x)$ can be Fourier-transformed by the methods above, the $(-\varkappa^2 D^2)/2!:$ term sticks out like a *sore thumb* (ST). One may regularize it using any available method; there is no reason, however, for the effective cut-off to depend on $\varkappa$. We shall call the Lagrangian L_I *finite* and, in contrast, L_{II} *renormalizable*, since L_{II} needs a subtraction constant of the conventional variety. At least one physical quantity cannot be computed within the theory so far as L_{II} is concerned. The ideal theory would of course be the one where there are no uncomputable, renormalization constants whatever.

6. – Enumeration of finite nonlocalizable theories.

Let us now turn to enumerate the cases when nonpolynomial theories are *finite* as opposed to being just *renormalizable* (*i.e.* still needing a finite number of constants unspecified and uncomputable within the theory). Unfortunately, all work so far on this problem [12] (papers by EFIMOV, FRADKIN, DELBOURGO, KOLLER and SALAM, TAYLOR and KECK) has concentrated on normally-ordered rational Lagrangians of the type

$$\mathscr{L} = :g\frac{\phi^n}{(1+\varkappa\phi^W)}:\,.$$

To summarize the results of these papers

a) $\mathscr{L}$ is *finite* if $n - w < 2$.

b) $\mathscr{L}$ is renormalizable provided $2 \leqslant n - w < 4$, *i.e.* there exist a finite number of ST's which cannot be computed in terms of g and $\varkappa$, but whose infinities can be absorbed into a renormalization of physical contents. Note the surprising inequality $n - w < 4$. One may naively have expected renormalizable theories to range over $n - w \leqslant 4$, just as is the case for polynomial theories.

c) For mixed theories like

$$L = {:}g\frac{\bar{\psi}\psi A}{(1+\varkappa\phi^W)}{:}\,, \tag{12}$$

where the Dyson weight of the polynomial part of L (*i.e.* $\bar{\psi}\psi A$) is $\leqslant 4$ (Dyson weight of a scalar field A is 1, of spinor field ψ is $\frac{3}{2}$), the theory is *renormalizable* for all $w > 0$.

d) We believe that if normal ordering is dropped (*i.e.* $D(0)$ is kept as an undetermined constant or alternatively $D(0)$ is re-expressed in terms of an uncomputable magnitude $\langle\phi\rangle$) the Lagrangian (12) gives a finite theory, provided $w > 2$. For the borderline case $w = 2$, the only ST's are those which correspond to vacuum-to-vacuum graphs.

e) For exponential (localizable) Lagrangians the situation is more favourable towards *finiteness* [13]. Lagrangians of the type ${:}g\phi^n \exp[\varkappa\phi]{:}$ are *finite*. The result is trivial for superpropagators of the second order. For higher-order superpropagators the proof of this assertion presents no difficulties whenever $n \leqslant 2$. For $n > 2$ the result is deduced by differentiating $(\exp[\varkappa\phi])$ with respect to $\varkappa$.

f) Less easy to prove is the finiteness of the mixed case ${:}g\chi^n \exp[\varkappa\phi]{:}$ for higher superpropagators. For $n \leqslant 2$ the proof goes through easily. For $2 < n \leqslant 4$ we believed these theories are finite, but a proof complete in details has not yet been constructed.

To summarize the result of this Section, *local* nonpolynomialities like $(\exp[\varkappa\phi])$ are superior to nonlocal ones in that their infinity-suppressing powers are unmeasurably superior. Roughly speaking, not only do superpropagators like $\exp[-\varkappa^2/x^2] \to 0$ when $x^2 \to 0$ from the appropriate direction; so also is the case for $(1/x^2)^n \exp[-(\varkappa^2/x^2)]$. Thus not only is $\exp[\varkappa\phi]$ finite; so is also $(\bar{\psi}\psi A)\exp[\varkappa\phi]$.

7. – Derivative couplings and the law of conservation of derivatives.

The chiral Lagrangian and the gravity-modified Lagrangian possess the property that the interaction Lagrangian contains as high powers of derivatives as the free Lagrangian. This is a new situation in physics and the first problem is to give the correct Feynman rules for such cases.

A number of authors have shown recently that for Lagrangians of the type

$$\mathscr{L} = \tfrac{1}{2}(\partial\phi)^2 v(\phi) + m^2\phi^2 ,$$

one needs to modify the Lagrangian by the addition of a term $-(i/2)\delta^4(0)\cdot$ $\cdot \log v(\phi)$, in order that one can consistently use the T^* product satisfying $\langle T^* \partial_\mu \phi d_\nu \phi\rangle = \partial_\mu \partial_\nu \langle T\phi(x)\phi(y)\rangle$. A detailed discussion of this is given in a paper by STRATHDEE and myself using the powerful path-integral methods, (*Phys, Rev.*, *D*, **2**, 2869 (1970), paper entitled « *On equivalent formulation of massive vector meson theories*). I shall not speak here of this aspect of derivative-containing theories.

What I shall be more concerned with is the development of a calculus of derivatives to generalize Gel'fand's calculus of products of singular functions $(1/x^2)^{z_1} \otimes (1/x^2)^{z_2}$.

Consider the following identity, which holds for all x except $x=0$:

$$\partial_\mu \left(\frac{1}{x^2}\right)^{z_1} \otimes \left(\frac{1}{x^2}\right)^{z_2} = \frac{z_1}{z_1+z_2}\, \partial_\mu \left(\frac{1}{x^2}\right)^{z_1+z_2} . \tag{13}$$

This formula holds also for $x=0$ in the sense that FT's of both sides agree when $0 < \operatorname{Re} z_1, z_2, z_1+z_2 < \frac{3}{2}$. For z_1 and z_2 outside this range, the two sides may differ at $x=0$ up to terms containing $\delta(x)$ and its derivatives.

We believe this formula and its generalizations offer the correct extension of Gel'fand-Shilov's calculus when derivatives occur. The important point about (13) is that the number of derivatives on both sides is conserved (*law of conservation of derivatives*). This ensures that the overall singularity from the line $(1/x^2)^{z_1+z_2}$ is not enhanced by the differentiations more on one side of (13) than on the other. To see the power and raison of this ansatz, consider the simplest generalization of (13) given by

$$\begin{aligned} &\partial_\mu \left(\frac{1}{x^2}\right)^{z_1} \otimes \partial_\nu \left(\frac{1}{x^2}\right)^{z_2} = \\ &\qquad = \frac{z_1 z_2}{(z_1+z_2)(z_1+z_2+1)} \left(\partial_\mu\partial_\nu + \frac{1}{2(z_1+z_2-1)}\,\partial^2\delta_{\mu\nu}\right)\left(\frac{1}{x^2}\right)^{z_1+z_2}, \\ &\qquad\qquad 0 < \operatorname{Re} z_1 z_2 , \qquad z_1+z_2 < 1 . \end{aligned} \tag{14}$$

The terms on the right contain two derivatives just as on the left. Now consider using (14) in a conventional photon self-energy calculation (zero-mass electrons with propagators $S(x) = \not{\partial}(1/x^2)$):

$$\Pi_{\mu\nu}(x) = \operatorname{Tr} \gamma_\mu \not{\partial}\left(\frac{1}{x^2}\right)\gamma_\mu \not{\partial}\left(\frac{1}{x^2}\right) = 2(\partial_\mu\partial_\nu - \delta_{\mu\nu}\partial^2)\left(\frac{1}{x^2}\right)^2 . \tag{15}$$

(We have continued the relation (14) outside the region of definition. This of course always needs care and we discuss this presently.) Now (15) correctly exhibits the transverse character of proton self-energy. Using even the most obtuse and old-fashioned of computing procedures, the FT of $\Pi_{\mu\nu}(x)$ equals

$$2(p_\mu p_\nu - \delta_{\mu\nu} p^2)\int \frac{d^4k}{(p-k)^2 k^2}\,.$$

This shows no sign whatsoever of generating any photon self-mass. The photon self-mass in conventional calculations is indecently manufactured by forgetting the ansatz regarding conservation of derivatives and by writing $\Pi_{\mu\nu}(x)$ in the form

$$\Pi_{\mu\nu}(x) = 2\partial_\mu\partial_\nu\left(\frac{1}{x^2}\right)^2 - 16\,\delta_{\mu\nu}\left(\frac{1}{x^2}\right)^3. \tag{16}$$

That is, in the second term on the right of (15), $\partial^2(1/x^2)^2$ is replaced by $8/(x^2)^3$ in most standard treatments. Since $(1/x^2)^3$ is more singular that $(1/x^2)^2$, it is alleged that the theory is giving a (quadratically infinite) photon self-mass.

Let us pursue this problem of gauge invariance a little further. (The problem of checking gauge invariances of a general type is, of course, the problem of knowing how to deal with derivatives.) We wish in particular to adopt a final form for the formula (14) outside the region $0 < \operatorname{Re} z_1 + z_2 < 1$ which should permit us, at the same time, to resolve the related problem of normal ordering in gauge theories.

To appreciate what the problem is, consider scalar electrodynamics with the Lagrangian

$$\mathscr{L} = ie(\phi^*\partial_\mu\phi - \phi_\mu\partial\phi^*)A_\mu + e^2A_\mu^2\phi^*\phi\,.$$

There are two photon self-energy graphs

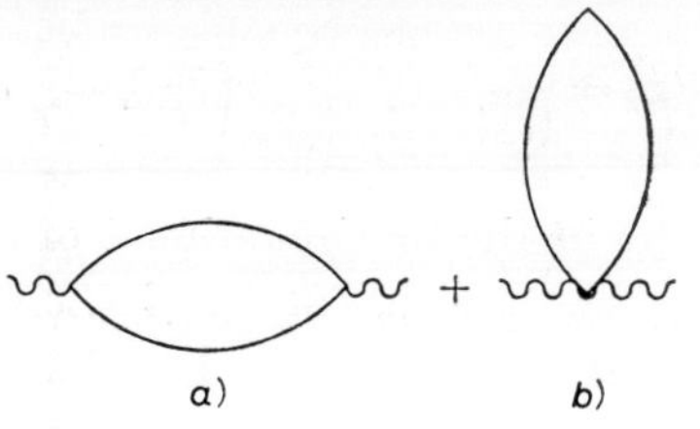

a) b)

The graph *b*) is a straight tadpole (we shall call this a tadpole of the first kind) with a contribution proportional to $e^2D(0)$. The important point is that the first graph *a*) (part of whose contribution can be written as $+e^2\int\partial^2D(x)\cdot D(x)\,d^4x$), also gives rise to a tadpolelike contribution when we replace $\partial^2D(x)$ inside the integral by $-\delta^4(x)$. Graphically what is happening is that one of the two internal lines of the graph a) shrinks, yielding from this seemingly two-point graph a structure very similar to graph b). (We shall call this « effective » tadpole a tadpole of the second kind.) The two tadpole contributions exactly cancel each other as they should in order to give zero photon self-mass. However, if we had normal-ordered the scalar electrodynamics Lagrangian and written it as

$$:ie(\phi^* \partial^\mu \phi - \phi \partial^\mu \phi^*) A_\mu : + : e^2 \phi^* \phi A_\mu^2 : ,$$

we would have eliminated the tadpole of the first kind without at the same time eliminating the tadpole of the second kind. (This is clear since such tadpoles arise from the two-point graph a) which is not directly touched by normal ordering.)

Ideally what we would like, then, is development of a calculus of derivatives outside the region $0 < \operatorname{Re} z_1, \operatorname{Re} z_2, \operatorname{Re} z_1 + z_2 < 1$, which should guarantee that contributions of tadpoles of the second kind—typically expressions like $D^z(x)\partial^2 D(x)$, $z \neq 0$, automatically reduce to zero, so that when used together with normal ordering (which eliminates tadpoles of first kind) we may never be faced with latent tadpoles of the second kind. This after all is—or should be—the ideal of a desirable mathematical formalism for use in physics—the formalism should automatically achieve the elimination, at all stages, of terms which we know (in this case from gauge invariance) never do make their appearance in the final answers.

To achieve this development we proceed as follows.

Analogously to (14), combine the factors D^z and $\partial_\mu \partial_\nu D^{z_1}$ into the form

$$D(x)^z \partial_\mu \partial_\nu D^{z_1} = \tag{17}$$

$$= \frac{z_1}{(z+z_1)(z+z_1+1)} \left((1+z_1)\partial_\mu \partial_\nu - \frac{z}{2(z+z_1-1)} \delta_{\mu\nu} \partial^2 \right) D^{z+z_1} ,$$

which in the case $z_1 = 1$ becomes

$$D(x)^z \partial_\mu \partial_\nu D(x) = \frac{2}{(z+1)(z+2)} \left(\partial_\mu \partial_\nu - \frac{1}{4} \delta_{\mu\nu} \partial^2 \right) D^{z+1} , \tag{18}$$

which is an identity except in the neighbourhood of $x_\mu = 0$, where it becomes ambiguous. We shall adopt this formula as a *definition* for all x_μ *except* in the neighbourhood of $z = 0$ where it needs to be elaborated. It is clear that (18) cannot be a satisfactory definition at $z = 0$ since the left-hand side assumes the well-defined form, $\partial_\mu \partial_\nu D(x)$, while the right-hand side assumes the equally well-defined form, $\partial_\mu \partial_\nu D(x) + \frac{1}{4}\delta(x)$, which is different.

To meet this difficulty we adopt the definition

$$[D(x)]^z \partial_\mu \partial_\nu D(x) = \lim_{\varepsilon \to 0} \frac{2}{(z+1)(z+2)} \left[\partial_\mu \partial_\nu - \frac{1}{4} \delta_{\mu\nu} \left(\frac{z}{z+\varepsilon} \right) \partial^2 \right] [Dx]^{z+1} ,$$

where ε is a positive number. (To see the raison for a factor like $z/(z+\varepsilon)$ set $z_1 = 1+\varepsilon$ in (16).) It is to be understood that the singularity at $z=-\varepsilon$ lies to the left of any z-contours if an integration over z is carried out and that the limit $\varepsilon \to 0$ is therefore to be taken *after* evaluating any Fourier transforms we may wish to take *after* translating the contour to the right of $z=0$. In this way one obtains a definition which is consistent at $z=0$ where (17) failed. For other values of z it coincides with (17).

Note now that, contracting the indices μ, ν, one finds

$$[D(x)]^z \partial^2 D(x) = 0\,, \qquad \text{for } z \neq 0\,.$$

This has the important consequence that all those tadpolelike graphs in the theory which arise from a consonance of terms like $D(x)^z \partial^2 D(x) = D(0)^z \delta(z)$, and which cannot be removed by the normal-ordering procedures, automatically vanish. Thus, in effect $D(0) = 0$ everywhere.

To summarise, we have made a beginning towards a calculus of derivatives —a generalization of Gel'fand-Shilov's calculus—which should help in resolving automatically the age-old gauge-invariance problems in normally ordered theories. Such problems should have been solved years ago but they were not. For chiral or gravity theories with up to 2nd-order derivatives in interaction Lagrangians these problems—purely mathematical distribution-theoretic problems, and with no physics in them—become very vicious. (Certainly, there is no relevance of these problems to ultraviolet infinities.) Our work is only a beginning but we do believe progress in this direction by those mathematically minded among us is essential to resolve what are essentially purely mathematical difficulties —how to write gauge-invariant matrix elements for visibly gauge-invariant Lagrangian theories. (See also a recent Trieste preprint (IC/71/38) by A. PATANI and G. LAZARIDES on this subject.)

8. – High-energy behaviour of localizable theories on the mass shell.

One of the hardest problems for nonpolynomial Lagrangians on the pure field-theory side is this. GLASER, MARTIN and EPSTEIN [14] in a fundamental paper have shown that all *localizable* theories should give Froissart-bounded high-energy dependence of S-matrix elements. The Volkov-Lehmann expansion [5] in the major constant does not achieve this. In this expansion *any given order* behaves like $\exp[|p^2|^\alpha]$ with $\alpha < \frac{1}{2}$. I believe this difficulty will be resolved when summations over the major coupling constants are performed. We already know that for conventional polynomial theories this type of summation—performed either directly by summing diagrams or by using Padé approximants or carried through indirectly using Bethe-Salpeter-type integral equations—alters the high-energy behaviour of individual diagrams. Specifically we know that even though individual diagrams behave like s^{α} (α constant), for the *sum* of ladder diagrams one finds a drastic change; one obtains the Regge behaviour $s^{\alpha(t)}$. Likewise I believe that a sum of a chain of superpropagators is likely to satisfy Glaser, Martin, Epstein results and to give Froissart-bounded high-energy dependence.

To see this in a simple example (worked out together with DELBOURGO, unpublished), consider $\mathscr{L}_{\text{int}} = g\,(\exp[\varkappa\phi - 1])$. The single-superpropagator contribution for two-particle scattering gives

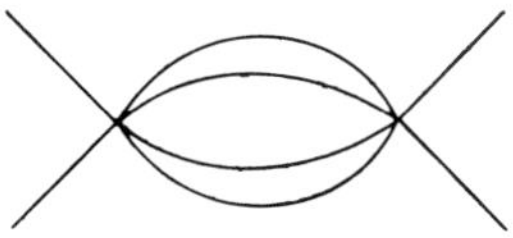

$$T(s) \approx ig^2 \exp[\varkappa^2 s]^{\frac{1}{2}}\,.$$

If we consider a linear chain of superpropagators with k initial and l final particles

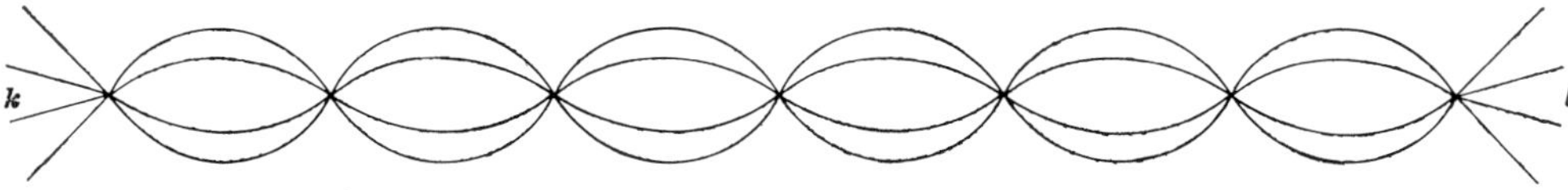

it is possible to write down an integral equation for the complete matrix element. Quite generally

$$F_{kl}(s) = F^0_{kl} + \sum_{n=1}^{\infty} F^0_{kn} \frac{D_n(s)}{\Gamma(n+1)} F_{nl}(s)\,,$$

where $D_n(s)$ is the Fourier transform of the nth power of the Feynman propagator in the massless case and F^0_{kl} is the contact (Born) term for the scattering which equals

$$F^0_{kl} = \left.\frac{\delta^{k+l}\mathscr{L}(\phi)}{\delta\phi^{k+l}}\right|_{\phi=0}.$$

(Graphically, the integral equation represents

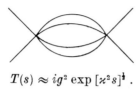

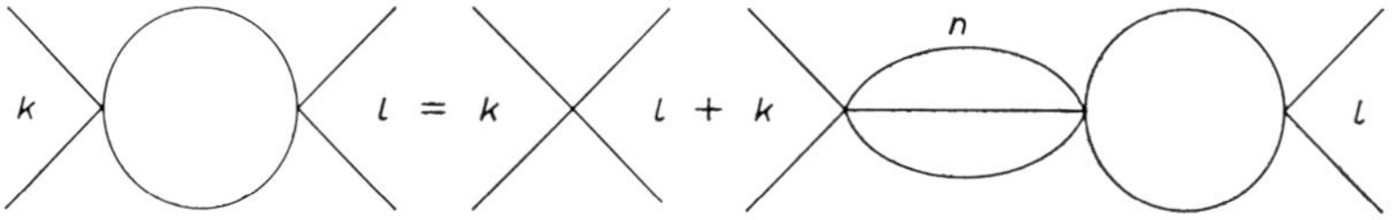

).

For the case of the Lagrangian $g(\exp[\varkappa\phi] - 1)$,

$$F^0_{kl} = g\varkappa^{n+l}$$

and

$$F_{kl}(s) = g\left(\varkappa^{k+l} + \sum_n \varkappa^{k+n}\frac{D_n(s)}{\Gamma(n+1)}F_{nl}(s)\right)$$

which solves to

$$F_{kl}(s) = g\varkappa^{k+l}\left[1-\sum_n \frac{g\varkappa^{2n}D_n(s)}{\Gamma(n+1)}\right]^{-1}.$$

Using Sommerfeld-Watson's transform method, together with

$$D_z(s) = \frac{1}{4\pi^2}\frac{\Gamma(2-z)}{\Gamma(z)}\left(\frac{-s}{16\pi^2}\right)^{z-2}, \qquad z\neq 2, 3, \dots,$$

one sums the series in the square bracket. The usual Meijer's G-function makes its appearance. For large s we obtain

$$F_{kl} = g\varkappa^{k+1}\frac{1}{1-c\exp[\varkappa^2 s/16\pi^2]^{\frac{1}{2}}},$$

which is indeed Froissart-bounded.

The summation for the exponential Lagrangian is particularly easy. The general case for other nonpolynomial Lagrangians is no more difficult and has been carried out with DELBOURGO. But clearly we cannot accept this type of summation as representing good physics because of unitarity difficulties.

The proper way to carry out the summation has been described by REDMOND and later refined by BOGOLUBOV, LOGUNOV and TAVKHELIDZE (see S. SCHWEBER, *Introduction to field theory*).

The method relies on computing the imaginary part of the linear chain and then writing a dispersion relation.

Clearly

$$\varrho_{kl} = \operatorname{Im} F_{kl}(s) \approx \frac{\exp[s\varkappa^2]^{\frac{1}{2}}}{|1+c\exp[\varkappa^2 s]^{\frac{1}{2}}|^2} \approx \exp[|s\varkappa^2|]^{-\frac{1}{2}}.$$

The dispersion integral is highly convergent and the resulting F_{kl} is Froissart-bounded.

In particular when $k=l=1$, we have the case of the two-point function; thus

$$\Delta(s) \approx \int\frac{\varrho(s')\,ds'}{s'-s},$$

which behaves like $1/s$ for large s in accordance with Lehmann-Källén bounds on the propagator.

A legitimate question which one may ask at this point is; if the propagator satisfies the Lehmann-Källen bound in these theories and behaves like a free propagator for large s, how has one obtained finiteness of matrix elements in such theories? To answer this recall the statement made in the introduction. It is a fallacy to bring to the nonpolynomial theories the notions we have inherited from polynomial Lagrangian theories. The principle of insertion of self-energy graphs in free lines, enunciated by DYSON, simply does not hold. Every graph in a nonpolynomial theory is a jumble of superpropagators (one superpropagator joining every two points) and no (self-energy) insertions can be isolated.

The Bogoliubov-Redmond procedure of summing linear chains explained above is rather simple. Till a better idea comes along, I would like to suggest that this procedure may be used for dealing, for example, with the nonpolynomiality in weak interactions. As stated earlier, the parts of the weak processes mediated by the spin-zero daughters of the W-mesons possess an in-built cut-off at inverse of the Fermi constant G_F^{-1}. Since W-meson theory is a local theory, the Glaser-Martin-Epstein theorem assures us that cross-sections must be Froissart-bounded. To estimate their size at GeV energies, the linear chain summation in Bogolubov-Redmond manner would at the very least save us from violating the exact result of local field-theories in respect of the high-energy behaviour of cross-sections.

If one were not considering linear chains of graphs, it would be worth remarking that the totality of graphs in nonpolynomial theories bear a striking resemblance to graphs in statistical mechanics. A graph of order g^n in the major constant is composed of $n(n-1)/2$ superpropagators joined together.

The external lines impinge at the n-vertices in all topological configurations but the inner heart of the graph is the same. In this sense the topology of the n-th order graphs (of order g^n) in nonpolynomial theories is much simpler than for polynomial theories. In a recent preprint from Maryland, FIVEL has exploited the analogy of such graphs to Mayer-Onsager graphs in statistical mechanics to carry through a summation in the major constant. I believe this is a line of approach which is likely to be very productive in finally resolving the problem presented by the high-energy behaviour of nonpolynomial theories.

PART II

I have already described the essential results which follow when we modify electrodynamics by incorporating in the Lagrangian a local nonpolynomial factor $\exp[\varkappa\phi]$. A Lagrangian like $e\bar{\psi}\gamma_\mu\psi A^\mu \exp[\varkappa\phi]$ is not gauge-invariant and cannot represent electrodynamics correctly and we should now like to consider the proper tensor gravity modification to electrodynamics rather than an *ad hoc* modification of the above type. Together with ISHAM and STRATHDEE [15] we have written two papers on this (one is already published in *Phys. Rev.* **3**, 1805 (1971)). This uses tensor gravity—constructing a complete tensor gravity superpropagator. Unhappily, in this paper we parametrized the fundamental gravitation field using a « rational » parametrization

$$g^{\mu\nu}(x) = \eta^{\mu\nu} + \varkappa\phi^{\mu\nu}(x)\,, \tag{1}$$

where $\phi^{\mu\nu}(x)$ is the physical graviton field whose asymptotic limits define the in and out states. The covariant field (the field with lower indices) $g_{\mu\nu}(x)$ is then given as the ratio of two polynomials in $\phi^{\mu\nu}$ of degree 3 and 4, respectively:

$$g_{\mu\nu}(x) = \frac{4\varepsilon_{\mu\alpha\beta\gamma}\varepsilon_{\nu\alpha'\beta'\gamma'}g^{\alpha\alpha'}g^{\beta\beta'}g^{\gamma\gamma'}}{\varepsilon_{\alpha\beta\gamma\delta}\,\varepsilon_{\alpha'\beta'\gamma'\delta'}\,g^{\alpha\alpha'}g^{\beta\beta'}g^{\gamma\gamma'}g^{\delta\delta'}}\,. \tag{2}$$

This parametrization is nonlocal. As stated earlier, there is the alternative (and by the mathematicians the more favoured) exponential parametrization

$$g^{\mu\nu} = (\exp\,[\varkappa\phi])^{\mu\nu}\,, \tag{3}$$

where $\phi^{\alpha\beta} = \phi^{\beta\alpha}$ are the basic interpolating fields. The covariant tensor $g_{\mu\nu}(x)$ is simple and is given by

$$g_{\mu\nu}(x) = (\exp\,[-\varkappa(\phi)])_{\mu\nu}\,. \tag{4}$$

Similarly the vierbein gravity field $L^{\mu a}$ can be parametrized as

$$L^{\mu a} = \left(\exp\left[\frac{\varkappa}{2}\phi\right]\right)^{\mu a}. \tag{5}$$

Since in gravity theory one always assumes that $\det g \neq 0$, it is clear from (1) that if $g^{\mu\nu}(x)$ is entire, so is $g_{\mu\nu}(x)$. (With exponential parametrization $\det g \neq 0$ is automatically guaranteed, even in quantum theory. With rational parametrization there is no such guarantee and this gives one reason why the exponential parametrization is so superior.)

With this parametrization

1) We can arrange the calculation so that *gauge invariance* is preserved.

2) We can see more clearly why it is the *tensor* rather than *scalar* gravity which is responsible for infinity suppression. This accords with one's physical intuition (emphasized particularly by WEISSKOPF) in that infinity suppression should come as a consequence of the light-cone fluctuations which are peculiar to tensor gravity and its (Schwarzschild-like) metrical aspects.

3) To exhibit this distinction between scalar and tensor gravity, we need a better comprehension of the role of equivalence theorems for field transformations. Since we have seen in Part I that localizable Lagrangians are ordinary, decent, unassuming types of Lagrangians, and since microcausality holds for operators in localizable theories, we can take over Borchers' results and assume that S-matrix equivalence theorems hold for localizable theories. *This will necessitate using exponential parametrization for gravity and not the usual rational parametrization.*

4) As shown in Part I (Sect. **6**), exponential nonpolynomiality is extremely potent in smoothing infinities. This will mean our problem in gravity-modified theories will not be making sure whether infinities can be regularized or not; it will rather be making sure if gauge invariance can be preserved.

1. – Gauge-invariant calculations in tensor gravity.

The gravity-modified Lagrangian for quantum electrodynamics may be written in the form

$$L_{\text{total}} = L_{\text{gravity}} + \frac{L^{\mu a}\bar{\psi}\gamma_a(\nabla_\mu - ieA_\mu)\psi + m\bar{\psi}\psi}{\det L} + \frac{g^{\mu\mu'}g^{\nu\nu'}F_{\mu'\nu'}F_{\mu\nu}}{\det L}\,. \tag{6}$$

$L^{\mu a}$ is the vierbein gravity field which in exponential parametrization will be written as $\exp[\varkappa\gamma_{\lambda c}h^{\lambda c}]$. $\gamma_{\lambda c}$ are 4×4 pseudosymmetric matrices and $h^{\lambda c}$ are the physical (quantized fields) describing tensor quanta. ∇_μ is the covariant derivative; $F_{\mu\nu}$ equals $\partial_\mu A_\nu - \partial_\nu A_\mu$, while the Einstein field $g^{\mu\nu}$ equals a bilinear product of vierbein fields $L^{\mu a}$, $L^{\nu a}$. Note that $\det L = \exp[\operatorname{Tr}(\varkappa h)]$. The simplicity of this expression makes work with exponential parametrization even simpler than with the rational parametrization we used in our previous papers so far as $\det L$ is concerned.

Now, a scalar gravity Lagrangian can be recovered from (1) by substituting

$$\left\{\begin{aligned} L^{\mu a} &= \exp[\varkappa\phi]\eta^{\mu a}, \qquad \text{where } \eta^{\mu a} = \begin{pmatrix} 1 & & & \\ & -1 & & \\ & & -1 & \\ & & & -1 \end{pmatrix} \\ g^{\mu\nu} &= \exp[\varkappa\phi]\eta^{\mu\nu}, \\ \det L &= \exp[4\varkappa\phi], \end{aligned}\right. \tag{7}$$

so that L_{total} reduces to the form

$$= L_{\text{gravity}} + [\bar{\psi}\gamma_\mu(\partial_\mu - ie_0 A_\mu)\psi]\exp[-3\varkappa\phi] + m_0\bar{\psi}\psi\exp[-4\varkappa\phi] + F_{\mu\nu}F_{\mu\nu}\,.$$

I want to show that scalar gravity does not suppress infinities. Note that, as physically expected, *the photon field does not interact with scalar gravity.* The important point is that massless electrons do not interact with scalar gravity. To see this let us make a field transformation:

$$\psi' = \exp\left[-\tfrac{3}{2}\varkappa\phi\right]\psi\,. \tag{8}$$

This will completely eliminate $\varkappa\phi$-coupling from the Lagrangian from all terms except the electron mass term. In the limit $m_0 = 0$ scalar gravity and (massless) electrons also get uncoupled.

Let us restate the result with a slightly different emphasis. Electromagnetic gauge invariance demands that any modification of the total electromagnetic Lagrangian should preserve the combination $(\partial_\mu - iA_\mu)$ intact. If it is a scalar modification, then it must clearly have the form

$$\mathscr{L}_T = v(\phi)\,\bar\psi\gamma_\mu(\partial_\mu - iA_\mu)\,\bar\psi\,.$$

However, for this form we can make the transformation

$$\psi \to \psi\sqrt{v(\phi)}\,,$$

and thereby remove the dependence of $\mathscr{L}_T$ on ϕ.

A scalar modification produces no regularization.

What is the next possible modification? Clearly the next simplest idea is to try a tensor modification like

$$L^{\mu a}\,\bar\psi\gamma_a(\partial_\mu - iA_\mu)\psi\,.$$

We have preserved in this way the sacred gauge-invariant combination $(\partial_\mu - iA_\mu)$. (The trick we have exploited is to delink the index a of the γ-matrix from the index μ of $\partial_\mu - iA_\mu$). Mathematically, a combination of the Fermi character of the electron and electromagnetic gauge invariance has forced us to a tensor rather then a scalar modification of electrodynamics.

As I said before, we have redone our calculations in a paper to be published (*Infinity suppression of gravity-modified electrodynamics*, II, Trieste preprint, to be submitted to the *Phys. Rev.*, IC/71/14).

We used the calculus of derivatives developed in the previous Section; we also needed the full graviton superpropagator:

$$\left\langle \frac{L^{\mu a}}{\det L}, \frac{L^{\nu b}}{\det L}\right\rangle,$$

$$L^{\mu a} = \left(\exp\left[\frac{\varkappa}{2}\phi\right]\right)^{\mu a},$$

where

$$\langle \phi^{\mu a} \phi^{\nu b} \rangle = \frac{(\eta^{\mu\nu}\eta^{ab} + \eta^{\mu b}\eta^{\nu a} - \eta^{\mu a}\eta^{\nu b})}{x^2}\,.$$

A completely general and beautiful method for obtaining such superpropagators has been worked out by ASHMORE and DELBOURGO (Imperial College preprint, *Journ. Math. Phys.*, to be published).

The final gauge-invariant result for the self-mass of the electron coincides with the result of the naive calculation of Part I, Sect. **2** (these lectures) up to terms of order $\alpha \log \varkappa^2 m^2$. (The difference comes in terms of order α.) We had to invent some new techniques of « kinking » and « cradling » of graphs to preserve manifest gauge invariance (these replace Dyson's techniques of « insertion » of graphs) and I shall refer you to the papers mentioned for details.

2. – Tensor gravity and curved space-time.

WEISSKOPF with his penetrating physical intuition has made the point (private communication) that infinity suppression in electrodynamics should be connected with the limiting frequency of a standing photon wave in the curved space-time around an electron. The wavelength of such a photon should be given by the Schwarzschild radius, *i.e.* its frequency k_{max} should be given by $1/R \approx 1/\varkappa^2 m$. It is this cut-off in (virtual) photon frequencies which in Weisskopf's view is really providing the inbuilt cut-off for quantum electrodynamics which we have exploited.

The late Prof. SCHIFF formalized for me the Weisskopf argument.

A Schwarzschild Universe round the electron is given by the metric

$$(\mathrm{d}s)^2 = (\mathrm{d}t)^2 \left(1 - \frac{2mG}{r}\right) - \frac{(\mathrm{d}r)^2}{1 - 2mG/r} - r^2[(\mathrm{d}\theta)^2 + \sin^2\theta (\mathrm{d}\phi)^2] \qquad (G = 8\pi\varkappa^2)\,.$$

With $\theta = \pi/2$ and $u = 1/r$ the path of a light ray is given by

$$\frac{\mathrm{d}^2 u}{\mathrm{d}\phi^2} + u = 3mu^2\,.$$

For $u =$ constant and $u = u_0 = 1/3mG$ we obtain a circular path which the ray travels in time $t = t_0 = 2\pi r_0/(1 - 2mG/r_0) = 6\sqrt{3}\pi m$. Thus the proper angular frequency of orbital revolution of a circularly orbiting photon is $1/3m$, *i.e.* the Weisskopf frequency k_{max} equals $1/3mG$. If this k_{max} is substituted in the old Weisskopf formula

$$\frac{\delta m}{m} \approx \alpha \int_0^{k_{max}} \frac{\mathrm{d}^3 k}{|k^2|^{\frac{3}{2}}}\,,$$

we recover our result for $\delta m/m$.

3. – Relation between the fine-structure constant and the Newtonian constant.

One final remark about the formula

$$\frac{\delta m}{m} \approx \frac{3}{4\pi} \alpha \log \varkappa^2 m^2 . \tag{9}$$

This term is the first term in an expansion which, with the gravity modification should go like

$$\frac{\delta m}{m} \approx \sum a_n(\alpha \log \varkappa^2 m^2)^r , \qquad r \leqslant n , \tag{10}$$

when higher orders in α are included. The important point is that the effective parameter in this expansion is $(\alpha \log \varkappa^2 m^2)$. This number has the surprising value near to unity ($\alpha \log \varkappa^2 m^2 \approx 100/137$). In an earlier paper we suggested- that nature probably intended the formulae (9) and (10) to be read back wards, that is to say, we might start with the assumption that all (or nearly all) electron self-mass is electromagnetic in origin, so that $\delta m/m \approx 1$. This may, in converse, determine $\alpha \log(\varkappa^2 m^2)$ and possibly help in understanding why this number is empirically so close to unity. In saying this one is of course making the assumption that the gravity-modified Maxwell equations keep holding up to 10^{18} GeV. Also that the muon self-mass is not predominantly electromagnetic in nature.

The problem of infinities in electrodynamics arose with Lorentz' classical electron theory some seventy years ago. WALLER investigated these infinities using Dirac's one-particle theory in 1930 and found that they persisted even after quantization. The modern formulation of the problem dates back to the famous 1934 paper of WEISSKOPF. There may be other solutions of the infinity problem; it has been suggested, for example, by JOHNSON and his collaborators that summations over the major constant α will regularize these. This may indeed be so. All we wish to point out is that there is in nature this powerful realistic regularizing effect of local-tensor gravity; that its effect is not small ($\approx \log \varkappa^2 m^2 \approx \alpha^{-1}$) and that the regularized answers are such that we visibly recover the old infinities when we take the limit $\varkappa \to 0$. The first calculations we published were finite but nongauge-invariant. We also worked with one particular *nonlocalizable* parametrization of gravity ($g^{\mu\nu}$ as the basic field, with $g_{\mu\nu}$ as the nonpolynomial subsidiary quantity). We did not understand ambiguities; neither the distribution-theoretic ones nor the Borel ambiguities. This is changed now. With Lehmann's ansatz and his work and with the *localizable* (exponential) parametrization of gravity, Borel ambiguities are completely gone and we are working with a field theory which is no longer mysterious—one may even call it orthodox. We are permitted to make field transformations at will. As we have shown, this allows us to exhibit explicitly the gauge invariance of gravity-modified electrodynamics. We believe at least one complete solution of the very long-standing infinity problem now exists within the context of an orthodox field theory. The infinity problem was intractable so long as bad mathematics was mixed up with the (missing) physics of gravity. (The best evidence of the missing physics is the seductiveness of the formula $\delta m/m \propto e^2 \log(\varkappa^2 m^2)$ where the limit gravity constant $\varkappa \to 0$ visibly generates back the infinity.) Once the missing physics has been supplied, there is hope that bad mathematics can at long last be tracked down and rectified.

PART III

Let us briefly review the status of inbuilt nonpolynomialities in some of the known field theories.

1. – Strong interactions.

As stated before, the $SU_2 \times SU_2$ chiral theories with

(1) $$\mathscr{L}_{\pi} = \operatorname{Tr} \partial S \partial S^{+} ,$$

(2) $$\mathscr{L}_{\pi\mathscr{N}}^{\text{inter}} = m_{\mathscr{N}} \bar{\psi} S'^{2} \psi ,$$

are intrinsically nonpolynomial in form. Here S and S' (in Gürsey's formulation) are given by

(3) $$S = \exp[i\tau\cdot\pi] , \qquad S' = \exp[i\lambda\gamma_5 \tau\cdot\pi] .$$

With the nonpolynomial methods developed these Lagrangians would give rise to finite matrix elements with an inbuilt cut-off at $4\pi\lambda^{-1} \approx 4\pi m_{\pi}$. Unhappily, in addition to pions (kaons and η's) there are other strongly interacting particles—notably the gauge 1^- and 1^+ particles—and the question arises: what can be done to compute the renormalization constants surviving in these theories?

Now there does exist a partially nonpolynomial formulation of gauge theories of massive spin-one particles which renders some—though not all—renormalization constants finite. This is the formulation due to BOULWARE (*Ann. of Phys.* **56**, 140 (1970)) and I shall describe it here to point out precisely what one may achieve and what problems are still left if one limits oneself to strong-gauge theories alone.

Consider a triplet of Yang-Mills fields described by

$$L_{\text{YM}} = \hat{\boldsymbol{W}}_{\mu\nu} \cdot \hat{\boldsymbol{W}}_{\mu\nu} + m^2 \boldsymbol{W}_\mu^2 , \tag{4}$$

where

$$\hat{\boldsymbol{W}}_{\mu\nu} = (\partial_\nu \boldsymbol{W}_\mu - \partial_\mu \boldsymbol{W}_\nu + 2if \boldsymbol{W}_\mu \times \boldsymbol{W}_\nu) . \tag{5}$$

The propagator of the W fields $(W_\mu, W_\nu)_+ = (g_{\mu\nu} + \partial_\mu \partial_\nu / m^2) \Delta(x)$ is highly singular and the theory as it stands is nonrenormalizable.

Following BOULWARE, let us now make a nonlinear Stückelberg-like transformation on the field variable $\boldsymbol{W}_\mu$. Write $W_\mu = \boldsymbol{W}_\mu \cdot \boldsymbol{\tau}$ and introduce two sets of fields A_μ and B, defined by the relation

$$W_\mu = S(B) A_\mu S^{-1}(B) + i/f S(B) \partial_\mu S^{-1}(B) . \tag{6}$$

Here $S(B)$ is a unitary matrix which could be taken in the Gürsey form as $\exp[iBf/m]$. Write $\mathscr{J}_\mu = (im/f) S \partial_\mu S^{-1}$. The net effect of (6) is to transform $\mathscr{L}_{\text{YM}}$ to a polynomial part $(\mathscr{L}_{\text{YM}}(A))$ and a nonpolynomial part of the form

$$\mathscr{L}_{\text{YM}}^{(W)} = (\hat{\boldsymbol{A}}_{\mu\nu} \cdot \hat{\boldsymbol{A}}_{\mu\nu} + m^2 \boldsymbol{A}_\mu^2) + (2m \boldsymbol{A}_\mu \cdot \mathscr{J}_\mu + \mathscr{J}_\mu \cdot \mathscr{J}_\mu) . \tag{7}$$

Now comes the important point. BOULWARE has shown that the two Stückelberg fields A_μ and B, in terms of which W_μ has been re-expressed, can be assigned normal propagators $(A_\mu, A_\nu)_+ = g_{\mu\nu} \Delta$ and $(B(x), B(0))_+ = \Delta(x)$ provided the conventional rules for writing the S-matrix corresponding to the Lagrangian (7) are supplemented by adding to (7) a term of the form $(if \boldsymbol{F} \times \partial_\mu \boldsymbol{F} \cdot \boldsymbol{A}_\mu)$. Here the triplet of $\boldsymbol{F}$ particles represents «fictitious» bosons of Fermi statistics first introduced into the theory by FEYNMAN who showed that the introduction of these bosons is needed to preserve unitarity of the S-matrix.

Consider now the final effective Lagrangian for the Yang-Mills field. It can be written in two parts, $\mathscr{L}_{\text{YM}} = \mathscr{L}_{\text{YM}}^{(1)} + \mathscr{L}_{\text{YM}}^{(2)}$:

$$\mathscr{L}_{\text{YM}}^{(1)} = \mathscr{L}_{\text{YM}}(A) + if \boldsymbol{F}^+ \times \partial_\mu \boldsymbol{F} \cdot \boldsymbol{A}_\mu + \partial_\mu \boldsymbol{F}^+ \cdot \partial_\mu \boldsymbol{F}, \tag{8}$$

$$\mathscr{L}_{\text{YM}}^{(2)} = 2im \mathscr{J}_\mu \cdot \boldsymbol{A}_\mu + \mathscr{J}_\mu \cdot \mathscr{J}_\mu . \tag{9}$$

$\mathscr{L}_{\text{YM}}^{(1)}$ is polynomial in form. $\mathscr{L}_{\text{YM}}$ is nonpolynomial with an inbuilt cut-off at about m/f. The contributions arising from the polynomial part of the Lagrangian $\mathscr{L}_{\text{YM}}^{(1)}$ are still, however, ultraviolet infinite in the traditional manner. To make them finite would need further *realistic* nonpolynomiality to be built into the theory. In the next Section we come back to a «realistic» provision of such nonpolynomiality using, for example, strong-gravity theory. Without this the traditional infinities arising from $\mathscr{L}_{\text{YM}}^{(1)}$ would survive.

To summarise: chiral Lagrangians possess inbuilt cut-off factors. Parts of Yang-Mills Lagrangians also possess such factors but there are other parts which are obstinately polynomial in form and give rise, if no further modification of these Lagrangians is made, to the traditional ultraviolet infinities. If an *ad hoc* procedure is adopted to regularize such infinities, there is no known way—as, in contrast, there probably is for nonpolynomial Lagrangians—to remove ambiguities.

2. – Weak interactions.

The discussion of the Yang-Mills field provides us with a model for weak interactions mediated by intermediate bosons $W_\mu^\pm$. Since nothing essential in the mathematics is altered even if we assume that the W-mesons from a gauge triplet $W_\mu^\pm$ and W_μ^0, we shall do so. (The physics is of course altered because of this introduction of neutral currents, but at this stage we are concerned with mathematical difficulties.) As is well known the Fermi constant and the constants f and m are related through the formula

$$G_F \approx \frac{f^2}{m^2}\,.$$

As before, make the Stückelberg split of the W-field into normal fields A_μ and B, using the transformation matrix $S(B)$, given by

$$S(B) = \exp\left[i\sqrt{G_F}\,B\right].$$

From the nonpolynomial part of the Lagrangian (analogous to $\mathscr{L}_{YM}^{(2)}$ of (9)) one can compute finite contributions to the renormalization constants. As explained before, these depend on $G_F^{-\frac{1}{2}}$.

The contribution from the polynomial part $\mathscr{L}_{YM}^{(1)}$, represented by a graph like

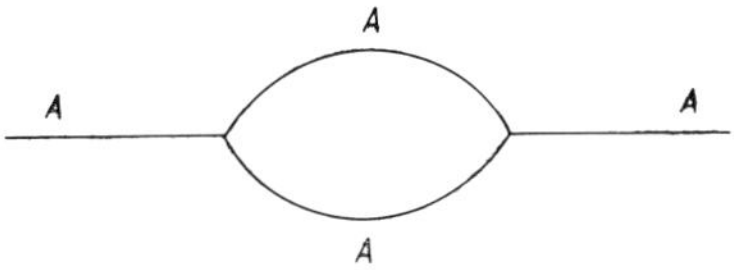

is, however, still unregularized and—as in the case of strong Yang-Mills theory—will, in accordance with the ideas of this lecture, need further (realistic) nonpolynomiality if we desire to assign to this contribution a well defined number.

To summarize, both for strong Yang-Mills theories and for weak interactions, whereas the spin-zero daughter contributions possess inbuilt cut-off factors (at a cut-off $4\pi m/f$), the spin-one particle contributions can still give rise to renormalizable infinities like those in conventional quantum electrodynamics. The question of course arises: is gravity the only force left which should provide the inbuilt cut-off for these remaining infinities? (Remember, if we are given two cut-offs operating on the same matrix element, the lower is the one which operates first. The higher cut-off is always the last court of appeal which one invokes only when all else has failed.)

Now so far as leptons are concerned, both for their weak as well as for their electromagnetic interactions, it would seem (forgetting the μ-e mystery) that gravity is the only known cut-off left. (Future, physics may show that there are other forces for leptons intermediate in coupling strength between weak forces and gravity. If they are nonpolynomial they will cut off matrix elements before gravity can.) For hadronic processes, however, (for strong as well as nonleptonic weak interactions) it seems that the effective cut-off mass is in any case much lower (compare the calculations of K_1-K_2 mass difference) possibly of the order of the nucleon mass. So for hadron physics we need a new universal (gravitylike) nonpolynomiality with a minor coupling constant $\approx m_N^{-1}$. Below we see how one may press hadronic F-meson interactions into service for this purpose. But before we consider this, I would like to conclude this Section with just one remark relevant to semi-leptonic weak interactions. It has been shown by SIRLIN (Trieste preprint 1968) that in a conventional intermediate vector boson mediated theory the radiative corrections to g_V in β decay are identical to the corresponding corrections for g_V in μ decay, so far as the (conventionally) logarithmically infinite factors are concerned.

Sirlin's result can be stated in a more familiar manner. The result simply is that the infinities in the purely hadronic part of the β-decay matrix element cancel themselves out.

If for leptonic weak interactions gravity is the only possible cut-off and none other interposes itself before, then Sirlin's result means g_V in β decay is corrected to the same extent (by gravity) as in μ decay, preserving the essential universality of the measured value of this constant.

3. – F-meson dominance gravity.

According to our present ideas, one of the fundamental forces of nature, electrodynamics, is mediated through two different mechanisms depending on whether we are considering leptons or hadrons. For lepton electrodynamics the present picture is that of a Dirac equation with photons interacting directly with muons and electrons. For hadrons, no such direct interaction is postulated. Instead the photon is pictured as interconverting into a (prescribed) mixture of the known 1^- strongly interacting particles (ρ^0, ϕ^0, ω^0), which themselves couple *strongly* to hadronic electric charge and which for this reason may be called « strong photons ».

Now nature has been prodigal in exactly the same manner with 2^+-particles. In addition to the massless graviton (with its obvious analogy with the massless photon), we know of at least three 2^+ massive strongly interacting particles f^0, $f^{0\prime}$ and A_2^0. It seems very natural that the analogy should carry further and that while leptons may interact directly with gravitons, so far as hadrons are concerned, it may be a mixture generically called F^0 of f^0, $f^{0\prime}$ (and perhaps other 2^+-objects which may be discovered) which provides the agency mediating gravity. For this to happen, it is mandatory that the F-meson in its strong interaction should couple to the hadronic stress tensor just as the graviton does to the lepton stress tensor.

We have constructed [16] a generally covariant theory of a universal strong coupling of F-mesons to hadronic stress tensors (with a coupling parameter $\varkappa_f \approx m_F^{-1} \approx 1$ GeV) and of the mixing of these particles to gravitons, on an analogy with ρ-γ mixing in electrodynamics. The formalism is elegant—as indeed everything where general relativistic invariance is concerned should be. The form of the final Lagrangian is simple; it consists of three pieces:

$$L^{(1)} = (-\det g)^{-\frac{1}{2}}\,[R(g) + L(\text{leptons})]\,,$$

$$L^{(2)} = (-\det f)^{-\frac{1}{2}}\,[R(f) + L(\text{hadrons})]\,,$$

$$L^{(3)} = \frac{M_f^2}{4K_f^2}(-\det f)^{-\frac{1}{2}}\,[\mathrm{Tr}\,(fg^{-1})^2 - (\mathrm{Tr}\,fg^{-1})^2 + 6\,\mathrm{Tr}\,fg^{-1} - 12]\,.$$

Notice the symmetry of $L^{(1)}$ and $L^{(2)}$ so far as f and g tensors are concerned. The lack of symmetry in $L^{(3)}$ (which is a sort of cosmological term) is a reflection of the physical lack of symmetry—in that the f field represents particles of mass M_f while the g field represents massless gravitons. (In the vierbein formalism, where

$$g^{\mu\nu} = L^{\mu a}L^{\nu}_{a}\,, \qquad L^{\mu a} = \eta^{\mu a} + \varkappa_g h^{\mu a}\,,$$

$$f^{\mu\nu} = \phi^{\mu a}\phi^{\nu}_{a}\,, \qquad \phi^{\mu a} = \eta^{\mu a} + \varkappa_f F^{\mu a}\,,$$

the physical fields are $h^{\mu a}$ and $F^{\mu a}$.)

It is clear what this design will achieve. Strong-interaction physics will now have a *universal* inbuilt cut-off from the nonpolynomiality of $(\det f)^{-\frac{1}{2}}$ at about $(\varkappa_f)^{-1}$. Lepton physics, or those parts of it included in $L^{(1)}$, will exhibit the inbuilt cut-off at $(\varkappa_g)^{-1}$ as before.

Let us reflect on the cut-off situation in the light of f and g gravitons and their interactions. Recall that in Sect. **2** a part of the Yang-Mills strong Lagrangian, even after the Stückelberg transformation, still remained obstinately polynomial in character. This part will now be multiplied by the factor $(\det f)^{-\frac{1}{2}}$. The same thing happens for weak W-mesons. Now if these are treated on a par with *leptons*, and their free (and self-interacting) Yang-Mills

Lagrangian is added to $L^{(1)}$, their self-mass will be proportional to $(f^2/\varkappa_g^2)+$ $+(f^2/G_{\text{Fermi}})$, *i.e.* these particles would each weigh some 10^{-5} g. It would seem more reasonable—if W-mesons exist at all—to class them with *hadrons*. Notice we are making the definite physical statement that W mesons interact *strongly* with F particles. Paradoxically this is in aid of making them *light* ($\delta m^2 \approx$ $\approx f^2/G_{\text{Fermi}}$) rather than too *massive* ($\delta m^2 \approx f^2/\varkappa_g^2$). (This conclusion is tentative to the extent that we do not know if quadratic self-mass infinities do arise for vector meson interactions at all. If they do not and if the self-mass of a vector boson is logarithmic, gravity would provide the appropriate cut-off.)

Of course, there is no question that the photon free Lagrangian must belong to $L^{(1)}$ and so must the weak ($J_\mu^{\text{lep}} W_\mu$) terms. On the other hand, terms of the form $m^2(\varrho_\mu^0 - A_\mu)^2$ giving the mixing of photons with the strong ρ^0-ϕ^0-ω^0 complex would belong to $L^{(2)}$ and so would the term $J_\mu^{\text{had}} W_\mu$. The mixed leptonic-hadronic weak processes thus acquire different cut-offs, depending on the company they keep. Clearly the interplay of the hierachy of the various inbuilt cut-offs λ_π^{-1}, $G_{\text{Fermi}}^{-\frac{1}{2}}$, $\varkappa_g^{-1}$ and $\varkappa_f^{-1}$, in prediction of reasonable magnitudes for the renormalization constants, when worked out fully, will provide severe and nontrivial tests of the ideas here expressed and also possibly explain some of the crazy numbers dotted around particle physics.

4. – The evidence for F-gravity.

Two f-particles (1260 and 1520 MeV) are known at present; a mixture of these two represents an SU_3 singlet. Should the F-field we have been talking about be identified with this mixed state or does the field correspond to some particle still undiscovered? (Since our F-field is proportional to the hadronic stress tensor, it should really transform in the SU_3 sense in the same manner as the strong Hamiltonian does.)

So far as any direct evidence on universal coupling of the known f-particles is concerned, it appears from an analysis of RENNER (see Miami Conference Report 1971) that f-boson couplings *are* universal. An indirect agreement (based on accepting ρ-f exchange degeneracy), however, leads one to think that f-nucleon coupling may be $\frac{3}{2}$ times larger than f-boson couplings. (The argument can be circumvented—as pointed out by G. O. FREUND (Imperial College preprint, 1971)—if exotics exist in the sense that nucleons are mixtures of three quarks, four quarks + one antiquark, 5 quarks + 2 antiquarks, ..., etc.) Alternatively one may assume that the F-meson we want is possibly a new particle of mass (1700 ÷ 2000) MeV which would lie on the pomeron trajectory (that is accepting that the pomeron is a Regge pole and also accepting present estimates of its slope). As has been pointed out by REMNER, JONES and SALAM (Miami Conference, 1971) one can then understand s-channel helicity conservation in «elastic» diffractive processes involving scattering of spin-0, $\frac{1}{2}$ and

1 particles (empirically associated with pomeron exchanges) as a consequence of the pomeron's possessing a stress-tensor coupling at its 2^1 recurrence.

5. – Exact static solutions of the F-gravity equations.

Even though the F-gravity theory was invented for the task of suppressing infinities, I would like to take F-gravity seriously in its own right as modifying conventional gravity for short distances. (It is easy to show that, notwithstanding the introduction of F-gravitons, the gravitational equivalence principle is not violated so that the Eötvös-Dicke experiment has nothing to say about F-gravitons.)

1) If we consider two hadronic particles it is clear (on account of f-g mixing) that in the linear approximation the static potential between them will be given by

$$V_{hh}(r) \propto -\frac{\varkappa_f^2}{\varkappa_f^2+\varkappa_g^2}\left[\varkappa_f^2\,\frac{\exp[-m_f r]}{r}+\varkappa_g^2\,\frac{1}{r}\right].$$

The corresponding potential between two leptons is

$$V_{ll}(r) \propto -\frac{\varkappa_g}{\varkappa_f^2+\varkappa_g^2}\left[\varkappa_g^2\,\frac{\exp[-m_f r]}{r}+\varkappa_g^2\,\frac{1}{r}\right].$$

There are, of course, no surprises here. The surprise comes for the hadron-lepton gravitational potential at distance $\approx 10^{-13}$ cm. This is given by

$$V_{lh} \propto -\frac{\varkappa_g^2}{\varkappa_f^2+\varkappa_g^2}\left[-\frac{\exp[-m_f r]}{r}+\frac{1}{r}\right].$$

Note that the $1/r$ singularity has disappeared. Layers of leptonic and hadronic matter do not attract (gravitationally) as strongly as had been assumed when they approach closer than 10^{-13} cm; alternate layers would produce partial shielding of conventional gravity.

2) One may except that the nonstatic gravitational potential between hadron and hadrons V_{hh} whould contain a short-range *repulsive* component of *the spin-two* f-graviton exchanges in second and higher orders. We have made no computations of this *repulsive gravity* so far and would appreciate help in finding out what the situation in higher orders in F-meson–nucleon coupling is.

3) For interactions of ordinary high-frequency gravitons with large concentrations of matter, we may expect to see surface effects familiar in the analogous ρ-γ mixing theory of hadronic electrodynamics. A high-frequency

g-graviton would convert into an f-graviton, through the mixing term which, on account of the short-range character of its force, would be absorbed predominantly at the *surface* of a (large) mass of matter, rather than penetrate into the inner layers. Very crudely speaking, gravitational effects may be expected to show $M^{\frac{2}{3}}$ rather than an M dependence where M is the mass of the large object. This weakening of gravity may have consequences in respect of the onset of gravitational collapse phenomena. (For the analogous ρ-γ mixing, photons of energies around 20 GeV empirically appear to exhibit a $Z^{0.9}$ dependence in their interactions with massive nuclei of charge Z.)

4) For use in hadronic particle physics, consider the F-meson–hadron equations neglecting the mass-mixing term as well as g gravity. Since the equations are exactly the same as Einstein equations (with $\varkappa_f$ replacing $\varkappa_g$; the relevant dimensions are now $(10^{-13} \div 10^{-14})$ cm rather than 10^{-33} cm). Since a certain number of exact solutions of the Einstein equations are known, we can take these over and attempt to use them.

Consider the Schwarzschild solution for a static point hadron as the source. The F-gravity field described by the Schwarzschild solution extends out to distances of the order of the Schwarzschild F-radius of the hadron; this being $2M\varkappa_f^2 \approx (10^{-14} \div 10^{-15})$ cm depending on the precise value of $\varkappa_f$.

COLLINS, DELBOURGO and WILLIAMS in London have considered the scattering of spin-zero meson in this Schwarzschild field of a static spinless hadron source. One can set up a «Schrödinger» equation from the Lagrangian

$$\mathscr{L} = \frac{f^{\mu\nu}\partial_\mu\phi\partial_\nu\phi}{\sqrt{\det f}},$$

where for $f^{\mu\nu}$ we substitute the Schwarzschild potential. On solving the equation (numerically) one sees (not surprisingly) a characteristic diffractive pattern of scattering. This work is being written up; it will also be extended for a spinning-hadron source (where the Kerr potential would be relevant). Its importance is that we are dealing with an exact F-meson potential. This is a new approach to scattering in hadron physics—the crucial assumption being that the f-mesic potential predominates over the other mesic potentials and that to write this f-mesic potential one may exploit the known exact solutions of EINSTEIN's equations. The method, of course, suffers from the weakness that these known exact solutions are static and also the mass of the F-meson has been neglected (though this later defect can perhaps heuristically be remedied by replacing M/r terms in the potential by a term like $M(e^{-m_f r}/r)$).

5) In a more speculative vein, consider F-gravity in its «metrical» aspects. For regions inside hadronic matter where g gravity may perhaps be neglected (to a good approximation), we are dealing with space-time of strong curvature. Let us for a moment take this aspect of F-gravity seriously.

Taking again the Schwarzschild or the Kerr solutions as describing a typical hadron, we could possibly translate the various « horizons ». Our relativistic friends have discovered as representing onion-ring like regions with different characteristics within the hadron. Thus the Kerr picture of a hadron (in F-gravity) would be an oblate object with a « red-shift horizon » at

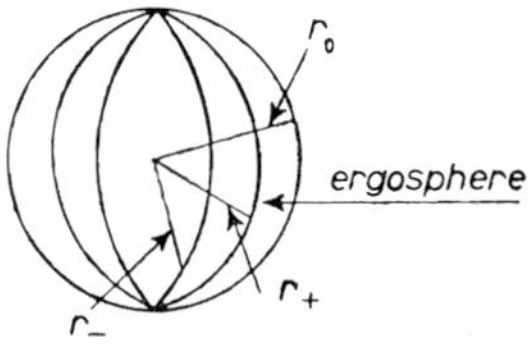

$$r_0 = m + \sqrt{m^2 - \frac{L^2 \cos^2 \theta}{m^2}}\,.$$

Inside is the ergosphere (the region of inelastic scattering) bounded by r^+ and r^0, where

$$r_+ = m + \sqrt{m^2 - \frac{L^2}{m^2}}\,.$$

Inside this is a region of capture; (recall that in classical theory r_+ is the one-way membrane). The capture region extends down to

$$r_- = m - \sqrt{m^2 - \frac{L^2}{m^2}}\,.$$

Between the origin and r_-, according to classical ideas, two-way propagation is possible (see R. PENROSE, Trieste Symposium on Contemporary Physics, 1968).

In quantum theory we have to reinterpret what classical relativity people have deduced on the basis of the study of classical geodesics. (For example, shadow scattering is something which classical study of geodesics will not automatically lead to). However, it is amusing that the condition for a « trapped surface » being formed ($r_- = r_+$) for an object of mass m and angular momentum L is given by $m = L/m$ (in units of $\varkappa_f$).

When charge is present, this reads

$$m^2 \varkappa_f^2 = \frac{L^2}{m^2 \varkappa_f^2} + Q^2\,.$$

A particle physicist will of course replace charge Q by unitary or isotopic charge ($Q^2 \Rightarrow I^2 = I(I+1)$ for the isotopic case) and L^2 by $J(J+1)$. The condition $r_+ = r_-$ then translates to a mass formula

$$m^2 \varkappa_f^2 = \frac{J(J+1)}{m^2 \varkappa_f^2} + I(I+1)\,.$$

The formula has a familiar look; it is an SU_4 like formula for a Regge hadronic trajectory. One had always hoped that SU_4 or SU_6 symmetries may some day emerge as dynamical symmetries. The analogy is with the case of the O_4 symmetry for the hydrogen atom, which emerges as a dynamical accident for the nonrelativistic Schrödinger equation with an $1/r$ potential. Could it be that the Kerr potential (with electric charge suitably generalized to unitary-charge in an SU_3 extended F-gravity scheme) may provide the dynamical input which yields aspects to a symmetry like SU_6 in a manner like the O_4 symmetry for the hydrogen atom with a $1/r$ potential in a Schrödinger equation?

Appendix

On the basis of their microcausality properties, Jaffe classifies field theories as localizable or nonlocalizable. Examples are:

A) *Jaffe-localizable*

$$L_{\text{int}} = g\,{:}\phi^n \exp[\varkappa\phi]{:} \qquad \text{(pure nonpolynomial)},$$

or

$$L_{\text{int}} = g\,{:}(\bar{\psi}\psi A)\exp[\varkappa\phi]{:} \qquad \text{(mixed Lagrangian)}.$$

The dots : : denote normal ordering.

B) *Jaffe-nonlocalizable*

$$L_{\text{int}} = g\,{:}\phi^n(1+\varkappa\phi)^{-w}{:}\,,$$

or

$$L_{\text{int}} = g\,{:}(\bar{\psi}\psi A)(1+\varkappa\phi)^{-w}{:} \qquad (w>0)\,.$$

Basically, Jaffe's distinction rests on the high-energy behaviour of the two-point spectral function $\varrho(p^2)$; it falls faster than $\exp[\sqrt{|p^2|}]$ for the localizable case and slower for the nonlocalizable. Jaffe shows that *localizable theories are* microcausal; the nonlocalizable ones are not. Also, by a rigorous theorem of Glaser, Epstein and Martin, S-matrix elements (on the mass shell) should exhibit Froissart high-energy boundedness for the localizable case. Far less is known about the nonlocalizable Lagrangians, apart from Steinmann-Taylor's results that, notwithstanding the possible breakdown of microcausality, the LSZ construction of the S-matrix can still be carried through—as also the proofs for CPT and spin-statistics theorems.

The Lehmann-Trute work referred to in the text shows that when derivatives are involved, the conditions for localizability become even more stringent. Thus for chiral theories they show that only the Gürsey exponential parametrization gives a localizable theory. This is a most beautiful result. Desirable nonpolynomial theories appear to belong to a very restricted class.

REFERENCES

[1] I. M. GEL'FAND and G. E. SHILOV: *Generalized Functions*, Vol. **1** (New York and London, 1964).

[2] W. GÜTTINGER: *Fortschr. Phys.*, **14**, 483 (1966).

[3] T. GUSTAFSON: *Ark. Mat. Astron. Fys.*, **34** A, No. 2 (1947).

[4] C. G. BOLLINI and J. J. GIAMBIAGI: *Nuovo Cimento*, **31**, 550 (1964); see also, E. R. SPEER: *Journ. Math. Phys.*, **9**, 1404 (1968); P. K. MITTER: Oxford preprint 18/70 (1970); F. CONSTANTINESCU and R. BLOOMER: University of Munich preprint (to appear in *Nucl. Phys.* (1971)).

[5] M. K. VOLKOV: *Commun. Math. Phys.*, **7**, 289 (1968); *Ann. of Phys.*, **49**, 202 (1968); *Commun. Math. Phys.*, **15**, 69 (1969); A. SALAM and J. STRATHDEE: *Phys. Rev. D*, **2**, 2869 (1970); H. LEHMANN and K. POHLMEYER: DESY preprint, DESY 70/26; M. KAROWSKI: *Commun. Math. Phys.*, **19**, 289 (1970). For a review of techniques in nonpolynomial Lagrangian theories see: A. SALAM: in *Proceedings of the VIII Coral Gables Conference on Symmetry Principles at High Energies* (ICTP, Trieste, preprint IC/71/3).

[6] B. W. LEE and B. ZUMINO: *Nucl. Phys.*, **13** B, 671 (1969).

[7] A. T. FILIPPOV: Dubna preprint (1968).

[8] G. V. EFIMOV: *Sov. Phys. JETP*, **17**, 1417 (1963); *Phys. Lett.*, **4**, 314 (1963); *Nuovo Cimento*, **32**, 1046 (1964); *Nucl. Phys.*, **74**, 657 (1965); ICTP, Trieste, Translation of Parts I, II and III of *Nonlocal quantum theory of scalar fields*, August 1969; CERN report No. Th. 1087 (1970); E. S. FRADKIN: *Nucl. Phys.*, **49**, 624 (1963); **76**, 588 (1966); R. DELBOURGO, A. SALAM and J. STRATHDEE: *Phys. Rev.*, **187**, 1999 (1969); A. P. HUNT, K. KOLLER and Q. SHAFI: Imperial College preprint ICTP/69/12.

[9] S. FELS: *Phys. Rev. D*, **1**, 2370 (1970).

[10] H. M. FRIED: *Phys. Rev.*, **17** A, 1725 (1968).

[11] J. G. TAYLOR: University of Southampton preprint, July 1970; E. KAPUŚCIK: ICTP, Trieste, Internal Report IC/69/132.

[12] R. DELBOURGO, K. KOLLER and A. SALAM: Imperial College preprint ICTP/69/10 (to be published in *Ann. of Phys.*); B. KECK and J. G. TAYLOR: University of Southampton preprint, April 1970.

[13] S. OKUBO: *Progr. Theor. Phys.* (*Kyoto*), **11**, 80 (1954); R. ARNOWITT and S. DESER: *Phys. Rev.*, **100**, 349 (1955); M. K. VOLKOV: *Commun. Math. Phys.*, **7**, 289 (1968); Q. SHAFI: Imperial College preprint ICTP/69/22 (to be published in *Phys. Rev.*).

[14] H. EPSTEIN, V. GLASER and A. MARTIN: *Commun. Math. Phys.*, **13**, 257 (1969).

[15] R. DELBOURGO, A. SALAM and J. STRATHDEE: *Lett. Nuovo Cimento*, **2**, 354 (1969); A. SALAM and J. STRATHDEE: *Lett. Nuovo Cimento*, **4**, 101 (1970); P. BUDINI and G. CALUCCI: *Nuovo Cimento*, **70** A, 419 (1970); A. SALAM: *Renormalization constants and interrelation of fundamental forces*, in *Lecture at the Kiev Conference* (Kiev, 1970) (ICTP, Trieste, preprint IC/70/106); C. J. ISHAM, A. SALAM and J. STRATHDEE: *Phys. Rev. D*, **3**, 1805 (1971).

[16] C. J. ISHAM, A. SALAM and J. STRATHDEE: *Phys. Rev. D*, **3**, 867 (1971); B. ZUMINO: CERN preprint and Brandeis Lecture Notes, June 1970.

An Informal Talk on Liquid Helium.

T. REGGE

Istituto di Fisica dell'Università - Torino

Liquid helium in the He-II phase is a remarkable physical system in that it shows quantum effects on a macroscopic scale.

It can be produced by cooling ordinary liquid He below $T = 2.18$ °K.

This transition point where the specific heat of helium shows a logarithmic singularity (by plotting C_V versus T one obtains a λ-shaped curve, hence the name λ-point).

Below the λ-point helium becomes superfluid, it flows throung capillaries without apparent viscosity and it exhibits a number of surprising properties (fountain effect, second sound, an anomalous and highly efficient heat transmission mechanism and others).

After many accurate experiments, notably by the Russian groups (Kapitza, Andronikashvili) a convincing explanation of these facts was given in 1943 by LANDAU through a modification and correct interpretation of a two-fluid model proposed by TITZA.

According to LANDAU the ground state of He-II is a single-quantum macroscopic state, which is degenerate if the number of particles in the vessel is not specified (this is related to the existence of the phonon as a Goldstone particle).

The excitation states are quasi-particle states which in the original theory of LANDAU were of two kinds:

A) Phonons with a dispersion curve $\varepsilon(\varrho) = cp(1 - \gamma p^2)$, where c is the speed of sound, ε and p the energy and momentum of the phonon. LANDAU assumed $\gamma > 0$.

B) Rotons of momentum around $|p_0| = 1.9$ Å$^{-1}\hbar$.

The dispersion curve is

$$\varepsilon(p) = \Delta + \frac{(|p| - p_0)^2}{2\mu},$$

with $\Delta \simeq 8.65$ °K.

Phonons explain the specific heat at the lowest temperatures. By neglecting their interaction we have $C_V = \text{const}\ T^3$ from a standard boson gas computation. Rotons enter at higher temperatures with a factor $\exp[-\Delta/kT]$ and play an important role near the λ-point.

More recent experiments by the Chalk River group, using inelastic neutron scattering experiments have shown that phonons and rotons are actually on the same excitation curve of the form indicated in Fig. 1.

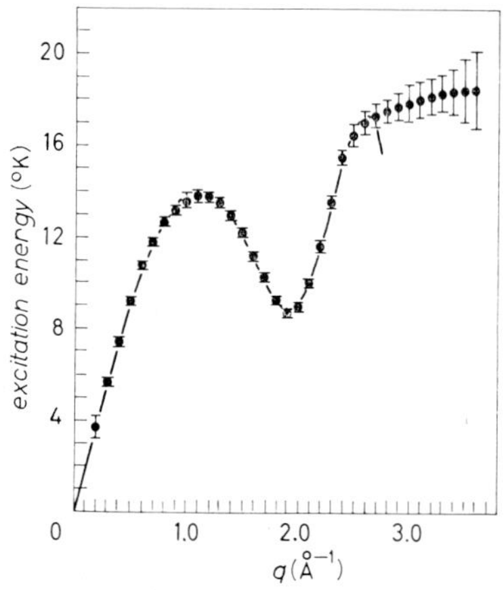

Fig. 1. – The dispersion curve in liquid He-II.

A third kind of excitation is provided by the quantized vortices of FEYNMAN-ONSAGER. I shall come back to them later.

At this point we may wonder why high-energy physicists should be interested at all in phenomena involving typical energies of the order 10^{-14} times those encountered in the accumulation ring experiments. To me it is enough to see that this field has produced physics of the highest caliber and it has influenced already particle physics in several ways (Josephson effect, Goldstone particles and general field-theoretical techniques). We should look at He-II as a particular model of field theory in which we may vary to some extent the relevant parameters and gain useful information.

In this direction a very important result was obtained by BOGOLUBOV, in He-II and in superconductivity, by introducing his celebrated transformation. He showed that a weakly interacting Bose gas has an excitation spectrum with a phonon structure; although the resulting dispersion curve is not realistic for He-II, this result is of the highest theoretical interest.

In particular, a Bogolubov gas shows superfluidity.

This property follows from the inability of a body moving through the fluid to lose energy in the only way it can do so, be radiating excitations into the fluid. Just as a charge in a refracting medium does not radiate unless it travels faster than light (Čerenkov effect) an impurity does not create phonons or any excitation unless it moves faster than the phase velocity of the excitation. From the dispersion curve we see that this lowest velocity occurs for rotons and is about 60 m/s. This is too high and experiments show that the critical velocity is actually lower by two orders of magnitude. Quantized vortices have been introduced in order to explain this discrepancy.

What is missing in the Bogolubov theory is a reliable treatment of phonon-phonon interaction. It can be reached to some extent by going through the machinery of many-body thermal Green's functions.

But this machinery is very complicated and most of the interesting results already follow from a naive application of the Landau quantum hydrodynamical equations and/or the Boltzmann diffusion equation.

If two phonons collide each phonon experiences a variation in the pressure of the medium induced by the other. His speed accordingly changes, a measure of this change and therefore of the strenght of the interaction is given by the Gruneisen constant

$$u = \frac{\mathrm{d}\ln c}{\mathrm{d}\ln p} \simeq 2.6\,.$$

This interaction means that He-II is to some extent an anharmonic medium, in fact much more so than most solids and fluids.

Data on He-II can be gathered through several experimental techniques, neutron probes, X-rays, attenuation of sound, specific heat, hot ions and there may be others.

All these give results which essentially confirm the Landau picture and with later theoretical work by FEYNMAN.

If we look at details however we see some disagreement.

None of these is significant per se, they are setting however a pattern which may require some interesting variation of the original picture.

I have stated that LANDAU ad all workers in the field have assumed $\gamma > 0$ in the dispersion. This condition ensures that phonons are stable against decay into softer phonons. Lately it has become clear that this condition is after all not so important and is probably false. It all started with a proposal by MARIS and MASSEY to the extent that a negative γ would explain the anomalous large attenuation of sound in He-II. If $\gamma > 0$, then it is impossible for a thermal phonon to absorb an acoustic phonon unless one assumes an energy width, if however $\gamma < 0$ this process (called the 3-phonon process) is much more favored, direct absorption takes place and the attenuation is larger. Also the specific heat data by PHYLLIPS and WATERFIELD show that the dispersion should be positive. The theoretical situation is in fact much more complicated than sketched here (KLEIN and WEHNER, RCA Zurich preprint) but the evidence seems now to be that the dispersion is in fact positive.

We remind that both the Bogolubov gas and the hard sphere gas treated by HUANG, LEE and YANG have positive dispersion.

Quite at the same time, FEENBERG proposed that the Van de Waals forces actually produce in the Bogolubov treatment an odd term in the expansion:

$$\frac{\varepsilon(p)}{cp} = (1 - \gamma p^2 + \sigma p^3 + \dots) .$$

In view of this, MOLINARI and myself tried to see whether a generic power expansion for the dispersion would fit the neutron and specific-heat data better than the conventional one in even powers. The surprising result is that the best fit is obtained by including a linear positive dispersion term.

I do not know how this term fares with the attenuation data, a crude computation shows that it may be even better than the quadratic version but more computations are needed.

A linear dispersion would imply a more drastic revision of the original scheme than we bargained for. It has the honest feature of being more easily testable than a quadratic one in the low-frequency range. Our fit proposes that this dispersion has the form (ν in Hz)

$$\frac{\varepsilon(p)}{cp} = 1 + 7.3 \cdot 10^{-13} \nu - 4.7 \cdot 10^{-24} \nu^2 .$$

The only experiment in this sense has been carried out in the $(20 \div 60)$ GHz region by ANDERSON and SABISKY at the RCA in Princeton.

He finds

$$\frac{\varepsilon(p)}{cp} = 1 + (6.3 \pm 0.7) \cdot 10^{-13} \nu \pm 8.10^{-25} \nu^2 .$$

The agreement in the linear term is quite impressive but obviously more experiments are needed before one draws the last conclusion.

A number of sharp objections have been raised against a linear dispersion. One is that the available phonon-phonon interaction is simply not enough to account in terms of self-energy for a linear term and that perturbative treatment on the Bogolubov frame fails to see it. Moreover, the idea is that $\varepsilon(p)$ is after all nonanalytic at $p = 0$ and that logarithmic terms should be included. On the other hand, in collaboration with BARUCCHI and PONZANO I have proposed a different approximation scheme (HEISENBERG 70th Birthday Festschrift) where the linear term appears.

It is clear that the theoretical situation is at the best terribly confused and that it will remain so till the dense interacting Bose gas will be under control.

Tied up to this fundamental difficulty is the theory of the λ-point transition. Also the generation of quantized vortices is an open question. Although these are interesting subjects, I have no time to go into them in details. I have no doubt that in solving them we must develop ideas of relevance to all fields of physics, including particle physics.

PROCEEDINGS OF THE INTERNATIONAL SCHOOL OF PHYSICS
« ENRICO FERMI »

Course XIV
Ergodic Theories
edited by P. Caldirola

Course XV
Nuclear Spectroscopy
edited by G. Racah

Course XVI
Physicomathematical Aspects of Biology
editel by N. Rashevsky

Course XVII
Topics of Radiofrequency Spectroscopy
edited by A. Gozzini

Course XVIII
Physics of Solids (Radiation Damage in Solids)
edited by D. S. Billington

Course XIX
Cosmic Rays, Solar Particles and Space Research
edited by B. Peters

Course XX
Evidence for Gravitational Theories
edited by C. Møller

Course XXI
Liquid Helium
edited by G. Careri

Course XXII
Semiconductors
edited by R. A. Smith

Course XXIII
Nuclear Physics
edited by V. F. Weisskopf

Course XXIV
Space Exploration and the Solar System
edited by B. Rossi

Course XXV
Advanced Plasma Theory
edited by M. N. Rosenbluth

Course XXVI
Selected Topics on Elementary Particle Physics
edited by M. Conversi

Course XXVII
Dispersion and Absorption of Sound by Molecular Processes
edited by D. Sette

Course XXVIII
Star Evolution
edited by L. Gratton

Course XXIX
Dispersion Relations and Their Connection with Causality
edited by E. P. Wigner

Course XXX
Radiation Dosimetry
edited by F. W. Spiers and G. W. Reed

Course XXXI
Quantum Electronis and Coherent Light
edited by C. H. Townes and P. A. Miles

Course XXXII
Weak Interations and High-Energy Neutrino Physics
edited by T. D. Lee

Course XXXIII
Strong Interactions
edited by L. W. Alvarez

Course XXXIV
The Optical Properties of Solids
edited by J. Tauc

Course XXXV
High-Energy Astrophysics
edited by L. Gratton

Course XXXVI
Many-Body Description of Nuclear Structure and Reactions
edited by C. Bloch

Information about Courses I-XIII may be obtained from the Italian Physical Society.

Course XXXVII
Theory of Magnetism in Transition Metals
edited by W. Marshall

Course XXXVIII
Interaction of High-Energy Particles with Nuclei
edited by T. E. O. Ericson

Course XXXIX
Plasma Astrophysics
edited by P. A. Sturrock

Course XL
Nuclear Structure and Nuclear Reactions
edited by M. Jean

Course XLI
Selected Topics in Particle Physics
edited by J. Steinberger

Course XLII
Quantum Optics
edited by R. J. Glauber

Course XLIII
Processing of Optical Data by Organisms and by Machines
edited by W. Reichardt

Course XLIV
Mulecular Beams and Reaction Kinetics
edited by Ch. Schlier

Course XLV
Local Quantum Theory
edited by R. Jost

Course XLVI
Physics with Storage Rings
edited by B. Touschek

Course XLVII
General Relativity and Cosmology
edited by R. K. Sachs

Course XLVIII
Physics of High Energy Density
edited by P. Caldirola and H. Knoepfel

Course IL
Fondations of Quantum Mechanics
edited by B. d'Espagnat

Course L
Mantle and Core in Planetary Physics
edited by J. Coulomb and M. Caputo

Course LI
Critical Phenomena
edited by M. S. Green

Course LII
Atomic Structure and Properties of Solids
edited by E. Burstein

Course LIII
Developments and Borderlines of Nuclear Physics
edited by H. Morinaga

Tipografia Compositori - Bologna - Italy